80

The analysis of starlight

One hundred and fifty years of astronomical spectroscopy

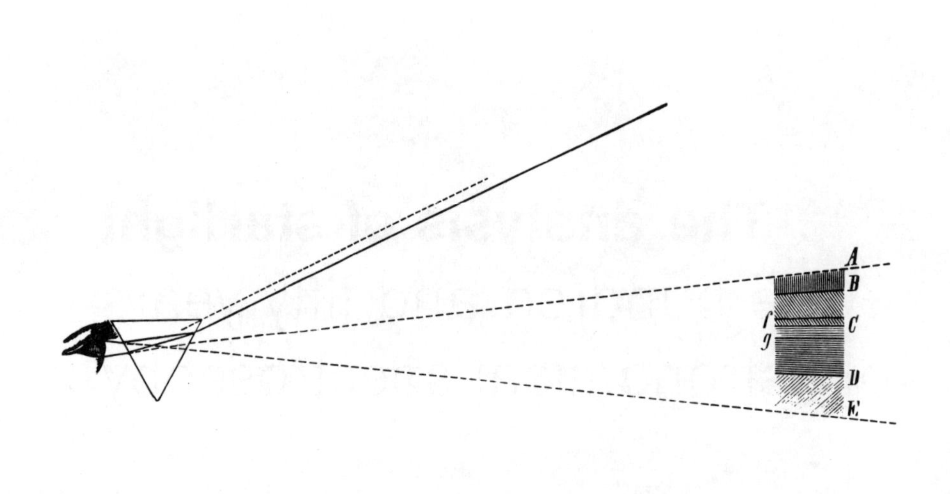
A
B
f
C
g
D
E

J.B. Hearnshaw

Senior Lecturer in Astronomy

University of Canterbury

Christchurch, New Zealand

The analysis of starlight

One hundred and fifty years of astronomical spectroscopy

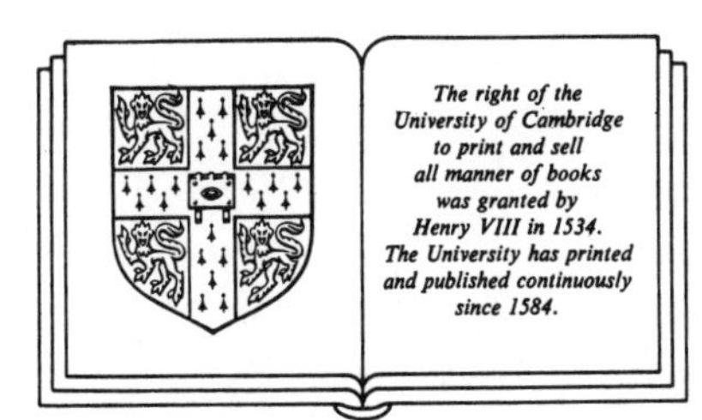

Cambridge University Press

Cambridge

New York Port Chester

Melbourne Sydney

Published by the Press Syndicate of the University of Cambridge
The Pitt Building, Trumpington Street, Cambridge CB2 1RP
40 West 20th Street, New York NY 10011, USA
10 Stamford Road, Oakleigh, Melbourne 3166, Australia

First published 1986
First paperback edition (with corrections) 1990

Printed in Great Britain at the University Press, Cambridge

British Library cataloguing in publication data

Hearnshaw, J.B.
The analysis of starlight: one hundred and
fifty years of astronomical spectroscopy.
1. Astronomical spectroscopy – History
I. Title
522′.67′09 QB465

Library of Congress cataloguing in publication data

Hearnshaw, J.B.
The analysis of starlight.
Includes indexes.
1. Astronomical spectroscopy – History. I. Title.
QB465.H43 1986 522′.67 85–21299

ISBN 0 521 25548 1 hard covers
ISBN 0 521 39916 5 paperback

TM

To Vickie, Alice and Edward.

Are not gross bodies and light convertible into one another; and may not bodies receive much of their activity from the particles of light which enter their composition?... The changing of bodies into light, and light into bodies, is very conformable to the course of Nature, which seems delighted with transmutations.

Newton, *Opticks,* Query 30; second edition (1717).

Contents

Preface

My main motivation for writing this book was an act of self-indulgence. As a hobby I enjoyed delving into the earlier literature of astronomical spectroscopy. As a practising observational astronomer, I found it especially refreshing to have a feel for the way the topic had developed, and to be able to glimpse at the lives of some of the early pioneers in stellar spectroscopy.

My hobby began in 1974 when I was at the Observatoire de Paris-Meudon on a fellowship. I frequently browsed in the excellent library there, and one day began reading the collected papers of the eminent English astronomer, Sir William Huggins, one of the founders of stellar spectroscopy. Huggins' lucid and eloquent papers and his many remarkable achievements provided the inspiration from which my interest developed further, to form the basis for this book.

However I had little time to pursue these interests very intensively until 1981, when an opportunity arose that allowed me to spend a year in Germany with the support of the Alexander von Humboldt Foundation, while on sabbatical from the University of Canterbury. I went to the Landessternwarte (State Observatory) in Heidelberg and resolved to spend most of my time there researching and writing a book on the development of stellar spectroscopy.

This book is not primarily intended for the science historian, nor is it a popular book for the layman, although I hope that readers in both these categories may find material here which is useful or interesting. Instead I have aimed at writing for the practising astronomer what is essentially an interpreted guide to the literature covering the development of observational stellar spectroscopy. For this reason I have placed a lot of emphasis on providing a good list of references to the original material. The primary sources are the papers in journals and observatory publications. I have generally avoided citing secondary references (with a few exceptions), such as modern commentaries by science historians.

A typical review paper in astronomy or any scientific discipline might devote an introductory paragraph to a brief historical summary, covering the subject over the past half century or so. The next several dozen pages might then discuss what has been achieved in the last five or at most ten years. This format is fine, and the review paper has an important rôle to play in today's world with its phenomenal growth in recent scientific literature. What I view less kindly is when that brief historical introduction is inaccurate and improperly researched. I know of several by distinguished authors with erroneous facts in the historical introductions; I trust that the readers of this book will not perpetuate such mistakes.

The method I have adopted here is just the reverse of the modern review article. For each topic I have deliberately emphasised the earlier references, so as to give precedence to whoever first entered a new field or established a new fact. On the other hand, for the more recent work up to 1965, I have generally been progressively briefer and more selective in the choice of material cited. I warn all readers that this is not a textbook of modern astrophysics; between 1965 and 1970 the treatment is not comprehensive and developments after 1970 are not covered here at all.

I found the Landessternwarte library in Heidelberg an excellent place to work on the manuscript of this book. For historical purposes it houses a collection of outstanding importance; in addition it is one of the least regulated libraries I have ever come across – no fixed library hours, no full-time librarian and no limits to loans either in time or in quantity of books. I spent a year in this paradise researching and writing about two-thirds of this manuscript. The remainder was written at the University of Canterbury in Christchurch between 1982 and 1985.

In addition to the libraries at the Landessternwarte and the University of Canterbury, I also referred to material in the libraries at the University of Heidelberg, the Physikalische Institute, Heidelberg, the Göttingen Observatory, the Observatoire de Paris and the Observatoire de Meudon, the Carter Observatory, Wellington, New Zealand, and the Niels Bohr Library at the American Institute of Physics in New York City. I am grateful to all these places for their kind assistance and for permission to use their excellent resources.

The scope of this book in the period it covers is discussed in section 1.1. The main emphasis is the century and a half from Fraunhofer's first observations of solar and stellar spectra to about 1965. The subject matter is observational stellar spectroscopy. Occasionally I have treated theoretical topics (for example model atmosphere theory) if these seemed indispensable to the discussion of the observations and their immediate interpretation. The development of solar physics has been covered only where this was essential for an understanding of stellar physics. Therefore some of the early nineteenth century work in solar spectroscopy is dealt with. On the other hand, later work in solar spectroscopy of real importance for the sun, such as Samual Pierpont Langley's infrared bolometry of sunlight, I have omitted altogether, as this work did not immediately lead to corresponding advances in stellar research. Similarly I have included some mention of the spectroscopy of gaseous nebulae, because it was generally supposed in the early days of spectral classification that nebular spectra could be incorporated into the classification of stellar spectra as part of a continuous sequence. The spectra of external galaxies I consider to comprise another topic of equal importance to stellar spectra, but whose historical development is best segregated from that of individual stars. It is not treated here at all.

A textbook in modern astronomy should only use the S.I. (Système International) units, though regrettably not all do. However, I found adherence to this rule inconvenient here. For example, many quoted passages refer to line wavelengths in Ångstrom units and not nanometres. I have therefore adopted the Ångstrom unit as the unit of wavelength. American and British astronomers until recently have almost invariably referred to the apertures of their telescopes in inches, so much so that appellations such as the Mt Wilson '100-inch telescope' have become familiar names rather than mere measures of a telescope's dimension. In view of this it seemed pedantic to change all the old and familiar references to telescopes in inches into metric units. I have therefore compromised and employed both imperial and metric units for telescopes in this book, whichever was the appropriate choice in each instance.

I have included numerous quotations from original works; for those that appeared originally in French or German, the translations are my own. I have included the dates of astronomers referred to in the text in many cases. Where a name is not followed by dates, this is because the dates appeared earlier on in the book, or because it is only a passing reference to that individual, or (in a few cases) because I was unable to ascertain the dates after a brief search.

Acknowledgements

I am grateful for the assistance of several people in the preparation of this book. Firstly, I wish to thank Dr Immo Appenzeller and Dr Bernhard Wolf at the Landessternwarte in Heidelberg for welcoming me as a guest in that institution. Professor Brian Warner at the University of Cape Town gave up his valuable time to read almost all the text and offered numerous suggestions which have resulted in major improvements. Miss E. Mistrik at the Landessternwarte, Mr A.C. Gilmore at Mt John University Observatory (University of Canterbury) and Mrs B. Cottrell (photographic section, University of Canterbury) skillfully undertook most of the photographic work for the illustrations. Mrs B. Bristowe at the University of Canterbury ably and willingly typed the manuscript and commendably kept her equanimity in spite of at least a thousand last-minute changes, corrections or additions to the text. She was assisted at times by Mrs G.M. Evans (University of Canterbury) and Mrs B. Schwander (Landessternwarte). Mrs S.E. Reynolds helped compile most of the material in the indexes.

I had helpful discussions on a variety of topics with Drs B. Baschek, U. Finkenzeller, D. Labs, K. Schaifers, H. Scheffler and B. Wolf (all in Heidelberg); and also with Dr G. Cayrel de Strobel in Paris. All these people, perhaps without realising it, have influenced the content of this book in one way or another. In addition the following made useful comments on parts of the manuscript, mainly relating to their own work: Drs L.H. Aller, H.W. Babcock, L. Biermann, D.H. DeVorkin, R.F. Griffin, W.W. Morgan, Y. Öhman, A. Przybylski, K. Aa. Strand, A. Unsöld, C. van't Veer-Menneret and K.O. Wright.

Many people have given assistance with the illustrations that appear here. Those who went out of their way to provide me with material that was otherwise unavailable to me were Mr L. Bartha (Budapest), Dr G. Cayrel de Strobel (Observatoire de Paris-Meudon), Mrs M. Dunham (Chocorua, New Hampshire), Drs O. Gingerich (Harvard–Smithsonian Center for Astrophysics), J.L. Greenstein (California Institute of Technology), M.L.

Hazen (Harvard College Observatory), Mrs K. Haramundanis (daughter of C.H. Payne-Gaposchkin), Drs W.A. Hiltner (University of Michigan), D.C. Morton (Anglo-Australian Observatory), Y. Öhman (formerly Stockholm Observatory) and K.O. Wright (formerly Dominion Astrophysical Observatory) as well as the staff of the Mary Lea Shane Archives of the Lick Observatory and of the Royal Dutch Academy of Sciences in Amsterdam.

I am grateful to the following for permission to make use of illustrations: L. Bartha, Fig. 4.22; A.H. Batten, Fig. 9.2; G. Cayrel de Strobel, Figs. 8.10 and 9.15; S. Chandrasekhar, Fig. 10.6; C. de Jager, Figs. 7.10 and 10.14; Mrs M. Dunham, Fig. 1.4; O. Gingerich, Fig. 7.7; J.L. Greenstein, Figs. 9.11, 9.12, 9.15, 9.16 and 10.12; R.F. Griffin, Fig. 6.22; K. Haramundanis, Fig. 7.9; M.L. Hazen, Fig. 11.1; G.H. Herbig, Fig. 6.12; W.A. Hiltner, Fig. 10.10; P.J. Ledoux, Fig. 9.16; D. Lynden-Bell, Fig. 6.6; Sir William McCrea, Fig. 10.4; W.W. Morgan, Figs. 8.13 and 8.14; D.C. Morton, Fig. 11.7; G.F.W. Mulders, Figs. 7.11, 10.7 and 10.8; Y. Öhman, Figs. 8.24 and 9.10; A. Unsöld, Fig. 10.11, and K.O. Wright, Figs. 7.18 and 10.13. I am also grateful to the following institutions or organisations for courtesy to reproduce the illustrations listed: the editorial office of Centaurus for Figs. 7.1, 7.2 and 7.4; Harvard College Observatory for Fig. 4.13, for Figs. 5.1 to 5.10 inclusive and for Figs. 7.9 and 9.17; the International Astronomical Union for Figs. 8.22 and 9.3; the Mary Lea Shane Archives of the Lick Observatory for Fig. 6.7; Mount Wilson and Las Campanas Observatories for Fig. 8.2; the *Publications of the Astronomical Society of the Pacific* for Figs. 6.5, 6.12, 6.18, 6.19, 6.21, 8.13, 8.21, 8.23, 9.13 and 10.11; the Royal Astronomical Society for Figs. 6.20, 8.1, 8.5, 11.5 and 11.6; the Royal Dutch Academy of Sciences, Amsterdam for Figs. 7.12 and 7.14; the Royal Society, London for Figs. 2.1, 3.3, 3.7 and 4.20; *Sky and Telescope* magazine for Fig. 9.9; *Sterne und Weltraum* for Figs. 7.6, 8.7, 10.1 and 10.2; *Vistas in Astronomy*, edited by A. Beer, published by the Pergamon Press, Oxford for Figs. 1.3, 1.5, 3.1 and 8.3; and the Yerkes Observatory for Figs. 1.1, 4.21, 6.16, 6.17 and 8.16.

The following individuals have kindly granted me permission to quote from their published papers, as referenced in the text: Drs H.A. Abt, L.H. Aller, H.W. Babcock, W.P. Bidelman, L. Biermann, E. Böhm-Vitense, P.S. Conti, M. de Groot, C. de Jager, D.H. DeVorkin, J.L. Greenstein, G.H. Herbig, C. Jascheck, P.C. Keenan, H.C. King, V. Kourganoff, W.J. Luyten, C.E. Moore(-Sitterly), W.W. Morgan, G.F.W. Mulders, Y. Öhman, D.M. Popper, G.W. Preston, A. Przybylski, A. Slettebak, K.Aa. Strand, A. Unsöld, C. van't Veer-Menneret, C.A. Whitney, O.C. Wilson, Sir Richard Woolley and Dr K.O. Wright. In addition, Mrs M. Dunham kindly gave me permission to quote from the American Institute of Physics Oral History Interview of Dr T. Dunham, Jr.

I am grateful to the editors of the following journals for their courtesy in granting permission to quote from them the passages which are referenced in the text: *Annalen der Physik*; the *Astrophysical Journal*, the *Astronomical Journal*, *Nature*, the *Observatory*, the *Philosophical Magazine*, *Publications of the Astronomical Society of the Pacific* and *Vistas in Astronomy* (edited by A. Beer, published by Pergamon Press, Oxford). The following institutions or societies kindly gave permission to quote from the publications given in each case in parenthesis: the Academy of Sciences of Paris (*Comptes Rendus de l'Académie des Sciences de Paris*); the American Astronomical Society (*Publications of the Astronomical and Astrophysical Society of America*); the British Astronomical Association (*Journal of the British Astronomical Association*); Dominion Astrophysical Observatory (*Publications of the D.A.O.*); Harvard College Observatory (*Annals*, *Circulars* and *Bulletin of H.C.O.*; also *Henry Draper Memorial Annual Reports* and H.C.O. Monograph No. 1, *Stellar Atmospheres*); Lick Observatory (*Publications of the Lick Observatory* and *Lick Observatory Bulletin*); the Department of Astronomy, University of Michigan (*Publications of the Astronomical Observatory of the University of Michigan*); the D. Reidel Publishing Co. (Utrecht symposium *The Solar Spectrum* ed. by C. de Jager and published by Reidel (1965)); the Royal Astronomical Society (*Monthly Notices* and *Quarterly Journal of the Royal Astronomical Society*); the Royal Society, London (*Philosophical Transactions* and *Proceedings of the Royal Society*); Sky Publishing Corporation (*Sky and Telescope*); the Utrecht Observatory (*Recherches Astronomiques de l'Observatoire d'Utrecht*). The American Institute of Physics kindly granted permission to quote from the Oral History Interviews of Drs I. Bowen, T. Dunham, Jr., J.L. Greenstein and W.W. Morgan.

In all cases the sources of direct quotations acknowledged above are explicitly referenced in the text.

1 Introduction to spectroscopy, spectroscopes and spectrographs

1.1 Introduction

Introduction One of the favourite quotations by astronomers is a passage by the French philosopher, Auguste Comte (1798–1857). The nineteenth lesson of his *Cours de Philosophie Positive* was one of several lessons dealing with the theory of knowledge in astronomy. It appeared in 1835. With reference to the stars, he wrote:

> We understand the possibility of determining their shapes, their distances, their sizes and their movements; whereas we would never know how to study by any means their chemical composition, or their mineralogical structure, and, even more so, the nature of any organised beings that might live on their surface. In a word, our positive knowledge with respect to the stars is necessarily limited solely to geometric and mechanical phenomena, without being able to encompass at all those other lines of physical, chemical, physiological and even sociological research which comprise the study of the accessible [i.e. terrestrial] beings by all our diverse methods of observation (1).

A little later he continued: 'I persist in the opinion that every notion of the true mean temperatures of the stars will necessarily always be concealed from us' (1).

These passages may be amusing in the light of present knowledge, and it seems probable that Comte was ignorant of Fraunhofer's investigations from about 1814 to 1823 in which he described the absorption lines in solar and stellar spectra (see sections 2.4 and 2.5). In any case the implications of Fraunhofer's spectroscopic work were far from apparent at that time, even in scientific circles. However, Comte has been much maligned, mainly by astronomers ignorant of his overall positivist philosophy. Immediately preceding the much quoted passage on chemical composition of the stars is the statement: 'Every research that is not finally reducible to simple visual observations is therefore necessarily disallowed in our study of the stars' (1). Comte therefore preached that true science was impossible if not based on direct observation or experiment (in the case of astronomy, on observation), a philosophy that scientists today should be happy to espouse.

In spite of my defence of Comte's views on the composition of the stars, which were admittedly erroneous yet pardonable, this book is largely a study of the subsequent investigations by many astronomers who reached the opposing view to Comte's by the analysis of starlight using the spectroscope or spectrograph. This is the story of how first the different chemical elements were identified in stars from their spectra, of how the temperatures and other physical properties of the outer layers of stars were measured, and finally of how the chemical composition of the stars themselves has been quantitatively determined.

The story described here spans approximately three centuries if the prismatic dispersion and analysis of sunlight by Isaac Newton in 1666 is taken as the starting point. Alternatively it encompasses one and a half centuries from the early investigations of Fraunhofer mentioned above, or just one century from the rebirth of stellar spectroscopy with work by Huggins, Secchi and others in 1863. Whichever starting point is adopted – and I prefer to take that of Fraunhofer, since he was the first stellar spectroscopist – the mid-1960s represent a significant anniversary for astronomical spectroscopy. For this reason 1965 has been chosen as the approximate limit of the material covered here; in a few cases I have found it necessary to continue the discussion up to about 1970, so as not to terminate some sections in mid-air; but any material included after 1965 is deliberately far from thorough or complete.

The remainder of this introductory chapter concerns some of the basic concepts in spectroscopy, which may be skipped by those already familiar with them. In addition, some historical comments on the development of astronomical spectroscopes and spectrographs are included.

1.2 Basic concepts in spectroscopy

1.2.1 Terminology The word 'spectrum' is defined in the *Oxford English Dictionary* as: 'the coloured band into which a beam of light is decomposed by means of a prism or diffraction grating' (2).* It was first

* It had been used earlier to signify a 'phantom' or 'apparition'. But the similar word 'spectre' is more normal in this context. Both date from the early seventeenth century, and both are derived from the Latin word 'spectrum' meaning either (*a*) an appearance or image, or (*b*) an apparition or spectre.

used in the English language with this meaning by Isaac Newton in 1671, in the first paper he sent to the Royal Society concerning the composition of white sunlight (see section 2.1). On the other hand the introduction of words such as spectroscopy, spectroscope and spectrograph was much more recent. All three date from the 1860s or 1870s. William Huggins in London, the pioneer of stellar spectroscopy, first used the first of these terms in 1870; while Henry Draper, the first person to successfully photograph a stellar spectrum in 1872, described his apparatus as a 'spectrograph' a few years later. On the other hand, a 'spectroscope' is for the visual instead of photographic observation of spectra. It is thus seen that stellar spectroscopists devised two of these three basic scientific terms at a time that the analysis of starlight was still in its infancy.

1.2.2 Wavelength and colour All spectroscopes or spectrographs employ a dispersing element consisting of one or more prisms or of a diffraction grating. Its function is to produce a spectrum by spreading the light into a one-dimensional strip according to a photon's colour or wavelength.

The link between the colour of light (a physiological concept) and its wavelength was demonstrated by Thomas Young in England. He demonstrated the wave nature of light from interference in 1801, and then used a simple diffraction grating, consisting of a series of parallel grooves in a glass plate, for measuring the wavelengths of sunlight of different colours (3). Spectra in four different orders were obtained from this apparatus and he showed how wavelengths could be obtained from the spacing of the grooves and the angle at which light of different colours was diffracted.

After Fraunhofer had rediscovered the absorption lines in the solar spectrum (see section 2.4) he introduced the idea of using spectral lines as wavelength standards. He used the orange line he named D in the solar spectrum and in 1821 measured a wavelength of 5888 Å*, using a simple grating consisting of parallel fine wires stretched in front of his telescope objective (4).

In the 1860s gratings of much improved quality became available and several attempts were made to establish an accurate wavelength scale for the solar spectrum. In this period E. Mascart (5), F. Bernard (6), L. Ditscheiner (7) and V.S.M. van der Willigen (8) all made wavelength measurements of absorption lines in the solar spectrum, generally to an

* In this book the Ångström unit ($1\,Å = 10^{-10}$ m) will be used for wavelength. It is no longer the favoured unit, but for historical purposes it is the most convenient, especially as its use occurs many times in quotations which are cited.

accuracy of a few Ångström units. However, all these contributions were eclipsed by the work of A.J. Ångström (1814–74) in Uppsala in 1868. He used a transmission grating to produce a drawing of the solar spectrum of unsurpassed quality showing about 1000 lines (9). Nine strong lines were used for an absolute wavelength calibration; Ångström obtained 5889.12 and 5895.13 Å for the close D-line doublet. In spite of the care of his measurements, this scale still contained a substantial systematic error of about 1 Ångström unit. Ångström probably suspected this himself, but the scale was not corrected until 10 years after his death by R. Thalén, his collaborator (10). Meanwhile the visual atlas of Ångström was extended to the ultraviolet by Cornu, and this work included a new ultraviolet wavelength scale with a comparable accuracy to Ångström's (11).

The uncertainty in the wavelength scale was essentially cleared up by H.A. Rowland in Baltimore in the last two decades of the century. He photographed the solar spectrum with a concave grating (12) and determined wavelengths to about 0.01 Å precision (13).

1.2.3 Continuous, emission and absorption spectra A spectrum can be regarded as a one-dimensional image in which the intensity of light can be analysed as a function of wavelength. The wavelength varies with position along the spectrum's length. The dispersion is a key parameter. High dispersion refers to the greatest degree of spreading of different wavelengths.*

Newton observed a continuous spectrum of sunlight in 1666. This is also the spectrum seen in a rainbow and from incandescent solids in the laboratory. The intensity of light in a continuous spectrum appears to change only slowly with wavelength. On the other hand, the spectrum of a candle flame was the best known example in the early nineteenth century of an emission spectrum. Here light occurs in bands at only certain discrete wavelengths, with at most a weak continuum or no light elsewhere in the spectrum. An emission spectrum of this sort was observed by Thomas Melvill as early as 1752 (14). William Wollaston was one of the first to study such flame spectra in more detail (15). The light from an electric spark also produces an emission spectrum with narrow lines. This was observed by Fraunhofer in 1824 (16) and detailed observations were made later by Wheatstone (17) and others.

Absorption-line spectra occur when light at discrete wavelengths is

* Dispersion is conveniently given in units of Å/mm. Strictly speaking this is the reciprocal dispersion, but here the notation will be abbreviated and the 'dispersion' will be referred to in Å/mm. Thus 'high dispersion' implies a small numerical value in Å/mm. For those forewarned of this common terminology, no confusion should arise.

absent from a continuous spectrum. The spectra of this type were observed in sunlight by Wollaston in 1802 and by Fraunhofer in 1814 (see Chapter 2) and were the first absorption spectra to be described. Subsequently, absorption spectra were produced in the laboratory, especially by Sir David Brewster, in the 1820s and 1830s, by the passage of light from a continuous-spectrum source through a cooler gaseous absorbing medium (18).

1.2.4 Resolution and resolving power of spectroscopes The concepts of wavelength resolution and resolving power are crucial in determining whether absorption lines are visible in the spectrum of sunlight or of a continuous source which has passed through an absorbing gas. The wavelength resolution $\Delta\lambda$ is the smallest resolvable spectral detail that can be discerned in a spectrum. It is a measure of the 'purity' of a spectrum, or the range in wavelengths present in any one spectral position. Low resolution will lead to the disappearance of narrow absorption lines. Resolving power is defined as a dimensionless ratio, $\lambda/\Delta\lambda$. For grating spectrographs it can be shown that this parameter is independent of the order of the diffraction and is often a more useful quantity to quote than the resolution. High resolving power in stellar spectroscopy signifies values of around several times 10^4; values as high as 10^5 are rarely achieved for other than solar spectra, nor are they required for most stellar investigations. On the other hand, low resolving powers of a few thousand are common for stellar spectral classification work, and would be typical for slitless spectra with an objective prism.

High resolving power can be achieved by employing dispersing elements of large physical size or by passing light through a narrow slit placed at the telescope's focal plane.* In this latter case, a collimator should be used after the slit to render the diverging rays parallel. William Swan in Edinburgh had employed a collimator in his spectroscope in 1856 (19). However, in this respect he may not have been the first. According to H. Schellen (20), William Simms in London had constructed a spectroscope with a collimator as early as 1848.

The use of both slit and collimator was employed by Bunsen and Kirchhoff in Heidelberg for their pioneering studies of laboratory and solar spectra, and also by Donati in Rome in 1860 (21). From about 1863, at the time that several pioneers began observing stellar spectra, the slit and collimator had become standard. Rutherfurd, Secchi and Huggins all used this arrangement in their first spectroscopic observations. In particular,

* The illumination of a grating at a large angle of incidence can also achieve high resolving power; modern échelle gratings make use of this property.

Lewis Rutherfurd in New York was a skilled instrumentalist, who criticised Donati's spectroscope for not having the slit at the focus of the telescope, and Secchi's for using multiple prisms of flint and crown glass that would have resulted in considerable light loss (22).

1.3 The development of the spectrograph and spectroscope design

1.3.1 Prism instruments in the nineteenth and early twentieth centuries The development of efficient slit and collimator spectroscopes in the mid-nineteenth century was indispensable to the rapid progress of stellar spectroscopy in the 1860s. This is the view of H.C. King in his text on *The History of the Telescope*:

> It is no exaggeration to say that the rapid rise of astrophysics was made possible by craftsmen like C.A. Steinheil and the Merz organization at Munich, John Browning, William Simms and, later, the Hilger brothers of London, and Howard Grubb of Dublin. These workers soon became acknowledged experts in the design and manufacture of high-grade prisms, spectroscopes, and auxiliary spectroscopic apparatus (23).*

Several other instrument-makers built stellar spectroscopes privately for their own use or for others at this time, notably Hoffmann in Paris, who constructed direct-vision instruments for Janssen (29), one of which was acquired by Secchi and used for his early stellar spectroscopy; and also Rutherfurd and Huggins constructed their own instruments for stellar spectroscopy. For stellar work it was necessary to broaden the stellar image on the slit using a cylindrical lens and this was used in nearly all stellar slit spectroscopes from 1860 onwards. However, such a lens was also used much earlier by Fraunhofer with his slitless objective prism spectroscope. Only when spectrography became established did Huggins dispense with this means of widening the spectrum, and instead he trailed the star image up and down the slit during the exposure (30). For this purpose he devised the technique of reflecting part of the stellar image obliquely into a small guiding telescope. This was later modified from using a small silvered mirror with an aperture immediately in front of the slit, to the use in 1893 of polished slit jaws made from speculum metal (31).

Henry Draper in New York State and William Huggins were the earliest

* Descriptions of early spectroscopes can be found in the following references: C.A. Steinheil (24); J. Browning (25); H. Grubb (26); L.M. Rutherfurd (22, 27); W. Huggins and W.A. Miller (28).

pioneers of spectrum photography. Initially Draper in 1872 had used a slitless spectrograph equipped with an ultraviolet transmitting quartz prism for his 28-inch reflector. By 1876 he had developed a slit instrument which also employed a portrait lens as a camera to focus the spectrum onto the plate (32). Huggins' experiments with spectrum photography also date from this time; his spectrograph was made by Adam Hilger in London and employed quartz lenses and an Iceland spar prism (33). It was mounted at the prime focus of his 18-inch reflector. A new instrument by Browning for the cassegrain focus in 1880 had two prisms and was easier to use (30) (see Chapter 4 for further details). An instrument of this type was obtained by Henry Draper on a visit to Europe in 1879, and used briefly when he resumed his stellar spectroscopy in the last few years of his life (34).

In the last two decades of the nineteenth century and the first of the twentieth, prismatic slit spectrographs became relatively commonplace at several major observatories. Apart from Huggins in London, they were developed most notably by Vogel at Potsdam (35), Campbell at Lick (36), Frost at Yerkes (37), Belopolsky at Pulkova, Küstner in Bonn, Newall at Cambridge (38) and J.S. Plaskett in Ottawa (39). Collimators and cameras were lenses, typically there were two or three flint glass prisms (useful in the range 3800–5000 Å) and the collimator-to-camera focal length ratio was generally two or three to one. This era in spectrograph design coincided with the period when new giant achromatic refractors were coming into service (such as those at Lick, Yerkes and elsewhere). However, prism spectrographs remained in widespread use at least until the end of the 1930s. An example of an advanced three-prism spectrograph is the cassegrain instrument designed by Walter Adams for the Mt Wilson 60-inch telescope in 1912. It could produce spectra at 5.2 Å/mm at Hγ in its highest dispersion mode (40).

The problems which arose in stellar spectrograph design at this time were the need to reduce flexure of the instrument attached to the moving telescope, the need to avoid temperature changes in the spectrograph, the problem of how to pass the ultraviolet starlight efficiently through a slit under visual guiding and the question of chromatic effects in the spectrograph lenses. Since the early photographic emulsions were sensitive in the ultraviolet and blue spectral regions, the fact that prisms gave very poor dispersion in the red and that most stars anyway seemed to have fewer interesting lines at longer wavelengths, meant that most observers were happy to record their spectra in the short wavelength photographic region of the spectrum. Flexure and temperature effects both prevented longer exposures unless they were carefully allowed for. At Potsdam Vogel rarely exposed for longer than 1 hour for these reasons. Wright at Lick paid

Fig. 1.1 The three-prism spectrograph on the Yerkes refractor, 1903.

careful attention to these problems, employing counterbalancing devices and a very stiff construction to reduce the flexure effects, and insulation of the whole spectrograph in a blanket and internal heaters to offset increased heat loss as the night got colder (41). Deslandres in Paris also had used a thermostat in the interior of his spectrograph in 1898 (42). He circulated water in the hollow walls of his spectrographs as thermal insulation, and proposed constructing spectrographs from 'invar', a nickel steel with a low thermal expansion.

Chromatic aberration in refractors proved a major drawback for photography and spectrography. Both Draper and Huggins had avoided these problems using reflecting telescopes, and this also allowed them to reach the ultraviolet region of the spectrum with quartz optics spectrographs. Deslandres in Paris was using the 1.2 m reflector for stellar spectrum photography from 1890 (43), and the Lick telescope installed in Chile in 1903 is another example of a reflector prominent in the early days of spectrography. However such instruments were the exception rather than the rule, and those refractors whose objectives were corrected for visual work were generally unsuited for spectrographic use unless further modified. Newall in Cambridge inserted an extra lens in his telescope 1.5 m in front of the focus to convert it to slit spectrography.

J. Hartmann at Potsdam was able to overcome the problem of chromatic aberration in spectrograph camera lenses by simply tilting the plate along the length of the spectrum, thus compensating for the change in the focal length with wavelength (44). This technique was widely copied elsewhere.

1.3.2 Early objective prism instruments The objective prism instrument, which had been used visually by Fraunhofer and by Secchi, came into its own from the 1880s when astronomical photography became an established technique. Here neither collimator nor slit is employed and one or more prisms of small apex angle are placed in front of the object glass of an astrographic telescope. The pioneering work of Edward Pickering at Harvard using objective prism spectrographs is the subject of Chapter 5. He began this work in 1882 with a 13° prism of 20 cm aperture. Objective prism spectrography was commenced at the Paris Observatory soon afterwards (45); here a 5° crown glass prism of aperture 21 cm was initially employed and the spectra of stars to ninth magnitude were recorded. Also in the nineteenth century Lockyer and McLean in England and von Gothard in Hungary undertook objective prism spectrography. Lockyer used objective prisms mounted on his 6- and 10-inch refractors (46) from about 1890, while McLean used a 12-inch Grubb astrograph with a 20° prism from 1895. More details are discussed in Chapter 4.

Objective prism spectrographs were fast and efficient but being slitless they usually gave low resolving power of several thousand or less. However, if three prisms were mounted, which would be justified only in excellent seeing, then quite high dispersion could be achieved. On the 13-inch Boyden telescope at Harvard's Arequipa station in Peru, three prisms gave spectra 7.43 cm in length from Hβ to Hε, an interval of nearly 900 Å

(47). At times up to four 15° prisms were used on the 11-inch Draper telescope at Harvard (48), but the long exposures limited such work to the brightest stars.

1.3.3 Early grating spectroscopes and spectrographs

Although diffraction gratings were used regularly for solar spectroscopy in the nineteenth century (see section 1.2.2), their use for stellar work at this time was very limited, due to their inefficiency which resulted from the available light being divided between several diffraction orders. Thus, even though gratings could give a nearly linear dispersion and were able to disperse light in the red part of the spectrum and did not suffer from the absorption of a long train of prisms, they were still little used. In 1875 Secchi received a grating spectroscope made by Rutherfurd in New York. Rutherfurd was producing reflection gratings diamond-ruled on speculum metal by about 1850 and his reputation for high quality gratings was considerable for the next three decades. Secchi commented: 'The spectra formed thus by interference possess a clarity and certainty which by far surpass those from prisms' (49). However, it is probable that he was referring to a laboratory or solar source; it is not certain if either Rutherfurd or Secchi attempted stellar observations with grating spectroscopes. If they did, it seems that they were unsuccessful.

Vogel at Potsdam is known to have attempted to use a grating (ruled by Wanscheff in Berlin) for stellar spectroscopy at Potsdam in 1881 (50). This is probably the first venture into stellar grating spectroscopy. However, the experiment was not successful, due to insufficient light in the spectrum.

The next experiments with gratings in astronomical spectroscopy were undertaken by Keeler at Lick from 1890 to 1891 (51). His visual spectroscope on the giant refractor had either three prisms or a Rowland grating ruled to 14 438 lines per inch (570/mm; see Fig. 6.7). He used this instrument in the grating mode to obtain a very accurate wavelength for the green emission line of the Orion nebula and of several planetary nebulae (see section 6.6). He also observed a number of stellar spectra with this instrument. He used the grating in the third or fourth order of diffraction which results in a high linear dispersion, which is an advantage for accurate wavelength determinations in an emission-line spectrum (where the lines are not dimmed by increased dispersion). Keeler's work was the first successful observation of grating spectra from a celestial source other than the sun. His success can be attributed to his use of a high quality Rowland grating, to his use of a large telescope, and to his restricting his observation either to emission-line nebulae or to very bright stars.

Several further tests with gratings for stellar work followed in the first decade of the nineteenth century. For example, W.S. Adams at Mt Wilson reported obtaining a blue spectrum of Arcturus with the assistance of Hale on the Snow solar telescope in 1905. As at Lick, a Rowland plane grating was employed. The exposure time was 23 hours over 5 nights (52). W.W. Campbell and S. Albrecht used a grating spectrograph at Lick in 1910; they observed the spectrum of Mars but the instrument appears not to have been used to any great extent thereafter (53).

J.S. Plaskett at the Dominion Observatory, Ottawa, can be regarded as the pioneer of successful stellar grating spectrography. In 1912 he obtained a grating from J.A. Anderson (1876–1959), Rowland's successor at Johns Hopkins University, and the pioneer of the 'blazed' grating which was claimed to concentrate half the diffracted light into the first order spectrum.* Plaskett compared prismatic and grating spectrograms of bright stars (54). The uniformity of dispersion and lack of ultraviolet absorption from the grating led to spectra uniformly exposed from the Hβ line to 3850 Å. He wrote:

> ... although the spectra obtained from the grating are disappointingly weak..., yet even under this handicap it can be used to advantage when the K line is required and if spectra of uniform intensity or of uniform dispersion are needed. It would be useful in the red end where prismatic spectra are so unduly compressed. If a grating giving twice the intensity could be obtained it would be superior even to single-prism dispersion for most work (54).

It was the potential for gratings in the red and infrared part of the spectrum, as noted by Plaskett, that motivated P.W. Merrill to begin experiments at Mt Wilson in 1922 (55). He cited the advantages that were anticipated: a longer length of spectrum of the correct exposure, a normal (i.e. uniform) dispersion, the ease of changing the spectral region under study (by tilting the grating), the small size of the spectrograph, and (for the first time) shorter exposures than with a prism. By 1924 he was employing a blazed 600 line/mm plane grating ruled at Mt Wilson by C. Jacomini who, with J.A. Anderson, had established the ruled grating section at that observatory in 1912 (see (56)). Together with R.F. Sanford, Merrill used this spectrograph to expose about 120 test spectrograms over the next 2 years. Flexure was a problem, but the instrument was found to be useful and efficient. The 1924 spectrograph was the prototype for Merrill's Mt

* The blazed grating is able to concentrate much of the light into one diffraction order by ruling with a specially shaped diamond which determines the groove profile.

Wilson cassegrain grating spectrograph of 1929 (Fig. 1.2) (55). This was the first modern and successful grating spectrograph in regular use in stellar astronomy. The elimination of flexure was a special feature of the design so as to allow for long exposures of up to 6 hours. The interchangeable

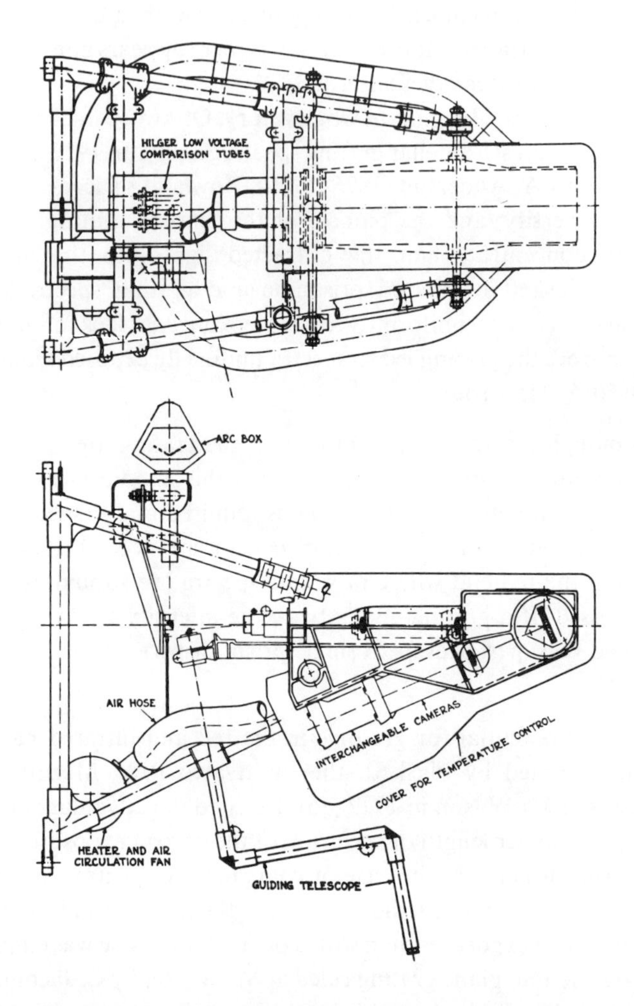

Fig. 1.2 Merrill's grating spectrograph at Mt Wilson, 1929.

cameras could provide three dispersions, of 111, 66 or 34 Å/mm. Some spectra were obtained on infrared sensitised plates to as far as 8700 Å and these included the first stellar observations of the three strong ionised calcium lines in the spectra of cooler stars.*

* The CaII infrared triplet lines are at 8498, 8542 and 8662 Å.

1.3.4 The development of coudé spectrographs The coudé configuration for a telescope is an optical arrangement devised by Maurice Loewy (1833–1907) at the Paris Observatory in the 1880s (57). Light reaches the coudé focus after two successive reflections from plane mirrors, which direct it to a fixed point. Here it can be recorded photographically or passed into a large spectrograph, which might be too massive for mounting on the moving telescope itself. This arrangement was used at Paris for a 27 cm coudé refractor, and then later on for a 60 cm instrument in which the converging coudé beam was directed up the polar axle to the northern pier of a two-pier equatorial mounting.

Loewy became director of the Paris Observatory in 1896 and he initiated the construction of the first coudé spectrograph on the 60 cm refractor in 1903. The construction was undertaken by P. Gautier and completed in 1907. This was a Littrow spectrograph, in which the light traversed the prism train twice after reflection. Three dispersions were available at 4, 8 or 20 Å/mm. Forty spectra of bright stars were photographed by M. Hamy with this instrument in 1907 for the purpose of obtaining radial velocities of the stars by the Doppler effect (58). Although the exposures were long, stellar spectra continued to be obtained with this instrument mainly by Hamy and P. Salet for several decades, except for a 6 year wartime break from 1914. The number of spectra obtained each year was, however, never very high.

The advantages of a stationary and thermally controlled environment for a large spectrograph of high resolving power, that could produce for stars what Rowland had already achieved for the sun, were recognised by George Ellery Hale, when he designed the 60-inch reflector at Mt Wilson with a coudé focus as one of several optical arrangements that could be chosen. This telescope made its first observations in December 1908 and the coudé spectrograph was completed by 1911. The coudé beam in this case was passed vertically down into an underground pit where the spectrograph was constructed.

The first spectra obtained with this coudé instrument were described by Walter Adams (59). The dispersion was very high, giving 1.4 Å/mm at 4300 Å and the plates were 43 cm in length. Good definition spectra of bright stars were obtained over all this length by Adams and Harold Babcock (1882–1968). A single large, dense, flint-glass prism was used in a Littrow mounting as dispersing element; in this arrangement a 15.2 cm lens served as both collimator and camera.

Undoubtedly the 60-inch coudé spectrograph served as the prototype for that of the 100-inch. This telescope was being planned even before the 60-inch had been completed. It was commissioned in 1918 but the coudé spectrograph was not installed by Adams until some years later, in 1925. Like the earlier coudé spectrograph on the 60-inch telescope on Mt Wilson, that on the 100-inch was a single-prism Littrow instrument; it could produce 2.9 Å/mm spectra at 4300 Å with a high resolving power of 70 000. A four-element Ross lens of 4.5 m focal length served as collimator and camera. Later a system with an aluminised parabolic mirror and a quartz prism was also installed to allow wavelengths between 3050 Å (short wavelength limit due to atmospheric ozone absorption) and 3900 Å (the short wavelength limit for the Ross lens and dense flint prism system). Walter Adams and Theodore Dunham used this spectrograph in 1938 to study the ultraviolet spectra of bright early-type stars (60).

Although undoubtedly successful for very high dispersion work on the brightest stars, the Mt Wilson coudé spectrographs were limited by the light absorption in the necessarily large prisms, and by the inability of large-aperture fast camera lens systems to achieve good image definition over a long spectrum. These limitations were dramatically overcome in the mid-1930s by Adams and Dunham. In the words of Ira Bowen: 'Suddenly, however, the picture was completely changed by the advent of the Schmidt camera and its various modifications and by the development of the blazed grating, which permits the concentration of 60–70 per cent of the light in one order' (61). In 1934 Dunham carried out the first test of a Schmidt camera to replace the Ross lens as a coudé camera (62). In Dunham's words:

> It was of course W. Baade, coming from Hamburg, who brought news of the... invention of the Schmidt plate, by Schmidt, in about 1933, or perhaps 1932,* when Baade came over... he told us about this extraordinary development of the Schmidt plate and the Schmidt camera for photographing stars... to anyone in my position, this struck instantly as a possibility for revolutionizing spectroscopy. You didn't have to use lenses now. The ordinary lenses of those days, for short focus cameras were perfectly horrible compromises. So we rubbed up this aspherical plate... and put it up there in a wooden mounting... And we got some gorgeous spectra out of it instantly (64).

The first Schmidt camera in the coudé spectrograph in 1934 was of 76 cm focal length and gave excellent definition spectra from the ultraviolet to the red (65). Soon afterwards it was replaced with 81 and 185 cm focal

* See Schmidt's 1932 paper (63). The Schmidt camera employs a special mirror and an aspherical corrector plate to eliminate aberrations for wide angle photography.

length Schmidts which were mounted on a completely redesigned steel girder framework in the coudé chamber in 1935, together with an off-axis parabolic mirror for collimation. In 1939 a third 2.85 m focal length Schmidt camera was added for spectra of the highest dispersion (Fig. 1.3).

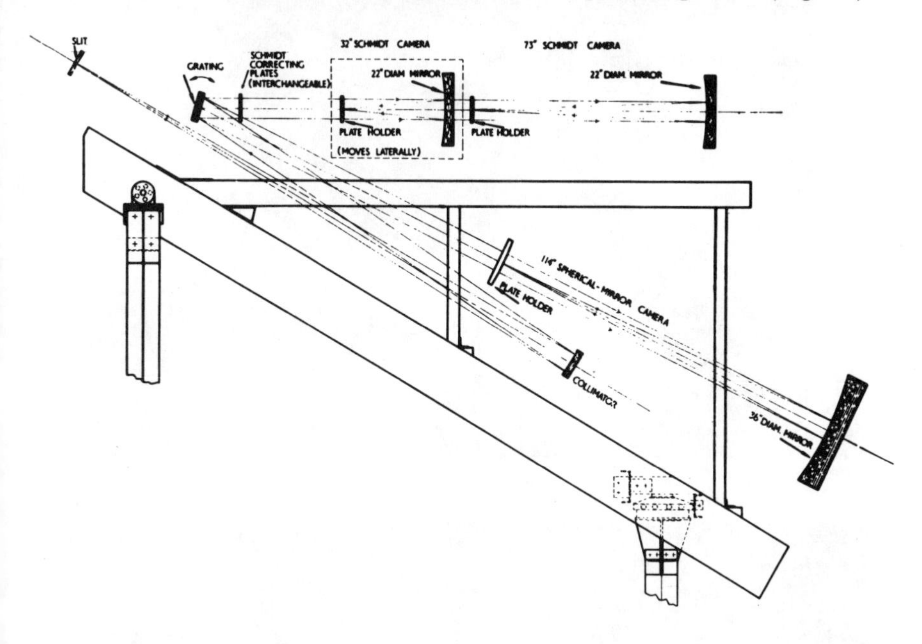

Fig. 1.3 The coudé spectrograph on the Mt Wilson 100-inch telescope showing the optical configuration from 1939.

In 1931 Adams and Dunham began their experiments with an Anderson plane grating to replace the prism in the coudé spectrograph. However, it was not until 1936 that Dunham acquired a grating from R.W. Wood at Johns Hopkins University. This 590 line/mm grating had a high blaze efficiency, peaking at 3500 Å for the second order of diffraction. The rulings were in aluminium which had been vacuum-deposited onto a pyrex base. The Wood grating gave high dispersion spectra at 10.5, 4.5 or 2.9 Å/mm, depending on the choice of camera. A resolving power of over 80 000 could be achieved with the largest camera at the highest dispersion. A sixth magnitude star, the variable Mira, was recorded in about 6 hours with this camera by Adams (66).

The use of the Wood grating and Schmidt cameras revolutionised high dispersion stellar spectroscopy from 1936. Several other coudé spectrographs copied the basic design features used at Mt Wilson. These were the coudé instruments built by W.A Hiltner for the 82-inch McDonald telescope in 1949 (see (61) for details), by Bowen for the 200-inch telescope

Fig. 1.4 Theodore Dunham, Jr, about 1930.

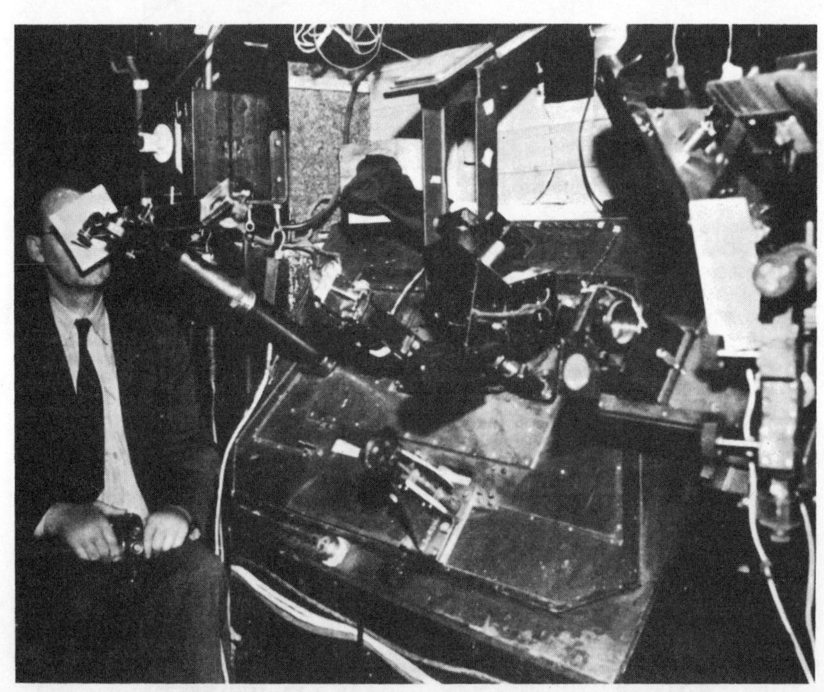

Fig. 1.5 Auxiliary equipment at the coudé focus of the Mt Wilson 100-inch telescope. Ira Bowen is guiding a star on the slit, while shielding his eyes with the mask from the glare of the iron comparison arc.

at Mt Palomar in 1952 (67), and then at Haute Provence, Radcliffe, Lick and Mt Stromlo Observatories over the next decade. From the late 1930s onwards blazed grating instruments increased in popularity while prism spectrographs went into decline. The clear advantages of linear dispersion, nearly uniform intensity (allowing the recording of long lengths of spectrum in one exposure), and the ease of changing the spectral region by tilting the grating, made the new coudé spectrographs the clear first choice for high dispersion work. On the other hand, the efficiency was not always very high compared with cassegrain instruments, because of losses at five of six mirrors and the grating before the light reached the photographic plate.

References

1. Comte, A., *Cours de Philosophie Positive*, II, 19th lesson (1835).
2. *Oxford English Dictionary*, vol. 10, Oxford University Press (1933).
3. Young, T., *Phil. Trans. Roy. Soc.* II, 12 (1802).
4. Fraunhofer, J., *Denkschriften der königl. Akad. der Wissenschaften*, München, **8**, 1 (1821).
5. Mascart, E., *Ann. Sci. de l'école normale supérieure*, **1**, 219 (1864); ibid., **4**, 7 (1866).
6. Bernard, F., *Comptes rendus de l'Académie des Sciences*, **58**, 1153 (1864).
7. Ditscheiner, L., *Wien Acad. Berichte*, **52** (II), 289 (1865).
8. van der Willigen, V.S.M., *Archiv du musée Teyler*, **I**, 1 (1869); ibid., **I**, 57 (1867).
9. Ångström, A.J., *Recherches sur le Spectre Normal du Soleil*, publ. W. Schultz (1868).
10. Thalén, R., *Nova acta reg. soc. sci. Uppsal.*, (3) **12**, 1 (1884).
11. Cornu, M.A., *Ann. Sci. de l'école normale supérieure*, (2) **3**, 421 (1874); *ibid.*, (2) **9**, 21 (1880).
12. Rowland, H.A., *Photographic Map of the Normal Solar Spectrum*, John Hopkins Press, Baltimore, Maryland (1887, 1888).
13. Rowland, H.A., Preliminary table of solar spectrum wavelengths, *Astrophys. J.*, a series of 19 articles from volumes **1** to **6** (1895–8). See also: ibid., *Amer. J. Sci.* (3) **33**, 182 (1887), and ibid., *Phil. Mag.* (5) **23**, 257 (1887).
14. Melvill, T., *Observations on Light and Colours*, Physical and Literary Essays, Edinburgh (1752).
15. Wollaston, W., *Phil. Trans. Roy. Soc.*, **92**, 365 (1802).
16. Fraunhofer, J., *Gilberts Ann.* **74**, 337 (1823). See also: ibid., *Bericht über die Arbeiten der königl. Akad. der Wissenschaften zu München vom Apr. bis Jun.*, p. 61 (1824).
17. Wheatstone, C., *Phil. Mag.* (3) **7**, 299 (1835).
18. Brewster, D., *Edinburgh Trans.*, **9** (II), 433 (1823) and **12** (III), 519 (1834). See also: ibid., *Phil. Mag.*, **2** (3), 360 (1833) and **8** (3), 384 (1836).

19. Swan, W., *Phil. Mag.* (4) **11**, 448 (1856); ibid., *Trans. Roy. Soc. Edinburgh*, **21**, 421 (1856).
20. Schellen, H., *Spectrum Analysis*, p. 230 (1872).
21. Donati, G.B., *Nuovo Cimento*, **15**, 292 (1862). See also: ibid., *Mon. Not. Roy. Astron. Soc.*, **23**, 100 (1863).
22. Rutherfurd, L., *Amer. J. of Science and Arts*, **36**, 154 (1863).
23. King, H.C., *The History of the Telescope*. Charles Griffin and Co., p. 284 (1955).
24. Steinheil, C.A., *Astron. Nachrichten*, **53**, 253 (1863).
25. Browning, J., *Mon. Not. Roy. Astron. Soc.*, **30**, 198 (1870); ibid., **31**, 203 (1871); **32**, 213 and 214 (1872).
26. Grubb, H., *Mon. Not. Roy. Astron. Soc.*, **31**, 36 (1871).
27. Rutherfurd, L.M., *Amer. J. of Science and Arts*, **39**, 129 (1865).
28. Huggins, W. and Miller, W.A., *Phil. Trans. Roy. Soc.*, **154**, 415 (1864).
29. Janssen, J., *Comptes Rendus de l'Académie des Sciences*, **55**, 576 (1862).
30. Huggins, W. *Phil. Trans. Roy. Soc.*, **171**, 669 (1880).
31. Huggins, W., *Astron. and Astrophys.*, **12**, 615 (1893).
32. Draper, H., *Amer. J. of Science*, (3) **18**, 419 (1879).
33. Huggins, W., *Proc. Roy. Soc.*, **25**, 445 (1877).
34. Draper, H., *Proc. Amer. Acad.*, **19**, 231 (1884).
35. Vogel, H.C., *Astrophys. J.*, **11**, 393 (1900).
36. Campbell, W.W., *Astrophys. J.*, **8**, 123 (1898).
37. Frost, E.B., *Astrophys. J.*, **15**, 1 (1902).
38. Newall, H.F., *Mon. Not. Roy. Astron. Soc.*, **65**, 608 (1905); ibid. **65**, 636 (1905).
39. Plaskett, J.S., *Report of the Chief Astronomer, Canada*, **1**, 161 (1909).
40. Adams, W.S., *Astrophys. J.*, **35**, 163 (1912).
41. Wright, W.H. *Publ. Lick Observ.*, **9**, 50 (1911); ibid., *Astrophys. J.*, **11**, 259 (1900).
42. Deslandres, H.A., *Bull. Astron.*, **15**, 57 (1898).
43. Deslandres, H.A. reported by E. Mouchez (director) in *Rapport Annuel sur l'Etat de l'Observatoire de Paris pour l'Année 1890* (1890).
44. Hartmann, J., *Zeitschrift für Instrumentenkunde*, **24**, 257 (1904).
45. Mouchez, E., *Rapport Annuel sur l'Etat de l'Observatoire de Paris pour l' Année 1885* (1885).
46. Lockyer, J.N., *Phil. Trans. Roy. Soc.*, **184**, 675 (1893).
47. Cannon, A.J., *Harvard Ann.*, **28**, (part 2), (1901).
48. Maury, A.C., *Harvard Ann.*, **28**, (part 1), (1897).
49. Secchi, A., *l'Astronomia in Roma*, p. 27, Rome (1877).
50. Vogel, H.C., *Zeitschrift für Instrumentenkunde*, **1**, 47 (1881).
51. Keeler, J.E., *Lick Observ. Bull.*, **3**, 161 (1894).
52. Adams, W.S., *Astrophys. J.*, **24**, 69 (1906).
53. Campbell, W.W. and Albrecht, S., *Lick Observ. Bull.*, **6**, 11 (1910).
54. Plaskett, J.S., *Astrophys. J.*, **37**, 373 (1913).
55. Merrill, P.W., *Astrophys. J.*, **74**, 188 (1931).
56. Babcock, H.D. and Babcock, H.W., *J. Optical Soc. America*, **41**, 776 (1951).
57. Loewy, M., *Comptes Rendus de l'Académie des Sciences*, **96**, 735 (1883).

58. Loewy, M., *Rapport Annuel sur l'Etat de l'Observatoire de Paris pour l'Année 1907* (1907).
59. Adams, W.S., *Astrophys. J.*, **33**, 64 (1911).
60. Adams, W.S. and Dunham, T., *Astrophys. J.*, **87**, 102 (1938).
61. Bowen, I.S., *Astronomical Techniques, Stars and Stellar Systems*, vol. 2, p. 34, edited by W.A. Hiltner, University of Chicago Press (1962).
62. Dunham, T., *Vistas in Astronomy*, **2**, 1223 (1956).
63. Schmidt, B., *Mitteilungen der Hamburger Sternwarte*, **7**, no. 36 (1932).
64. Dunham, T., *Oral History Interviews*, American Inst. of Physics, p. 49 (1977).
65. Dunham, T., *Phys. Rev.*, **46**, 326 (1934).
66. Adams, W.S., *Astrophys. J.*, **93**, 11 (1941).
67. Bowen, I.S., *Astrophys. J.*, **116**, 1 (1952).

2 The analysis of sunlight: the earliest pioneers

2.1 Isaac Newton and the composition of sunlight

The story of solar, and hence also of astronomical spectroscopy, began in 1666 when the young Isaac Newton (1642–1726) wrote these famous words:

> I procured me a Triangular glass-Prisme, to try therewith the celebrated Phaenomena of Colours. And in order having darkened my chamber, and made a small hole in my window-shuts, to let in a convenient quantity of the Suns light, I placed my Prisme at his entrance, that it might thereby be refracted to the opposite wall (1).

The quotation is from Newton's first paper which he sent in 1672 to the Royal Society. As he himself notes, the phenomenon was already well-known and views on the nature of colour by Descartes, Grimaldi, Hooke and others had already been published. The key feature distinguishing Newton's repeat of the experiment was probably his large distance from the prism to the screen or far wall (2). Newton tells us he used a wall 22 feet from the prism, so allowing sufficient space for the colours to separate out and to give a clear spectrum. 'Comparing the length of this coloured spectrum with its breadth, I found it about five times greater; a disproportion so extravagant, that it excited me to a more than ordinary curiosity of examining, from whence it might proceed' (1).

Newton found that the white sunlight was composed of a mixture of light of different colours. He eliminated prism irregularities as a possible cause for the colours, by combining the light again with a second prism so as to reproduce a white solar image. And hence he concluded that white light is 'ever compounded, and to its composition are requisite all the aforesaid primary Colours, mixed in due proportion' (1). To each colour Newton associated a refrangibility, or refractive index, which he was able to deduce by applying Snell's law of refraction.

> And so the true cause of the length of that Image was detected to be none other, then that Light consists of Rays differently refrangible... To the same degree of Refrangibility ever belongs the

> same colour, and to the same colour ever belongs the same degree of Refrangibility. The least Refrangible Rays are all disposed to exhibit a Red colour... The most refrangible Rays are all disposed to exhibit a deep Violet Colour (1).

His interpretation of the prismatic colours provoked a bitter controversy, notably with Hooke who was the acknowledged expert in optics in England at that time. This may indeed account for the delay of more than three decades before Newton published the first edition of his *Opticks* (3) in 1704, the year following the death of Hooke, in which all his optical experiments were described again in a single work.

Newton's discovery of the composite nature of white sunlight was undoubtedly a major breakthrough, though it was hardly exploited at all until almost a century and a half later. Instead his remarks on telescope design in the same paper may have had an unfortunate negative influence on astronomy, as Newton incorrectly concluded that the bothersome coloured or chromatic images of refracting (i.e. lens) telescopes were unavoidable. Only a reflecting telescope could be successfully made achromatic, he maintained. It wasn't until nearly the middle of the eighteenth century that Chester Hall found that achromatic refractors were indeed possible, a discovery that was exploited by John Dollond in London.

Also unfortunate is that Newton narrowly failed to discover the dark bands, or absorption lines, in the solar spectrum. This failure arose from his use of a circular aperture a quarter of an inch in diameter instead of a narrow slit through which to pass the sunlight. This gives a relatively impure spectrum with each colour overlapping considerably the adjacent colours, thus blurring the detail which can be perceived in the spectrum. He did subsequently observe the solar spectrum with a narrower slit corresponding to an angular width of 6 arc min, still not small enough to observe the absorption lines. Newton's observation of the spectrum of Venus, in which the light was collected by a lens, is not so widely known. He placed a prism just before the focus and found Venus' image was 'drawn out into a long and splendid line', but the spectrum was nevertheless still continuous (4).

If in the solar experiment he had instead used either a collimator lens or a slit 1 mm wide, the dark lines in the solar spectrum would probably (depending on his prism quality) have become visible (5), and the whole course of solar and laboratory spectroscopy might have been different. Indeed, the resolving power ($\lambda/\Delta\lambda$) in Newton's first spectrum cannot have been better than about ten, and his realisation that this could be improved with a narrower slit was still inadequate to achieve the necessary resolving power of a few hundred to see the absorption lines clearly.

Thus Newton's work produced more controversy than the basis for further developments. The fact that the next major advance in analysing sunlight did not come till practically the nineteenth century is in itself testimony of how little influence Newton had on his immediate followers in the domain of sunlight analysis. The early history of solar spectroscopy consists of a few islands, represented by a handful of famous names, with long intervals of time intervening. Not until the 1860s were the intervals of no productivity finally eliminated.

2.2 Invisible rays in the solar spectrum: Thomas Young and the measurement of wavelength The next advance after Newton came from another giant of astronomy, William Herschel (1738–1822). In 1800 Herschel investigated the heating power of rays in the solar spectrum, by allowing different parts of the spectrum produced by a prism to fall on a sensitive thermometer. He found that red light raised the temperature by 7° F above the surroundings, green by $3\frac{1}{2}$ °F, violet by only 2 °F (6). What is more, Herschel found the maximum heating effect of the solar spectrum fully $1\frac{1}{2}$ inches (38 mm) beyond the red end of his 4-inch long visible spectrum. Here the temperature increase was 9 °F (7). There was no heating effect, however, beyond the violet end. Herschel noted that the maximum visible brightness of sunlight lay in the green or yellow, so the maximum intensity of the solar heat rays was well displaced from that for visible light. In two further papers (8) Herschel found that heat rays obey the same laws of refraction and reflection as light, though the suggestion that radiant heat and light might be essentially the same phenomenon did not come till later, from J.B. Biot (1774–1862) in 1813 (9), and was not generally accepted until the mid-nineteenth century.

Although Herschel had found no heating effect in the solar spectrum beyond the violet end, the discovery of these rays followed soon afterwards, by the German scientist R.W. Ritter (1776–1810). He found a blackening effect on silver chloride due to ultraviolet solar light (10). His experiment preceded the discovery of photography using silver salts by some 38 years.

At the same time, Thomas Young (1773–1829) in England was investigating the phenomena of interference and diffraction of light and these experiments clearly demonstrated the wave nature of light. He used a 500 grooves/inch diffraction grating to achieve the first measurement of the wavelengths of light of different colours (11). His values are in good agreement with later work. The visible spectrum stretched from 6750 Å in the far red to 4240 Å at the violet limit for the eye. From now on wavelength was the fundamental quantity determining what we see as colour. Although Newton himself was always reluctant to accept a wave theory for light, he surely would have been pleased with this elegant single parameter to specify colour. However Newton had recognised an important point, namely that the unaided human eye cannot distinguish between a monochromatic colour (light of only one wavelength) and the same visual colour sensation induced by a mixture of different wavelengths.

2.3 William Wollaston and the discovery of the solar line spectrum

The first few years of the nineteenth century were unusually productive ones in experimental spectroscopy. In 1802, just 2 years after giving up a successful medical practice, the English chemist and mineralogist William Wollaston (1766–1828) published a paper with the title: 'A method of examining refractive and dispersive powers by prismatic reflection' (12). In this work he described the production of a solar spectrum in which he saw dark gaps between the primary colours. He thought these narrow lines of darkness were just that, gaps that for some reason separated the colours. His diagram showed five easily visible dark lines which he labelled A to E and two fainter ones (f and g) (Fig. 2.1). The description of his discovery was almost an incidental note to the rest of the paper. Clearly Wollaston himself completely failed to recognise the importance of the dark lines, nor did his results arouse any immediate interest from others. The identifications of Wollaston's dark lines are not all certain. His B is sodium absorption; his D, in the blue part of the spectrum, is due mainly to the CH molecule, and his E is due to ionised calcium absorption in the far violet (13).

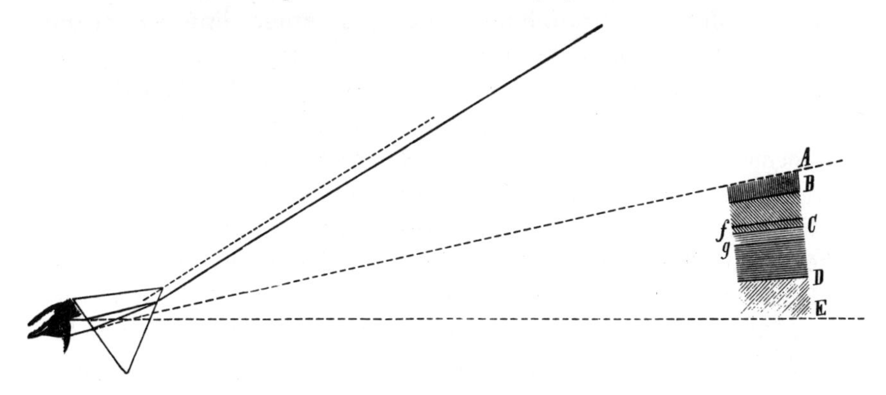

Fig. 2.1 Wollaston's solar spectrum.

Despite Wollaston's failure in interpretation of the lines, his experimental success came from using a slit aperture which he described thus:

> If a beam of day-light be admitted into a dark room by a crevice 1/20 of an inch broad and received by the eye at a distance of 10 or 12 feet, through a prism of flint glass, free from veins, held near the eye, the beam is seen to be separated into the four following colours only, red, yellowish green, blue and violet... The line A that bounds the red side of the spectrum is somewhat confused, which seems in part owing to want of power in the eye to converge red light. The line B, between red and green, in a certain position of the prism, is perfectly distinct; so also are D and E, the two limits of violet. But C, the limit of green and blue, is not so clearly marked as the rest; and there are also, on each side of this limit, other distinct dark lines, f and g, either of which, in an imperfect experiment, might be mistaken for the boundary of these colours (12).

In these inauspicious circumstances was astronomical line spectroscopy born. Wollaston had recognised the importance of a narrow slit, had discovered and described seven solar absorption lines, and he first used the word 'line' in the spectroscopic context, which has remained. It was, however, left to a far greater genius than Wollaston independently to rediscover the solar dark line spectrum, to examine and draw it in far greater detail, and to relabel the lines with a nomenclature which, at least for some of them, is still in common use. This genius was Joseph Fraunhofer (1787–1826).

2.4 Joseph Fraunhofer and the solar line spectrum

Fraunhofer came from humble parentage in Straubing near Munich and had very little formal education, having lost both parents when he was eleven. In 1807, at the age of 20, he was hired by the Mathematical Mechanical Institute Reichenbach, Utzschneider and Liebherr, a firm founded in 1804 for the production of military and surveying instruments, for which high quality optical glass for lenses was essential. The optical works of the firm were outside Munich, at a disused monastery in Benediktbeuern, where Fraunhofer received his training from a Swiss named Pierre Guinand. Guinand's considerable reputation rested on his skill in the production of relatively large and optically pure pieces of crown and flint glass. However, due to a clash of personalities, Guinand resigned his contract in 1814 and returned to Switzerland, and at this time the whole

Fig. 2.2 Joseph Fraunhofer.

firm passed into the hands of Utzschneider and Fraunhofer. About forty people were employed at the Benediktbeuern glassworks (see (14) and (15) for more details).

The success of this famous early glass factory lay in the production of optical crown and flint glass free from bubbles and veins. The technique of stirring the molten glass was discovered by Guinand, and developed by Fraunhofer. The use of these glasses enabled Fraunhofer to construct achromatic optical instruments of hitherto unsurpassed quality, and this was undoubtedly a key factor to his successful pioneering work in solar spectroscopy. Fraunhofer embarked on a careful examination of the optical properties of his glass, so as to measure the refractive index and dispersion. His work on the solar spectrum can therefore be seen as the means to Fraunhofer's end goal of perfecting optical instruments, for he realised that accurate refractive indices must be measured in monochromatic light. For, having rediscovered the solar absorption lines, he saw that the lines defined the precise wavelength of the light far better than the mere sensation of colour to the human eye.

Fraunhofer observed the solar spectrum using a telescope of 25 mm aperture taken from one of his theodolites. A prism was mounted in front of the objective and this enabled him to focus a relatively pure spectrum for direct visual inspection through the eyepiece. His introductory words are almost reminiscent of those used by Newton:

Fig. 2.3 Interior of the Fraunhofer glassworks at Benediktbeuern.

In a shuttered room I allowed sunlight to pass through a narrow opening in the shutters, which was about 15 seconds broad and 36 minutes high, and thence onto a prism of flint glass, which stood on the theodolite... The theodolite was 24 feet from the window, and the angle of the prism measured about 60°... I wanted to find out whether in the colour-image [i.e. spectrum] of sunlight, a similar bright stripe was to be seen, as in the colour-image of lamplight. But instead of this I found with the telescope almost countless strong and weak vertical lines, which however are darker than the remaining part of the colour-image; some seem to be nearly completely black (16).

Fraunhofer convinced himself that the lines in no way represent colour boundaries, as the same colour is found on both sides of a line with only a gradual and continuous colour change throughout the spectrum. Ten of the strongest lines were labelled with the letters A, a, B, C, D, E, b, F, G and H from the far red to the limit of the eye's vision in the violet. The last letter was used for the pair of strong violet lines that we now know are due to absorption by calcium. He noted that A was very near the red limit of the spectrum, but he was still able to see some red light beyond this feature. He showed the D feature to be composed of two close dark lines which exactly coincide with the bright lines emitted by lamplight, while b consists of three very strong lines, amongst the strongest in the solar spectrum. The G feature was also found to be composite, consisting of 'many lines clustered together, among which several stand out through their strength. The two stripes at H are the most extraordinary; they are both almost completely the same and consist of many lines; in their middle is a strong line which is very black' (16).

Between the lines B and H, Fraunhofer observed 574 fainter lines and was able to give precise positions for some 350 of these in his drawing of the solar spectrum (Fig. 2.4). In this figure he also indicated by the curve the approximate apparent intensity distribution of the light in the spectrum as judged by the eye. In a second paper he was able to measure the wavelengths of the Fraunhofer lines, as the solar absorption lines have

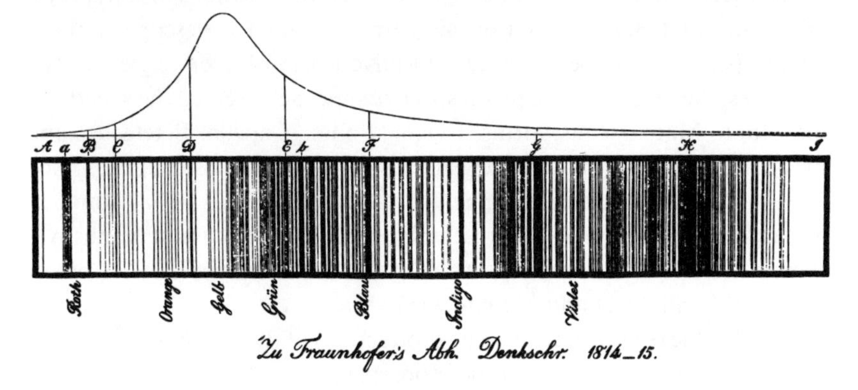

Fig. 2.4 Fraunhofer's solar spectrum.

since been called (17). For this he used a diffraction grating consisting of a large number of equally spaced thin wires placed before the objective of the telescope. He observed a series of spectra in several orders which he interpreted quantitatively in terms of the wave theory of light so as to measure the wavelengths of the solar lines. In the production of diffraction gratings he was one of the early pioneers.

Fraunhofer did not attempt to explain the origin of the dark solar lines. He knew they were intrinsic to the nature of sunlight, and not any instrumental effect. He restricted himself to careful and accurate observation rather than the speculation that characterised the work of some other spectroscopists over the next four decades.

2.5 Planetary and stellar spectra observed by Fraunhofer

Fraunhofer's spectroscopic work did not stop at the sun. Half a century ahead of his time, he initiated the science of planetary and stellar spectroscopy. With his theodolite telescope he observed the spectra of Venus, Sirius and other first magnitude stars. For Venus he wrote: 'I have seen the lines D, E, b, F perfectly defined... I have convinced myself that the light from Venus is in this respect of the same nature as sunlight' (16). For stars Fraunhofer found something surprisingly different:

> I have seen with certainty in the spectrum of Sirius three broad bands which appear to have no connection with those of sunlight; one of these bands is in the green, two are in the blue. In the spectra of other fixed stars of the first magnitude one can recognise bands, yet these stars, with respect to these bands, seem to differ among themselves (16).

In 1823 Fraunhofer was able to report new observations of the spectra of the stars and planets (18). For this purpose he used a telescope of about 10 cm aperture equipped with an objective prism of apex angle 37° 40′. With this, the first objective prism spectroscope, he observed the spectra of the Moon, Mars, Venus, Sirius, Castor, Pollux, Capella, Betelgeuse and Procyon: 'The spectrum of Betelgeuse (α Orionis) contains countless fixed lines which, with a good atmosphere, are sharply defined; and although at first sight it seems to have no resemblance to the spectrum of Venus, yet similar lines are found in the spectrum of this fixed star in exactly the places where the sunlight D and b come' (18).

In 1819 the optical section (lens production) of the instrument firm was shifted to Munich. Fraunhofer therefore spent most of his time there, going only on occasions to Benediktbeuern where the glass works were kept on.

He died of tuberculosis when only 39, and it is likely he would have made further outstanding contributions to spectroscopy and telescope design had he survived a normal lifespan. From 1826 the optical section of the firm was directed by Georg Merz, who had been a pupil of Fraunhofer's since 1808. When Utzschneider died in 1839, Merz purchased the entire business with the financial assistance of his friend Mahler. High quality refracting telescopes were produced under Merz, and the firm continued to operate the Benediktbeuern glass works until about 1884, when they were finally closed (14). However, the optical firm continued in the Merz family in Munich until 1903. Although Fraunhofer had died in his thirties, at the height of his powers, the legacy he left behind in the form of this famous instrument company was nevertheless an important factor in the promotion of astronomy in the nineteenth century, not least in stellar spectroscopy.

References

1. Newton, I., *Phil. Trans. R. Soc.*, **6** (no. 80), 3075 (1672).
2. Kuhn, T.S., Newton's Optical Papers, Published in *Isaac Newton's Papers and Letters on Natural Philosophy*, edited by I.B. Cohen. Harvard Univ. Press (1958).
3. Newton, I., *Opticks* 1st ed. 1704; 4th ed. 1730. Republished by Dover Books (1952).
4. Newton, I., in letter to Henry Oldenburg, secretary of the Royal Society, dated 13 April 1672. See also Newton's *Lectiones Opticae*, written 1669–70, published London (1729).
5. Kayser, H., *Handbuch der Spectroscopie*, Vol. I, p. 4 Hirzel-Verlag, Leipzig (1900).
6. Herschel, F.W., *Phil. Trans. R. Soc.*, **90**, 255 (1800).
7. Herschel, F.W., *Phil. Trans. R. Soc.*, **90**, 284 (1800).
8. Herschel, F.W., *Phil. Trans. R. Soc.*, **90**, 293 and 437 (1800).
9. Biot, J.B., *L'Institut* (1813) and *Gilberts Ann.*, **16**, 376 (1814).
10. Ritter, J.W., *Gilberts Ann.*, **7**, 527 (1801) and ibid., **12**, 409 (1803).
11. Young, T., *Phil. Trans. R. Soc.*, **92**, 12 (1802).
12. Wollaston, W.H., *Phil. Trans. R. Soc.*, **92**, 365 (1802).
13. Brewster, D., *Rep. British Assoc.*, p. 308 (1832).
14. Chance, W.H.S., *Proc. Phys. Soc.*, **49**, 433 (1937).
15. Seitz, A., *J. Fraunhofer und sein optisches Institut.* Springer-Verlag, Berlin (1926).
16. Fraunhofer, J., First as lectures to the Munich Academy of Sciences in 1814 and 1815, printed in: *Denkschriften der Münch. Akademie der Wissenschaften.*, **5**, 193 (1817) and also in: *Gilberts Ann.*, **56**, 264 (1817). His work was published in English in: *Edinburgh Phil. J.*, **9**, 296 (1823) and ibid **10**, 26 (1824).
17. Fraunhofer, J., *Denkschriften der Münch. Akademie der Wissenschaften.*, **8**, 1 (1821).
18. Fraunhofer, J., *Gilberts Ann.*, **74**, 337 (1823).

3 The foundations of spectral analysis: from Fraunhofer to Kirchhoff

3.1 The beginnings of spectral analysis: the work of Sir John Herschel We consider now the development of spectral analysis from where Fraunhofer left it until about 1860. This date marks the culmination of several decades of slow and often frustrating efforts by many workers mainly in Germany, Great Britain and France, with the eventual explanation by Kirchhoff and Bunsen of the origin of the Fraunhofer lines, and of how spectral lines can be used for chemical analysis. The early progress following Fraunhofer was slow. However, the reasons for this are clear enough in retrospect, as several formidable obstacles that barred the way forwards had to be overcome. For example, the bright double yellow–orange line (named by Fraunhofer as R when seen in emission) was found in the spectra of all flames, regardless of what substances might be introduced into the flame. The earliest reports of studying flame spectra go back to the eighteenth century with experiments in England by Thomas Melvill (d. 1753) (1) and by George Morgan (1754–98) (2).

The investigation of the spectra of flames was taken up from the early 1820s by Sir John Herschel (1792–1871). He wrote in 1823: 'The colours thus communicated by the different bases to flame afford, in many cases, a ready and neat way of detecting extremely minute quantities of them' (3). Herschel was therefore one of the first to suggest that flame colours could be used for chemical analysis. Owing, however, to the ubiquitous presence of the R light in all his flames, he failed to demonstrate that each one substance always gives rise to its own unique and characteristic emission-line spectrum, and that the orange light was in all cases due to minute sodium impurities in his samples. According to Kayser (4), the progress of spectroscopy was held up 30–40 years because of the extreme ease of producing the bright orange emission from minute quantities of sodium impurities.

Fig. 3.1 Sir John Herschel.

3.2 Sir David Brewster and spectral analysis Another stumbling block to progress was the belief that the different colours imparted to flames by various substances were in some way due to the interference of light waves. No doubt this idea originated from the well-known phenomenon of the colours produced by thin films, such as oil on a water surface or in soap bubbles. One of the leading optical experimenters in Britain was the Scotsman, Sir David Brewster (1781–1868) who adhered to the interference theory of colours (5), mentioning it even in his later writings (e.g. (6)) after the basis for the correct explanation for spectral lines had already been established by others.

Brewster enjoyed an enormous reputation as a scientist. He was a keen observer, a most productive writer (of more than 300 scientific articles), and a great populariser of scientific discovery. He is best remembered for his work on the polarisation of light. However, his name became associated with two further misconceptions which confused the issues in spectroscopy. He believed that all spectral colours were a mixture of red, yellow and blue (7). These were the three basic colours which were found in different proportions in each part of a pure continuous spectrum, a result which

Fig. 3.2 W.H. Fox Talbot.

complicated the relation between refractive index and colour and contradicted results established by Newton.

In addition, Brewster studied the absorption lines produced when light was passed through nitrous acid gas analysed with a prism (7,8). He compared these lines produced by nitrous acid in the laboratory with the Fraunhofer solar absorption lines, and believed that many of the 2000 or so of the former also appeared in the solar spectrum. He wrote: '... the same absorptive elements which exist in nitrous acid gas exist in the atmospheres of the sun and the earth' (8).

3.3 Fox Talbot and the spectra of flames

At the same time as Sir John Herschel, another English scientist, W.H. Fox Talbot (1800–77) was also investigating the spectra of flames. Fox Talbot was an English country squire who dabbled in mathematics, physics, chemistry, astronomy, botany, archaeology and literature (9). He is best known for his discoveries in photography, which he announced in 1839 practically simultaneously with Daguerre in France. In 1826 Fox Talbot wrote:

> The flame of sulphur and nitre contains a red ray, which appears to me of a remarkable nature.... This red ray appears to

> possess a definite refrangibility, and to be characteristic of the salts of potash, as the yellow ray is of the salts of soda, although, from its feeble illuminating power, it is only to be detected with a prism. If this should be admitted, I would further suggest that whenever the prism shows a homogeneous ray of any colour, to exist in a flame, this ray indicates the formation or the presence of a definite chemical compound... If this opinion should be correct and applicable to the other definite rays, a glance at the prismatic spectrum of a flame may show it to contain susbstances which it would otherwise require a laborious chemical analysis to detect (10).

Together with Herschel, Fox Talbot was therefore one of the first to suggest that spectral observations could be used for chemical analysis. He continued these ideas in a later article on the flame spectra of strontium and lithium salts (11). However, he seems not to have been convinced of the validity of his own suggestion, for he notes that 'the yellow rays may indicate the presence of soda, but they nevertheless frequently appear, where no soda can be supposed to be present' (10). Like Herschel, he was therefore led astray by the presence of sodium impurities in his samples. This problem persisted for many years and was only finally solved by the Scotsman William Swan (1818–94), about 1856 (12, 13), when he recognised the need for very pure samples if their composition was to be deduced by spectral analysis. Swan is also still remembered for his work on the flame spectra of carbon and hydrocarbon compounds. The molecular carbon spectral bands, seen in the spectra of cool stars in absorption, are named after him. In addition, he was one of the first to employ a collimator in his spectroscope to increase the purity of the spectrum.

3.4 Further progress in studying the solar infrared by J. Herschel, Fizeau and Foucault One area where progress was more rapid was the study of heat rays, or infrared radiation, from the sun. William Herschel's pioneering experiment was soon repeated by Sir Henry Englefield (1752–1822), who employed a blackened bulb for his mercury thermometer and focussed the solar spectrum with a lens behind the prism, so as to increase the heating effect (14). Following the development of the thermocouple by Seebeck in 1823, further studies of the invisible solar heat rays followed, most notably from the Italian, M. Melloni (1798–1854). He continued studies of heat from about 1831 to his death, making use of the thermopile to explore the heat distribution in the solar spectrum. He showed that heat rays have a wavelength, in the same way as light rays do, and indeed that both forms of radiation are similar phenomena (15, 16).

Sir John Herschel employed a novel method to detect infrared radiation from the sun (17). He focussed the solar spectrum onto a sheet of black paper moistened with alcohol. The heat from the sun dried the alcohol giving a clearly visible strip corresponding to the infrared spectrum. Herschel found this to be discontinuous, with a series of four broad dark bands separating the peaks of infrared radiation (Fig. 3.3). The result seems to have been doubted for some time, as later repeats of the experiment by

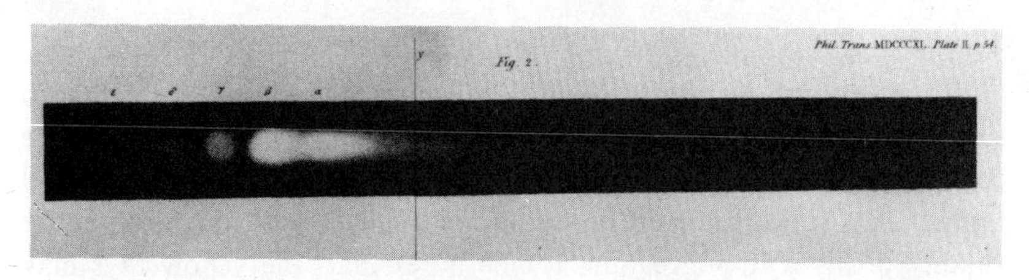

Fig. 3.3 Sir J. Herschel's infrared absorption bands in the solar spectrum.

Lord Rayleigh (18) and by J.W. Draper (19) gave null results. However, subsequent work by Langley at the Smithsonian Institution in 1882 confirmed the presence of broad infrared absorption bands in the solar spectrum due to water vapour absorption in the earth's atmosphere. It seems likely that Herschel had observed these bands (known since as ρ, Φ, ψ, Ω) and detected the infrared spectrum to a wavelength of about 2 μm.

The same bands were found a few years later, in 1847, by Armand Hippolyte Fizeau (1819–96) and Léon Foucault (1819–68) (20). They used

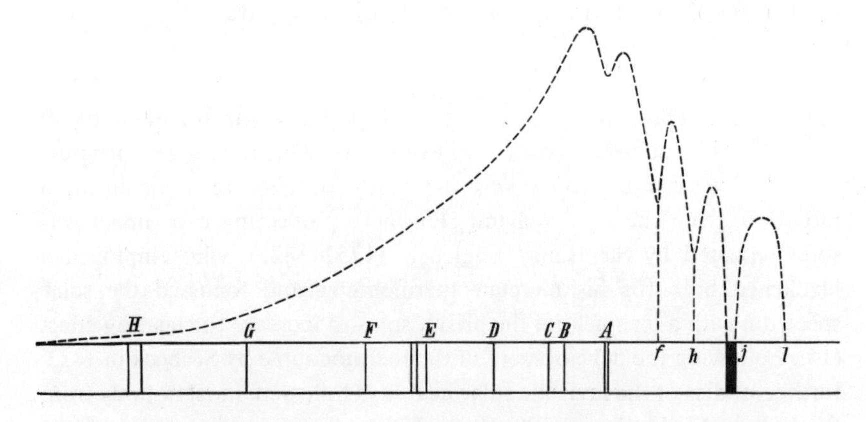

Fig. 3.4 Infrared absorption bands in the solar spectrum observed by Fizeau and Foucault.

an ultrasensitive alcohol thermometer to explore the sun's heat distribution in the infrared spectrum. They labelled the central troughs found f, h and j, with the end of the heat radiation occurring abruptly at a point 1 (Fig. 3.4). Fizeau measured the wavelengths of these infrared solar bands (21). The strongest (j) was at 14 450 Å, and the spectrum termination (1) at 19 400 Å

3.5 Edmond Becquerel and solar spectrum photography

The first experiments in photography by Fox Talbot and Daguerre in 1839 were quickly applied to photographing the solar spectrum. Sir John Herschel was one of the first pioneers in this field (17), though he failed to detect any Fraunhofer lines in his photographically recorded spectra.

The first successful application of photography to the solar line spectrum was undertaken by Edmond Becquerel (1820–91) in France. He came from an illustrious family of physicists, as both his father and his son were professors of physics at the Musée de l'Histoire Naturelle in Paris, as was also Edmond Becquerel himself. In his experiment he passed sunlight through a narrow slit, then dispersed it with a high quality flint glass prism and finally focussed the spectrum on a screen about 1 metre distant using a lens. Onto the screen he mounted a Daguerrotype plate, and was successfully able to record a solar spectrum showing the Fraunhofer lines (22). First he found the lines between F and H, later, with longer exposures, beyond the visible limit in the ultraviolet, and for extremely long exposures even in the green, yellow and red spectral regions to just beyond the A band. Over three decades later Sir Norman Lockyer described Becquerel's achievement thus:

> The result was that on June 13, 1842, Becquerel did what I may venture to call a stupendous feat. He did what has never been done since so far as I know. He photographed the whole solar spectrum with nearly all the lines registered by the hand and eye of Fraunhofer. I do not mean merely the blue end of the spectrum, as you may imagine, but the complete spectrum from the 'latent light' [ultraviolet] to the extreme red end (23).

The most important consequence of this work was the discovery of numerous absorption lines in the ultraviolet solar spectrum, which extended far beyond the visible limit. Becquerel continued Fraunhofer's nomenclature by labelling the strongest ultraviolet lines with letters I to P (in order of decreasing wavelength), with the apparent end of the spectrum being designated Z (Fig. 3.5).

Fig. 3.5 The first photograph of the solar spectrum, by E. Becquerel.

3.6 The photographic solar spectrum of J.W. Draper J.W. Draper (1811–82) repeated Becquerel's work in the United States (24) in 1843. Although born in England, Draper was at the time professor of chemistry in New York. Like Becquerel, he also had a famous son, Henry, who later was to become one of the pioneers in stellar spectroscopy. Draper's spectrum, which he published as a drawing (Fig. 3.6) extended further into the red than Becquerel's, and he recorded three absorption lines beyond the Fraunhofer A band, which he labelled α, β and γ in order of increasing wavelength. In the ultraviolet he estimated at least 600 lines lay between H and P. Draper had independently named the strongest ultraviolet lines with letters L to P. He also designated K the shorter wavelength line of the pair that Fraunhofer had called H.

3.7 Sir George Stokes and the fluorescent ultraviolet solar spectrum Also deserving mention are the experiments of Sir George Stokes (1819–1903) on the ultraviolet solar spectrum. Born in Ireland, Stokes spent most of his working life in Cambridge where he was professor of mathematics for over 50 years. He was also secretary of the Royal Society for more than 3 decades, and then president 5 years – certainly an illustrious record. He is remembered for his work in hydrodynamics, elasticity, waves and optics. In optics he studied the phenomenon of fluorescence, in which certain media such as fluorspar or quinine sulphate solution emit blue light when irradiated by ultraviolet light.

Stokes used fluorescence as an ingenious method of investigating the invisible solar ultraviolet line spectrum. His results were published in 1852 (25) in a paper described rather extravagantly by Lockyer as 'this magnificent paper, one of the crowning glories of the work of the century' (23). In this work Stokes focussed a solar spectrum onto a quinine sulphate solution. A blue fluorescent light was emitted, always of longer wavelength than the exciting solar radiation, which could be detected almost to the G band at one limit, and also far beyond H into the ultraviolet. Stokes wrote: 'Some of the fixed lines less refrangible than H were very plain, and beyond H a good number were visible'. The lines were observed as dark planes

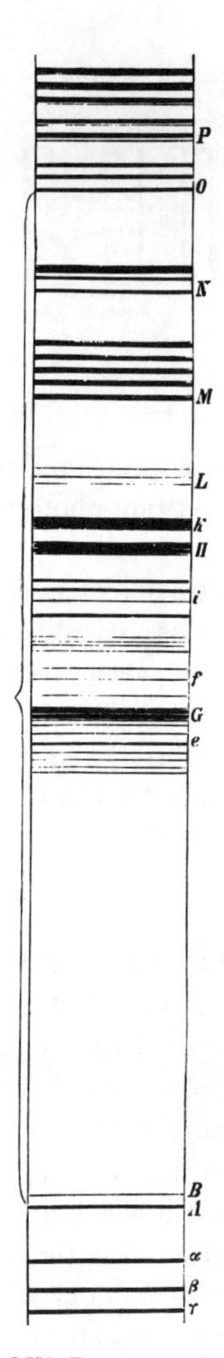

Fig. 3.6 J.W. Draper's solar spectrum.

interrupting the mass of fluorescent light from the body of the fluid. He was thus able to sketch the ultraviolet solar spectrum, and recorded in his diagram (Fig. 3.7) '32 fixed lines or bands more refrangible than H'. He gave the letters l, m, n, o, p to the stronger ones. He was unable to identify his lines with those in Draper's drawing. But in a footnote to his paper he concluded that Becquerel's O is Stokes' n, and Becquerel's P is probably his own o. Stokes recognised that the use of quartz optical components in the spectrograph (which, unlike glass, were transparent to ultraviolet light)

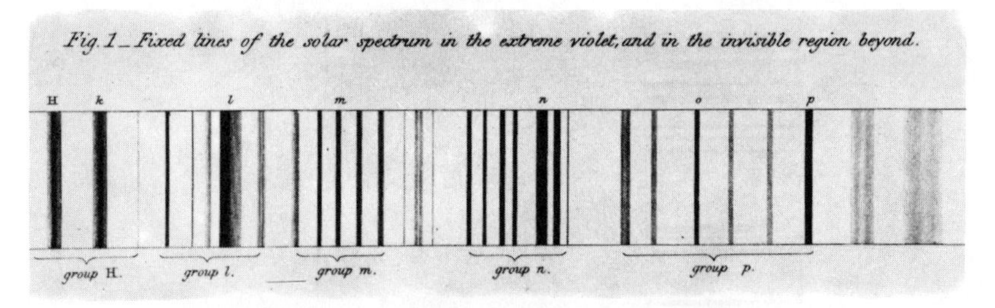

Fig. 3.7 G.G. Stokes' solar ultraviolet fluorescent spectrum.

allowed one to probe further into the ultraviolet spectrum. This fact was also exploited by Sir William Crookes (1832–1919) in photographing the solar ultraviolet spectrum (26).

Meanwhile the first wavelengths of solar ultraviolet lines were measured by E. Esselbach (1832–64), a German whose scientific career was terminated at an early age due to an accident when working on the telegraph in India. He also extended the recorded ultraviolet spectrum beyond Stokes' P to two further strong lines he called Q and R. He claimed once to photograph a line beyond R which he named S. For R he measured 3090 Å; for S, 3040–3050 Å (27).

3.8 The relationship between heat, light and 'chemical rays' Throughout the first half of the nineteenth century, there was no general consensus on the nature either of 'heat rays' (infrared) or of the 'chemical rays' (ultraviolet) which had been found to come from the sun. J.B. Biot in France, as early as 1813, was one of the first to suggest they might both be similar in nature to light (28), but Brewster always opposed this idea, even though it was known that both types of invisible ray could be reflected and refracted in the same way as light. We have already mentioned Melloni, who was one of the strong advocates for Biot's ideas, especially as he had measured the wavelength of heat radiation. To Melloni we should add the name of J.W. Draper. In 1847, Draper published an excellent paper (29), in which he successfully sets out:

(1) To determine the point of incandescence of platinum and to prove that different bodies become red-hot at the same temperature.
(2) To determine the colour of the rays emitted by self-luminous bodies at different temperatures. This is done by the only reliable method – analysis by the prism.

Fig. 3.8 Gustav Kirchhoff.

(3) To determine the relation between the brilliancy of the light emitted by a shining body and its temperature.

He carried these experiments out by observing the prismatic spectrum of an incandescent platinum strip whose temperature was measured by its expansion.

Draper, who himself had earlier doubted the identity of the natures of ultraviolet and light rays, now became convinced that all radiation was identical in nature except for its wavelength. For heat he stated: 'I cannot here express myself with too much emphasis on the remarkable analogy between light and heat which these experiments reveal'. Moreover, he reaffirmed the Newtonian view that 'to a particular colour there ever belongs a particular wavelength, and to a particular wavelength there ever belongs a particular colour', and took the opportunity of criticising Brewster for thinking otherwise, with his theory of three primary colours coexisting at any one refrangibility.

Brewster's views, though influential, had evidently also been controversial for he was sharply rebuked by George Airy, the Astronomer Royal (30) for believing that an absorbing medium can change the colour of light of a given refrangibility (that is, at a given place in the spectrum), an observation on which his theory of the composite nature of light had rested. In hindsight, Brewster's indignant reply (31) to the Astronomer Royal seems very inept, if not amusing, when we consider how far from the truth his reasoning had led him. He wrote in his defence:

> When a philosopher examines, and pronounces an opinion upon the researches of others, especially upon those which competent judges have recognized as sound, he is bound to repeat the identical experiments which he challenges, with similar apparatus and similar materials; to state the differences which he observes, to inquire into its causes by which such discrepancies have arisen, to establish his own views by new and effective experiments, and to publish his researches in vindication of his charges against a fellow-labourer in science. Mr Airy, however, has not done this; but, as we shall presently see, has followed a course which is unusual in the history of science as it may be injurious to its progress.

He goes on: 'I will hazard the supposition that the Astronomer Royal cannot distinguish colours, and is a genuine specimen of an idiopt [one who is colour-blind]'. On Newton, Brewster states: 'Newton committed a mistake, if mistake is a proper term, when he asserted that to the same refrangibility always belongs the same colour' (31).

3.9 The origin of the Fraunhofer lines

If the relationship between heat, light and 'chemical' rays was a source of controversy, then so also were the uncertainties caused by the Fraunhofer absorption lines themselves. Although most observers clearly distinguished between an absorption spectrum such as that from the sun or a cool gas, and an emission spectrum produced by a hot gas, such as in a flame or a spark, this was not always so. A complex molecular emission spectrum observed in a flame can resemble an absorption spectrum, as it is not obvious if the lines are the bright peaks or the spaces between them. Both Fox Talbot and Draper, for example, appear at times to have been confused by this similarity.

More interesting is the speculation on the cause of the solar absorption-line spectrum. Fox Talbot (32) had suggested that the absorption of a cool gas was due to a discordance of the light's frequency of vibration with the natural frequency of vibration of the gaseous molecules, which, although almost the opposite to the truth, was at least a step forward from the idea that the lines are caused by interference. At any rate, it seemed clear that the

Fraunhofer lines were due to absorption by a cool gas, and Herschel (33) had early on considered that this gas might be either the solar atmosphere or the earth's.

One of Brewster's positive contributions to solar spectroscopy was his finding that some of the Fraunhofer lines had strengths that varied with the season and with the sun's proximity to the horizon (8). These lines were mainly in the red and yellow, and he called them 'atmospheric lines', as he correctly believed the terrestrial atmosphere to be their cause. That most of the Fraunhofer lines must be solar in origin seems however to have escaped his and Herschel's attention, as Fraunhofer himself had observed stellar spectra in which some of the lines of the solar spectrum were weak or absent, although the terrestrial atmosphere of course always intervened.

The Scotsman, J.D. Forbes (1809–68) went further, and maintained that all the lines were terrestrial. He observed the lines from sunlight at the edge of the solar disk (34) during an annular eclipse in May 1836. He found them no stronger than from the sun as a whole, even though the light had presumably traversed a greater path through the supposed solar atmosphere when it emanated from the limb.

Meanwhile Draper in the United States confirmed Brewster's observations concerning the strengthening of some of the red Fraunhofer lines as the sun neared the horizon (24) and the whole issue was the subject of a thorough investigation by Piazzi Smyth (1819–1900) in 1856 (35). Piazzi Smyth was sent by the British Admiralty to Teneriffe to conduct spectroscopic observations of the sun from altitudes ranging between 8000 and 10 000 feet. He too found some red lines much stronger when the sun was low, and others which only appeared at that time. On the other hand, many solar lines were unchanged with the height of the sun in the sky. Also, when the sun was highest, he couldn't see as far as the A line, but then the violet end was brighter and clearer.

Brewster and his colleague J.A. Gladstone (1827–1902) rediscussed the whole issue in 1860 (6), and concluded 'the origin of the fixed lines in the solar spectrum must therefore still be considered an undecided question', even though the terrestrial origin was favoured. This paper was published after the work of Bunsen and Kirchhoff had appeared, in which a solar origin was ascribed to the majority of the Fraunhofer lines.

3.10 A key observation by Foucault as a step towards understanding the Fraunhofer spectrum

The problem of the origin of the Fraunhofer lines is one that was solved bit by bit over many decades. Much of the credit for a clearer understanding certainly goes to Kirchhoff and Bunsen. However the origins go back to Fraunhofer himself, since he early on had observed the coincidence of the solar D lines with the bright double line occurring in the spectra of flames (36). Although this coincidence was generally known for several decades, a key observation by Foucault in 1849 (37) was an important step towards an explanation. Foucault observed the bright orange emission lines from a carbon arc, and, in order to demonstrate their exact coincidence with the solar D lines, he passed sunlight through the arc to obtain two spectra superposed. He found to his surprise that the D lines were still in absorption and now stronger than in the solar spectrum without the arc. In another experiment, he reflected light from one of the glowing charcoal points back through the arc and then to his spectroscope. The light from the glowing charcoal gives a continuous spectrum, and the D lines were now seen in absorption in this spectrum, just as in the sun, whereas using arc light alone they had been in emission. Foucault did not speculate on the causes of what he found. However, it is clear that he demonstrated that a given medium could produce either emission or absorption lines, depending on whether its own light is observed, or light from another hotter source passing through it.

3.11 Kirchhoff announces the presence of sodium in the sun Foucault's work was largely ignored until Gustav Kirchhoff (1824–87) independently repeated it 10 years later (38). Kirchhoff noted that the light from a source giving a continuous spectrum, when passed through a powerful salt flame, gave the D lines in absorption. He added the important note that the absorbing medium must be cooler than the source of continuous light, otherwise emission is observed. In this paper Kirchhoff wrote:

> I conclude furthermore that the dark lines of the solar spectrum which are not produced by the earth's atmosphere, result from the presence of that substance in the luminous solar atmosphere which produces in the flame spectrum bright lines in the same place... The dark D lines in the solar spectrum allow one therefore to conclude, that sodium is to be found in the solar atmosphere (38).

This important observation was followed up by two significant developments, one theoretical, the other observational. The theoretical step forward was the understanding of the relation between the emissive and absorptive powers of any body, including gases. At a given temperature and wavelength, the most emissive bodies are also the most absorptive, a result which applies both to a continuous spectrum (for example heated glass radiates less than soot), and to spectral lines.

3.12 The emission and absorption of radiation: the theoretical work of Balfour Stewart and Kirchhoff That the most emissive bodies are also the best absorbers was first deduced in 1858 by the Scotsman, Balfour Stewart (1828–87), who was then an assistant in Professor Forbes' laboratory at the University of Edinburgh. His first paper (39) applied only to heat, but this was extended later, in 1860, to include light (40). The second paper stressed the similarity between these two different types of radiation. However his 1860 work was subsequent to Kirchhoff's own publication (41) on this subject, which treated the theory far more rigorously. The statement that in thermodynamic equilibrium the radiant energy of any wavelength emitted by a body equals the radiant energy absorbed by that body is therefore known as Kirchhoff's Law of Emission and Absorption. This law helped explain the reversal of the D lines in the flame and arc spectra, and was essential much later for the quantitative interpretation of the solar spectrum.

3.13 Further laboratory work in the analysis of flame, arc and spark spectra Kirchhoff's name is also associated with the observational development which now followed. For some time laboratory work in flame, arc and spark emission line spectroscopy had been going on. Masson in France, Miller in England, Alter in the United States, van der Willigen in Holland, Swan in Scotland and Plücker in Germany were all active in the 1850s in studying the lines emitted by different luminous gaseous bodies. We should also mention that in England Wheatstone had reinitiated the study of spark spectra as early as 1835, although the first observations of sparks had been made even earlier by Fraunhofer.

For example, A.J. Ångström showed that spark spectra contained some lines characteristic of the metallic electrodes, while others were due to the air being present with all electrodes (42). He recognised the equivalence of emission and absorption lines and was one of the first to explore systematically the solar absorption line spectrum for coincidences with the spark emissions due to air. Although he thought he found some coincidences, the agreement was poor. He wrote in 1855: 'A complete correspondence, however, between the solar and electric spectrum is the less to be expected, as the lines in the former, as is generally assumed, are not only due to the action of the atmosphere, but also to be referred to the action of the sun itself'. By this time the fact that each element or compound produced its own characteristic spectrum was well established, and the feasibility of chemical analysis this afforded was recognised. J. Plücker's (1801–68) work on the discharge spectra of gases at low pressure in Geissler tubes is also of note (43). He identified the Fraunhofer F line with a bright hydrogen line he called Hβ, while the solar C line was practically coincident with a red hydrogen line Hα, and G in the sun was not far from Hγ (this last identification was however erroneous).

3.14 Bunsen and Kirchhoff: chemical analysis of the solar spectrum

It was this solid background in laboratory spectroscopy established in the 1850s that enabled Robert Bunsen (1811–99) and his colleague Kirchhoff, who were both then at Heidelberg, to continue this work so effectively and then apply it to the analysis of the solar spectrum. Their paper entitled *Chemical analysis by spectral observations* (44) carefully recorded the spectra of lithium, sodium, potassium, calcium, strontium and barium salts in flames and sparks. The paper then continues:

> ... it is evident that the same mode of analysis must be applicable to the atmospheres of the sun and of the brighter fixed stars. A modification must, however, be introduced in respect to the light which heavenly bodies themselves emit. In a memoir published by one of us, 'On the Relation between the Coefficients of Emission and Absorption of Bodies for Heat and Light' [(41)], it was proved from theoretical considerations that the spectrum of an incandescent gas becomes reversed (that is, the bright lines become changed into dark ones) when a source of light of sufficient intensity, giving a continuous spectrum, is placed behind the luminous gas. From this we may conclude that the solar spectrum, with its dark lines, is nothing else than the reverse of the spectrum which the sun's atmosphere alone would produce. Hence, in order to effect the chemical analysis of the solar atmosphere, all that we require is to discover those substances which, when brought into the flame, produce lines coinciding with the dark ones in the solar spectrum.

This suggestion was carried out in Kirchhoff's monumental work which now followed in two parts, the *Investigations of the solar spectrum and the spectra of the chemical elements* (45), which can be regarded as a landmark

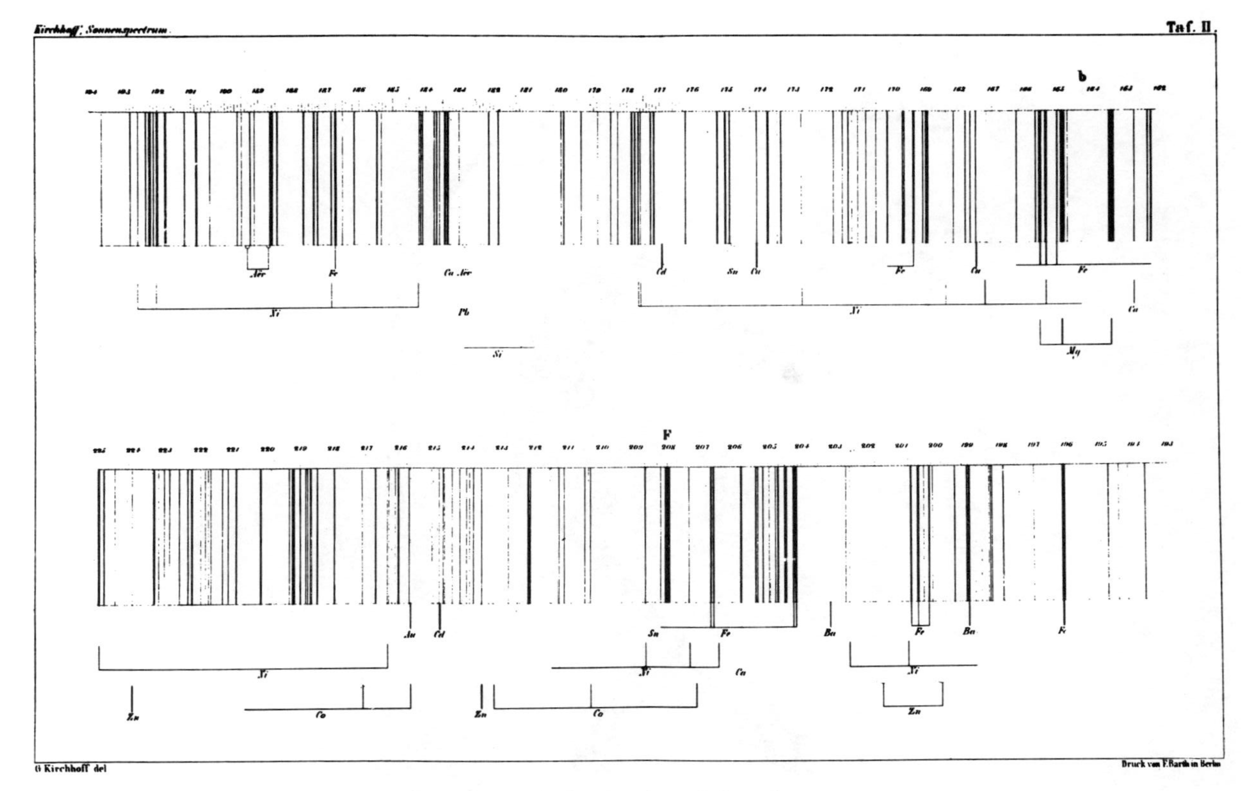

Fig. 3.9 Part of Kirchhoff's drawing of the solar spectrum.

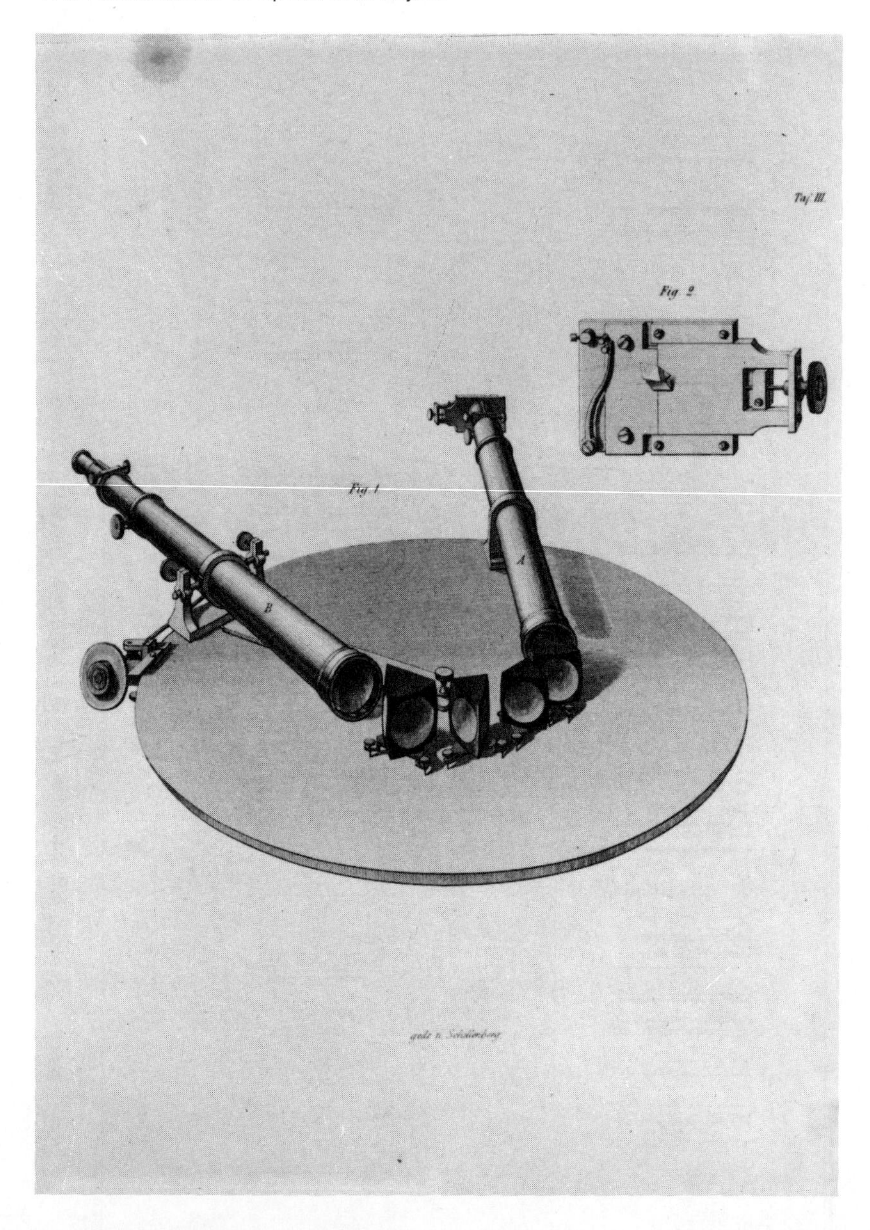

Fig. 3.10 Kirchhoff's spectroscope.

in our qualitative understanding of the Fraunhofer spectrum. Kirchhoff carefully drew the solar spectrum and compared this with the spark spectra of thirty different elements (Fig. 3.9). For this purpose he used a 4-prism spectroscope by Steinheil equipped with a collimator to give exceptional spectral purity (Fig. 3.10). The apparatus allowed simultaneous viewing of both spark and solar spectra for the most accurate comparison of line positions. Kirchhoff concluded that iron, calcium, magnesium, sodium, nickel and chromium must be present in the cooler outer layers of the sun, giving the Fraunhofer absorption. The elements cobalt, barium, copper and zinc were probably present, while the remaining elements studied were not found. For iron, the best observed element, he found exact coincidences for as many as sixty lines, virtually excluding the possibility that this might have arisen by chance. Kirchhoff also discounted Forbes' argument that the Fraunhofer lines could not be solar in origin; Forbes had believed that a thick solar absorbing layer would produce centre-to-limb spectral line variations if the lines originated in the solar atmosphere. He had looked for such variations but found none (34).

3.15 Reactions to the work of Kirchhoff and Bunsen

The work of Kirchhoff and Bunsen excited widespread scientific and popular interest. The concept of a chemical analysis of a distant heavenly body such as the sun aroused the admiration and wonder of many. However, it also provoked considerable controversy over the question of the priority for the discovery of spectral analysis and for the explanation of the dark solar lines. For example, Lord Kelvin wrote to Kirchhoff claiming that Sir George Stokes had already done much of the work which Kirchhoff had published. His letter is worth quoting:

> Prof. Stokes mentioned to me at Cambridge some time ago, probably about ten years ago, that Prof. Miller had made an experiment testing to a very high degree of accuracy the agreement of the double dark line D of the solar spectrum with the double bright line constituting the spectrum of the spirit lamp burning with salt. I remarked that there must be some physical connexion between two agencies presenting so marked a characteristic in common. He assented and said he believed a mechanical explanation of the cause was to be had on some such principles as the following: Vapour of sodium must possess by its molecular structure a tendency to vibrate in the periods corresponding to the degree of refrangibility of the double line D. Hence the presence of

> sodium in a source of light, must tend to originate light of that quality. On the other hand vapour of sodium in an atmosphere round a source must have a great tendency to retain in itself, i.e. to absorb and to have its temperature raised by light from the source of the precise quality in question. In the atmosphere around the sun, therefore, there must be present vapour of sodium, which, according to the mechanical explanation thus suggested, being particularly opaque for light of that quality, prevents such of it as is emitted from the sun from penetrating to any considerable distance through the surrounding atmosphere. The test of this theory must be had in ascertaining whether or not vapour of sodium has the special absorbing power anticipated. I have the impression that some Frenchman did make this out by experiment, but I can find no reference on that point.
>
> I am not sure whether Prof. Stokes' suggestion of a mechanical theory has ever appeared in print. I have given it in my lectures regularly for many years, always pointing out along with it that solar and stellar chemistry were to be studied by investigating terrestrial substances giving bright lines in the spectra of artificial flames, corresponding to the dark lines of the solar and stellar spectra (46).

Stokes himself partially repudiated Lord Kelvin's letter to Kirchhoff, saying that he never had thought that a gas absorbs the same wavelength as it emits: 'I have never attempted to claim for myself any part of Kirchhoff's admirable discovery, and cannot help thinking, that some of my friends have been overzealous in my cause' (47).

Meanwhile Ångström believed a large share of the credit should go his way (48). He carried out a similar analysis of the solar spectrum to Kirchhoff's, but this was published in full in 1862, the year following the first part of Kirchhoff's work.

As a result of the controversy over priority, Kirchhoff felt obliged to defend himself, which he did in 1863 in an article 'On the history of spectral analysis' (46). He rightly points out that even Fox Talbot had considered the possibility of chemical analysis by spectroscopy, but no one prior to himself had either conclusively proved it was feasible, let alone actually applied the method systematically to the sun. Similarly, of Balfour Stewart's work on absorption and emission, Kirchhoff describes the conclusions reached as being in the realm of hypothesis, unlike his own work which constituted a proof.

Nor was Sir David Brewster silent on the question of priority. In 1866 he wrote: 'The great interest created by the important researches of Messrs. Kirchhoff and Bunsen, on what has been called spectral analysis, makes it desirable that the work of those who preceded them is not forgotten . . . I can remark in favour of my friend Mr. Fox Talbot, who has yet earlier rights pertaining to this important research' (49). Most of Brewster's article is taken up, however, in making sure that his own claim cannot be foregotton.

A more objective summary of the developments is given by Pritchard, President of the Royal Astronomical Society, in 1867 (50). It is clear that Kirchhoff did more than anyone to unravel the mystery of the Fraunhofer lines. He went further than his contemporaries, but only after the groundwork had been laid by others.

References

1. Melvill, T., *Physical and Literary Essays*. Edinburgh (1752).
2. Morgan, G.C., *Phil. Trans. R. Soc.*, **75**, 190 (1785).
3. Herschel, J.F.W., *Edin. Trans.*, **9**, (II), 445 (1823).
4. Kayser, H., *Handbuch der Spectroscopie*, Vol. I, p. 14, S. Hirzel Verlag. Leipzig (1900).
5. Brewster, D., *Phil. Mag.*, (3) **21**, 208 (1842).
6. Brewster, D. and Gladstone, J.H., *Phil. Trans. R. Soc.*, **150**, 149 (1860).
7. Brewster, D., *Edin. Trans.*, **12** (I), 125 (1834).
8. Brewster, D., *Phil. Mag.*, (3) **8**, 384 (1836).
9. Obituary, W.H., Fox Talbot, *Nature*, **16**, 523 (1877).
10. Fox Talbot, W.H., *Brewster's J. of Sci.*, **5**, 77 (1826).
11. Fox Talbot, W.H., *Phil. Mag.*, (3) **4**, 112 (1834).
12. Swan, W., *Edin. Trans.*, **21** (III), 411 (1857).
13. Swan, W., *Phil. Mag.*, (4) **20**, 173 (1860).
14. Englefield, H.C., *R. Inst. J.*, **1**, 202 (1802).
15. Melloni, M., *Ann. chim. et phys.*, **48**, 385 (1831).
16. Melloni, M., *Comptes Rendus*, **35**, 165 (1852).
17. Herschel, J.F.W., *Phil. Trans. R. Soc.*, (I), 1 (1840).
18. Rayleigh, Lord, *Phil. Mag.*, (5) **4**, 348 (1877).
19. Draper, J.W., *Phil. Mag.*, (5) **11**, 157 (1881).
20. Fizeau, H. and Foucault, L., *Comptes Rendus*, **25**, 447 (1847) A fuller account is given in: *Ann chim. et phys.*, (5) **15**, 363 (1878).
21. Fizeau, H., *Société philomatique*, 11th Dec. (1847) p. 108. See also *Ann. chim. et phys.*, (5) **15**, 394 (1878).
22. Bequerel, A.E., *Bibliothèque universelle de Genève*, **40**, 341 (1842).
23. Lockyer, J.N., *Nature*, **10**, 109 (1874).
24. Draper, J.W., *Phil. Mag.*, (3) **22**, 360 (1843).
25. Stokes, G.G., *Phil. Trans. R. Soc*, (II), 463 (1852).
26. Crookes, W., *Bull. Photograph. Soc. London*, **2**, 292 (1856).
27. Esselbach, E., *Poggendorff's Ann.*, **98**, 513 (1856).
28. Biot, J.B. *L'Institut* (1813). See also *Gilbert's Ann.*, **16**, 376 (1814).

29. Draper, J.W., *Phil. Mag.*, (3) **30**, 345 (1847).
30. Airy, G.B., *Phil. Mag.*, (3) **30**, 73 (1847).
31. Brewster, D., *Phil Mag.*, (3) **30**, 153 (1847).
32. Fox Talbot, W.H., *Phil. Mag.*, (3) **7**, 113 (1835).
33. Herschel, J.F.W., *A Treatise in Astronomy*, p. 212. London (1833).
34. Forbes, J.D., *Edin. Trans.*, (II), 453 (1836).
35. Piazzi Smyth, C., *Phil. Trans. R. Soc.*, (II). 465 (1858).
36. Fraunhofer, J., *Denkschriften der Münchner Akademie der Wissenschaften*, **5**, 193 (1817) (See Chap. 2, ref. (16)).
37. Foucault, L., *L'Institut* (1849) p. 45. See also: *Ann. de chim. et phys.*, (3) **68**, 476 (1860) and: *Phil. Mag.*, (4) **19**, 193 (1860).
38. Kirchhoff, G., *Monatsberichte Berliner Akad.* p. 662 (1859). See also: *Phil. Mag.*, (4) **19**, 93 (1860).
39. Stewart, B., *Edin. Trans.*, **22**, 1 (1858).
40. Stewart, B., *Proc. R. Soc.*, **10**, 385 (1860).
41. Kirchhoff, G., *Monatsberichte Berliner Akad.* p. 783 (1859). See also: *Poggendorff's Ann.*, **109**, 275 (1860).
42. Ångström, A.J., *Svensk. Vetensk. Akad. Handl.*, p. 327 (1852). See also: *Phil. Mag.*, (4) **9**, 327 (1855).
43. Plücker, J., *Poggendorff's Ann.*, **107**, 497 (1859).
44. Kirchhoff, G. and Bunsen, R., *Poggendorff's Ann.*, **110**, 160 (1860). See also: *Phil. Mag.*, (4) **20**, 89 (1861) and *Phil. Mag.*, (4) **22**, 329 (1861).
45. Kirchhoff, G., Part I: *Abhandl. d. Berliner Akad.*, p. 63 (1861); p. 227 (1862). Part II: ibid., p. 225 (1863).
46. Kirchhoff, G., *Poggendorff's Ann.*, **118**, 94 (1863). See also: *Phil. Mag.*, (4) **25**, 250 (1863) (in these articles Kirchhoff publishes the letter from Lord Kelvin).
47. Stokes, G.G., *Nature*, **13**, 188 (1876).
48. Ångström, A.J., *Phil. Mag.*, (4) **24**, 1 (1862).
49. Brewster, D., *Comptes Rendus*, **62**, 17 (1866).
50. Pritchard, C., *Mon. Not. R. Astron, Soc.*, **27**, 146 (1867).

4 Early pioneers in stellar spectroscopy

4.1 Stellar spectroscopy before 1860 Fraunhofer first observed stellar spectra in 1814. Using his 2.5 cm aperture theodolite telescope he found three broad stripes in the spectrum of Sirius (1). Nine years later with his 10 cm refractor he described the lines he saw in Sirius, Castor, Pollux, Capella, Betelgeuse and Procyon (2). The main result from this work was that stars have dark absorption lines in their spectra, yet that the lines present differ from star to star. Sirius for example, with its three strong lines, was quite dissimilar to sunlight, while Betelgeuse displayed countless lines in its spectrum, some of which corresponded in position to the solar lines (see Chapter 2).

It is perhaps remarkable that the first pioneer to explore line spectra of any source at all systematically should have included stellar spectra in his observations. After Fraunhofer, no significant work was undertaken in stellar spectroscopy for 40 years. It is also surprising that these decades that saw so much activity in solar and laboratory spectroscopy should have seen practically no continuation of the spectroscopic work on stars that Fraunhofer had initiated.

Fraunhofer's 1823 paper describes his objective prism mounted on the 10 cm telescope. One of the few references to stellar spectroscopic observations in the intervening four decades came in 1838 from J. Lamont (1805–79), who was then director of the Royal Observatory in Munich (3). Lamont set up Fraunhofer's apparatus again, and observed spectra of some of the brightest stars. Then he placed a prism between the objective and the eyepiece of the large Munich refractor, and he was able to obtain intense spectra of fourth magnitude stars which showed dark lines with great clarity. However this research appears not to have been pursued further.

The next attempt to repeat and extend Fraunhofer's results came from the Scottish scientist William Swan, whose other spectroscopic work in the laboratory was mentioned in the last chapter. In May 1856 Swan built a spectroscope for the express purpose of accurate measurements of line positions in stellar spectra. He wrote: 'I therefore resolved, as soon as I had leisure, to attempt a series of observations on the spectra of stars' (4). As it happens, he observed Mars on 16 May when 'the spectrum was more brilliant then I anticipated', rushed off his rather sketchy paper to the *Philosophical Magazine* the following day and appears then to have had no more leisure for pursuing stellar spectroscopy further.

In Paris, the Italian astronomer I. Porro (1795–1875) observed Donati's comet of 1858 spectroscopically and tried to compare its spectrum with that of Arcturus (5). He used a large flint glass objective prism mounted in front of his 6 cm refractor. He wrote: 'Directed towards Arcturus, this apparatus gave a perfect spectrum readily measurable, but it wasn't the same with the comet. The low intensity of its light didn't allow me to use a narrow slit, and its angular dimensions prevented me, with a wide slit, from having a pure enough spectrum to measure the Fraunhofer lines'. He was the first, but certainly not the last, astronomer to be caught by the ever-present problem of wanting both sufficient spectral intensity and resolution.

4.2 Stellar spectroscopy: a new beginning G.B. Donati (1826–73) in Florence then took up the study of stellar spectra in 1860, and is generally regarded as the person who reset the ball rolling following Fraunhofer, although his results were of no great significance. He used a large 41 cm chromatic objective lens to collect the light (Fig. 4.1). A cylindrical lens was placed near the focus to widen the spectrum and the light then entered a single-prism spectroscope equipped with a rotatable viewing telescope to measure line positions. Donati published the description of the spectra of some fifteen stars, noting the presence of a green-blue line (in fact Hβ) in many of them (6) (Fig. 4.2). The positions he gave for other lines were, however, discordant with Huggins' measurements of a few years later. An English account of Donati's work is to be found in the *Monthly Notices of the Royal Astronomical Society* (7) for 1863.

The year 1863 is certainly a memorable one in stellar spectroscopy. Apart from Donati's work, four other astronomers published their first accounts of stellar spectroscopic observations in that year. They were Rutherfurd in New York, Airy at Greenwich, Huggins in London and Secchi in Rome. In addition to these four pioneers, we know that C.A. von Steinheil (1801–70) built a stellar spectroscope in 1862 in Munich, where he was professor of astronomy. This instrument was described in the *Astronomische Nachrichten* in the following year (8), and then sent to G.W. Lettsom (1805–87), the British consul in Montevideo who was also an amateur astronomer. Lettsom mentioned using the instrument on his 9-

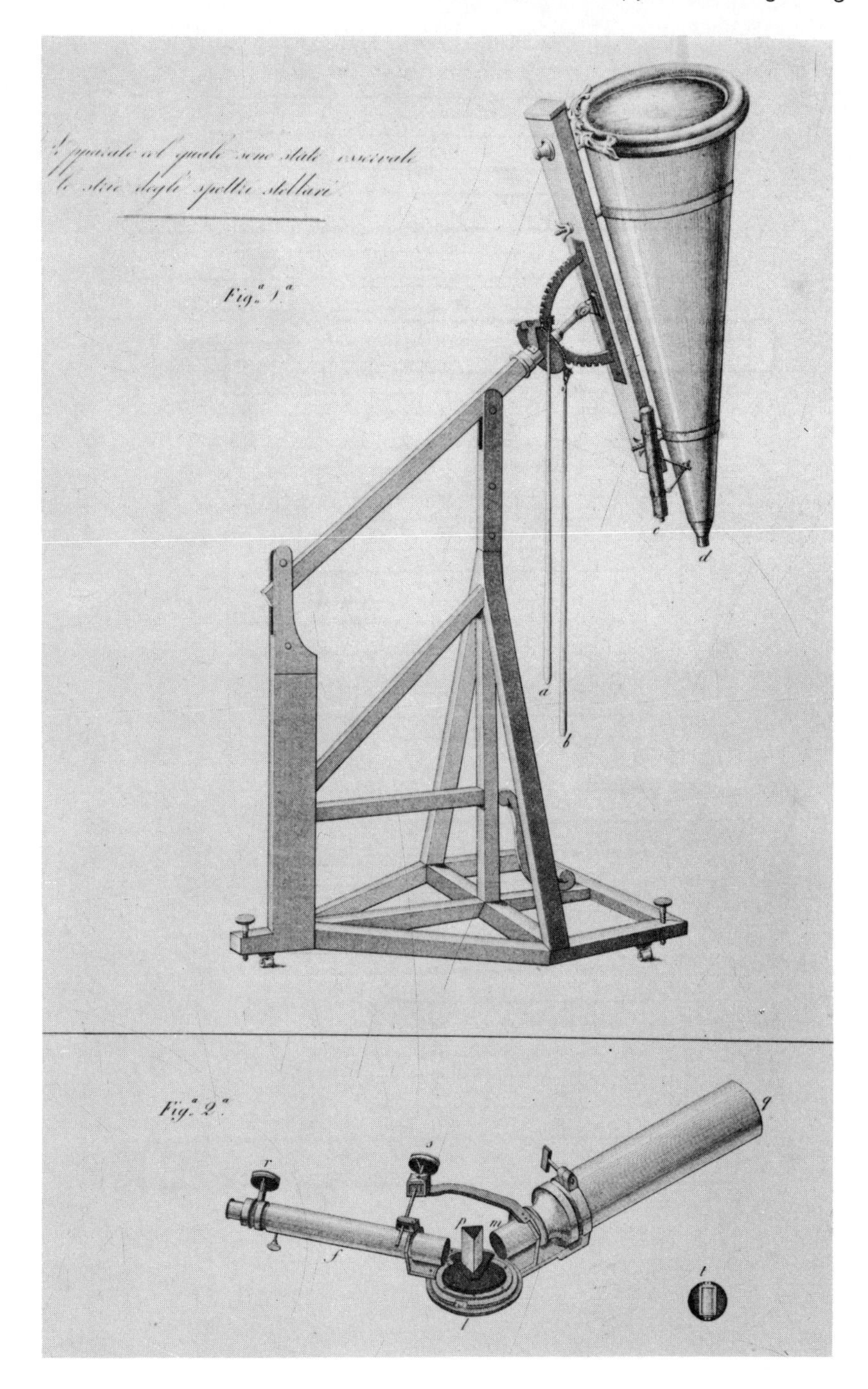

Fig. 4.1 Donati's telescope and spectroscope, 1860.

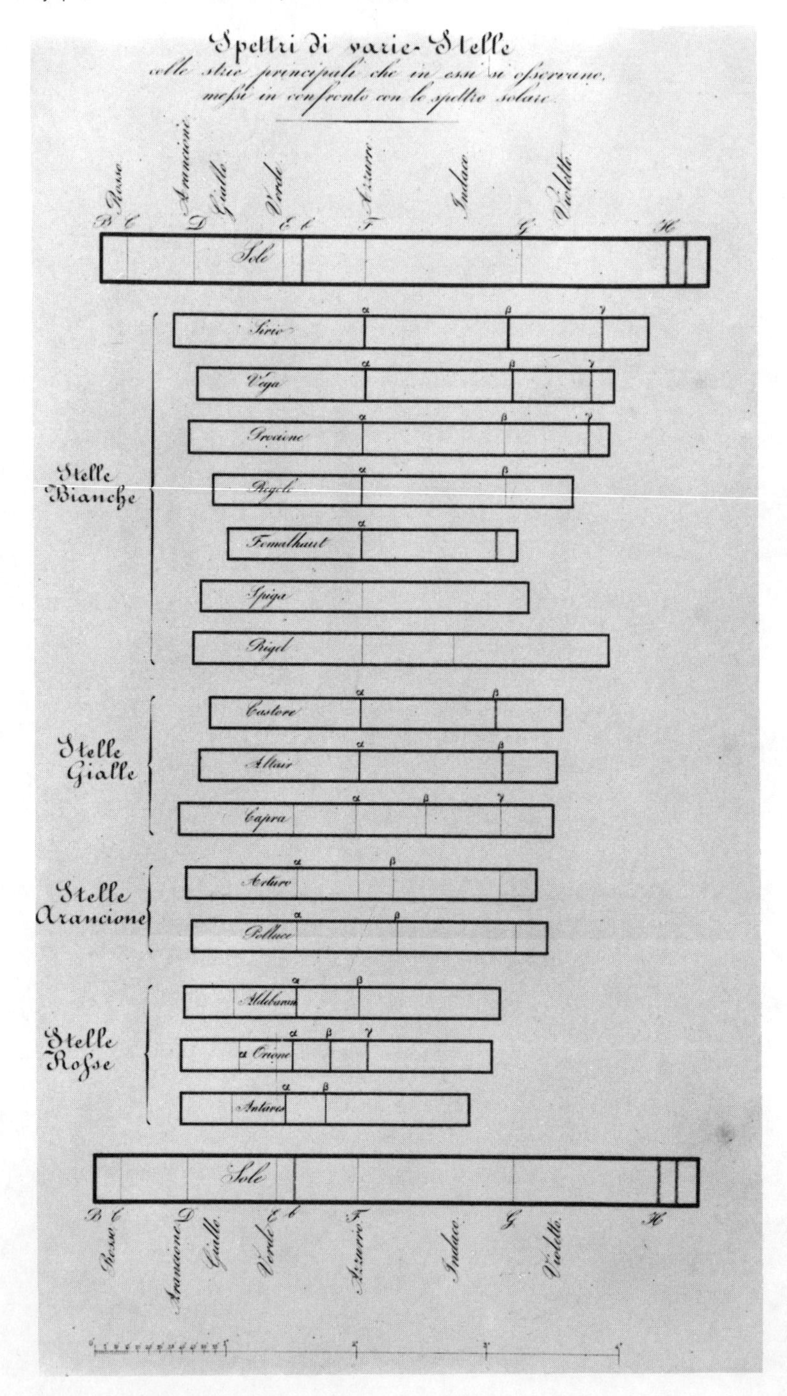

Fig. 4.2 Donati's drawings of stellar spectra.

inch refractor in 1864 (9) so, in a sense, he was the founder of stellar spectroscopy in the southern hemisphere. However his results were of little scientific value, and he does not even mention that lines were ever seen in his spectra. Jules Janssen (1824–1907) was also known to be interested in astronomical spectroscopes in 1862. He had several built by Hoffmann in Paris (10), but he appears not to have used them for other than solar spectroscopy at the time.

Of the four spectroscopists who did publish useful stellar results in 1863, Lewis M. Rutherfurd (1816–92) may have been the first to enter the field. He was an American amateur astronomer who had practised law in Massachusetts until 1850, when he moved back to his home state New York where he founded on observatory from his private means. He equipped this with an $11\frac{1}{4}$-inch Fitz refractor. His pioneering work in celestial photography and the manufacture of diffraction gratings is also still remembered. Rutherfurd's first paper (11) was dated 4 December 1862 and published in the January issue of *Silliman's American Journal.* William Huggins (1824–1910), assisted by William Allen Miller (1817–70), on the other hand did not send his first paper to the Royal Society until 19 February 1863 where it was read on the 26th of that month (12). In a footnote dated 21 February they added that they had seen Rutherfurd's paper and then hastened to mention that they had observed many more stars than the three reported on, and that this work had been proceeding over the previous 12 months.

Father Angelo Secchi (1818–78) began his work in December 1862, the month he received his Hoffmann spectroscope from Paris. His first publication in stellar spectroscopy appeared in a Collegio Romano bulletin (13) and was repeated in *Comptes rendus* (14) and *Astronomische Nachrichten* (15) for a wider distribution. The German version cited is dated 19 February 1863. George Airy's (1801–91) results were reported to a meeting of the Royal Astronomical Society in London on 10 April 1863 (16).

It is certainly remarkable that all these pioneers should have been working practically simultaneously and independently on similar problems. If we are to believe the Huggins and Miller footnote, they were the first in the field after Donati, but this is hardly doing justice to Rutherfurd who published first, and Huggins' later writings do not support successful observations being made before the end of 1862. Furthermore, Rutherfurd almost certainly completed the observations for his first paper before Secchi had started his.

Fig. 4.3 Lewis Rutherfurd.

4.3 Lewis Rutherfurd To Rutherfurd we owe the first attempt to classify the spectra of the stars into different groups. His scheme had three classes which he described as follows (11):

> The star spectra present such varieties that it is difficult to point out any mode of classification. For the present, I divide them into three groups: first, those having many lines and bands and most nearly resembling the sun viz., Capella, β Geminorum, α Orionis, Aldeberan, γ Leonis, Arcturus, and β Pegasi. These are all reddish or golden stars. The second group, of which Sirius is the type, presents spectra wholly unlike that of the sun, and are white stars. The third group, comprising α Virginis, Rigel &c., are also white stars, but show no lines; perhaps they contain no mineral substance or are incandescent without flame.

On the MK system in use today it is easy to identify Rutherfurd's groups as respectively late-type G, K and M stars; late B to early F stars with strong hydrogen lines; and early-type stars without clearly visible hydrogen lines in low resolution spectra, such as early B stars or supergiants where the hydrogen lines are narrow.

Rutherfurd improved his spectroscope by incorporating the facility for simultaneous viewing of a comparison spectrum and he experimented with carbon disulphide liquid prisms (17). He was clearly skilled at instrumentation, as his third paper shows, in which he described and criticised the spectroscopes of Donati, Airy and Secchi (18).

4.4 Early spectroscopy at Greenwich The Astronomer Royal, George Airy, also described the initiation of a program in stellar spectroscopy to the Royal Astronomical Society in 1863 (16). A simple single-prism spectroscope without either a slit or a collimator was built at the Royal Greenwich Observatory, and the positions of some of the lines were recorded by his assistant Mr Carpenter, for nineteen bright stars. The quality of the sketches presented was quite low, and the program to record line positions was apparently not pursued. However, from this work grew the extensive Greenwich program to measure the motions of stars in the line of sight using the Doppler effect (see section 6.4).

Fig. 4.4 Father Angelo Secchi

4.5 Angelo Secchi and spectral classification More significant to the progress of stellar astrophysics was the entry of Father Angelo Secchi into the new world of stellar spectroscopy. Secchi began his training in the church and became a Jesuit priest. However, he abandoned this career in favour of physics, in which he had also been trained. In 1848 he travelled to the United States where he had been offered a professorship in physics at Georgetown College near Washington. In 1850 Secchi returned to Italy where he became director of the Roman College Observatory (Collegio Romano). His interests spanned meteorology, terrestrial magnetism, sunspots and the solar chromosphere, double stars and comets as well as spectroscopy.

Within 2 years of his return to Rome, Secchi had founded a new observatory at the Collegio Romano, thanks to generous support from Pope Pius IX. The principal instrument was a 24 cm equatorially mounted refractor by Merz, which Secchi praised for its excellent optical definition (19) (Fig. 4.5).

Fig. 4.5 Secchi's refractor at the Collegio Romano Observatory.

Secchi embarked on his spectroscopic observations of the planets and stars in December 1862. He appears to have been inspired to undertake this work by Janssen's description of his pocket Hoffmann spectroscope (10) that same year, whereupon Secchi ordered a similar instrument for himself (Fig. 4.6). Janssen also brought his own instrument to Rome with him and Secchi says 'I begged him to be willing to mount it on the great equatorial

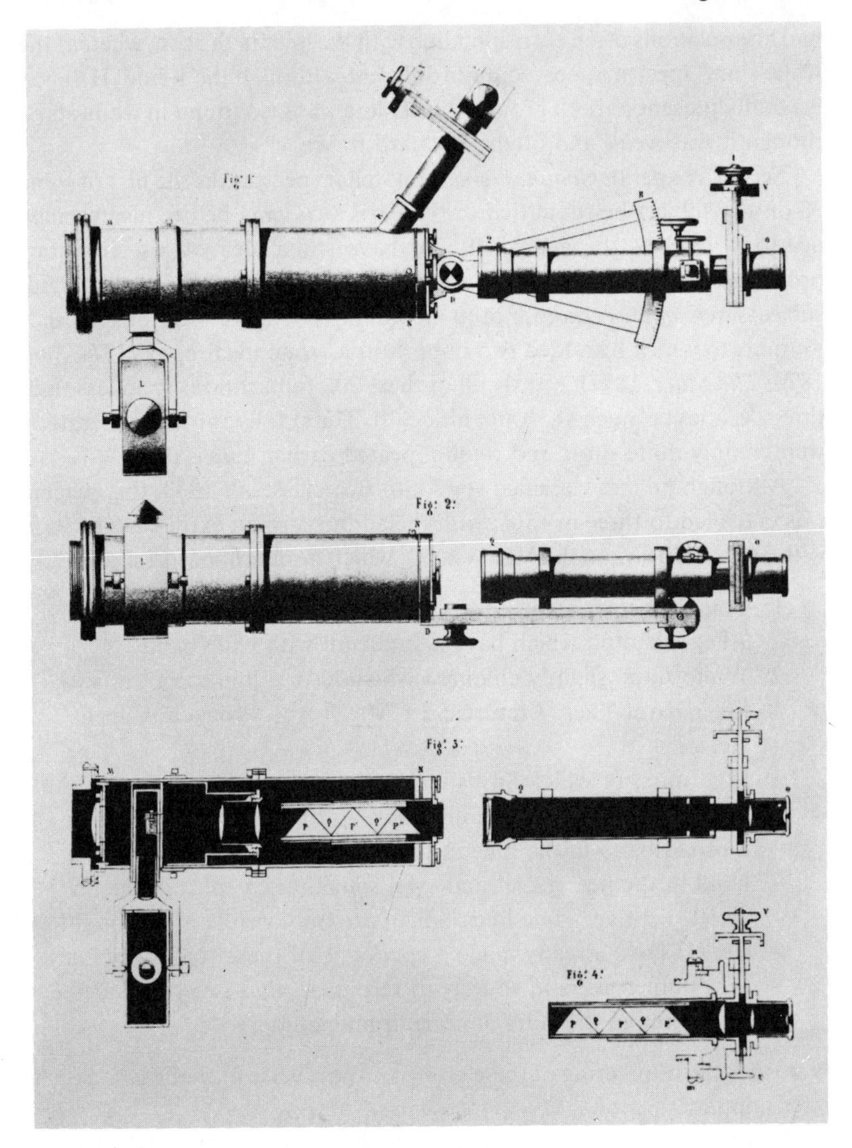

Fig. 4.6 Secchi's first direct vision spectroscope.

telescope, and we were astonished by the magnificent results that we obtained at the first attempt' (15).

His paper to the French Académie des Sciences (14) divided the stars observed into two groups: (a) yellow or red stars including α Ori, α Tau, Algol, β Peg, Arcturus and β UMi, and (b) white stars with few lines such as Sirius, Rigel, β Sco, Castor, ζ UMa, ε UMa, α Lyr, δ Ori. The coloured stars had line positions often corresponding with the lines in the sun, whereas the white stars' spectra appeared quite different, although the F line (Hβ) was generally present in both groups. The D line was also found in white stars, though it was weak and often very hard to see.

Secchi's paper in *Comptes rendus* on stellar spectra was the first of some 33 on the subject he submitted to the Paris Academy before his untimely death in 1878 at the age of 59. To have some idea of his enormous productivity we should remember he published a total of 186 papers on all subjects in *Comptes rendus*, not to mention over 500 publications in other journals to which he added two important astronomical books (*The Sun*, 1870; *The Stars*, 1877). Nearly all of these 700 publications span less than three decades between 1850 and his death. Thankfully for the reader, they were mainly quite short and often repeated earlier material.

Although he had classified stars into two classes in 1863, this scheme was extended to three in 1866, with the addition of an extra class of stars with bands (type M on the MK system), which he described as follows (20):

> 1: Coloured stars, which have as their prototype α Ori, α Sco, β Peg etc., and which have a spectrum with wide bands.
> 2: White stars, slightly coloured which have a line spectrum with fine narrow lines: Arcturus, α UMa, β Aql, Procyon, Capella etc.
> 3: Blue stars, of which Sirius is the type star, and also Vega, α Aql etc. These are the most numerous, and have the characteristic of a broad band in the blue in the position of f,* another broad band in the near violet and even sometimes a third in the far violet, with very fine lines, which are only visible in the brightest stars. I have already made a catalogue of these stars, which are very numerous, and so easy to recognise, that one can find the broad line in the stars of seventh magnitude.

Note that the numbering of these groups is the reverse of that which Secchi later adopted.

Later that year Secchi gave a more detailed account of his classification scheme (21). Class I stars of the Sirius type have hydrogen absorption lines, but a few like γ Cas have these lines in emission. Yet later he was to put the emission-line stars γ Cas and β Lyr into a separate group, class V (22). He also noted that in Orion there are a number of white stars with narrow lines, including one in the place of F (Hβ), and without the usual wide bands. The Orion stars are early B-type stars which Secchi grouped together in class I.

* Secchi here used f for Fraunhofer's symbol F, the Hβ line.

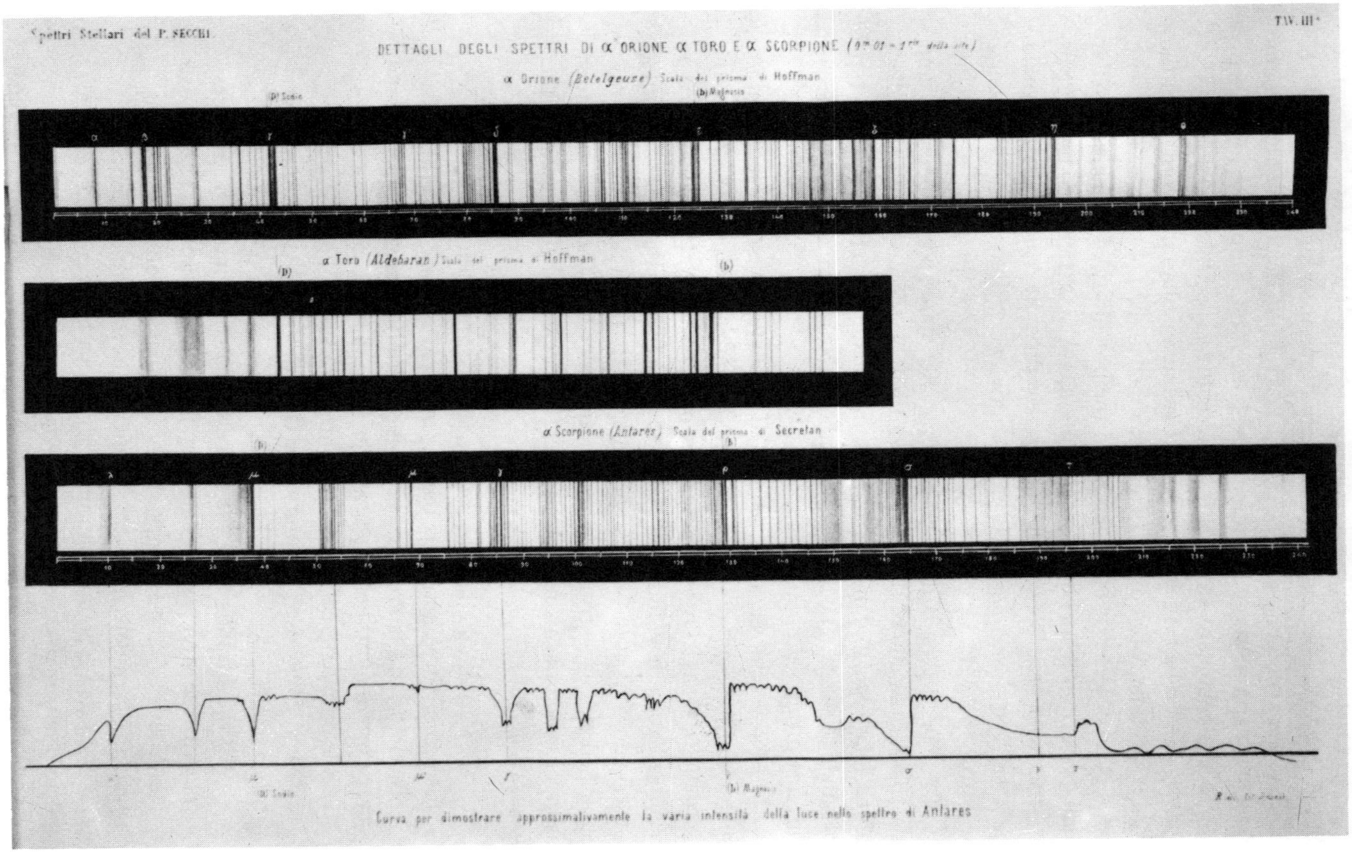

Fig. 4.7 Secchi's drawings of the spectra of Betelgeuse, Aldeberan and Antares, 1867.

The spectra of slightly coloured stars of class II were found to resemble closely the solar spectrum, and the main Fraunhofer lines such as B, D, b, E, F, G were generally visible. As for Class III (the Betelgeuse type), many irregular variable stars, including Betelgeuse itself, were to be found here.

In this important paper, Secchi presented a spectral type catalogue for 209 stars, of which 95 were of class I (white or blue) including 11 in the Orion sub-type, 94 of class II (yellow, solar) and 20 of class III (red, with wide bands). Here he also mentioned the possibility of measuring stellar radial velocities from line displacements, a task first attempted by Huggins the following year. He concluded, in hindsight quite realistically: 'It would be absurd to want to exhaust in this initial work, the new field of stellar spectroscopy' (21).

By 1867 Secchi had observed some 500 stars spectroscopically and classified about 400 of these (23). About half were blue–white, nearly half of the solar type, and only a small number belonged to the α Herculis type of banded spectra. The red long-period variable star Mira was, however, in this last group.

In 1868 Secchi was surprised to find a new type of stellar spectrum for some faint red stars (24). It was quite dissimilar to those red stars of the third type such as α Orionis, α Herculis, β Pegasi and Antares. The new type belonged to a few faint red stars always fifth magnitude or fainter. Observations of such objects had been greatly facilitated by the recently published Schjellerup catalogue of red stars (25). The most prominent member found in this new group was Lalande 12 561. He described their spectra thus (26):

> The essential character of this type is to exhibit a spectrum formed of three luminous bands separated by dark intervals. The brightest band is in the green; it is generally strong, divided up and very broad. Another much weaker band is in the blue but often visible only with difficulty. The third band is in the yellow and extends redwards. This one alone is further subdivided into several other bands.
>
> All these bands have the property that their intensity increases towards the violet side where they terminate abruptly. On the other hand, towards the red they show a gradual degradation until reaching complete darkness. There is thus a complete opposite between this type and the third; for in this latter not only are the bright bands twice as numerous, but in addition they show a maximum in intensity on the red side and a minimum on the violet. The two spectra are not therefore a variation of just one type, but are clearly due to totally different substances.

> The weakness of the light, which means I am unable to use a slit spectroscope, prevents me from determining rigorously the substances that produce these phenomena; one might say however that there is here a considerable analogy with the reversed spectrum of carbon.

The reversed spectrum of carbon is of course the laboratory emission spectrum which Secchi correctly identified. His description of bright bands is confusing. In reality the spectra show absorption bands separated by bright spaces. The absorption band heads are then on the longer wavelength, red side of each band. He was able to list seventeen faint stars from Schjellerup's catalogue, between magnitude 5.8 and 8.5 which belonged to the new class IV. With this work he had discovered the carbon stars, one of his crowning achievements.

Spectral classification was certainly the lasting monument that Secchi left behind him. He classified at least 4000 stars, including most of the northern naked-eye objects plus a number of fainter stars down to eighth magnitude. Apart from his classification work, a few other results should be mentioned. He confirmed that the strong lines in his type I stars were due to hydrogen, by direct comparison of the α Lyrae spectrum with the emission from a hydrogen-filled Geissler tube (26, 27). He studied the spectrum of the long-period variable star R Geminorum (28) and wrote: 'The spectrum is one of those rare examples where the hydrogen lines are bright! It's the third I've found in the sky. It shows as well other spectral bands, the main ones corresponding to the dark bands in the spectrum of α Orionis'. For Sirius he noted the hydrogen lines were exceptionally broad, and correctly anticipated that 'this could lead one to estimate the considerable pressure that the gas possesses in the atmosphere of this star' (29).

Particularly interesting is Secchi's defence of his classification when Lockyer suggested he merely had adopted Rutherfurd's scheme. In 1872 he replied to Lockyer in *Comptes rendus* (30):

> With regard to stellar spectra, I should put right an opinion which seems to be spreading among scientists. Mr Lockyer in a review of my book *The Sun* (*The Academy* no. 53, Aug. 1872, p. 289) supposes that I have taken over the division of stellar types made by Mr Rutherfurd in 1863. My friend Mr Schellen inadvertently fell into the same error; he even added a note on this subject in his translation of *The Sun*.
>
> Now a glance at Rutherfurd's classification and at mine proves there is nothing in common between the two. Mr Rutherfurd

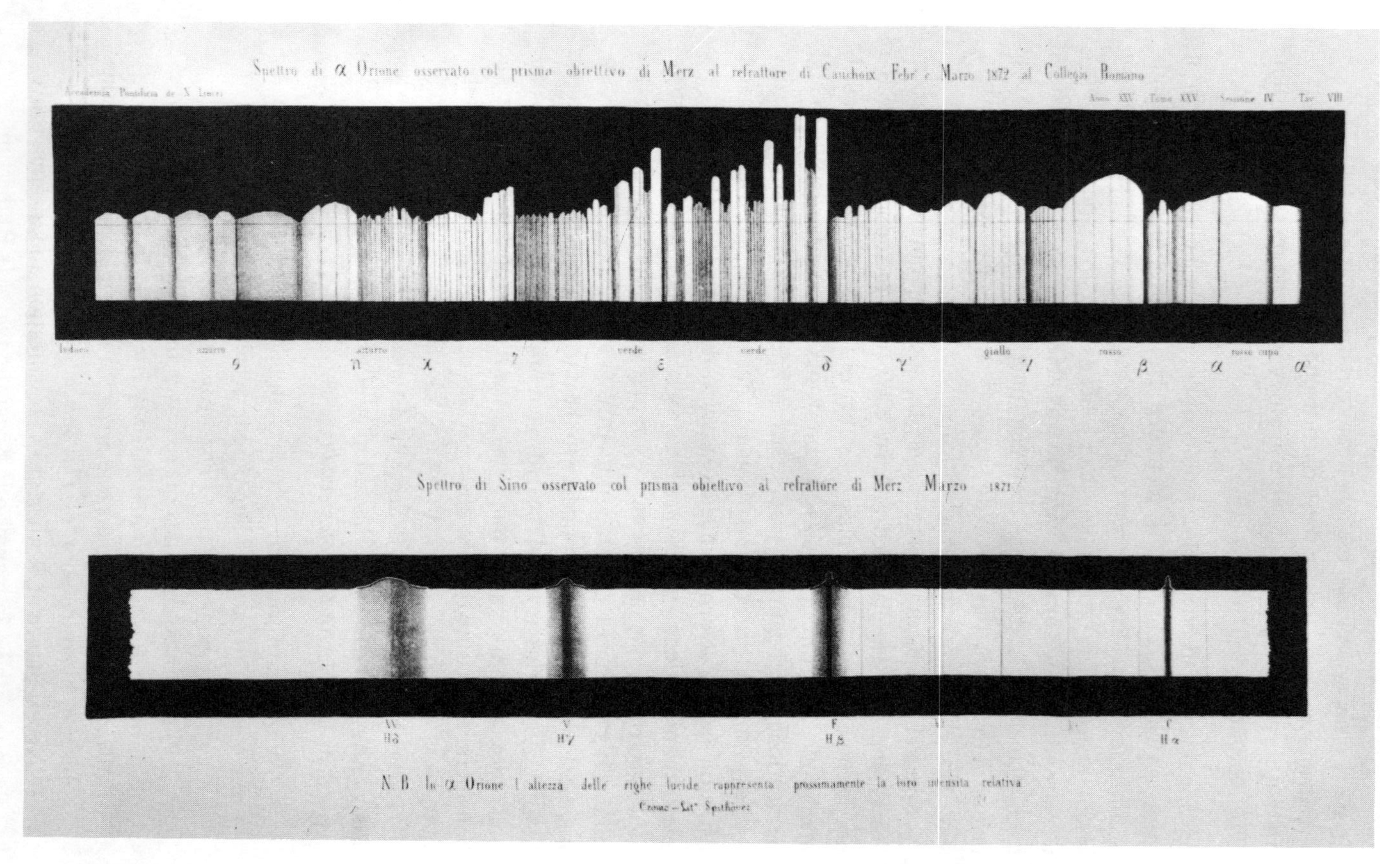

Fig. 4.8 Secchi's drawings of the spectra of Betelgeuse and Sirius, 1872.

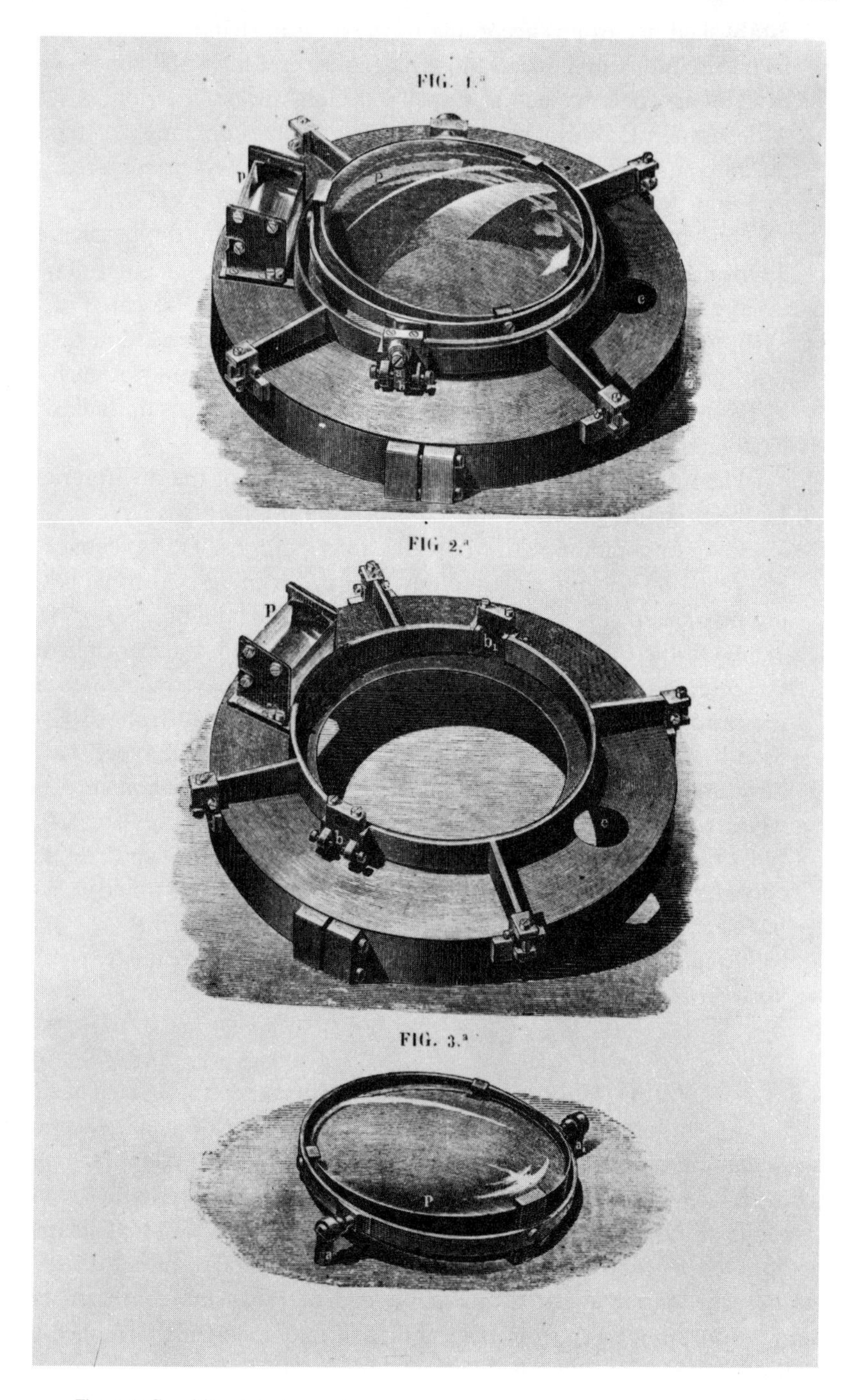

Fig. 4.9 Secchi's objective prism.

established his groups according to the natural stellar colours, including in his first group those having several lines and zones and resembling our sun, such as Capella, β Geminorum, α Orionis, Aldeberan, γ Leonis, Arcturus and β Pegasi. It is obvious that this group comprises all the bright stars which are yellow or coloured, and that those of my second type, like β Geminorum, Capella, γ Leonis etc. are mixed up there with those of my third type such as α Orionis, β Pegasi etc. The second of these groups is the first of my types, and I had already classified it as such in my research of 1863, which was at the same time as that of Mr Rutherfurd. The third group of Mr Rutherfurd consists of stars with no lines, such as Rigel and α Virginis. Now it's known that these stars do indeed have lines and belong to different types.

The result is then that neither by the method of the division nor by the principle governing the classification do my types have anything in common with Mr Rutherfurd's groups. This scientist has relied on general stellar colour, and this is in agreement with my first classification published in the bulletin of the observatory in 1863. On the other hand, I have based my work on the principle of the column-type bands, the constancy of which I have recognised in the third and fourth types, and which is quite different from that used by Mr Rutherfurd. The differentiation between the types that I have undertaken does not belong to the American astronomer, who, in fact, has never claimed it. In order to firmly establish these types, it is necessary to examine all the first magnitude stars and a considerable number of second magnitude, which Mr Rutherfurd has never done, but which I accomplished in the years that followed, and published in 1865 and in 1866, as the Academy well knows.

4.6 William Huggins and stellar composition If Secchi made important strides forward with his system of stellar spectral classification, even more important was the work of Sir William Huggins who can fairly claim to be the founder of stellar astrophysics. His illustrious career is remarkable in that it was pursued entirely as an amateur and without the benefit of any university education in natural science. He began his observations in spectroscopy at the end of 1862, and continued his active pioneering work from then until well into his eighties. He died in 1910, aged 86. Huggins devoted his time in 1856 to astronomy and supported himself by his own private means from then on. In that year he

Fig. 4.10 Sir William Huggins.

purchased a house at Tulse Hill on the outskirts of London and installed an observatory in his garden (31) which he equipped with a 5-inch Dollond equatorial. Two years later he acquired an 8-inch Alvan Clark objective glass which was installed in a new telescope, and this instrument he used for his early spectroscopic observations (Fig. 4.11).

His early astronomical work was confined to drawing the surfaces of the planets and sun. How he became involved in spectroscopy Huggins himself related in an autobiographical article in the *Nineteenth Century Review* (32):

> I soon became a little dissatisfied with the routine character of ordinary astronomical work, and in a vague way sought about in my mind for the possibility of research upon the heavens in a new direction or by new methods. It was just this time, when a vague longing after newer methods of observation for attacking many of the problems of the heavenly bodies filled my mind, that the news reached me of Kirchhoff's great discovery of the true nature and chemical composition of the sun from his interpretation of the Fraunhofer lines.
>
> This news was to me like the coming upon a spring of water in a dry and thirsty land. Here at last presented itself the very order of work for which in an indefinite way I was looking – namely, to

Fig. 4.11 Huggins' 8-inch Clark-Cooke refractor.

extend his novel methods of research upon the sun to the other heavenly bodies.

Huggins tells of how he attended a meeting of the Pharmaceutical Society in January 1862 at which his friend and neighbour William Allen Miller, the professor of chemistry at King's College, London, was lecturing on spectrum analysis. After the meeting the two returned to Tulse Hill together, and it was at this time that Huggins proposed a collaborative effort to apply the Kirchhoff method of prismatic analysis to the stars. They built a spectroscope consisting of two 60° flint-glass prisms, which received the light from a collimator (Fig. 4.12). The spectrum was broadened by a cylindrical lens and observed by a small rotatable viewing telescope whose position was controlled by a micrometer to enable accurate line positions to

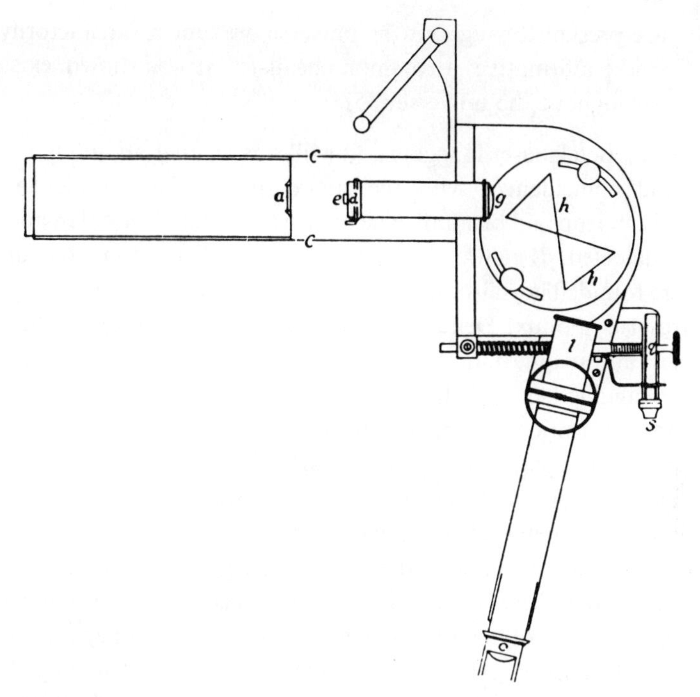

Fig. 4.12 Huggins' first stellar spectroscope.

be recorded (33). The spectra were about one-third of an inch in length from Fraunhofer A to H, so the mean dispersion was about 430 Å/mm.

Their first paper using this instrument gave drawings of the principal lines in Sirius, Betelgeuse and Aldeberan with their positions relative to the solar lines (12). However, two major publications followed in 1864. The first recorded the lines in the spark spectra of air, nitrogen, oxygen and twenty-five other elements, using a six-prism spectroscope (34). The second paper described the spectra of some fifty stars, with some three or four in considerable detail, by comparing the stellar absorption line positions with the bright lines from a comparison spectrum (33).

For Aldeberan some 70 line positions were measured, and the presence of the elements sodium, magnesium, hydrogen, calcium, iron, bismuth, thallium, antimony and mercury in this star was reported, though some of these were based on three or four lines only, with the result that the last four elements were misidentified. In a footnote added to the paper in 1909, Huggins wrote:

> One important object of this original spectroscopic investigation of the light of the stars and other celestial bodies, namely to discover whether the same chemical elements as those of our earth

are present throughout the universe, was most satisfactorily settled in the affirmative; a common chemistry, it was shown, exists throughout the universe (35).

For Betelgeuse, Huggins and Miller measured about eighty lines and found coincidences with the spark spectra for sodium, magnesium, calcium, iron and bismuth. The hydrogen C and F lines (Hα and Hβ) were not detected. βPeg was similar to Betelgeuse. For it too, no hydrogen lines were found. They also described the lines present in Sirius, Vega, Capella, Arcturus, Pollux, Deneb and Procyon. All these showed the sodium D lines, and magnesium, hydrogen and iron lines were found in several.

Huggins believed the red stars owed their colour to the presence of many absorption lines mainly in the blue part of the spectrum, and not to the lower temperature than white stars, which had few lines, an idea that Sir David Brewster had also advanced a few years earlier. He argued that the fact that the metals were in a gaseous state in stellar atmospheres testified to the high temperature of red stars like Betelgeuse. Although these ideas on temperature were wrong, it is nevertheless now well-known that the blocking effect of absorption lines does play a secondary role in determining stellar colours. Here we have one of the earliest references to an effect which a century later was to be crucial for the understanding of the ultraviolet excesses of metal-poor stars.

He also concluded that these observations showed different stars had different compositions and hence 'the composition of the nebulous material [from which the stars condensed] must have differed at different points', a wrong conclusion, though a natural one to make at that time. The significance of this early work was clearly not lost on Huggins and Miller when they concluded their monumental paper with a comment on the elements needed for living organisms (33):

> It is remarkable that the elements most widely diffused through the host of stars are some of those most closely connected with the constitution of the living organisms of our globe, including hydrogen, sodium, magnesium and iron.... These forms of elementary matter, when influenced by heat, light and chemical force, all of which we have certain knowledge are radiated from the stars, afford some of the most important conditions which we know to be indispensable to the existence of living organisms such as those with which we are acquainted.

We can see the influence of Darwin's *Origin of Species*, published in 1859, impinging on astronomy.

In August 1864 Huggins turned his attention to the nature of the nebulae. These fuzzy patches of faint light in the sky were widely held to be composed of unresolved stars, in which case their spectra would be essentially stellar. In his first paper on this subject he found eight nebulae with bright lines in emission, quite unlike the spectrum of any other celestial object (36). The general feature of all the bright emission-line nebulae in the visible spectrum was four (or sometimes three) lines at wavelengths 5007 and 4959 Å and two lines of hydrogen corresponding to Hβ and Hγ. The first two green lines are the famous 'nebulium' lines. Huggins at first believed the 5007 Å line may be due to nitrogen, as the spark spectrum of this gas showed two bright lines at 5001 and 5005 Å, though he later had to abandon this idea after careful wavelength measurement confirmed the discrepancy (37).

The great nebula in Orion was observed by Huggins soon after the original nebular observations (38) and this remained a favourite object with him over several decades. By 1868 some seventy nebulae had been studied spectroscopically, with about one third showing emission lines, the remainder a continuous or star-like spectrum. Later, when studying the ultraviolet spectrum of the Orion nebula photographically, he found a bright emission line at 3730 Å (39).

Huggins and Miller correctly concluded in their first discussion of the bright-lined nebulae that these objects 'must be regarded as enormous masses of luminous gas or vapour' which would therefore never be resolved into stars. However, the problems of identifying the green nebulium and strong ultraviolet lines, and of explaining why some nebulae gave solar-like spectra was for the future. Both were finally solved in the 1920s, by Bowen, who recognised oxygen at very low densities as the source of the unidentified lines (see section 8.3), and by Hubble who established the extragalactic nature of those nebulae giving star-like spectra.

Having thus pioneered in the first work in stellar chemistry and in nebular spectroscopy, Huggins was now fortunate to be able to record another first in 1866. This year a new star or nova appeared in May in the constellation of Corona Borealis. It was first seen from Ireland on 12 May and Huggins heard the news by the 16th when it was as bright as second magnitude. He observed it spectroscopically with Miller, the last time the two worked together. T Coronae Borealis (T CrB) thus became the first nova to be studied with the spectroscope (40). The visible spectrum showed five bright emission lines including Hα, Hβ and Hγ, and in addition there was an absorption spectrum on a weak continuum, containing at least two strong red lines as well as many weaker ones including the D lines. The spectrum of T CrB was also observed at Greenwich by Stone, Carpenter and Airy (41) from 19 May, and by C.J.F. Wolf (1827–1918) and G.A.P. Rayet (1939–1906) in Paris (42), who were able to confirm Huggins and Miller's results of the apparent superpositions of two spectra. Meanwhile the nova was fading rapidly, and by the beginning of June 1866 was already down to ninth magnitude.

Nova Coronae of 1866 was not the only nova that Huggins became associated with. In December 1891 Nova Aurigae erupted. It was discovered in January, and the spectrum was studied in considerable detail by Huggins and several others, such as Campbell, Vogel and Belopolsky. It too showed the characteristic emission-line spectrum with absorption lines displaced to the short wavelength side, most noticeably for the hydrogen lines which were photographed into the ultraviolet (43). The D line also showed this double emission and absorption profile. In August 1892 the nova went into a nebular stage, showing the characteristic nebular spectrum with the two green lines, especially 5007 Å, dominating the spectrum. Huggins thought he could resolve this line into a number of components (44) and was therefore reluctant to accept the nebular interpretation, as the lines of true gaseous nebulae were single and sharp.

4.7 Wolf and Rayet and their emission-line stars The first decade of stellar spectroscopy turned up a variety of rare stars with emission lines. We have already mentioned such dissimilar objects as γCas, βLyr, R Gem and T CrB, which today we recognise respectively as a rapidly rotating Be star, a close binary, a long period Mira-type variable of spectral type M, and a nova near maximum light.

In 1867 another type was added by the French astronomers Wolf and Rayet, that would immortalise their names in the Wolf–Rayet stars. In this year they reported finding three faint, eighth-magnitude stars* in the constellation of Cygnus with several bright bands (that is, very broad emission lines) on a continuous background (45). One of the bands in the blue was particularly intense. The spectra were described more fully by the German astronomer H.C. Vogel (1841–1907) who showed that the blue band in the first star was centered at about 4680 Å as against 4640 Å in the other two (46). Huggins' contribution was a paper in 1890 in which he agreed with Vogel's result about one star having a longer blue-band wavelength, and he dismissed any suggestion that any of the stars' bands might be due to hydrocarbon molecules (47).

We now know that Wolf and Rayet's first star belongs to the nitrogen

* These stars were BD + 35°4001, + 35°4013, + 36°3956, also known as HD 191765, 192103, 192641.

subsequence showing the broad 4686 Å feature of ionised helium as well as other emission lines mainly of ionised nitrogen. The other two stars are in the Wolf–Rayet carbon subsequence, and show a broad feature at about 4650 Å due to ionised carbon lines (C III and C IV), as well as other bands of ionised carbon, oxygen and helium (see section 9.2).

4.8 Huggins' later work: comets and the Doppler effect

Huggins' work on comets also deserves mention as he was among the pioneers in cometary spectroscopy, although Donati (48) had earlier found three bright emission bands in comet I (1864). Secchi and Huggins both entered this field in 1866, when they observed Tempel's comet. However it was 2 years before two much brighter periodic comets appeared, Brorsen's Comet (comet I, 1868) and Winnecke's (comet II, 1868) for which Huggins was able to confirm the presence of three bright bands which came from the coma. He showed they corresponded to the same bands observed on burning olefiant gas (hydrocarbons) in the laboratory (49).

The spectra of comets will not be discussed in detail here except to note that later Huggins was the first to photograph a cometary spectrum (comet 1881b) (50) in which he found ultraviolet bands at 3883 and 3870 Å that he identified with the cyanogen bands already found in the laboratory by Liveing (1827–1924) and Dewar (1842–1923).

The early 1870s were fortunate years for William Huggins. The Oliveira bequest had left £1350 to the Royal Society and this sum went towards the construction of a 15-inch Grubb refractor and an 18-inch reflector, to be interchangeable on a single mount. The Society erected these telescopes at Tulse Hill on indefinite loan to Mr Huggins. A new 18-foot diameter observatory dome was erected to house them, and the whole installation was completed by February 1871.

It was with the 15-inch telescope that Huggins soon made a pioneering effort to measure stellar radial velocities, using the principle of the Doppler effect, which was predicted to give small displacements in the positions of the spectral lines for stars moving rapidly away from or towards the observer along the line of sight. The possibility of measuring stellar radial velocities had been suggested both by Secchi (21) and by Huggins and Miller (33). Huggins' first observations followed in 1868 when he compared the position of the F (Hβ) line of Sirius with the same line in a low-pressure hydrogen discharge tube mounted in front of the object glass. He wrote:

> I am certain that the narrow line of hydrogen, though it appeared projected upon the dark line of Sirius, did not coincide

> with the middle of that line, but crossed it at a distance from the middle, which may be represented by saying that the want of coincidence was apparently equal to about one-third or one-fourth of the interval separating the components of the double line D (51).

He actually measured the shift to be 1.09 Å to longer wavelengths, corresponding to a recession of 41.4 mile/s, which was reduced to a heliocentric value of 29.4 mile/s after allowing for the earth's motion about the sun. The value was in error both in its amount and the direction of motion (the currently accepted value is about -8 km/s or -5 mile/s) but at least the principle was correct and Huggins deserves the credit for being the first to attempt to apply the method to the stars.

In spite of the painstaking care required to measure the minute shifts of spectral lines that this work demanded, Huggins re-embarked on this research in 1872 with the new 15-inch telescope. His paper of that year (52) to the Royal Society presented results for thirty stars with a probable error estimated to be about 5 mile/s. The observations included Sirius with a revised heliocentric radial velocity of 18–22 mile/s. The still considerable error only goes to emphasise the intrinsic difficulty of visual measurements of Doppler shifts.

4.9 Henry Draper, Wm Huggins and spectrum photography In 1875 Huggins married Margaret Murray (1848–1915) of Dublin. She was 24 years her husband's junior, but one of the most successful husband-and-wife partnerships in astronomy was the result. She became his untiring companion at the telescope, and much of their subsequent research was thenceforth published jointly. Soon after the marriage a new endeavour occupied their efforts, that of spectrum photography. Huggins and Miller had attempted to record the spectra of Sirius and Capella on the old wet collodion plates in 1863, but no lines, only a continuum had been recorded (33). The first successful stellar spectrogram was recorded in 1872, not by Huggins, but by Henry Draper (1837–82) in the United States (53). He placed a quartz prism in front of the focus of his 28-inch reflector at his private Hastings-on-Hudson Observatory in New York State, so as to photograph the spectrum of Vega. He found a series of strong dark lines probably due to hydrogen:

> In the photographs of the spectrum of Vega there are eleven lines, only two of which are certainly accounted for, two more may be calcium, the remaining seven, though bearing a most suspicious

Fig. 4.13 Henry Draper.

> resemblance to the hydrogen lines in their general character, are as yet not identified. It would be worthwhile to subject hydrogen to a more intense incandescence than any yet attained, to see whether in photographs of its spectrum under those circumstances any trace of these lines, which extend to wavelength 3700, could be found.

The two lines he could account for were presumably Hγ and Hδ; the remaining seven were the ultraviolet Balmer series lines of hydrogen, which had been observed for the first time, 7 years before they were to be studied in the laboratory. It was an important new step into the world of ultraviolet stellar spectrography.

Draper went on to photograph the spectra of other bright northern stars, including Arcturus, Capella and Altair, noting that the first two resembled the solar spectrum whereas Altair's spectrum resembled Vega's. Unfortunately he became embroiled in a controversy in the last years of his life, after claiming in 1877 to discover emission lines of oxygen in the blue part of the photospheric solar spectrum that he had photographed (54).

This appeared to be an especially interesting result, as not only were emission lines previously unknown in the photospheric spectrum, but the known elements producing the Fraunhofer lines were metallic. Draper's claim was quickly attacked in England, and especially vehemently by Norman Lockyer who claimed that Draper's bright lines were simply the peaks between the crowded absorption lines (55). This assertion of Lockyer's was shown conclusively to be correct by John Trowbridge and Charles Hutchins, physicists at Harvard University (56), but only several years after Draper had died prematurely in 1882.

By the mid-1870s new dry plates were becoming available. They were more sensitive than the wet collodion and the light-sensitive silver bromide salts were held in a gelatine emulsion on the glass. The ease of handling these plates, especially for the long exposures encountered in stellar spectrography, made them vastly superior to the old method. Huggins was the first to use the dry plates when he took up stellar spectrography again in 1876 (57). For this purpose, the 18-inch reflector was the ideal telescope. It was equipped with a new ultraviolet-transmitting spectrograph with quartz lenses and an Iceland spar prism. This instrument was mounted at the prime focus of the reflector, and he observed the star on reflecting slit jaws with a small telescope which looked up through the hole in the primary mirror. Huggins was able to dispense with the cylindrical lens and instead introduced the technique of trailing the star along the slit so as to widen the spectrum. With this apparatus he recorded seven strong lines of hydrogen in Vega, five of them in the ultraviolet.

Huggins published an extensive account of his work on the photographic spectra of stars in 1880 (58). He described the advantages of dry plates thus:

> At the early stages of these experiments I used wet collodion, but I soon found how great would be the advantages of using dry plates. Dry plates are not only more convenient for astronomical work, being always ready for use, but they possess the great superiority of not being liable to stains from draining and partial drying of the plates during the long exposures which are necessary even with the most sensitive plates. I then tried various forms of collodion emulsions, but finally gave up these in favour of gelatine plates, which can be made more sensitive.

Huggins photographed the spectra of Sirius, Vega, α Aquilae, α Virginis, α Cygni, η Ursae Majoris, Arcturus, Rigel, Betelgeuse, Aldeberan and Capella and gave tables of line wavelengths. For the white stars he now found twelve very strong lines from Hγ into the ultraviolet and noted the likelihood that all these were due to hydrogen, which was confirmed at about the same time by Hermann Wilhelm Vogel (1834–98) in his laboratory in Berlin (59) (not to be confused with H.C. Vogel, famous for his work in photographing stellar spectra). In an interesting footnote to the 1880 paper Huggins speculates on whether the different stellar spectra form an evolutionary sequence, starting with Sirius and Vega for the youngest objects and with red stars such as Betelgeuse at the end of the evolutionary process.

Huggins soon went on to apply the photographic method to comets (60) and nebulae (39). He was unable to identify the strong ultraviolet 3730 Å emission line found on the Orion nebula spectrogram. We now know it is due to the forbidden 3726 and 3729 Å lines of singly ionised oxygen.

In 1896 Huggins was already an old man, but his research went on unabated. In this year he constructed a new ultraviolet spectrograph for the 18-inch telescope, this time to be mounted more conveniently at the cassegrain focus behind the hole in the primary mirror. It incorporated two Iceland spar prisms and quartz lenses, and this was the instrument used for the publication of *An Atlas of Representative Stellar Spectra* by Sir William and Lady Huggins (Huggins was knighted in 1887), being volume I of *Publications of the Sir Wm Huggins Observatory* (61). The photographic spectra presented ranged between 3300 and 4870 Å (Fig. 4.14).

For most astrophysicists Sir William Huggins is regarded as the founder of stellar spectroscopy. His career was long and illustrious and it is easy to see how his discoveries captured the popular imagination of the day, and why his name is still regarded with such a high esteem a century later. The value of his research was fully recognised in his lifetime, as the awards of FRS (1865), KCB (1887) and the OM (one of the twelve original members) (1902) testify. He also served as president of the Royal Astronomical Society (1876–78) and of the Royal Society (1900–5). The two Royal Society telescopes were returned to the Society in 1908, and they were then presented to the University of Cambridge and erected there before Huggins' death in 1910. Lady Huggins survived Sir William by 5 years.

4.10 Hermann Carl Vogel

Of the early pioneers in stellar spectroscopy, the Italians Porro, Donati and Secchi all died in the 1870s. So too did the German astronomer H.L. d'Arrest (1822–75) who had established spectroscopic research during his directorship of the Copenhagen Observatory. In the United States, Rutherfurd didn't pursue

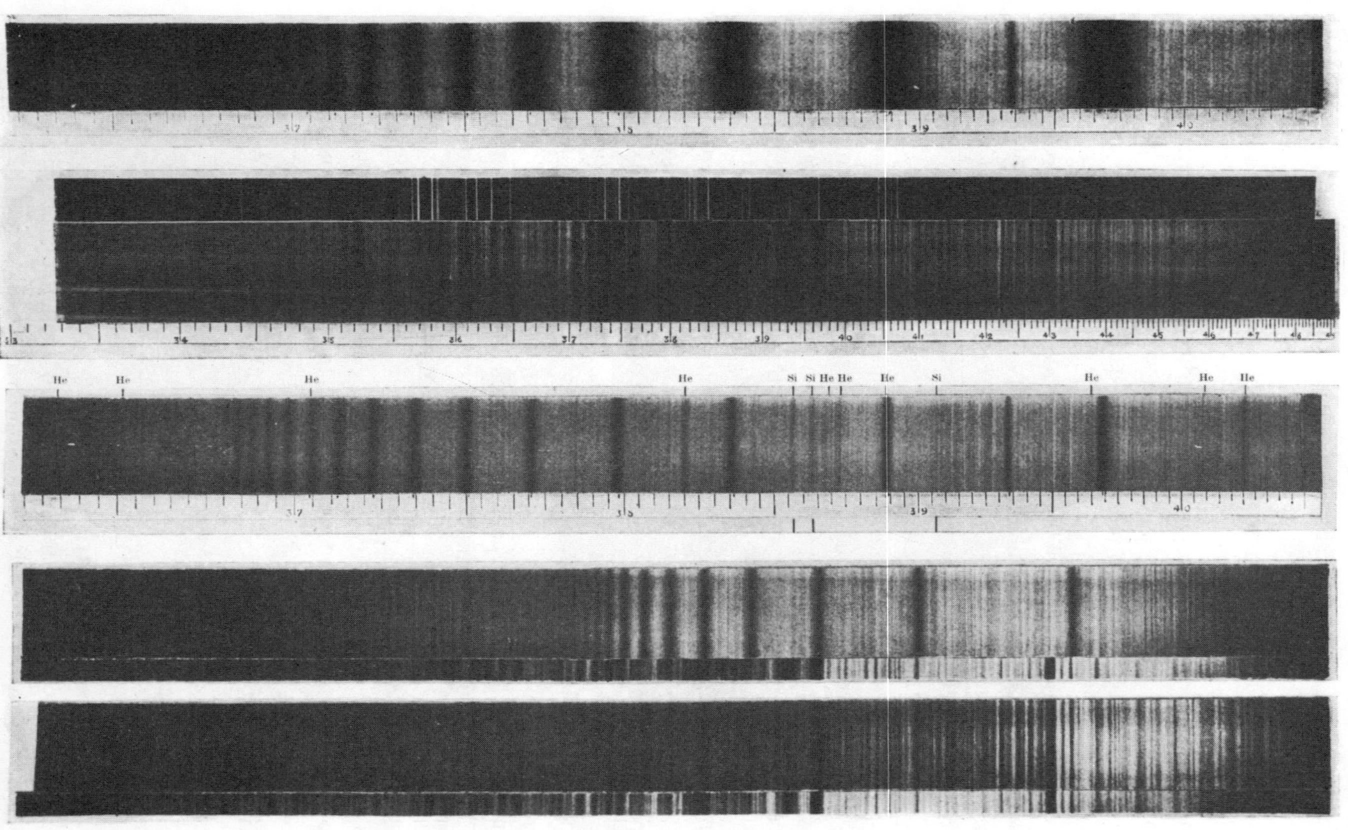

Fig. 4.14 Stellar spectra photographed by Sir Wm Huggins. From top to bottom Vega, Arcturus, Rigel, β Cygni (blue star), β Cygni (yellow star).

Fig. 4.15 Hermann Carl Vogel.

his stellar spectroscopic research after his early début, and Henry Draper died in the same year as his father, 1882. In Paris, Wolf and Rayet contributed very little to stellar spectroscopy apart from their one famous discovery of 1867. Airy never involved himself closely with stellar spectroscopy, but the Greenwich program of visual radial velocity measurements continued into the early 1890s, the observations being mainly undertaken by E.W. Maunder (1851–1928). By the end of his first two decades of stellar spectroscopy (i.e. 1883) Huggins was therefore already the undisputed leader, and before the end of the century he was the grand old man of stellar spectroscopy. Meanwhile, new workers were entering the field and, of these, the German researchers at Potsdam headed by Hermann Carl Vogel were the most prominent, followed soon afterwards by the Lick Observatory astronomers in California under James Keeler (1857–1900).

Vogel's entry into spectroscopic research began in 1870 when he was appointed director of a new private observatory at Bothkamp near Kiel owned by the Kammerherr F. von Bülow. The observatory possessed an 11-inch refractor, at the time the largest in Germany. Here, with the assistance of W.O. Lohse (1845–1915) he embarked on a program of spectroscopic observations of the sun, stars, nebulae, planets and comets. As early as 1871 Vogel had experimented with dry plates for solar photography, and it

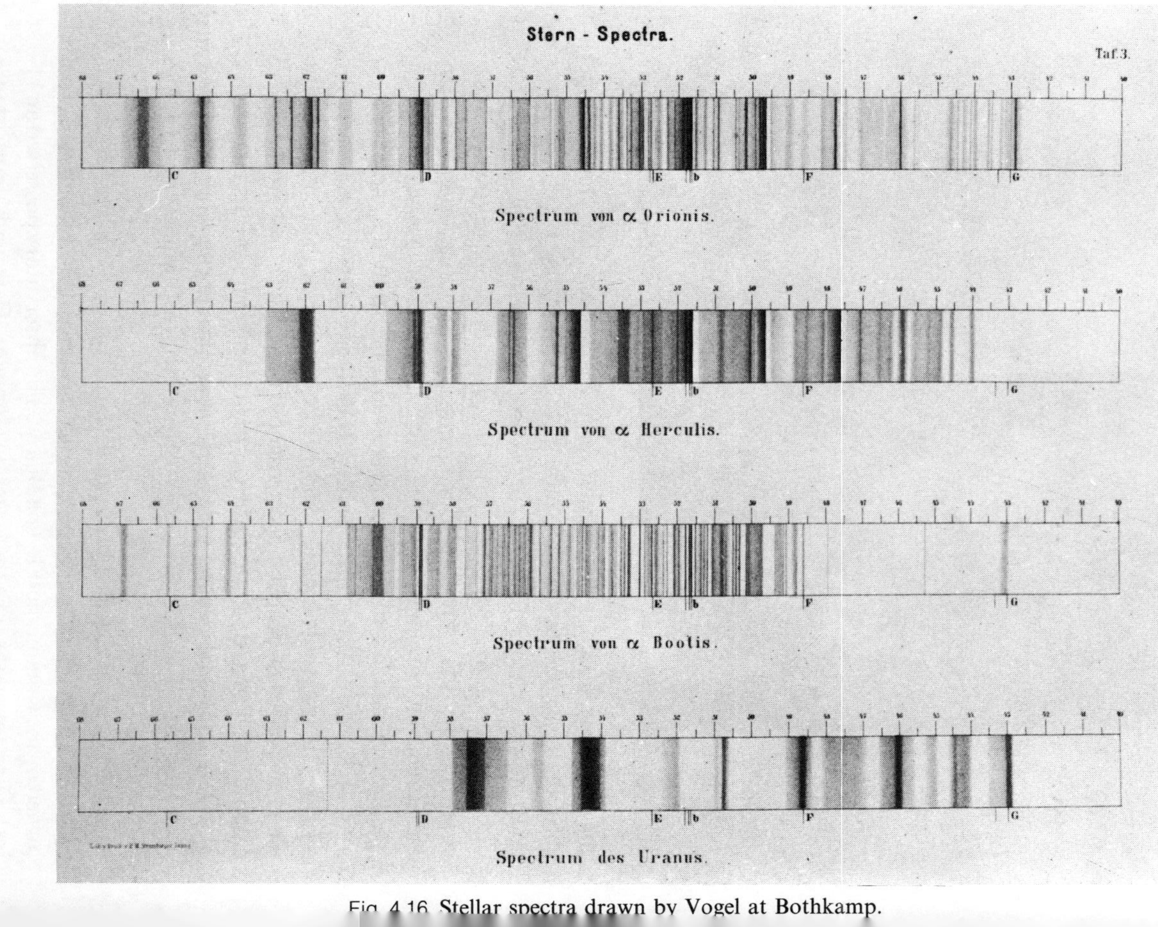

Fig. 4.16 Stellar spectra drawn by Vogel at Bothkamp.

was the later application of this process to stellar spectra that was, in due course, to make him famous. He stayed at Bothkamp only 4 years, but in this short time his reputation in spectroscopy was established. He catalogued some 2600 lines in the solar spectrum and measured their wavelengths, though this work was soon superceded by the more exhaustive study of H.A. Rowland.

In 1873 Vogel and Lohse embarked on a spectral survey of all stars in the declination zone -10° to $+20^\circ$ down to magnitude 4.5. The aim was to classify their spectra using a new classification scheme that Vogel had devised (62). Vogel's system resembles Secchi's, but his intention was to put more emphasis on the stars' evolutionary states as deduced from their spectra, though in reality such an ambition was grossly premature. The survey in this form was never completed, because of Vogel's departure from Bothkamp in 1874 to take up a post at the newly established astrophysical observatory at Potsdam near Berlin. However the classification scheme was used quite widely, alongside Secchi's, for the remainder of the nineteenth century.

In Vogel's own words, his scheme ran as follows:

> The only rational classification of the stars by their spectra might be achieved, if one starts from the point of view, that in general the evolutionary state of the respective heavenly bodies is mirrored in their spectra. Three quite distinctly separated classes can then be distinguished, namely:
>
> 1. Stars whose incandescence [presumably Vogel means temperature] is so considerable that the metallic vapours contained in their atmospheres can only exercise a small absorption, so that either no or only weak lines are to be seen in the spectrum (the white stars belong here).
> 2. Stars similar to our sun which contain metals in their surrounding atmospheres as shown by strong absorption lines in the spectrum (yellow stars), and finally,
> 3. Stars whose incandescence is so much lower that association of the material of which their atmospheres are composed can take place and which, as recent studies have shown, are always characterised by more or less broad absorption bands (red stars).
>
> On looking at spectra of Secchi's 3rd and 4th types, we find they belong in this case to the above-mentioned third class and are only differentiated by the ordering of the dark bands in the spectrum, or in other words the difference is solely to be sought in the different compositions of the atmospheres surrounding the luminous bodies. For this reason it seems to me that Secchi's 4th

Table 4.1. *Summary of Vogel's first spectral classification of 1874*

Vogel Class	MK equivalent	Secchi class	Typical stars
Ia	A	I	Sirius, Vega
Ib	O, B	I (Orion)	β, γ, δ, ε Ori
Ic	Be	I (later, V)	γCas, βLyr
IIa	F5 to K5	II	Capella, Arcturus, Aldeberan
IIb	novae, WR stars Mira variables with bright lines	—	T CrB, WR stars observed in Cygnus, R Gem
IIIa	M	III	αHer, αOri, βPeg
IIIb	C (carbon stars)	IV	78, 152, 273 Schjellerup

type shouldn't be left as a separate class, even though these are readily distinguished from his 3rd type by inspection.

Vogel then goes on to define his three spectral classes, each of which is split up into two or three subclasses. Table 4.1 gives the correspondence with Vogel's classification and modern spectral types on the MK system, while Appendix II gives a translation of Vogel's original description in full. He concluded his discussion with the remark: 'I would like to believe that the classification so arranged will last for a considerable time...' (62). However, this last hope was soon to be dashed. Vogel had to revise his classifcation in 1895 following the discovery of terrestrial helium and the cataloguing of its spectrum. And then, as we will see, the far more numerous objective prism classifications at Harvard were soon to swamp all other schemes through the large number of stars classified.

After moving to Potsdam, Vogel at first worked under the directions of a three-man committee (Kirchhoff was one of the members) in establishing the equipment for the astrophysical observatory. The building was completed in 1879 and was originally to be only a solar observatory. However, the extension of activities to the wider field of stellar astrophysics was soon approved, and Vogel then became the director in 1882. Celestial spectroscopy was from the start one of the major activities. The principal telescope was a 30 cm Schröder refractor, and with this instrument Vogel was able to continue his spectroscopic research. Two larger telescopes were however installed during Vogel's directorship. The first was a double refractor (on a single mount) with twin 32 cm photographic and 24 cm visual objectives, which was erected in 1889. Ten years later there followed a second larger double refractor with 80 cm (photographic) and 50 cm (visual) apertures.

Initially there were three astronomers appointed to Potsdam (Vogel, Lohse and G. Spörer (1822–95)). In the years 1877–81 D.H.G. Müller

Fig. 4.17 Vogel's spectrograph on the 30 cm refractor at Potsdam.

(1851–1925), P.F. Kempf (1856–1920), J. Wilsing (1856–1943) and J. Scheiner (1858–1913) joined the staff, and nearly all these astronomers were to leave their mark on the development of astronomical spectroscopy. A major undertaking in the early years of the observatory was the continuation of the spectroscopic Durchmusterung that had been begun at Bothkamp. Vogel undertook this work with Müller using the 30 cm telescope. The earlier program was greatly extended and the aim was to classify the spectra of all northern stars to magnitude 7.5 using a small Zöllner spectroscope. The first publication of this program covered 4051 stars in the declination zone -1° to $+20^\circ$ (63). As well as classifying the spectra on the Vogel classification scheme devised earlier at Bothkamp, they made visual estimates of stellar colour on a seven-point scale from white to red.

4.11 The discovery of helium

In 1895 Vogel had to make a substantial revision of his stellar classification scheme. The reasons for this go back to 18 August 1868, and form an important chapter in the story of stellar astronomy. On that day a total solar eclipse took place. The passage of the moon in front of the sun's disk was observable only from India and the Malay peninsula and a number of expeditions, including two from Britain and two from France, were despatched to observe it. With the moon precisely obscuring the bright solar disk, the tenuous gases immediately above the apparent solar surface could be observed spectroscopically, and it was here that the prominences were known to occur. Lt J. Herschel (1837–1921) (64), the youngest son of Sir John Herschel, J.F. Tennant (1829–1916) (65), Janssen (66) and Rayet (67) all observed the line emission from a solar prominence, the last mentioned recording nine bright spectral lines whose positions relative to the Fraunhofer absorption lines he listed as corresponding to: B, D, E, b (2 lines), F, G, a green line between b and F, and a blue one near G. In fact B should have been C (Hα) and hydrogen was the dominant element observed, accounting for three lines. What concerns us for the moment however is the line ascribed by Rayet to the sodium line D, but which Lockyer showed had a wavelength smaller than either of the D lines (68). Lockyer's observations were made in October 1868 using a technique that first Janssen (on the day after the eclipse) and then, independently, Lockyer had discovered for observing prominence spectra at times outside eclipse by placing the spectrograph slit tangential to the solar limb. The new line was at 5876 Å and named D_3 to distinguish it from D_1 and D_2 of sodium. It corresponded to none of the Fraunhofer lines, nor to any line of a known terrestrial element, so Lockyer and his colleague, Professor E. Frankland

Fig. 4.18 Sir Norman Lockyer.

(1825–99), chose the name 'helium' for what they correctly supposed was a new element which at that time had not been isolated on earth.

The discovery of terrestrial helium from the lead uranate mineral Clèveite by W. Ramsay (1852–1916) in 1895 was a major breakthrough (69). Ramsay boiled the mineral with weak sulphuric acid and obtained a gas he claimed to be a mixture of argon and helium. The latter of these was identified by the presence of the long-sought-for D_3 line at 5876 Å when it was analysed spectroscopically in a discharge tube. Wavelengths for six helium lines were measured by T.R. Thalén (1827–1905) from a gas sample prepared by P.F. Clève at Uppsala (70) and the results reported even before Ramsay had announced his discovery. Meanwhile Lockyer also studied the laboratory spectrum of the gas he had named (71). He showed that several other chromospheric lines could be ascribed to the new element, as well as a number of absorption lines in certain Orion stars such as Bellatrix (γ Orionis, MK type B2), including both 4472 and 4026 Å which were particularly prominent in the laboratory.

Next followed a detailed study of the Clèveite gas spectrum with precise wavelengths in the ultraviolet to the far red by C.D.T. Runge (1856–1927) and L.C.H.F. Paschen (1865–1947) (72). They grouped seventy-four of the lines into six different series whose wavelengths followed regular formulae, somewhat similar in structure to the well-known Balmer series of hydrogen for which the formula had been announced only a decade earlier (73). Six series for a single element was then unprecedented, especially as many of the alkali metals gave lines in only three series. They concluded that Clèveite gas must in reality be two separate elements, a view supported by Lockyer (71). These two gases were named helium (giving lines such as 7066 Å, D_3 and 4472 Å) and parhelium – or as Lockyer preferred to name it, asterium (with 6678, 5016 and 4922 Å being the prominent lines). This conclusion was supported by the changes in the relative strengths of the helium and parhelium spectra, both as observed in the laboratory and when observed astronomically. For example, the chromospheric helium spectrum tended to dominate that of parhelium, whereas in Nova Aurigae and gaseous nebulae, parhelium gave the more intense spectrum. However, the two-element theory for Clèveite gas later had to be abandoned. In reality there is only one substance, helium, which presents two independent neutral atomic spectra corresponding to triplet (electron spins aligned parallel) or singlet (antiparallel) states.

4.12 Vogel's second classification

We now resume the story of Vogel's work in stellar classification with his confirmation of Lockyer's identification of helium absorption lines in ten of the Orion stars (including β, γ, δ, ε, ζ Orionis) and fourteen other white stars, as well as nineteen helium lines which he found in the emission spectrum of the binary star β Lyrae (74).

With the discovery of many stars with helium absorption a revision of the spectral classification was needed, and Vogel revised his earlier scheme for class I stars by creating a new subdivision for those showing helium. His class Ia was now subdivided into Ia_1, Ia_2 and Ia_3. These all showed strong hydrogen lines, no helium and progressively stronger metallic lines in going to the third subtype. On the MK system these types run from A0 to early F and form a natural link to Vogel's second class, whose definition was unaltered.

The helium stars were placed in class Ib which Vogel described as follows:

> Ib. Spectra in which, besides the still dominant hydrogen lines, the lines of Clèveite gas appear, and above all the lines λ 4026, λ 4472, λ 5016, and λ 5876 (D_3), (The strongest line in the violet λ 3889, is so nearly coincident with Hζ that it is not a reliable criterion of the presence of the lines of Clèveite gas in star spectra).

Class Ib thus now differed from the class of this name in the first Vogel classification which had been reserved for white stars with the hydrogen lines missing. The old class Ib included O and early B stars, and also late B stars only if the hydrogen lines were unusually narrow, as in the spectrum of Rigel (MK-type B8Ia), and thus not readily discernible. The new class Ib included O stars and all B stars (on the MK system) with helium lines in absorption.

Type Ic was similar to that of Vogel's first scheme, except now he divided it into Ic_1 and Ic_2. The former had only hydrogen in emission (γ Cas); the latter had emission lines of hydrogen, helium and other metals, as seen in one component of the star β Lyr which Vogel had just shown to be a spectroscopic binary. The variable star P Cygni, was also assigned to class Ic_2. Its unusual spectrum had been discovered by E.C. Pickering (1846–1919) at Harvard (75) and then studied in more detail by J.E. Keeler at Lick (76) who found it to have three bright lines ascribed by Vogel to helium (see section 11.2.1).

The spectra of 528 mainly class I stars were photographed and classified by Vogel and Wilsing on the revised scheme (77). Of these, one hundred Ib helium stars were recorded, though not all of them for the first time, as objective prism classifications by Pickering were already underway at Harvard. In addition, the line positions and strengths were carefully measured for 130 of these stars, making it one of the most detailed investigations in stellar spectrography of its time.

4.13 Vogel and photographic radial velocity determinations

If Draper and Huggins were the first pioneers in stellar spectrography, the Vogel was the person who perfected the photographic art so that measurements on stellar spectra with hitherto unattainable precision became a standard, even if laborious, procedure. Vogel began stellar spectrography at Potsdam in 1888. His first paper (78) gave the photographic results for Sirius, Procyon, Rigel and Arcturus (Fig. 6.3). Radial velocity measurements were from the outset one of his major aims, and in the following year Vogel's first radial velocities, for five stars (α Aur, α Tau, α UMi, α Per, α CMi) were published (79). The accuracy of these final results was 5–10 km/s, but this was greatly improved in a classical paper with the collaboration of Scheiner in 1892 (80), where an accuracy of better than 3 km/s was achieved for the radial velocity from a single plate. This was about an order of magnitude improvement over what had been achieved visually by Huggins. The radial velocity of Sirius now became about −8 km/s, very close to the value accepted nearly a century later. Vogel was for the first time clearly able to show the effect of the earth's orbital motion on the results. The velocities relative to the earth obtained for

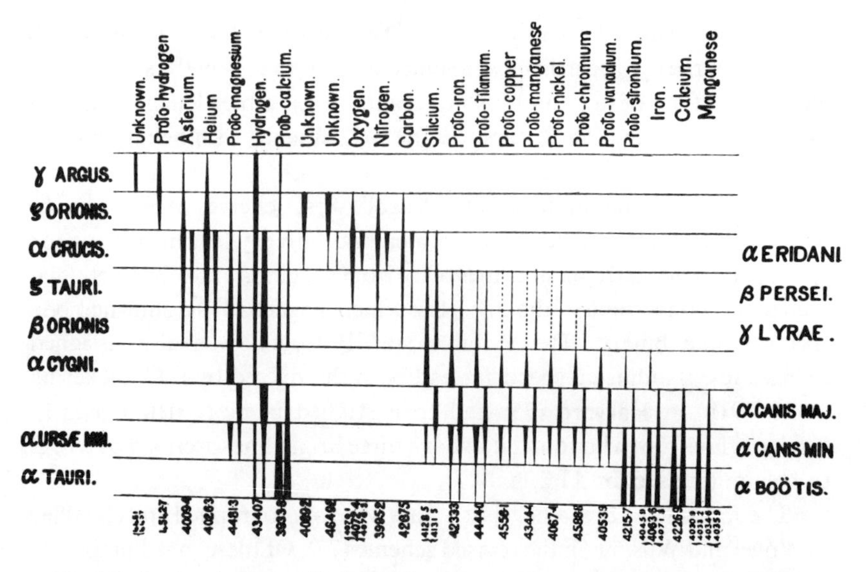

Fig. 4.19 Lockyer's map of chemical substances present in stars of different temperatures.

a given star from different plates showed considerable scatter, which could be up to ± 30 km/s for stars near the ecliptic. However, when applying the standard formula which allows for the earth's motion, the scatter in the heliocentric radial velocities so obtained was reduced to no more than the above-mentioned observational scatter, thus elegantly demonstrating the validity of the whole Doppler effect method.

Vogel's velocities were obtained from the small shifts of stellar lines in the neighbourhood of $H\gamma$. Most of his work involved a comparison with the hydrogen spectrum photographed from a narrow Geissler tube mounted in front of the slit in the converging stellar light beam of the telescope. The plates were analysed with a travelling microscope. Generally he compared the positions of several stellar lines with the corresponding lines in a solar spectrogram, which served as an intermediate standard. At other times he experimented by dispensing with the hydrogen as the primary comparison, and used an iron arc instead, which allowed a direct comparison with the positions of iron lines in stellar spectra (81).

With this work Vogel and his collaborators acquired a reputation for precision radial velocities of the highest quality, and with an accuracy still regarded as respectable a century later. Another important development also arose out of this research. In the early 1870s Vogel was attempting radial velocity measurements from visual observations made at Bothkamp. One of the objects he observed was Algol which had been long suspected of being an eclipsing binary, since Goodricke had found its brightness variations a century earlier. The results for velocity variations, due to the motion of one of the stars in its orbit, were then negative, but Vogel came back to Algol in 1888, and the following year he was able to publish the velocity results from six spectra showing variations that correlated with the phase of the light curve (82). Algol was thus shown to be a spectroscopic binary, consisting (as we know today) of a luminous B dwarf orbiting in about 69 hours a fainter K subgiant companion which is losing mass to the hotter star. As the B star dominates the light, only its lines are seen. This was the second spectroscopic binary to be found, and confirmed the suspected compound nature of Algol. From his variable radial velocities, Vogel estimated the size of the orbit and the masses of the two close stars.

Another single-lined spectroscopic binary, Spica (α Virginis), which is not eclipsing, was found soon afterwards (83). Neither Algol nor Spica, however, were the first spectroscopic binaries to be found. Also in 1889 Edward C. Pickering had discovered at Harvard periodic doubling in the spectra of ζ Ursae Majoris (84) and then soon afterwards Miss Maury found the same phenomenon for β Aurigae. The discovery for ζ UMa was announced to a meeting of the National Academy of Sciences in Philadelphia on 13 November 1889, the month before Vogel sent his Algol paper to *Astronomische Nachrichten*.

Radial velocity work and especially the study of spectroscopic binaries and the application of photography to stellar spectroscopy will long be remembered as Vogel's main achievements. He was active at Potsdam up to his death in 1907. From 1895 to 1899 he gave a lot of time to the construction and planning of the 80 + 50 cm double refractor. In 1892 he was one of several observers to make a detailed photographic study of the spectrum of Nova Aurigae (85), which he did once more in 1901 for the very bright Nova Persei (86). His classification scheme for stellar spectra had less lasting impact than his other research, not so much because of any inherent deficiencies, but more because the objective prism classifications at Harvard from 1890 embraced far more stars in both hemispheres (leading to their widespread use and eventual official adoption by the International Astronomical Union) than either the Secchi or Vogel systems.

4.14 Norman Lockyer and the meteoritic hypothesis

In our discussion of stellar classification schemes devised in the nineteenth century, we will defer the story of the work at Harvard until Chapter 5. Here we must, however, mention the work of Sir Norman Lockyer (1836–1920). Lockyer was a self-made 'entrepreneur' in astronomy. He had no university education, and his hobby was at first pursued in his spare time during his employment as a clerk in the British War Office. Lockyer had the capacity for very hard work, and also a considerable flamboyance in his dedication to the popularisation of astronomy in general and his theories in particular. He set up his first observatory in his garden at Wimbledon with a $6\frac{1}{4}$-inch refractor in 1862. Three years later he acquired a Herschel–Browning spectroscope. His scientific hobby was soon noticed by his War Office superiors with the result that he was given various scientific administrative posts by the government, as well as a grant to acquire a larger spectroscope to observe the solar eclipse of 1868. As it happened, this instrument was not ready in time for the eclipse, but Lockyer (87) used it to observe the spectra of prominences in the absence of the eclipse, as Janssen had done before him (66). His resulting joint discovery of the D_3 line of helium made him famous, and this was soon followed by his coining of the term 'chromosphere' to refer to the hot layer of hydrogen completely surrounding the sun above the photosphere (68). In 1879 the Solar Physics Observatory in South Kensington was established by the government, and Lockyer became the first director. In 1881 he was also given a professorship at the Royal College of Science, and for most of the 1880s he was able to devote himself entirely to solar studies.

Not until about 1890 did Lockyer become extensively involved in stellar spectroscopy, using a 2-prism spectroscope attached to his 30-inch reflector, as well as an objective prism on his 6- and 10-inch refractors. His main observational results were published in 1893 by the Royal Society, and present a discussion of the spectra of 171 stars from 433 spectrograms mainly taken with the 6-inch objective prism (88). In this paper Lockyer developed further his scheme of stellar evolution which he had first put forward 5 years earlier (89). It was unfortunately characteristic of the way he tackled research that the theorising came first and the observations second.

The basis for Lockyer's spectral classification was to support his theory of evolution known as the meteoritic hypothesis. The universe was supposed to be full of streams of meteorites which, on colliding, were heated and vaporised giving first gaseous nebulae or comets. The vapours of these bodies then condensed to form young stars which contracted further to give a rise in their surface temperature. Eventually the loss due to radiation from the star was just balanced by the heating due to contraction, at which point the stars were at their highest temperatures. Thereafter the life of a star entered a prolonged phase of cooling, ending in final extinction. At any given temperature there were therefore stars on ascending and descending branches, corresponding to objects of increasing and decreasing temperatures.

Lockyer's classification was based on being able to estimate qualitatively the relative temperatures of the stars from their spectra and also on being able to differentiate between stars on the ascending and descending branches. The first part he did quite well by today's standards. That is, he assigned Secchi's types III and IV as the coolest and the hydrogen-line stars such as Vega as the hottest (89, 90). Lockyer's theoretical considerations for the second part of the task remain unclear, although he was quite explicit on the observational criteria that distinguish the stars on the ascending and descending branches. Whereas both Lockyer and Vogel claimed to base their classifications on the principles of stellar evolution, Lockyer placed the Vogel class IIIa (i.e. M stars) as the youngest, Vogel class IIIb, the carbon stars, as among the oldest. Without any detailed appeal to supporting evidence, Lockyer asserted:

> There is now, however, no doubt whatever that Vogel's class IIIa represents stars in which the temperature is increasing, and with conditions not unlike those of nebulae – that is to say, the meteorites are discrete, and are on their way to form bodies of Class II and Class I by the ultimate vaporisation of all the meteoric constituents. There is also no doubt that the stars included in Class IIIb have had their day; that their temperature has been running down, until owing to reduction of temperature they are on the verge of invisibility brought about by the enormous absorption of carbon in their atmospheres.
>
> I pointed out in the year 1886 that the time had arrived when stars with increasing temperatures would require to be fundamentally distinguished from those with decreasing temperatures, but I did not then know, that this was so easy to accomplish as it now appears to be (90).

In the earliest version of his evolutionary scheme (89), Lockyer made use of the notation of Vogel's first classification (Fig. 4.20) but the following year he used his own notation with the Roman numerals I to VI for six groups which he considered to be in an evolutionary sequence. Cool stars such as Betelgeuse were in group I and carbon stars were in group VI, while group IV stars (e.g. Vega) were the hottest (90).

In the final version of his evolutionary scheme, published in 1900, Lockyer arranged the stars in a number of groups which were named after the characteristic star in each. For example the first group in the evolutionary sequence following the nebular stage was the Antarian, the Secchi type III star Antares being the leading example. The ascending temperature series with their present-day MK types was represented by the

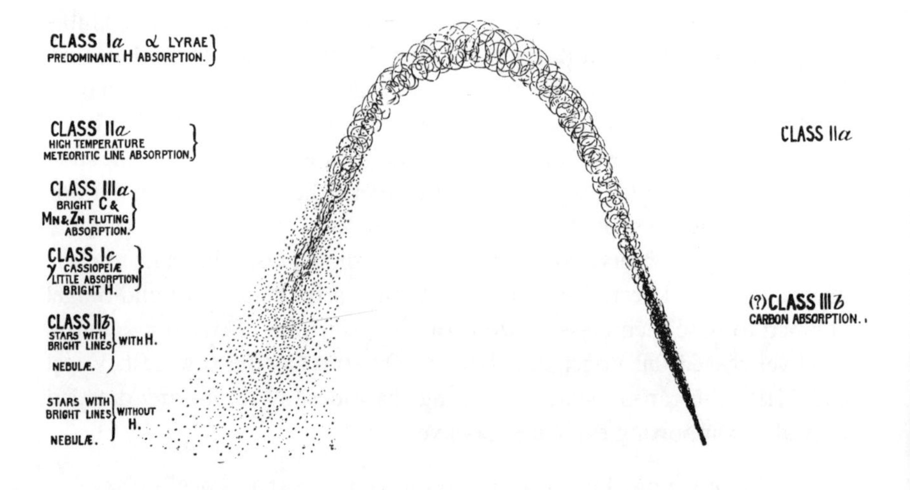

Fig. 4.20 Lockyer's temperature curve for stellar spectral evolution

stars; Antares (M1 Ib), Aldebaran (K5 III), Polaris (F8 Ib), αCygni (A2 Ia), Rigel (B8 Ia), ζ Tau (B2 IVp), β Crucis (B0.5 IV). At the temperature peak he now placed ε Orionis (B0 Ia) and then γ Velorum (WC7 + O7) as the type examples. Then on the descending or cooling branch were Achernar (B5 IV), Algol (B8 V), Markab = α Peg (B9.5 III), Sirius (A1 V), Procyon (F5 IV), Arcturus (K2 IIIp), 19 Psc (C6). This arrangement of the evolutionary sequence is from Lockyer's book *Inorganic Evolution* (91). Like Vogel, he had earlier modified his classification scheme to incorporate the helium stars (92) and correctly recognised that these were indeed hotter than, for example, Vega.

Lockyer's temperature criteria were based on his laboratory observations of arc and the hotter spark spectra from which he concluded that elements are dissociated by increased temperature into proto-elements, as he termed them, which showed different spectral lines that he termed 'enhanced', meaning they were stronger at the higher temperatures.

> We have then to face the fact that on the dissociation hypothesis, as the metals which exist at the temperature of the arc are broken up into finer forms, which I have termed protometals, at the fourth stage of heat (that of the high tension spark) which gives us the enhanced spectrum; so the proto-metals are themselves broken up at some temperature which we cannot reach in our laboratories into other simpler gaseous forms, the clèveite gases, oxygen, nitrogen and carbon being among them.
>
> Does the story end here? No there is still a higher stage; as the

> clèveite gases have disappeared as the arc lines and enhanced lines did at the lower stages; the raw form of hydrogen to which I have before called attention and which we may think of as 'proto-hydrogen', makes its appearance (91).

Lockyer's proto-hydrogen was simply ionised helium, the lines of which were discovered by Pickering in ζ Puppis in 1896 (93) and at first mistaken for a form of hydrogen.

As we have seen, Lockyer persistently glossed over his reasoning in assigning rising or falling temperatures to stars of various spectral characteristics. Throughout his working he resorts to woolly and repetitive rhetoric which would be entirely unacceptable in scientific publications of today, and his style is reminiscent of the unproductive theorising half a century earlier of Sir David Brewster. Whatever his methods, what is certain is that he distinguished the spectra of giants and supergiants from those of dwarfs. The former were on his ascending branch. The latter, including the sun, were supposed to be cooling, and the observational data were listed as follows (92):

> .... it may be stated that stars at about the same temperature, as judged by the iron lines on the ascending side of the curve, differ from those on the descending side.
>
> (1) In greater continuous absorption in the violet or ultraviolet, especially at the lower stages of temperature.
> (2) In the relative thinness of the hydrogen lines at the higher stages of temperature.
> (3) In the greater intensity and thickness of the metallic lines whether of low or high temperature.
> (4) In the relatively greater thickness of the lines of clèveite gases at those stages of temperature in which they appear.

Just as Lockyer's dissociation theory predated the ionisation theory of Saha (1920) and even J.J. Thomson's discovery of the electron (1897), so did his two stellar evolutionary branches anticipate Henry Norris Russell's division of stars into dwarfs and giants in 1910 and Miss Maury's classification based on apparent line width. His meteoritic theory and scheme of stellar evolution were never generally accepted and frequently attacked (see for example the note by Sir Arthur Schuster (94)). However his work on the enhanced lines illustrated Lockyer's unusual scientific insight, in spite of his unorthodoxy. His self-confident style and unorthodox theorising may have caused some friction with his great English contemporary Sir William Huggins, as correspondence in *Nature* would indicate (95, 96). As the founder of *Nature* in 1869 and its editor for the next

50 years, Lockyer had a ready-made outlet for publicising his theories. If he had his critics, he also had admirers and his share of recognition. The awarding of a Royal Society fellowship (1869), a professorship (1881) and a knighthood (1897) are proof enough.

Lockyer's last years ended on a somewhat sour note. He was compulsorily retired from the South Kensington Observatory in 1913 when a committee recommended the resiting of the observatory in Cambridge as part of that university. Lockyer however objected to such a move and, although now powerless to prevent it, he instead established a private observatory in Sidmouth, Devon, which was maintained financially entirely by its private benefactors. When Lockyer died in 1920 his son Major W.S.J. Lockyer took over as director, and objective prism spectroscopy was the main activity throughout the interwar years, when the observatory was named the Norman Lockyer Observatory. It came under University of Exeter administration in 1948.

4.15 New southern emission-line stars: Herschel, Ellery, Pechüle, Copeland So far we have outlined the main developments in stellar spectroscopy up until 1895 with the exception of the work at Harvard and Lick. The four main figures were Secchi, Huggins, Vogel and Lockyer. Significant work by other astronomers should not, however, be ignored. Leaving aside the application of the Doppler effect which will be discussed more fully in Chapter 6, the work by astronomers such as d'Arrest, Dunér, Copeland, Pechüle, Espin, Konkoly and McLean in cataloguing stellar spectra, especially of very red stars of Secchi types III and IV, and of bright-line stars, deserves some mention.

The stars and nebulae of the southern hemisphere, as might be expected, were only occasionally observed with the spectroscope in the nineteenth century. Some rich prizes were soon found by those few observers who took their spectroscopes south. Lt J. Herschel, for example, was in Madras for the 1868 solar eclipse; throughout much of that year he made occasional observations of the spectra of southern nebulae, including the great nebula in Argus ($= \eta$ Carinae) (97). The installation of the Great Melbourne Telescope of 48 inches aperture in Australia in 1867 should have opened the way for an extensive investigation of southern stellar spectra. It is with this instrument that A. le Sueur observed the spectrum of the 30 Doradus nebula in the Large Magellanic Cloud as well as of η Carinae (98). The observation of 30 Dor marks the beginning of extragalactic spectroscopy. As for η Car he described the spectrum in more detail in a second article (99):

> The spectrum of this star is crossed by bright lines... The most marked lines I make out to be, if not coincident with, very near to C, D, b, F and the principal green nitrogen line.* There are possibly other lines, but those mentioned are the only ones manageable. The yellow (or orange?) line in the star has not yet received sufficient attention; it is however very near D...; at present it cannot be said whether the line may not be slightly more refrangible than D... Owing to the faintness of the spectrum no dark lines are made out; one in the red is strongly suspected.

This description is interesting as it is probably the first account of reliable stellar spectroscopy in the southern skies, if Lettsom's observations of 1864 are not considered (9). The director of the Melbourne Observatory, R.L.J. Ellery (1827–1908) later carried out a southern spectroscopic survey with the help of another assistant, P. Baracchi, using a McLean spectroscope on the 48-inch telescope (100). The spectra of 200 stars were described, but the quality of the results was poor. With such a large telescope the practical limit for spectroscopic work was only about fifth magnitude, which tends to confirm the well-known fact that this telescope was largely a failure.

Meanwhile an English amateur astronomer who was based in India as an engineer to the Public Works in Madras, E.H. Pringle (1844–82), pursued stellar spectroscopy in his spare time. He found the very bright Wolf-Rayet star γ Argus ($=\gamma$ Velorum) in January 1872. As he correctly mentions, the professor of astronomy in Rome, L. Respighi (1824–89) had also found the same spectrum from Madras the previous month (101). Pringle wrote: 'The three principal bright lines, two in the yellow and one in the blue, were very distinct'; two other emission lines were also suspected, at each end of the visible spectrum (102).

The early discovery of such a bright emission-line star in the south was certainly an important event, though the far southern declination resulted in continued relative neglect. Two expeditions to the Caribbean to observe the transit of Venus across the sun's disc in December 1882 also took stellar spectroscopes with them, and were hence able to continue the observations of γ Velorum as well as of other southern stars. One of these expeditions was led by C.F. Pechüle (1843–1914) from the Copenhagen Observatory. He set up his base on the island of St Croix, and in December and January he surveyed 568 southern star spectra, going as faint as about fifth or sixth magnitude with a 6-inch refractor. Pechüle gave a description of the

* This presumably refers to the 'nebulium' line at 5007 Å, which is in fact due to ionised oxygen.

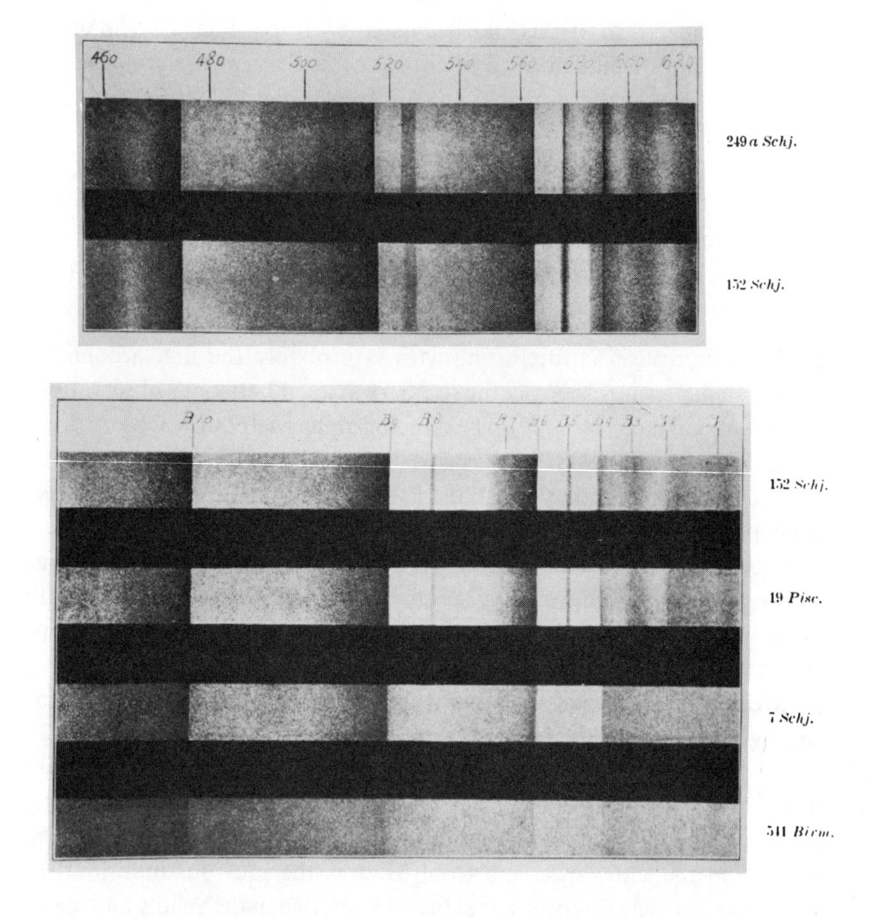

Fig. 4.21 Spectra of carbon stars photographed by Vogel (above) and by Dunér (below).

spectrum of γ Velorum which confirmed the presence of two close yellow emission lines and of a blue one, all very bright (103).

The other expedition was led by Ralph Copeland (1837–1905) the Astronomer Royal for Scotland, who travelled to Peru and Bolivia after observing the Venus transit in Jamaica. He established an observing station on the shores of Lake Titicaca at an altitude of 4000 m in Bolivia (104). His description of the γ Velorum spectrum is quite poetic:

> Its intensely bright line in the blue, and the gorgeous group of three bright lines in the yellow and orange, render its spectrum incomparably the most brilliant and striking in the whole heavens. To a great extent it was the extraordinary beauty of this spectrum (which, as I have since learned, was first seen by Respighi in 1871)

that led me to devote a considerable part of my time to more or less systematic sweeps of the neighbourhood of the Milky Way.

Copeland's South American journey was highly productive as he found no fewer than five further Wolf–Rayet stars* at that time, though all were relatively faint, between visual magnitudes 7.5 and 9.8 After his return he added a sixth to this list of Wolf–Rayet discoveries when he found emission lines in BD +37° 3821 in Cygnus (105). Pickering incorrectly claimed this last discovery as his own the following year (75). However three other stars of the Wolf–Rayet type were his own,† as well as the unusual emission line star P Cygni. This brought the total of known Wolf–Rayet stars by the mid-1880s in Cygnus alone to seven, and in the whole sky to thirteen, though γ Velorum was easily the brightest. Pickering's discovery of the peculiar spectrum of P Cygni, which had possibly been a nova about 1600 and shows the characteristic double emission and violet-displaced absorption lines, was followed 2 years later by the observations of Maunder at Greenwich (106), who seems to have been unaware of the earlier mention of the spectrum at Harvard.

4.16 The spectra of red stars: d'Arrest, Dunér, Espin The study and classification of red stars with banded spectra which Secchi had initiated attracted interest from other observers too, no doubt because of their relative rarity and because of their frequent association with red variable stars. The director of the Copenhagen observatory, H.L. d'Arrest, contributed four papers on red star spectra during the last 2 years of his life (107). He classified about 100 red stars, but only four were of Secchi's type IV, the carbon stars.

N.C. Dunér (1839–1914) at Lund Observatory in Sweden undertook a detailed study of Vogel's class III with the 24.5 cm Merz refractor (108). He classified and described the spectra of 297 class IIIa (= M type) stars and 55 IIIb (carbon) stars, noting that 273 Schjellerup (or 19 Piscium) was one of the finest examples of the latter. Moreover he showed that carbon stars have a distinct tendency to be found at small galactic latitudes (or in the Milky Way) because, as we now recognise, they are mainly younger stars of the disc population in our Galaxy. Dunér was later to become well-known for his spectroscopic study of solar rotation. Both these researches were undertaken with the greatest care and precision which earned him a justly high reputation.

* They were CD −41° 11041, CD −47° 4504, CPD −57° 5981, CPD −58° 2546 and CPD −60° 2578.

† They were BD +36° 3987, +35° 3953 and +38° 4010.

Fig. 4.22 Nicholas von Konkoly, about 1900.

The comparative rarity of carbon stars can be judged from a catalogue by the English amateur, the Rev. T.E. Espin (1858–1934) published in 1889, and based on the observations of Dunér, Pechüle, Konkoly and himself (109). Only 113 stars are given of which 29 were south of the equator. Over fifty were Espin's own discoveries, undertaken with his $17\frac{1}{4}$ inch reflector at his private observatory at Wolsingham, Durham. It is remarkable that Espin with this instrument was visually observing the spectra of stars in one case as faint as magnitude 11.5, and regularly to tenth magnitude (110), probably fainter than any other observer at the time, and fully 100 times better than the performance of the much larger Melbourne telescope! In his observations of red star spectra he found a number of long-period variables of Secchi type III but with bright hydrogen lines, including R Leo, R Hya, χ Cygni, R And and S Cas (111).

Meanwhile the study of the spectra of red stars was advanced further by a new catalogue by F.J.C. Krüger (1864–1916) in Kiel (116). He gave a list of 2153 red stars together with their published spectral types on the Secchi and Harvard systems and colour estimates on a sixteen-point scale.

4.17 S Andromedae: the first supernova spectrum In Hungary, N. von Konkoly (1842–1916) was at this time also classifying spectra at his private observatory at O'Gyalla. He extended Vogel and Müller's spectral survey of stars to magnitude 7.5 southwards by observing in the zone from the equator to $-15°$ (112). Konkoly was also one of several astronomers to make the first observations of a supernova spectrum (113). This was in 1885 and the star was S Andromedae in the great Andromeda galaxy. In fact Vogel observed S And on 1 September of that year, which preceded Konkoly by 3 nights (114). Of course Konkoly and Vogel were ignorant of both the facts that it was not a nova like those that had been seen earlier in Corona Borealis and Cygnus, but the quite different and far more energetic phenomenon of a supernova, and that they were observing the first spectrum of an extragalactic star. Vogel had found the spectrum continuous but Konkoly reported four broad emission lines in the red, yellow, green and blue, two of which he judged to be hydrogen lines ($H\alpha$ and $H\beta$; see section 9.14). Konkoly's compatriot, E. de Gothard (1857–1909), also contributed to spectral classification from his private Herény Observatory, but his published results were not extensive (115).

4.18 Spectrum photography in the 1890s: McLean, Scheiner, Sidgreaves F. McLean (1837–1904) was another English amateur who took up stellar spectroscopy at his private observatory at Tunbridge Wells in Kent. In 1895 he undertook an objective prism survey of all northern stars down to magnitude 3.5 with his 12-inch Grubb astrograph and 20° prism (117). The survey was completed in the southern hemisphere when he visited the Cape Observatory in 1897 but, as this work came after the Draper Memorial catalogue at Harvard (which went much fainter), it was only of minor importance. However McLean, soon after Vogel, was one of the first to identify the stars with neutral helium absorption lines (118), and he devised a new system of spectral classification based on this. His scheme resembled Secchi's except that he divided Secchi's type I into three groups (I, II, III) corresponding to B, A and F stars respectively. His types IV, V and VI were then G and K stars, M stars and carbon stars. In this respect, his classification was thoroughly up-to-date and not inferior to Vogel's revised scheme or to Pickering's, but its lack of use by others ensured its early demise.

One of the more remarkable of McLean's discoveries was that of oxygen in the helium stars, particularly in β Crucis (MK type B0.5 IV). He found several lines in these stars which correspond to Thalén's spark spectra for oxygen (119). This discovery was treated with considerable scepticism at the time, although the lines are now recognised as being from ionised oxygen (O II).

McLean was a solitary worker who nevertheless made some munificent donations, including a new 24-inch photographic telescope for the Cape Observatory. Sir David Gill (1843–1914) used this instrument to measure more precise wavelengths in the emission-line star η Carinae from an objective prism spectrum (120). If McLean was an amateur who applied photography to stellar spectroscopy from his own private means, Scheiner at Potsdam was a professional who had the resources of one of the leading observatories of the day to draw on. His detailed photographic study of the spectra of fifty stars, in which he measured line wavelengths to six figures (to one-hundredth of an Ångström unit), was one of the most exhaustive stellar-wavelength studies in the nineteenth century (121). Father W. Sidgreaves (1837–1919) was another who successfully made use of photography. He directed the observatory at Stonyhurst College which belonged to a Jesuit establishment in Lancashire, England. As an example of his work, his investigation of the spectrum of Nova Aurigae in 1892 gave an extensive list of emission and absorption line wavelengths in this star, measured to the nearest Ångström unit (122).

In the nineteenth century much of stellar spectroscopy was no more than the collection and classification of data. It could hardly be otherwise, for how could interpretation play a significant role, considering that J.J. Thomson and the electron, Albert Einstein and quantum theory, Niels Bohr and the electronic theory of atomic structure and M.N. Saha's theory of ionisation were all in store for the future? Without these powerful new methods, theorising in astrophysics was of limited value, as for example Lockyer's meteoritic hypothesis shows only too clearly.

References

1. Fraunhofer, J., *Gilberts Ann.*, **56**, 264 (1817) See ref. 16, Chap. 2.
2. Fraunhofer, J., *Gilberts Ann.*, **74**, 337 (1823).
3. Lamont, J., *Jahrbuch der königl. Sternwarte bei München für 1838*, pp. 190, 191 (1838).
4. Swan, J., *Phil. Mag.*, (4) **11**, 448 (1856).
5. Porro, I., *Comptes Rendus*, **47**, 873 (1858).
6. Donati, G.B., *Nuovo Cimento*, **15**, 292 (1862).
7. Donati, G.B., *Mon. Not. Roy. Astron. Soc.*, **23**, 100 (1863).

8. von Steinheil, C.A., *Astron. Nachrichten*, **53**, 253 (1863).
9. Lettsom, W.G., *Mon. Not. Roy. Astron. Soc.*, **24**, 217 (1864).
10. Janssen, J., *Comptes Rendus*, **55**, 576 (1862).
11. Rutherfurd, L.M., *Americ. J. of Sci. and Arts*, **35**, 71 (1863).
12. Huggins, W. and Miller, W.A., *Proc. R. Soc.*, **12**, 444 (1863).
13. Secchi, A., Memoria sugli spettri prismatici della luce de' corpi celesti. *Bull. meteorologico dell' Osservatorio del Collegia Romano*. Vol. II (1863).
14. Secchi, A., *Comptes Rendus*, **57**, 71 (1863).
15. Secchi, A., *Astron. Nachrichten*, **59**, 193 (1863).
16. Airy, G.B., *Mon. Not. Roy. Astron. Soc*, **23**, 188 (1863).
17. Rutherfurd, L.M., *Americ. J. of Sci. and Arts*, **35**, 71 (1863).
18. Rutherfurd, L.M., *Americ. J. of Sci. and Arts*, **36**, 154 (1863).
19. Secchi, A., *Astron. Nachrichten*, **41**, 109 (1855).
20. Secchi, A., *Comptes Rendus*, **63**, 364 (1866).
21. Secchi, A., *Comptes Rendus*, **63**, 621 (1866).
22. Secchi, A., *'die Sterne'*, p. 70 et seq. (1877).
23. Secchi, A., *Comptes Rendus*, **64**, 345 (1867).
24. Secchi, A., *Comptes Rendus*, **66**, 124 (1868).
25. Schjellerup, H.C.F.C., *Astron. Nachrichten*, **67**, 97 (1866).
26. Secchi, A., *Comptes Rendus*, **67**, 373 (1868).
27. Secchi, A., *Comptes Rendus*, **64**, 774 (1867).
28. Secchi, A., *Comptes Rendus*, **68**, 361 (1869).
29. Secchi, A., *Comptes Rendus*, **71**, 252 (1870).
30. Secchi, A., *Comptes Rendus*, **75**, 655 (1872).
31. Huggins, W., *Mon. Not. Roy. Astron. Soc.*, **16**, 175 (1856).
32. Huggins, W., *The Nineteenth Century Review* (June, 1897).
33. Huggins, W. and Miller, W.A., *Phil. Trans. R. Soc.*, **154**, 413 (1864).
34. Huggins, W., *Phil. Trans. R. Soc.*, **154**, 139 (1864).
35. The Scientific Papers of Sir William Huggins (*Publications of Sir Wm Huggins Observatory*, Vol. II) W. Wesley & Son (1909).
36. Huggins, W. and Miller, W.A., *Phil. Trans. R. Soc.*, **154**, 437 (1864).
37. Huggins, W. and Lady Huggins, *Proc. R. Soc.*, **46**, 40 (1889).
38. Huggins, W., *Proc. R. Soc.*, **14**, 39 (1865).
39. Huggins, W., *Proc. R. Soc.*, **33**, 425 (1882).
40. Huggins, W., *Mon. Not. Roy. Astron. Soc.*, **26**, 275 (1866). See also: Huggins, W. and Miller, W.A., *Proc. R. Soc.*, **15**, 146 (1866).
41. Stone, E.J., *Mon. Not. Roy. Astron. Soc.*, **26**, 292 (1866).
42. Wolf, C. and Rayet, G., *Comptes Rendus*, **62**, 1108 (1866).
43. Huggins, W. and Lady Huggins, *Proc. R. Soc.*, **50**, 465 (1892).
44. Huggins, W. and Lady Huggins, *Proc. R. Soc.*, **51**, 486 (1892).
45. Wolf, C. and Rayet, G., *Comptes Rendus*, **65**, 292 (1867).
46. Vogel, H.C., *Publ. Astrophys. Observ. Potsdam*, **4**, 17 (1883).
47. Huggins, W. and Lady Huggins, *Proc. R. Soc.*, **49**, 33 (1890).
48. Donati, G.B., *Astron. Nachrichten*, **62**, 375 (1864).
49. Huggins, W., *Proc. R. Soc.*, **16**, 386 (1868).
50. Huggins, W., *Proc. R. Soc.*, **33**, 1 (1881).
51. Huggins, W., *Phil. Trans. R. Soc.*, **158**, 529 (1868).
52. Huggins, W., *Proc. R. Soc.*, **20**, 379 (1872). See also *Mon. Not. Roy. Astron. Soc.*, **32**, 359 (1872).

53. Draper, H., *Americ. J. of Sci. and Arts*, **18**, (Series 3) 419 (1879).
54. Draper, H., *Proc. Amer. Philosoph. Soc.*, **17**, 76 (1877). See also: *Amer. J. of Sci. and Arts*, **14**, (Series 3) 89 (1877).
55. Lockyer, J.N., *Phil. Mag.*, **6**, 174 (1878). See also: Christie, W.H.M., *Mon. Not. Roy. Astron. Soc.*, **38**, 473 (1878) for further criticisms of Draper's work.
56. Trowbridge, J. and Hutchins, C.C., *Amer. J. of Sci.*, **34**, 270 (1887). An account of the whole controversy and its resolution is given by Plotkin, H., *J. History Astron.*, **8**, 44 (1977).
57. Huggins, W., *Proc. R. Soc.*, **25**, 445 (1877).
58. Huggins, W., *Phil. Trans. R. Soc.*, **171**, 669 (1880).
59. Vogel, H.W., *Sitzungsberichte d. Königl. Preuss. Akademie d. Wissenschaften zu Berlin* (1879) p. 586 and ibid. (1880) p. 192.
60. Huggins, W., *Proc. R. Soc.*, **34**, 148 (1882).
61. Huggins, W. and Lady Huggins, 'An Atlas of Representative Stellar Spectra' (*Publ. of the Sir Wm. Huggins Observatory*, Vol. I) W. Wesley & Son (1899).
62. Vogel, H.C., *Astron. Nachrichten*, **84**, 113 (1874).
63. Vogel, H.C., and Müller, G., *Publ. Astrophys. Observ. Potsdam*, **3**, 127 (1883).
64. Herschel, J., *Proc. R. Soc.*, **17**, 116 (1869).
65. Tennant, J.F., *Mon. Not. Roy. Astron. Soc.*, **28**, 245 (1868).
66. Janssen, J., *Comptes Rendus*, **67**, 838 (1868).
67. Rayet, G., *Comptes Rendus*, **67**, 757 (1868).
68. Lockyer, J.N., *Proc. R. Soc.*, **17**, 131 (1868) and *Phil. Trans.*, **154**, 425 (1869).
69. Ramsay, W., *Proc. R. Soc.*, **58**, **65**, and **81** (1895).
70. Clève, P.F., *Comptes Rendus*, **120**, 834 (1895).
71. Lockyer, J.N., *Proc. R. Soc.*, **58**, 67, 113, 166 and 192 (1895).
72. Runge, C. and Paschen, F., *Sitzungsberichte d. Königl. Preuss. Akad. der Wissenschaften zu Berlin* (1895) pp 639 and 759. See also *Astrophys. J.*, **3**, 4 (1896).
73. Balmer, J.J., *Ann. der Physik*, **25**, 80 (1885).
74. Vogel, H.C., *Sitzungsberichte d. Königl. Preuss. Akademie d. Wissenschaften zu Berlin* (1895) p. 945. See also: *Astrophys. J.*, **2**, 333 (1895).
75. Pickering, E.C., *Nature*, **34**, 439 (1886).
76. Keeler, J.E., *Astron. Astrophys.*, **12**, 361 (1893).
77. Vogel, H.C. and Wilsing, J., *Publ. Astrophys. Obs. Potsdam*, **12**, 1 (1899).
78. Vogel, H.C., *Sitzungsberichte d. Königl. Preuss. Akademie der Wissenschaften zu Berlin* (1888) p. 397.
79. Vogel, H.C., *Astron. Nachrichten*, **121**, 241 (1899).
80. Vogel, H.C., *Publ. Astrophys. Observ. Potsdam*, **7**, 1(1889). See also *Mon. Not. Roy. Astron. Soc.*, **52**, 87 (1891) and **52**, 541 (1892) for English version.
81. Vogel, H.C., *Sitzungsberichte d. Königl. Preuss. Akademie der Wissenschaften zu Berlin* (1891) p. 533.
82. Vogel, H.C., *Astron. Nachrichten*, **123**, 289 (1890).
83. Vogel, H.C., *Astron. Nachrichten*, **125**, 305 (1890).
84. Pickering, E.C., *Amer. J. of Science*, **34**, 46 (1890).

85. Vogel, H.C., *Astron. Astrophys.*, **12**, 896 (1893).
86. Vogel, H.C., *Astrophys. J.*, **13**, 217 (1901).
87. Lockyer, J.N., *Comptes Rendus*, **67**, 949 (1868).
88. Lockyer, J.N., *Phil. Trans. R. Soc.*, **184**, 675 (1893).
89. Lockyer, J.N., *Proc. R. Soc.*, **43**, 117 (1887).
90. Lockyer, J.N., *Proc. R. Soc.*, **44**, 1 (1888).
91. Lockyer, J.N., *Inorganic Evolution*, Macmillan & Co. (London) (1900).
92. Lockyer, J.N., *Proc. R. Soc.*, **61**, 148 (1897).
93. Pickering, E.C., *Astrophys. J.*, **4**, 369 (1896).
94. Schuster, A., *Proc. R. Soc.*, **61**, 209 (1897).
95. Lockyer, J.N., *Nature*, **55**, 304 and 341 (1897).
96. Huggins, W., *Nature*, **55**, 316 (1897).
97. Herschel, J., *Proc. R. Soc.*, **16**, 417 (1868).
98. le Sueur, A., *Proc. R. Soc.*, **18**, 222 (1870).
99. le Sueur, A., *Proc. R. Soc.*, **18**, 245 (1870).
100. Ellery, R.L.J., *Mon. Not. Roy. Astron. Soc.*, **49**, 439 (1889) and ibid. **50**, 66 (1890).
101. Respighi, L., *Comptes Rendus*, **74**, 516 (1872).
102. Pringle, E.H., *Mon. Not. Roy. Astron. Soc.*, **34**, 267 (1874).
103. Pechüle, C.F., *Expédition danoise pour l'observation du Passage de Vénus 1882* (Copenhagen, 1883).
104. Copeland, R., *Copernicus*, **3**, 193 (1884).
105. Copeland, R., *Mon. Not. Roy. Astron. Soc.*, **45**, 90 (1885).
106. Maunder, E.W., *Mon. Not. Roy. Astron. Soc.*, **49**, 300 (1889).
107. d'Arrest, H.L., *Astron. Nachrichten*, **84**, 263 and 364 (1874); ibid. **85**, 250 (1875); ibid **86**, 53 (1875).
108. Dunér, N.C., Sur les étoiles à spectres de la 3ème classe... *Kongl. Svenska Vetenskaps-Akademiens Handlinger*, Vol. **21**, no. 2, Stockholm (1884).
109. Espin, T.E., *Mon. Not. Roy. Astron. Soc.*, **49**, 364 (1889).
110. Espin. T.E., *Mon. Not. Roy. Astron. Soc.*, **54**, 100 (1894).
111. Espin, T.E., *Astron. Nachrichten*, **121**, 143 and 351 (1889); ibid. **123**, 31 and 143 (1890).
112. von Konkoly, N., *Publ. der Beobachtungen des Astrophys. Observatioriums zu O'Gyalla,* Vol. **8**, Part 2 (1887).
113. von Konkoly, N., *Astron. Nachrichten*, **112**, 286 (1885).
114. Vogel, H.C., *Astron. Nachrichten*, **112**, 283 (1885).
115. de Gothard, E., *Mon. Not. Roy. Astron. Soc.*, **43**, 421 (1883).
116. Krüger, F., *Publ. der Sternwarte in Kiel*, Vol. **8**, (1893).
117. McLean, F., *Phil. Trans. R. Soc.*, **191**, 127 (1898).
118. McLean, F., *Mon. Not. Roy. Astron. Soc.*, **56**, 428 (1896).
119. McLean. F., *Proc. R. Soc.*, **62**, 417 (1898).
120. Gill, Sir D., *Mon. Not. Roy. Astron. Soc.*, **61**, [66] (Appendix no. 4.) (1901).
121. Scheiner, J., *Publ. Astrophys. Observ. Potsdam*, **7**, 167 (1895).
122. Sidgreaves, W., *Mem. Roy. Astron. Soc.*, **51**, 29 (1892).

5 Spectral classification at Harvard

5.1 E.C. Pickering at Harvard College Observatory

So far only occasional reference has been made to the work in stellar spectroscopy at Harvard College Observatory in Cambridge, Massachusetts. However, the developments that took place there from 1885 are so important that they merit a separate chapter. The action took place over a period of four decades from this date, and five actors filled the leading rôles. Four of these were women. The part played by Professor Edward C. Pickering in the development of Harvard stellar spectroscopy was, however, the most significant. Pickering came from a prominent New England family and his brother (W.H. Pickering (1858–1938)) was also a physicist and astronomer of some note (he was an assistant professor in astronomy at Harvard from 1887). The fact that Edward Pickering was appointed to a chair as Professor of Physics at the Massachusetts Institute of Technology (MIT) at the age of only 22 shows that his scientific abilities were already manifest in comparative youth. His research at MIT was mainly in the field of optics, although astronomy was also an interest, as he took part in solar eclipse expeditions in both 1869 and 1870.

After 9 years as an MIT professor, Pickering was appointed director of Harvard College Observatory where he took up his duties in February 1877. Apparently the appointment of a physicist provoked some criticism, as several able astronomers were also candidates. However, a more suitable choice, in hindsight, would be hard to imagine, as Pickering was an energetic observer, administrator and fund raiser. He was fully aware of the importance of the new developments in spectroscopic astrophysics and he combined this awareness with his most remarkable characteristic, the ability to undertake routine projects of data collection on a scale hitherto unprecedented in spectroscopic astronomy. He initiated a vast program of visual photometry at Harvard, and personally recorded about 1 400 000 observations of stellar magnitudes at the telescope. He also soon recognised the importance that photography would play for the future of astronomy, and the regular photographic patrolling of the sky with wide-angle cameras was begun under his direction. Many of these patrol plates

Fig. 5.1 Edward Pickering in 1891.

were barely examined before going into storage, which his critics saw as both wasteful of effort and extravagant. But Pickering foresaw the need to have a permanent record of the sky for the discovery of new novae, asteroids and variable stars. The last of these became one of his favourite fields of investigation, and the fact that he found and catalogued 3435 new variable stars points to the value of the patrol plates even within his lifetime. The record of the sky stretching back about a century (the program was commenced in March 1885) is now a unique one of inestimable and continuing value.

5.2 Mrs Draper and the Henry Draper Memorial It is in the field of objective prism stellar spectroscopy that Pickering made his most important contribution to astronomy. And it is here that Mrs Anna Palmer Draper (1839–1914) entered the scene at Harvard. Her husband, Henry Draper, had died aged 45 in 1882. At his private observatory in Hastings-on-Hudson he had in 1872 been the first to photograph successfully a stellar spectrum (Chapter 4). Mrs Draper now wished to establish a fitting memorial to her husband. Rather than attempt to have Henry Draper's spectroscopic research continued at his private observatory, she chose to approach Edward Pickering for the establish-

Fig 5.2 Mrs Anna Palmer Draper.

ment of a stellar spectroscopy program at Harvard, to be funded by her endowments. These gifts continued over many years during her lifetime, and included a large bequest to the observatory when she died in 1914; in all they amounted to several hundred thousand dollars. It should be mentioned that Mrs Draper's gifts to Harvard astronomy were not the only substantial ones during Pickering's directorship. The Paine Fund of about $400 000 in 1886, the Boyden Fund of $230 000 in 1887, Miss Bruce's donation of $50 000 in 1889 and Professor Pickering's personal contribution of more than $100 000 (from his own pocket) all helped ensure that the ambitious programs Pickering initiated could continue on a grand scale.

Pickering described the establishment of the Henry Draper Memorial in the following words:

> It would have been difficult, if not impossible, to find anyone who, without years of practice, could have attained the skill and experience of Dr Draper in the various departments required for successfully solving this problem [of stellar photographic research]. But great advances had been made in photographic processes, and new methods of investigation had been discovered. Accordingly, an arrangement was made by Mrs Draper for continuing research on a liberal scale at the Harvard College Observatory, under the name of the Henry Draper Memorial (1).

Fig. 5.3 Harvard College Observatory in 1887. The telescopes are: left dome, 13-inch Boyden refractor; centre foreground, 28-inch Draper reflector; right dome, 11-inch Draper refractor; centre background, 15-inch refractor; wooden hut to immediate left of 15-inch, 8-inch Bache refractor.

This was early in 1886 and it was natural that the main effort was to be with the photographic objective prism. Objective prisms had been used by both Fraunhofer and Secchi, but Pickering had realised their value in conjunction with photography for the simultaneous recording of several, or in some cases, hundreds of spectra in a single exposure. The first Harvard experiments in objective prism spectrography were in May 1885, and a regular program was commenced in October of that year. The telescope used for this work, known as the Bache telescope, was an 8-inch photographic doublet of focal ratio f/5.6 giving a 10° square field on photographic plates. A 13° prism was provided for the spectrographic program. The spectra of sixth magnitude stars could be recorded in as little as 5 minutes.

5.3 Williamina Fleming and the Draper Memorial Catalogue

The first major spectroscopic program soon became established as part of the Henry Draper Memorial. In this program the spectra of 10 351 stars, nearly all north of −25°, were photographed on 633 plates containing 28 266 spectra suitable for classification. The plates were each 20 × 25 cm and all stars brighter than seventh magnitude could be studied on them. Few brighter than sixth in the northern sky were omitted. The observing program was complete by January 1889. From this store of data Pickering wished not only to classify and describe the spectra, but also to estimate the photographic stellar magnitudes, from the photographic density in the neighbourhood of 4320 Å. The task of examining the spectra was assigned to Scottish-born Mrs Williamina P. Fleming (1857–1911) who had, since 1881, been employed as a copyist and computer (non-electronic kind) at the observatory. She found that her new duties were to occupy her fully for the next 4 years from the commencement of the Draper Memorial program. As well as classifying over 26 000 spectra and estimating the photographic magnitudes, the work also entailed recording the shortest wavelength hydrogen line visible, estimating the strength of the calcium K absorption and ascertaining the presence or absence in the spectrograms of the F line (Hβ) at or near the long wavelength cutoff of the emulsion.

The Draper Memorial Catalogue of 10 351 stars was published by Pickering in 1890 (2). It was the most extensive and detailed spectral classification catalogue of the nineteenth century. The system of spectral classification used had been devised for the Draper Memorial program, and represents the forerunner of the system for the later Henry Draper Catalogue. The basis was to take the four Secchi spectral types, and to divide them further into thirteen types as follows:

Secchi type	Draper Memorial types
I	A, B, C, D
II	E, F, G, H, I, K, L (J not used)
III	M
IV	N

In addition, type O was used for Wolf–Rayet spectra with bright lines, P for planetary nebula spectra, and Q for objects otherwise unclassified by the letters A to P. Types O and P together were sometimes referred to at Harvard as spectra of the fifth type. Only one O-type object appeared in the catalogue (the Wolf–Rayet star CD − 23° 4553 = HR 2583), and no carbon stars of type N were bright enough to be classified. Type A stars had broad hydrogen absorption lines, B stars were mainly of the ‘Orion’ type, described first by Secchi, with neutral helium lines, though of course at this time the helium spectrum had not been studied in the laboratory. C stars had apparently double lines (probably due to an instrumental fault), and D represented Secchi type I stars with emission, such as γ Cas. In Secchi’s type

II, E signified only the F, H and K lines were visible, while types F and G showed progressively more and stronger lines. Type H resembled type F except the spectrum was weak below 4310 Å while I stars had this property as well as additional lines. Bright bands were seen in type K (though presumably these in reality were continuum peaks between CN bands in cooler stars) and L was reserved for stars with any other peculiarities not clearly specified.

Fig. 5.4 Mrs Williamina Fleming.

Mrs Fleming again used the Draper Memorial Catalogue classification for her later work on the spectra of stars in clusters (3). However, this was published in 1897 after the discovery of terrestrial helium and the cataloguing of its spectrum (see Chapter 4). The 1897 work incorporated several modifications, including the explicit assigning of class B to stars with the characteristic helium absorption lines at 4026 and 4472 Å, the dropping of class C from the classification, and the use of class F ahead of E for stars intermediate in type between Secchi's I and II, in which the H and K lines are of about equal strength. The sequence now ran A, B, F, E, G, H, M, N, O.

The Draper Memorial Catalogue was undoubtedly a significant milestone in the development of spectral classification. However, it was also grossly inadequate, mainly because, for the relatively large number of sixteen classes, the criteria defining the classes were not adequately quantitatively specified. Some classes turned out to be redundant and could be rejected, while others had the potential for further subdivision with more precise defining criteria.

5.4 Establishment of the Boyden Station at Arequipa, Peru

At the time of publication of the Draper Catalogue, important developments were taking place in the observatory's instrumentation. In 1879 Mr U.A. Boyden left $230 000 for the establishment of a mountaintop observatory, and this sum was transferred to Harvard in 1887 by the Boyden trustees. Pickering used these funds in 1890 to establish the Boyden Station of Harvard College Observatory in Arequipa, Peru at an altitude of over 8000 feet (2440 m) (Fig. 5.5). The 8-inch Bache telescope had been taken out of service at Cambridge in 1889, and was now erected at Arequipa, at a latitude of 16° south of the equator.

Fig. 5.5 The Boyden Station at Arequipa, Peru.

In addition the 13-inch Boyden refractor was also provided from the fund to equip the new observatory. This telescope was unusual in having an achromatic objective that could be adjusted for either photographic or visual work by reversing the crown-glass component of the doublet.

The departure of the Bache telescope from Cambridge left a hole which the munificence of Mrs Draper was able to fill:

> The removal of the Bache telescope interrupted several researches which would otherwise have been carried on with it and

for which it is especially suited. Accordingly Mrs Draper has provided a second instrument of nearly the same dimensions. This instrument has been kept at work throughout nearly every clear night since September 27, 1889 (4).

Her generosity did not stop with the new 8-inch Draper telescope. Henry Draper's 11-inch Alvan Clark refractor was loaned to Harvard in 1886 and this formed an ideal complement in Cambridge to the 13-inch Boyden telescope in Peru for higher dispersion spectrography of the brighter stars. It was modified for photographic work by the use of a correcting lens. Later the 11-inch refractor became the outright property of Harvard by gift. Henry Draper's 28-inch reflector, with which some of the first pioneering work in stellar spectrum photography had been undertaken, was also acquired by Harvard and erected in Cambridge.

5.5 The Maury classification

From the inception of the Henry Draper Memorial, Pickering had envisaged not only a general survey of spectral types, as published in the Draper Catalogue, but also a more detailed study of the brighter stars. For this purpose the 11-inch Draper telescope was used for an investigation of 681 bright stars north of $-30°$. Four 15° prisms were available. If all were mounted, the mean reciprocal dispersion (between Hβ and Hε) was 11 Å/mm. With a single prism it was 45 Å/mm, and this was used for classification purposes. The examination of about 4800 plates was the task assigned to Miss Antonia C. Maury (1866–1952) who was employed as a research associate at the observatory from 1888 until 1935. The study of the spectra of the bright northern stars was her first major work and was to occupy her until at least 1895. Antonia Maury was the niece of Henry Draper and grand-daughter of J. W. Draper, so it was only fitting that she should be engaged on the work of her uncle's memorial.

Pickering decided to give Miss Maury a free choice in devising a classification scheme suited to the task in hand. Since the resolution of these higher dispersion spectra was greater, a more detailed classification scheme than that used in the Draper Memorial Catalogue was felt to be justified. Miss Maury's classification consists of twenty-two groups represented by the Roman numerals I to XXII. Groups I to V were stars of the Orion type with increasing strength of the hydrogen lines from I to V. Although the classification was published in 1897 (5) the work had been practically completed by the time the helium spectrum was catalogued. A supplementary note was added to the work discussing the helium lines found in these

Fig. 5.6 Miss Antonia Maury.

groups. Groups VII to XI were of Secchi type I with progressively weaker hydrogen lines and stronger metallic lines while group VI was intermediate in type between the Orion objects and group VII. Group XII formed a link between Secchi's first and second types, while XIII to XVI had progressively weaker hydrogen and stronger metallic lines of the solar type. The stars of Secchi's third type were placed in groups XVII to XX, for which the spectral bands increased progressively in strength. Additionally in group XX the hydrogen lines were seen in emission, as in Mira variables around maximum light. Finally two groups completed the classification, XXI for carbon stars (Secchi type IV) and XXII for Wolf–Rayet stars.

So far, the Maury classification resembled that of Mrs Fleming, the sequence from I to XX corresponding to finer divisions of the Draper Catalogue letters B, A, F, G, K, M. However Miss Maury was the first to place the Orion or B stars ahead of those with the strongest hydrogen lines (A stars). She justified this with three lines of evidence: (a) '...by the gradual falling off of the more refrangible rays in successive groups...', (b) 'the comparative simplicity of the Orion spectra and the increasing complexity shown throughout the series' and (c) 'the prevalence of the Orion type in great nebulous regions, as in Orion and the Pleiades, indicates very emphatically that stars of this type are in an early stage of development' (5). The last point implicitly indicates that she had an evolutionary sequence in mind, though neither this, nor the possibility that her classification might represent a temperature sequence, is elaborated.

The remarks for some of the stars in Miss Maury's catalogue are of interest, for they show that she had noticed peculiarities in several spectra now classified as Ap (peculiar A-type) or Am (metallic-line A-type). Thus for the Ap star α CVn (Group VIII) she wrote: 'This star...has marked peculiarities. Thus the K line is extremely faint, and the lines 4131.4 and 4128.5 [of ionised silicon] have greater intensity than in any other stars except those of Division c in Group VIII. Some of the fainter lines...differ in intensity from the corresponding lines of the stars in Division a'. And for the metallic line star τ UMa (Group XII) she wrote: 'From the number and intensity of its solar lines it should be classed in Group XII. The width of the lines K and H, however, and the intensity of the hydrogen spectrum, would place it in Group X' (5).

5.6 Antonia Maury's 'collateral divisions' based on line width

The most original feature of Miss Maury's work was the further subdivision of the groups, as defined by the letters a, b, and c, based on the appearance of the lines, whether hazy or sharp. The sharp-line stars also had some lines of unusual strength. The principal division was called 'Division a' and contained 355 of the 681 stars. The lines were clearly discernible and of average width. Division b comprised

> stars in the spectra of which all the lines are relatively wide and hazy...the fainter lines becoming altogether imperceptible, and the total number of lines which can be seen is therefore often comparatively small. The relative intensity of those which are visible, however, remains about the same as in those of Division a' (5).

Presumably many of these stars were either rapid rotators or double-lined spectroscopic binaries in which the two components of each line were imperfectly resolved. The second of these possibilities is explicitly mentioned by Miss Maury; the first not so, although the effect of rotation in 'washing out' spectral lines by the Doppler effect had first been proposed by Captain W. de W. Abney (1843–1920) in 1877 (6). However, this theory was not at the time generally accepted (see Chapter 6). The division b stars were found exclusively in groups I to X and numbered 91 out of a total of 304 stars in these ten groups. This clustering of the stars with hazy lines in those of Secchi's first type (including the Orion subtype) is the first reference to the distribution of rotational velocities among spectral types, which was only fully explored by Shajn, Struve and Elvey about 1930.

Miss Maury's division c stars had unusually narrow lines of hydrogen and, for the Orion stars, of helium as well. Moreover, some of the metallic lines were unusually intense, giving relative intensities of various metallic lines quite different to that found in the sun (spectral type XIVa). In fact, some of the lines in later type c stars were frequently absent altogether in the solar spectrum. Miss Maury described these stars with the comment: 'In general, Division c is distinguished by the strongly defined character of its lines' (5). Division c was used comparatively rarely by Miss Maury. Eighteen stars had this classification in groups III to XIII while a further seventeen were classified ac, being intermediate between a and c divisions.

In Europe, the harshest critics of the Harvard classification schemes came from supporters of Vogel's system. Naturally these included most of the Potsdam spectroscopists, and also the Uppsala astronomer Dunér. Of the former group, Scheiner was the most vocal, attacking Miss Maury's classification very severely, especially on account of its use of so many different classes (7). Dunér, who had earlier praised Vogel's second scheme so warmly (8) went on to deliver the harshest attack on Antonia Maury's work:

> Still much worse [than the Pickering-Fleming classification] is the situation regarding Miss Maury's system consisting of 154 classes. For here the comparison with Vogel's classes shows the most frightful confusion, and one is forced to wonder where it will lead to, if everyone who works on stellar spectra also introduces a new classification. In fact, if one believed that every barely perceptible difference in a stellar spectrum justifies setting up a new spectral class, then one would do better to put every star into a class of its own. For as spectral analyses are carried out, with greater detail, the clearer it is that no two stellar spectra are absolutely identical (9).

Although Lockyer had also taken the first steps in distinguishing high and low luminosity stars in his classification which used the so-called enhanced lines (Chapter 4), Miss Maury's 'collateral divisions' were far more concise. The fact that the narrow lines in the c stars were associated with high intrinsic luminosity was first proposed in 1905 by E. Hertzsprung (1873–1967) at Potsdam (10). He was able to show statistically that the division c stars had smaller proper motions than those of division a, due to their on average greater distance, and therefore they must be assigned a higher luminosity. 'In other words the c stars are at least as luminous as the Orion stars... The hypothesis of the collateral series is suitable for explaining the principal observation, that among stars brighter than fifth apparent visual magnitude, the c and Orion stars shine the brightest' (10). Hertzsprung's important discovery is confirmed by modern MK luminosity classes for Maury's eighteen stars of division c; twelve of these are supergiants of class Ia, whereas the remaining six are class Iab or Ib supergiants. In the intermediate division ac (seventeen stars), six are bright giants (MK luminosity class II) and eight are Ib supergiants. There are no Ia supergiants, two of class Iab and only one star is classified as a giant (class III). We can conclude that her c and ac divisions represent very reliable classifications for detecting high luminosity stars.

With Miss Maury's two-dimensional classification the number of types which could have been defined was 110, from 22 groups and five divisions (a, b, c, ab, ac). In practice such a fine scale was not required for her rather small stellar sample and, in any case, some types (such as later groups with division b) simply were found not to occur. On the other hand, she sometimes assigned stars to a group halfway between two defined by Roman numerals, in which case the numeral of the earlier group was given in italics. For these reasons, the actual number of distinct types used was only 48, out of the possible 110. If the stars in intermediate groups, and those objects designated with a P (for peculiar) are included, then the number of types used rose to 74. R.H. Curtiss in 1932 wrote that:

> the Maury classification of stellar spectra is without question the most complete, thorough, and comprehensive achievement of its kind by any one investigator in this field. That it did not find wider application was due to its appearance of excessive detail, and to the greater acceptability of the more flexible Draper Classification as revised and developed by Miss Cannon (11).

Curtiss' comments may seem a little harsh as the MK system now in use is considerably more complex than Miss Maury's. Her system could doubtless have evolved to satisfy all our present requirements had it been more widely accepted after its introduction.

One further feature of the Maury classification also deserves mention. She placed in her group I stars such as ι Orionis and S Mon which today carry O-type MK classifications as they show ionised as well as neutral helium lines in absorption. Other O stars (MK system) with emission lines were however collected together with the Wolf–Rayet stars in Group XXII. Both Pickering (12) and Miss Maury (5) considered that Group I stars may be related to those in Group XXII and, through them, to the planetary nebulae. In this respect they showed some foresight. At the time Miss Maury wrote her classification description, specific mention is not made of Pickering's discovery of ionised helium lines in ζ Puppis in 1896 (13), which however he believed to be due to hydrogen. Lines such as 4200 and 4542 Å are characteristic of the ζ Puppis stars, and they were also found by Miss Maury in her group I objects, which differed from Group XXII in the absence of emission lines.

Miss Maury's work in volume 28 of the *Harvard Annals* (5) was her main contribution to spectral classification. Her other interest for many years was spectroscopic binaries, of which she discovered the second at Harvard (β Aurigae) in 1889. The binary β Lyrae was for a long time a special interest. She published a discussion of its spectrum in 1933 (14), and she continued to study the spectrum even after her retirement from Harvard in 1935, when she became curator of the Draper Park Museum at Hastings-on-Hudson, formerly the site of her uncle's observatory.

5.7 Ionised helium lines and the Pickering series

The story of the ionised helium lines just alluded to is nearly as intriguing as that of neutral helium. Following his discovery of the so-called Pickering series of lines in ζ Puppis, Pickering then showed (15) that the wavelengths corresponded to the formula found for lines of the hydrogen series by J.J. Balmer (1825–98) (16), provided half-integral values of the index n were allowed. The Balmer formula is $\lambda_n = An^2/(n^2 - 4)$ where $A = 3646.1$ Å (the series limit) and n is since known as the principal quantum number of the upper level. A similar formula to fit the Pickering series was also proposed shortly afterwards by H. Kayser (1853–1940) (17). The lines with half-integral n in the Balmer or Kayser formulae were given the notation by Vogel of Hα' (10124 Å, $n = 2\frac{1}{2}$), Hβ' (5412 Å, $n = 3\frac{1}{2}$), Hγ' (4542 Å, $n = 4\frac{1}{2}$) and so on. Pickering decided the lines were due to hydrogen, becuase the series formula 'is so closely allied to the hydrogen series, that it is probably due to that substance under conditions of temperature or pressure as yet unknown' (18).

It wasn't until 1912 that one of Lockyer's pupils, Alfred Fowler (1868–1940), showed that the Pickering series lines could be readily produced in the laboratory from a mixture of hydrogen and helium (19). At that time Fowler still subscribed to Pickering's view that their origin was hydrogen. But Niels Bohr (1885–1962) soon showed that the lines were indeed due to helium at high temperatures (20). As we now recognise, the helium is in its ionised state in which one electron has been stripped from each atom leaving an ion resembling the atom of hydrogen, except for twice the nuclear charge and four times the nuclear mass. The higher charge gave rise to the apparently half-integral quantum numbers, the higher mass to a violet shift, predicted to be of about 2 Å, of the helium lines of integral n in the Balmer formula relative to the Balmer lines of hydrogen. (Because they are blended in stellar spectra with the normal hydrogen lines, these integral-n lines were not detected by Pickering.) This violet shift was demonstrated by L.C.F. Paschen in 1916 (21). Fowler himself confirmed

Bohr's findings very soon afterwards (22).

One of the lines of ionised helium which is seen in ζ Puppis and similar stars in emission is the 4686 Å line, which however is not part of the Pickering series. Sir Norman Lockyer also observed this line in the solar chromosphere spectrum of 1898 (23) and in his report to the Royal Society commented on its close position to a blue emission line he had once recorded in the laboratory from a helium tube. This tentative identification was discussed in more detail in 1905 by Lockyer and another of his former students, F.E. Baxandall (1868–1929) (24), but unfortunately they were unable to repeat the laboratory conditions which had produced the line in question. They thus narrowly missed being the first to catalogue the ionised helium spectrum and solve the enigma of the 'additional hydrogen lines'.

5.8 Annie Cannon and the Harvard classification of 1901

Miss Maury's spectral work was exclusively north of −30°, and the remaining quarter of the sky was therefore still awaiting a detailed investigation of its brighter stars. The first plates for this southern survey had been exposed as early as 1891 using the 13-inch Boyden telescope at Arequipa. The work continued until 1899 using one, two or three objective prisms. By this time, 5961 plates had been taken with spectra of 1122 stars, mainly in the little-explored southern sky. The mean dispersion of these Boyden telescope spectra between Hβ and Hε was 40 Å/mm with one prism, and this was used to classify 813 of the stars. Only forty-one of the brightest stars were studied at 12 Å/mm with all three prisms.

The work of the southern classifications of bright stars was assigned to Miss Annie Jump Cannon (1863–1941), who must rank as among the most dedicated of astronomers of all time and certainly one of the most illustrious from the female ranks. Annie Cannon entered Harvard College Observatory in 1896 as an assistant. Although she had graduated from Wellesley College a decade earlier, she didn't turn to astronomy until after her mother died in December 1893. Two years at graduate school at Wellesley and Radcliffe Colleges then gave her the necessary astronomical background for her Harvard career.

In her classification scheme (25) Annie Cannon chose not to adopt that used by Miss Maury, but to take over the existing Draper Catalogue system devised by Mrs Fleming, and to bring it up-to-date so as to take into account both the Orion lines in the B stars and the Pickering series lines in the O stars. The only Draper letters used were O, B, A, F, G, K, M (in that order), together with P for planetary nebulae (only one observed), and Q

Fig. 5.7 Miss Annie Cannon.

for three peculiar stars with bright lines. There were no carbon stars in the catalogue, so the letter N was not used.

The main development in Miss Cannon's work was the decimal subdivision of the types from B to M. In her notation B signified what was later to be designated by B0, and K would today be named K0. The intermediate types were represented by B5A, F2G etc., meaning respectively halfway from B to A, or a fifth of the way from F to G, but in no case were all nine intermediate subtypes found necessary. Within classes O and M the decimal interpolation could not be used, but these classes were extrapolated using an alphabetical subdivision from Oa to Oe, and from Ma to Md, although no Mc stars were included in the catalogue, and Md was reserved for Mira variable stars with Balmer-line emission. Pickering and Mrs Fleming had also subdivided the O stars in 1891, but in their case only three subdivisions were defined (12). As in the Draper Catalogue, all Miss Cannon's O stars have emission lines. Following Miss Maury's tentative suggestion (5), she now placed them all ahead of the B stars, and she linked type Oe to type B with the intermediate type Oe5B (Miss

Maury's group I) which had no emission, but still had some of the Pickering series lines (from ionised helium) and 4686 Å in absorption. Classes Oa, b and c would today be classified as Wolf–Rayet stars, and Od (typical star ζ Puppis) and Oe (e.g. 29 CMa) as MK system O stars with some lines in emission (in particular ionised helium 4686 Å and doubly ionised nitrogen 4634 Å) but the Pickering series in absorption. The link between the Wolf–Rayet stars and other O stars was in the common presence of the Pickering series lines, which appear in emission in types Ob and Oc.

Miss Cannon thus firmly established the now accepted temperature sequence from Wolf–Rayet stars, through the other O stars, to B then A and beyond, a sequence which Miss Maury had first suggested in 1897, and which Lockyer had in 1900 also implicitly adopted (23), though of course without using the Harvard notation. She justified this order, without referring to temperature, in the following way:

> Stars whose spectra are of the fifth type are called O in the Draper Catalogue. In the present classification five subdivisions of Class O are represented by the letters Oa, Ob, Oc, Od, and Oe. These subdivisions depend upon the varying intensities of the bright bands named above, and the various combinations of these bands with other lines and bands, bright or dark. It is between spectra of Class Oe and Class B that the few spectra with wholly dark lines above referred to as falling outside of the series from B to Mb appear to belong. They are therefore designated by the symbol Oe5B. Spectra of Class Oe5B differ mainly from the spectra of Class Oe in having 4685.4 dark, and in the presence of the dark line 4649.2 [a blend of ionised carbon lines] instead of the bright band 4633; they also differ mainly from the spectra of Class B in the greater intensities of the additional hydrogen lines [Pickering series, He II], and of the line 4685.4 [also He II]. These spectra, in combination with those of Class Oe, appear to establish the position of spectra of the fifth type as preceding those of the Orion type. The letter O has been placed, therefore, in this classification, before the letter B instead of after the letter M (25).

According to Miss Cannon, the sequence represented an evolutionary one, probably proceeding from O to M, though a progression from M to O was not excluded.

In practice, the 1901 Cannon classification was almost as complex as Miss Maury's. As R.H. Curtiss has noted (11), Miss Cannon employed 48 line strength criteria, as against Miss Maury's 49 (of which 35 were in common), to determine her spectral types. She did not use any subdivisions

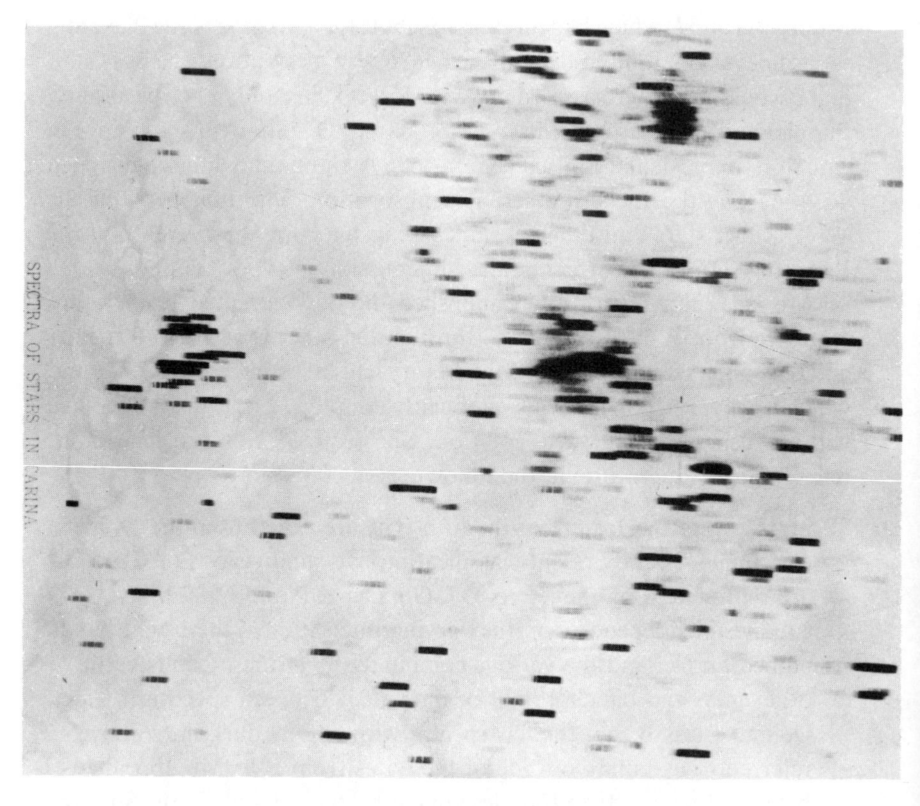

Fig. 5.8 Objective prism spectra of stars in Carina.

based on line width, but reserved such observations to the remarks on individual stars. The total number of separate types actually used from Oa to Md was 31 (including P and Q), exactly the same as employed by Miss Maury in her sequence from I to XXII, if her nine intermediate groups are also included.

The remarks in the catalogue contain some interesting comments on stars with abnormal spectra. The most important of these was the observation of A stars with unusually strong lines either of silicon (Si II, 4128 and 4131 Å) or of strontium (Sr II, 4078 and 4216 Å). A few of the peculiar silicon stars had already been noted by Miss Maury in the northern hemisphere (5). As a class, the peculiar A stars were not studied and classified in detail until the 1930s by Morgan (27). Examples of silicon stars described by Miss Cannon are ν For, τ^9 Eri and α Dor. For ν For she commented: 'The spectrum is peculiar in the faintness of the line K, which is barely seen on plates of normal exposure, and in the intensity of the double line 4128.5 and 4131.4, which together are three or four times as intense as any other line in the spectrum except those of hydrogen' (25). Examples of strontium stars are ξ Phe, θ^1 Mic and ι Phe. She described the spectrum of θ^1 Mic: 'The spectrum is peculiar. The faintness and haziness of the line K, which is not more than 0.1 as intense as Hδ, combined with the intensity of the solar lines suggest a composite type... 4077.9 is unusually intense, being 0.3 or 0.4 as intense as Hδ, and is sharply defined' (25).

In addition, Miss Cannon noted peculiarities in other A stars which in the MKK classification (28) of 1943 were recognised as metallic-line stars (type Am).* These stars have weak calcium lines but other metallic lines are generally as strong as in F stars, the hydrogen lines corresponding in strength to an intermediate type. For example, δ Nor was classified as A3F (peculiar) with the comment:

> This spectrum is very peculiar, and suggests a composite type. While the lines of hydrogen are as intense as the typical star [of type A3F], τ^3 Eridani, and the line K is not more than 0.7 or 0.8 as intense as the line H, the solar lines resemble those of Class F5G... The solar lines are nearly equal in intensity to those of α Canis Minoris of Class F5G, except that 4227.0 and 4415.3 are only about 0.5 as intense as in α Canis Minoris (25).

Following the MKK definition of the metallic-line stars in 1943, δ Nor was not recognised as such until 1960 (29).

The 1901 classification also contained useful tables, for example the wavelengths of thirty-four lines in the early B stars α Crucis, θ Carinae and β Crucis, and of 55 emission lines or 'bands' in γ Velorum, some of which were several tens of Ångström units broad and blended. The majority of lines tabulated for all four stars was of unidentified origin.

5.9 Annie Cannon's classifications of 1911

The Henry Draper Memorial was continued with the 1912 publication of *Harvard Annals*, volume 56. One of the aims was the classification of further northern stars omitted in 1897 from Miss Maury's catalogue. Some of these were as bright as fourth magnitude. Miss Cannon now classified a further 1477 stars on her system of 1901, and this completed the survey north of $-20°$ for all stars brighter than fifth magnitude. All the spectra were photographed with one prism on the 11-inch Draper telescope (30). The only differences in notation were the dropping of the second letter in the subclasses (e.g. B2A became B2), and the use of class N, which in 1901

* The Morgan–Keenan–Kellman classification is discussed in section 8.8.

had not been required. The description of the 29 classes was given in much more concise and condensed form than before. In addition, class Md was used (though not described) and peculiar objects noted as 'Pec.' instead of Q. Among those in this last category was the variable carbon star R Coronae Borealis, with a spectrum in some respects resembling a K star (H and K lines were present), but with many lines anomalous. There was a mention of Nova Aquilae No. 2 which erupted in 1905, and also of the peculiar star υ Sagittarii. This star was known to be a spectroscopic binary (31), and Miss Cannon described the variable composite spectrum as sometimes resembling β Orionis, at other times α Cygni or ε Aurigae.

In the same volume of *Harvard Annals*, Miss Cannon continued the classifications of fainter stars to fifth magnitude between $-20°$ and $-30°$, and to sixth magnitude south of $-30°$ (32). A total of 1688 southern stars were classified, mainly from 13-inch Boyden telescope plates taken with a single objective prism. In this catalogue 'according to the suggestion of Dr Hertzsprung, the Classes B, A, F, G and K, are called B0, A0, F0, G0 and K0. In a few doubtful cases, the numeral is omitted.' In the two catalogues of 1912, Miss Cannon recorded 18 stars 'with the additional series of hydrogen' (i.e. He II) as well as 64 new peculiar A stars of the silicon or strontium types. Tables were also given of 23 newly discovered B stars with emission lines (similar to γ Cas) and by combining the three catalogues of Miss Cannon with that by Miss Maury, a compilation of 91 high-luminosity stars (corresponding to Maury divisions c or ac) was presented.

With these four catalogues (5,25,30,32), Edward Pickering, Antonia Maury and Annie Cannon had completed a major undertaking. Nearly 5000 stars had been classified over the entire sky in the most detailed systems yet devised, and large numbers of stars with unusual spectra, as judged by emission lines, or by abnormal line strengths or breadths, were catalogued for the first time. Although the total number of stars was only half that of the earlier Draper Catalogue, the quality of the classifications was now far superior, and this allowed detailed statistical analysis to be undertaken on the space distribution and frequency of occurrence of objects of different spectral types. It seemed that Pickering's ambition of solving the secrets of stellar astrophysics by the mass accumulation of great volumes of data was to be realised. In 1912 Pickering undertook an analysis of the galactic distribution of different spectral types, albeit using data taken mainly from the Draper Catalogue, but supplemented with more recent classifications by Mrs Fleming in the southern hemisphere from Bache telescope plates (33). By counting the stars of given types in different galactic latitude zones, he concluded: 'These figures show very clearly that the spectra of Classes A and B are more numerous in the Milky Way than outside of it, and that the maximum point is a little south of the Galactic Equator'. The era of exploring galactic structure by means of stellar spectroscopy had begun.

5.10 The 1910 meeting of the International Solar Union and the spectral classification questionnaire

An especially valuable document for the history of stellar spectroscopy arose out of the proceedings of the fourth meeting of the International Union for Cooperation in Solar Research at Mt Wilson Observatory in September 1910. A Committee on the Classification of Stellar Spectra was formed at that time with F. Schlesinger (1871–1943) of Allegheny Observatory as secretary. The committee resolved to distribute widely a questionnaire to those astronomers active in stellar spectroscopy. The following five questions were asked (34):

(1) ... the Draper Classification is the most useful that has thus far been proposed. Do you concur in this opinion? If not, what system do you prefer?
(2) In any case, what objections to the Draper System have come to your notice and what modifications do you suggest?
(3) Do you think it would be wise for this committee to recommend at this time or in the near future any system of classification for universal adoption? If not, what additional observations or other work do you deem necessary before such recommendations should be made?...
(4) Do you think it desirable to include in the classification some symbol that would indicate the width of the lines, as was done by Miss Maury in *Annals of the Harvard College Observatory*, Vol. 28?
(5) What other criteria for classification would you suggest?

Replies were received from twenty-eight prominent spectroscopists in seven countries, mainly in the United States, with their comments on the Draper classification, as defined by Miss Cannon in 1912 (30). These were published in the *Astrophysical Journal* (34) and are summarised here in tabular form (Appendix III). Twenty-four replies were broadly favourable to the Draper classification (question 1), and only one (J. Scheiner) was unequivocally against. Of course the most developed of the rival schemes was Vogel's second classification of 1895. Loyalty to his former colleague almost certainly played a role in Scheiner's reply. Lockyer was one of the most notable omissions from the questionnaire's distribution. According to DeVorkin (35) the demise of his spectral classification scheme can be partly attributed to his 'effective exclusion from the process of the committee. Unlike Lockyer, Pickering was at the centre of the proceedings. The way he grasped leadership left little doubt that his system would be strongly favoured'. Another 'elder statesman', Dunér, known for his unswerving support for the Potsdam spectral classification, was also left out of the consulting process.

There was less unanimity in the replies to the other questions. Eleven were opposed to the official adoption of any classification scheme for universal use, while fourteen were more or less in favour, but most of these with qualifications added. Miss Cannon, Mrs Fleming and Professor Pickering all voted not to adopt the Draper classification officially at the present time, while Miss Maury abstained on this point. Most of the respondents found faults in the Draper classification, mainly with the notation rather than any astrophysical deficiency. Only Henry Norris Russell believed the non-alphabetical order of the letters to be advantageous: 'This helps to keep the novice from thinking that it is based on some theory of evolution' (34). Seventeen replies indicated that the incorporation of a line width parameter (question 4) was desirable or definitely necessary. Four were against such a parameter, mainly because the classification would become unduly complicated. Various additional classification criteria were proposed for question 5, but the most frequently made comment, in eight of the replies, was the need for the yellow and red spectral regions also to be taken into account.

Overall, the reply that showed the most insight came from Karl Schwarzschild (1873–1916) at Potsdam. His discussion of the number of independent stellar variables required to determine completely a star's spectral type is worth quoting:

> The Draper classification represents the spectra as a function of one variable which one can call 'evolutionary state'. In Miss Maury's scheme, a second variable, the line width, has also been added, and this is, according to Hertzsprung's investigations, a physically very important criterion. It is indeed possible that one needs yet more variables, in order to represent all the fine variations of the spectra. I would like to suppose, however, that the number of variables is limited, and perhaps does not exceed the number three. That is to say, I wouldn't readily believe that the distribution of the elements in the stars is accidental, but I would much prefer to accept that the mix of the stellar material out of different elements is everywhere the same, or at most it depends on the age of the stars in a regular

way. If this latter assumption is right, then ultimately the spectrum of a star might depend on nothing other than its mass, its age and its temperature (energy content) (34).

As a result of this questionnaire and the replies received, or perhaps in spite of the wide spectrum of opinion represented in the replies, the committee at the fifth meeting of the International Solar Union in Bonn in 1913 resolved:

(1) That the Committee on the Classification of Stellar Spectra be asked to secure by cooperation the material necessary for the establishment of a system that can be recommended for permanent and universal adoption; and
(2) That, pending the establishment of such a system, the use of the Draper classification be recommended in the form described in Vol. 56, p. 66 of the *Annals of the Harvard College Observatory*; except that hereafter, in accurate classification, a zero be added to letters not followed by other numerals and that the absence of any numeral be taken to indicate only a rough classification (36).

The Draper classification of 1912 therefore received temporary official recognition at the Bonn meeting. However, the temporary nature of the adoption became permanent at the first General Assembly in Rome of the newly formed International Astronomical Union (IAU) in May 1922 (37). At this time the Draper classification was reaffirmed as the official system, though with numerous modifications in the form of additional symbols (see Chapter 8). As will be seen, there were pressing reasons for doing so, as the new Henry Draper Catalogue was then nearly complete. The sheer weight of numbers of the Harvard classifications virtually ensured the extinction of any rival scheme and left the IAU Commission 29 for spectral classification with little real choice. The decisions taken at the 1922 meeting nevertheless proved to be a fortunate and notable milestone in the development of spectral classification.

5.11 Mrs Fleming's work on stars with peculiar spectra

If Antonia Maury and Annie Cannon made the greatest advances in developing systems of spectral classification, we should not forget Mrs Fleming. From 1899 till her death in 1911 she was given the position of curator of astronomical photographs at the observatory. Much of her time was devoted to cataloguing stars with peculiar spectra (here interpreted in a fairly broad sense). After the completion of the Draper Catalogue, the 8-inch Bache telescope was installed in Peru and the similar Draper instrument at Harvard. Both these telescopes continued taking low dispersion objective prism spectra (mainly at mean dispersions of 130 Å/mm or 450 Å/mm) of fainter stars. In addition, the Arequipa station acquired a new and larger telescope in 1895, the 24-inch Bruce photographic refractor, which was made possible by a $50 000 donation to Harvard by Miss Catherine Bruce of New York in 1889. The spectra recorded by all three of these instruments were now examined by Mrs Fleming, and the results of her extensive work on peculiar spectra were published posthumously in 1912 (38).

The record she left behind on peculiar stars is an especially valuable one for the history of stellar spectroscopy. Not only are complete lists given of various types of peculiar star discovered at Harvard and elsewhere up to about 1911, but each object has its discoverer, year of discovery and literature reference cited, so the reader can readily see the enormous growth in the discovery of unusual stellar spectra from the inception of the Draper Memorial program onwards. The peculiar stars are listed under novae (including supernovae)*, gaseous nebulae, O stars (with emission lines), hydrogen emission line stars of types A or B, spectroscopic binaries, variable stars of the Algol, β Lyrae and short and long period types, N stars, R stars, Oe5 stars (with the Pickering series, but no emission lines) and other miscellaneous unclassifiable objects. The statistics from the paper are impressive and worth quoting to appreciate the growth of this branch of astronomy around the turn of the century.

5.11.1 Novae From the establishment of the Henry Draper Memorial to 1911 nineteen novae were discovered, fifteen of them at Harvard and ten of these spectroscopically by Mrs Fleming from their characteristic emission-line spectra. Nova Persei in 1901 was exceptionally bright and Pickering made an extensive survey of its changing spectrum using plates exposed from 22 February, when it was still in rising light, to 30 October 1901, using the 11-inch Draper telescope (39).

5.11.2 Gaseous nebulae 151 objects are listed which are non-stellar as shown by emission lines on a weak continuous background. Of these 104 were discovered independently at Harvard, 71 of them for the first time, 33 being repeat discoveries.

* Supernovae were not recognised as distinct from novae until 1934 (see Chapter 9).

5.11.3 O stars These included Wolf–Rayet stars as well as other emission-line stars with the Pickering series lines. As mentioned in Chapter 4, 13 O stars were known in 1884 (all of the Wolf–Rayet type). Mrs Fleming listed 108 in 1912, of which 94 were discovered at Harvard from 1886, including 21 in the Large Magellanic Cloud, one in the Small Cloud and the remainder in the Milky Way. Of these 94 from Harvard, Mrs Fleming accounted for 91 of the discoveries, a truly impressive record.

5.11.4 A and B stars with emission lines Mrs Fleming listed 92 objects with Balmer emission. Only γ Cas was known in this category prior to the Henry Draper Memorial program.* Of the remainder, eighty-six were Harvard discoveries.

5.11.5 Spectroscopic binaries Although the first spectroscopic binaries were only discovered in 1889 at Harvard and at Potsdam, 306 were known by 1910. Mrs Fleming listed forty-seven of spectral type A with double lines, which can be detected most readily from objective prism spectra. Only seven were Harvard discoveries. The skill of the Lick and Yerkes astronomers such as Campbell and Frost with their large refractors had cornered the market in new discoveries (see Chapter 6).

5.11.6 Variable stars Large numbers of variables were discovered at Harvard on the 40 000 direct sky plates taken by each of the 8-inch telescopes in Cambridge and Arequipa. Most of these stars, if their spectral type was unknown, were the subject of a spectroscopic study by Mrs Fleming. 134 Algol eclipsing variables, 13 eclipsing variables of the β Lyrae type, 168 short period variables (diverse types), and 629 long period variables (mainly Mira stars) are listed. Mrs Fleming made a special study of this last group and discovered 187 of those in her list. Most of these stars show emission in the Balmer lines near maximum light, as first discovered in 1869 by Secchi for R Gem (41). Such objects were therefore classified Md, and 389 were listed by Mrs Fleming. However, she extended this classification further on a decimal scale Md1 to Md10 depending on the absorption-line intensity. Unfortunately she left no description of her subclasses, so Pickering had to omit this when he published the work posthumously. Use was also made of class Mc which she had employed

* One star, HR 1536, was listed in 1887 by Konkoly as Vogel class Ic? (40), but recent classifications are a normal F8 or G0 dwarf. As Mrs Fleming incorrectly gives the date as 1883 for Konkoly's work, it appears that two objects in her list were known prior to the start of the HD memorial.

earlier in 1898 (42), although this subdivision was not used by Miss Cannon until 1918, in the Henry Draper Catalogue.

We note here that the fluted spectral bands characteristic of all M stars, including the Mira stars, were identified by the English astronomer A. Fowler in 1904. Fowler was able to demonstrate that the bands were due to titanium or titanium oxide:

> It will be seen that eight of the ten bands recorded by Vogel and Dunér [for Antarian stars] agree within the possible limits of error with the flutings of titanium, and it is to be noted also that the only one of the principal titanium flutings, which is not represented in the stellar spectrum, is out of range in the extreme red... It will be seen... that the details of the titanium flutings are reproduced with remarkable fidelity in stellar spectra, and more especially in *o* Ceti ... The chief substance concerned in the production of the flutings is titanium, or possibly a compound of that element with oxygen (43).

5.11.7 N stars Of the 267 objects listed as having at one time or another been reported to be of class N (Secchi type IV), 142 had this classification confirmed at Harvard. Of these, 65 were Harvard discoveries, nearly all of them by Mrs Fleming. The Rev. T.E. Espin at his private Wolsingham Observatory claimed 123 of the 267 discoveries, a remarkable achievement for an amateur (see Chapter 4). Many of Espin's stars were however too faint to be classified on the low dispersion Harvard plates.

5.11.8 R stars A new type of stellar spectrum was announced by Pickering in 1908 (44), and sometimes called by him a sixth type of stellar spectrum, so extending Secchi's notation. The characteristics were absorption bands resembling the carbon stars, but with far more blue light, usually extending to the H and K lines in the far violet. As Pickering wrote: 'An examination of the enlargements [of the spectrograms] shows these spectra cannot be classed as fourth type stars, since they contain so much blue light, and stars having spectra of the fourth type are commonly regarded as red stars' (44). Since, in the Draper Catalogue (2), all the letters up to Q had already been used, the new class of carbon star was designated R. Pickering listed 51 stars, of which the brightest were of visual magnitude seven. Most of these objects had been observed earlier at Harvard and noted as being peculiar. For example in 1896 Pickering had first noted four stars then classified as N with 'rays of much shorter wave length than ordinary fourth type stars' (45). In *Harvard Annals* Vol. 56, Mrs Fleming extended this list to 61 stars of type R. Of these, two were observed first by Espin (who classified them as Secchi type IV), one by Pickering, and the remainder were Mrs Fleming's discoveries.*

A detailed study of ten R stars was made at the Detroit Observatory of the University of Michigan in 1915 by W. Carl Rufus (1876–1946) using the 95 cm Brashear reflector and a slit spectrograph. The exposures were frequently long (7 or 8 hours). An extensive list was made of wavelengths in the blue region, and the R stars were placed in an evolutionary sequence linking the G and N stars, which ran parallel to the G, K, M sequence. The carbon star sequence joined this latter at about K0 (46). The evidence for placing R ahead of N seems to have been based mainly on stellar colour rather than the line spectrum. Rufus' scheme reversed the implicit order used by Mrs Fleming who had classified S Cen as N5R (38), i.e. as intermediate between N and the later type R (on Miss Cannon's 1901 notation).

5.11.9 Oe5 stars These stars, showing the Pickering series of 'additional lines of hydrogen' (i.e. He II) in absorption, but no lines in emission, were 25 in number in Mrs Fleming's list. All but one were Harvard discoveries. The typical star was τ CMa.

5.11.10 Stars with peculiar spectra Twenty-nine peculiar stars not classified under any of the above groups completed the paper. Three (R Gem, π^1 Gru, T Cam) were later to be classified as S stars (which contain zirconium oxide bands), although this type was not defined by Miss Cannon until 1923 (47), following an IAU recommendation of the previous year (37). Four were peculiar emission line objects, including η Carinae. Two were the variable carbon stars R CrB and RY Sgr which were tentatively classified as classes Gp and N respectively. The list also included the eclipsing binary, β Lyrae.

The spectrum of R CrB was first described by Espin in 1891 (48) who found it to be sometimes continuous (no lines or bands), and at other times with bands resembling Secchi's type IV. Miss Cannon described the spectrum as resembling a K star, though an earlier reference by her mentions 'a number of bright and dark lines. Among the bright lines there

* It is interesting to note the following passage in Mrs Fleming's 1912 paper (38): 'In Harvard Circular 145 a class of spectrum was described by Mrs Fleming which she had previously regarded as a subdivision of the fourth type'. But Harvard Circular 145 carries only the name of Pickering, and the article makes only one brief mention of Mrs Fleming. Evidently it was then a common practice at Harvard for female research assistants to do much of the tedious work, but the director to reserve for himself the right of publication.

are two of slightly shorter wave length than Hδ and Hγ. The spectrum is peculiar and appears to be identical with that of RY Sagittarii' (30). The light curve of R CrB and similar variables is characterised by sudden plunges of five or six magnitudes at irregular intervals, followed by a slower recovery to maximum. R CrB frequently spends a year or more steadily shining at maximum light of about sixth magnitude.

5.12 Emission-line stars catalogued by Annie Cannon

A few years later Miss Cannon published a list of stars with emission lines (49) which supplemented much of Mrs Fleming's work. She divided the emission-line stars into six types (Table 5.1).

Table 5.1 *Miss Cannon's emission-line stars (1916)*

Emission-line type	Number listed
P (nebulae)	150
O (Wolf–Rayet and other ionised He stars)	107
P Cygni type	10
Novae (including supernovae)	20
Emission-line B stars (including β Lyrae)	99
Md	364
Total	750

The list actually included slightly fewer stars in some categories than Mrs Fleming's list in 1912 (38). The paper proposed a subdivision of the gaseous and planetary nebulae on a scale from Pa to Pf based mainly on the presence or otherwise of the ultraviolet lines 3726 and 3729 Å (of ionised oxygen) or of the ionised helium 4686 Å line. The two ultraviolet lines were strong in the earlier types, and the blue helium line in the later ones, while Pf was supposed to form a continuous link to the O stars which also have 4686 Å in emission. This scheme was also employed by Miss Cannon in the Henry Draper Catalogue on which she had been working since 1911. In spite of that, it never became widely adopted for nebular classification by other astronomers. The Cannon classification did however bear some relation to nebular excitation; the low excitation objects were labelled Pa, the highest Pf. Only three of the 117 nebulae studied by Miss Cannon (33 listed were not studied) were later recognised as gaseous nebulae (all classified as type Pb). The others are all planetary nebulae. Two years later the Cannon scheme for nebulae was criticised by W.H. Wright

(1871 – 1959) at the Lick Observatory (50), who devised his own classification based on the intensities of the 4686 Å and 3869 Å ([Ne III]) lines.

Miss Cannon also attempted to classify the spectra of novae based on seventeen such stars studied spectroscopically at Harvard. She had a subdivision running from a to e, but the small sample seems to have rendered this scheme not particularly useful. She included an update of her earlier discussion (25) of η Carinae, including the remark that the emission line spectrum presented many similarities to Nova Aurigae of 1892.

5.13 The Henry Draper Catalogue: program initiated by Pickering and Miss Cannon Even before volume 56 of *Harvard Annals,* with its eight classic papers on various aspects of the spectroscopic program under the Henry Draper Memorial, was finally printed in 1912, Pickering had initiated a new large program of spectral classification that was to dwarf all the work hitherto undertaken at Harvard and elsewhere. The Henry Draper (HD) Catalogue (not to be confused with the 1890 Draper Catalogue of Stellar Spectra (2)) had its origins going back into the nineteenth century, as the program of securing objective prism plates, mainly with the 8-inch Bache and Draper telescopes, had been continuing ever since these instruments were established in Arequipa and Cambridge in 1889.

However 11 October 1911 marks the day when Miss Cannon started classifying the spectra of 225 300 stars which together make up the Henry Draper Catalogue. The classifications, all undertaken by Miss Cannon, were essentially complete by 30 September 1915, in less than 4 years. Since this interval contains about 1000 working days, it is clear Miss Cannon was able to classify stars at a sustained average rate of well over two hundred stars a day, or some 30 an hour. Much of the routine work in organising the catalogue was undertaken by several assistants, the average number being five. Pickering's aim in committing the observatory to this new undertaking can be inferred from his opening remarks in the catalogue's preface (51): 'In the development of any department of astronomy, the first step is to accummulate the facts on which its progress will depend. This had been the special field of the Harvard Observatory. An attempt is made to plan each investigation on such a scale that it will not be necessary to repeat it shortly...' The Henry Draper Catalogue was no exception to this rule.

The telescopes used were principally the 8-inch Draper camera in Cambridge and the 8-inch Bache camera at Arequipa. However, the results for 4287 bright stars classified earlier by Miss Cannon (25, 30, 32), using in

Fig. 5.9 'Pickering's Harem', 1913. Annie Cannon is at the rear,the second to the right of Professor Pickering.

the main either the 11-inch Draper (Cambridge) or the 13-inch Boyden (Peru) telescopes, were taken over for the new work without alteration. A total of 1409 southern Bache plates were used (going as far north as $+10^\circ$ in some cases), together with 709 northern eight-inch Draper plates. Miss Cannon also reclassified all the spectra used by Miss Maury for her 1897 study of bright northern stars (5) with the 11-inch Draper telescope. In addition some of the fainter northern stars were observed with the 16-inch Metcalf instrument, for which the achromatic objective had been figured and donated by the Massachusetts amateur astronomer, the Rev. J.H. Metcalf.

Pickering wished to include in the catalogue the classification for as many stars as possible with spectra recorded on the objective prism plates. In practice, the limiting magnitude for southern stars was about one magnitude fainter than in the north (52). The northern stars were complete down to about eighth magnitude, although some as faint as 9.5 were included. The reasons for the discrepancy between the hemispheres are due both to the better seeing (i.e. image quality) at Arequipa and also to the fact that on the Bache telescope the best prism was one giving a mean dispersion of 400 Å/mm, while on the 8-inch Draper telescope the best prism gave more dispersed spectra at 160 Å/mm. Higher dispersion inevitably results in a brighter limiting magnitude (as the light is more spread out on the plate), other things being equal. Thus the majority of the HD Catalogue spectra employed the Bache camera with a 5° prism, or the Draper camera with a 13° prism.

The classification scheme of the HD Catalogue included the usual P, O, B, A, F, G, K, M, R, N types in that order. The nebulae (P) were divided into six, as in *Harvard Annals*, vol. 76 (49). The only other differences in the classification not appearing in Miss Cannon's 1912 work were the inclusion of type Mc (as well as Ma, b, d used earlier) as Mrs Fleming had done, the subdivision of the R stars into R0, 2, 5 and 8, which were now all placed ahead of N, and type N was also subdivided into Na and Nb. The practice of using decimal subdivisions for all the letter classes except the terminating ones (P, O, M, N) was thus retained. The Mc stars are those with the strongest titanium oxide molecular bands, and those of vanadium oxide are also prominent (although these latter were not identified until 1936 by N.T. Bobrovnikoff (53); P.W. Merrill (54) confirmed the vanadium oxide identification). The four R subtypes corresponded to increasingly less ultraviolet light below 4240 Å. Table 5.2 is taken from the review by Curtiss (11) and shows a comparison of the principal spectral classification schemes that had been devised up to and including that of Miss Cannon in the HD Catalogue.

Table 5.2 *Comparison of the principal stellar spectral classifications*

Secchi type	Vogel class (1895)	McLean division	Lockyer genus	Pickering class[a]	Maury group	Cannon class[b]
(V)	—	—	—	P	—	P
(V)	IIb	(Ia)	Argonian	O	XXII	Oa
(V)	IIb	(Ia)		O	XXII	Ob
(V)	IIb	(Ia)		O	XXII	Oc
(V)–I*O*	IIb	(Ia)		O	XXII	Od
(V)–I*O*	IIb	(Ia)		O	XXII	Oe
I*O*[c]	Ib	Ia		B	I	Oe5
I*O*	Ib	Ia	Alnitamian	B	II	B0
I*O*	Ib	Ia		B	III	B1
I*O*	Ib	Ia	Crucian	B	IV	B2
I*O*	Ib	Ia	Taurian	B	IV	B3
I*O*	Ib	Ib		BA	V	B5
I*O*–I	Ib	Ib	Algolian	BA	VI	B8
I*O*–I	Ib	Ib	Rigelian	BA	VI	B9
V	Ic1, Ic2	—	Crucian	D	L	Oe5p-B9p
I	Ia2	II	Markabian	A	VII	A0
I	Ia2	II	Sirian	A	VIII	A0
I	Ia2	II	Cygnian	A	VIII	A2
I	Ia2	II		AF	IX	A2
I	Ia3	III		AF	IX–X	A3
I	Ia3	III		AF	X	A5
I	Ia3	III		F	XI	F0
I	Ia3	III		F	XI–XII	F2
I–II	Ia3–IIa	III	Procyonian	FG	XII	F5
II	IIa	IV	Polarian	G	XIII	F8
II	IIa	IV		G	XIV	G0
II	IIa	IV		GK	XIV–XV	G5
II	IIa	IV	Arcturian	K	XV	K0
II	IIa	IV		K	XV-XVI	K2
II–III	IIa–IIIa	IV-V	Aldebarian	KM	XVI	K5
III	IIIa	V	Antarian	Ma	XVII	M0 (Ma)
III	IIIa	V		Ma	XVIII	M0 (Ma)
III	IIIa	V		Mb	XIX	M3 (Mb)
III	IIIa	V		(Mc)	—	M6,5 (Mc)
III	IIIa	V		Md	XX	Md
—	—	—		—	—	S
IV	IIIb	VI			XXI (?)	R0
IV	IIIb	VI			XXI (?)	R3
IV	IIIb	VI			XXI (?)	R5
IV	IIIb	VI			XXI (?)	R8
IV	IIIb	VI	Piscian	Na	XXI	N0 (Na)
IV	IIIb	VI			XXI	N3 (Nb)
IV	IIIb	VI			XXI	Nc
—	—	—	—	—	—	Pec
—	—	—	—	—	—	Con

[a]In collaboration with Mrs Fleming.
[b]The Draper Classification.
[c]Secchi's Orion subtype.
Adapted from R.H. Curtiss (11).

5.14 Publication of the HD Catalogue The Henry Draper (HD) Catalogue was published between 1918 and 1924 in nine volumes of the *Harvard Annals* (51, 55–62). The costs of the program were borne largely by the bequest left by Mrs Draper when she died in December, 1914 (63). However, Pickering himself took over the publication costs of the first volume, and Mr George R. Agassiz of Boston is recorded therein as contributing the salaries of two of Miss Cannon's assistants. The effort required to see these volumes of the annals through to publication occupied Pickering for much of the time in the closing years of his life. By now he was an old man, and he lived to see only the first three volumes of the Catalogue in print. He died on 3 February 1919 after 42 years as director of the observatory, and thereafter Miss Cannon was left to supervise the printing of the final six volumes. The publication of the first two volumes was reviewed in 1920 by H.F. Newall (1857–1944), the professor of astrophysics in Cambridge (England) (64):

> The publication of the first volume of the new Henry Draper Catalogue... was issued in the autumn of 1918, and marked the first step in the last stage of a piece of work of colossal magnitude, carried out with all the precision of a masterly organisation. Thirty years ago the Draper Catalogue of the spectra of 10 498 stars seemed a huge undertaking, successfully accomplished by Professor E.C. Pickering; but it is altogether dwarfed by the new Henry Draper Catalogue, which will contain a specification of the types of spectrum of more than 20 times the number of stars, in a classification approximately 10 times as detailed as that adopted in the earlier work. The second volume was distributed in the autumn of 1919, and contains the stars from pole to pole in the range of R.A. $4^h0^m.0$ to $6^h59^m.9$. Miss Cannon and her devoted assistants have fulfilled the expectation which Professor Pickering expressed in his preface to the first volume under the date 1917 December 13, and the whole work is finished, except for the printing, in 1920 January. Miss Cannon's work, carried out with unfailing enthusiasm, has included the classification, the revision, and the supervision of the whole. It has involved six years of very strenuous application on her part with the aid of several assistants, the average number being five. The title pages of both volumes bear the names of Annie J. Cannon and Edward C Pickering, as joint authors; and in congratulating Miss Cannon on the completion of her task, we would express our deep regret that her stimulating collaborator has not lived to receive our renewed homage on the completion of this great work.

The eighth volume of the HD Catalogue (61) saw two changes to the classification. One was the introduction of type Nc, being carbon stars with little or no light shortwards of Hβ, the brightest part of the spectrum being entirely in the orange to red region. More important was the introduction of a new class designated S, with typical stars π^1 Gruis and R Geminorum. S-type stars were first described in 1922 at the first IAU meeting in Rome (37), in a report of the Committee on Spectral Classification, which was chaired by Walter Adams of Mt Wilson Observatory. The report described the new type as follows:

> Their spectrum in the region λ4 500 to 4 700 is of a most complicated nature, and appears to consist of both absorption and emission lines, with absorption bands present at about 4650 and 6470. Most of the stars belonging to this type are long period variables, and show bright hydrogen lines. The type may represent a third branch of the main spectral sequence, cognate with the K5-M and R-N branches. The letter S is suggested for this type (37).

However, the credit for the discovery of stars of type S belongs to Miss Cannon, even though she did not publish a brief description of their spectral characteristics until the following year (61). No doubt the IAU report was compiled from her comments, and even the first volume of the HD Catalogue gives an account of seven long-period variables with peculiar spectra at a time before the S designation had been introduced.

> Several spectra which have hitherto been called Md1, or Md2 in which Hβ is the strongest bright line, are found to be peculiar and are designated Pec. in Table I. The variable stars R Andromedae, U Cassiopeiae, S Cassiopeiae, R Lyncis, R Canis Minoris, T Geminorum, and R Cygni may be given as examples. These spectra do not show the titanium bands having bright edges at 4762, 4954, and 5168 as in all divisions of Class M, but more nearly resemble the spectrum of π^1 Gruis, which may be placed in a subdivision of Class R. assuming some peculiarities (51).

As it happens, her suggestion concerning π^1 Gruis was a poor one, as all S stars have molecular absorption bands with heads on the violet sides, as do also the M stars. The peculiarities of one of the S stars, R Cygni, was also described by T.E. Espin (65) and by W.H. Wright (66), the latter author showing for example the presence of a broad red band with a head at 6468 Å which is referred to in the IAU report. S stars were first mentioned in Harvard publications in a note by Harlow Shapley (1885–1972) (67), who had succeeded Pickering as observatory director. However, the most detailed early analyses were due to Paul W. Merrill (1887–1961) at Mt Wilson, who in 1923 first published an identification of the S star bands as due to zirconium, or one of its compounds, probably its oxide (68). F.E. Baxandall in England had suggested the same thing in an earlier private communication to Merrill. At this time Merrill was able to list 22 members of the new class, and he increased the list to 31 stars by 1927 (69). By this time the zirconium oxide bands had been firmly established as the dominant absorption features in the red, but many S stars showed titanium oxide, characteristic of M stars, as well. The relative strength of the bands of the titanium and zirconium oxides varied from star to star, and some objects therefore appeared to be intermediate between the two types. The relatively sharp dichotomy between carbon and M stars was not repeated in comparing M with S.

The final and ninth volume of the HD Catalogue (62) was printed in 1924. The total cost of the whole project was then, since the time the first plates had been exposed, about a quarter of a million dollars (63), which must today seem quite a modest sum for a project that extended over 13 years. At almost exactly 1 dollar per spectral classification, it was even a bargain.

5.15 The Henry Draper Extension

The publication of the ninth volume was by no means the end of Miss Cannon's work in spectral classification for the Henry Draper Memorial. In 1923 Shapley planned an extension of the catalogue (the HDE) 'to balance the survey for fainter stars, and especially to explore the more important Milky Way fields for stars considerably beyond the present limits of magnitude' (52). By 'balancing the survey', Shapley was referring to the relatively brighter limiting magnitude for northern stars when compared to the southern. Shapley had estimated that Harvard plate material already exposed contained the spectra of about a million faint stars so far unclassified. Miss Cannon therefore had a wealth of material for further classifications in selected areas of the Milky Way. For example, the spectra for the first Milky Way field in Cygnus that she worked on as part of the extension were all contained on a single objective prism plate covering 80 square degrees and taken with the 10-inch Metcalf telescope. This instrument had a 10-inch triplet Metcalf objective and was installed in Peru, but at times was used for observing as far north as $+45^{\circ}$. The spectra had a mean dispersion of 400 Å/mm, and 4490 were recorded on this plate alone. Miss Cannon included 3992 of them in the Henry Draper Extension (part 1) (70), and the limiting magnitude for completeness was just below eleventh, with some still fainter stars, a few even below twelfth magnitude, also included.

The HD Extension (HDE) classifications were ostensibly on the same system as used before, though as Shapley noted, the HDE types overall tended to average 0.8 decimal subtypes earlier. For the G stars, the HDE types were as much as three decimal subtypes earlier, while for M stars this systematic trend was reversed (the HDE types for M stars averaged 1.6 decimal divisions later than in the main catalogue). Shapley concluded these differences were satisfactorily small (52). All together the HDE (70) contained 46 850 spectral classifications and appeared in six parts, the last being dated July 1936 for a field in the Large Magellanic Cloud (LMC).

Miss Cannon continued classifying stellar spectra up till her death in 1941. Her latest work appeared first in the tercentenary volume of the *Harvard Annals* (71) (published to mark the founding of Harvard College in 1637) for another LMC field. This field together with a further six was then published in a second volume of the HDE in 1949 (72). The format in these volumes was changed to charts, which saved time in measuring the coordinates of faint stars not appearing in any positional catalogue. 86 932 stars appear on the HDE Charts (with HDE numbers) and nearly all these were classified by Miss Cannon, plus a few by Mrs Mayall who supervised the production and publication of the charts following Miss Cannon's death. Most of the plates for the charts were exposed on the 10-inch Metcalf triplet, either at Arequipa or, later, at its new site near Bloemfontein in South Africa.

In addition to the 359 082 stars classified in the HD Catalogue, Extension and Charts, Miss Cannon classified about 36 700 spectra without HD designations. In all, over 395 000 stars were classified by her between 1896 and 1941. If the 4 years up to 1915 devoted to the main part of the HD Catalogue were her crowning achievement, the overall picture of nearly half a century devoted with single-mindedness to the sole cause of spectral classifications is remarkable by anyone's standards. The final posthumously published volume of her work (72) was as much a memorial to herself as to Henry Draper. It was entitled the *Annie J. Cannon Memorial Volume of the Henry Draper Extension.*

5.16 Statistical analysis of the HD data: Shapley and galactic structure

The vast store of data in the HD Catalogue was used for further detailed statistical studies, notably by Shapley and Miss Cannon. For example, the distribution of stars as a function of spectral type and magnitude was studied in 1921 (73). The use of spectral

Fig. 5.10 Harlow Shapley.

classifications to confirm the existence of a belt of early type stars inclined at a small angle to the Milky Way plane was undertaken in 1922 (74), and the galactic distribution of stars in different spectral classes was studied also in 1922 (75) and continued the following year (76). A special study of the galactic distribution of M stars followed in 1923 (77) and of F stars in 1925 (78), and the space density of stars of different spectral types was analysed in 1923 (79). In this last paper Shapley concluded 'for every Class B star that is in the stage of development represented by the stars in the Orion and Scorpius clusters, there are about five giant M stars and seventeen hundred dwarf stars like our Sun. This conclusion should be of some significance in considerations of stellar evolution'.

Shapley's studies of galactic structure using the Henry Draper spectral classifications in the early 1920s came at a time when our concepts of the nature and size of the Milky Way and of its relation to the universe as a whole were undergoing profound change. At a famous debate with Heber Curtis in April 1920, Shapley had argued in favour of the Milky Way being a very large galactic system of stars with the sun well displaced from the centre. He also maintained our Galaxy is the centre of the universe, with the spiral nebulae (such as M31 in Andromeda) being within it or nearby (80). Up till the mid-1920s, both these issues were hotly contested by astronomers, with Shapley playing a leading role. It is therefore important to see his galactic structure studies from Harvard spectroscopy in this context.

Both the issue of the Galaxy's dimensions and of the relation of the spiral nebulae to the Galaxy were solved in the mid-1920s. The work of B. Lindblad in Sweden from the analysis of stellar radial velocities supported the concept of a rotating Galaxy of stars with the sun well out from the centre (81); Edwin Hubble's (1889 – 1953) photographic resolution of stars in M31 and M33 showed these spiral nebulae to be distant galaxies of stars similar to our Milky Way (82).

From the early 1920s the course of astrophysics also took a new turn, with the application of ionisation theory and of atomic physics to the interpretation of stellar spectra. Henry Norris Russell at Princeton and Miss Cecilia Payne and Donald H. Menzel at Harvard are three of those who played major roles in the new astrophysics. What could therefore be more appropriate than to end this chapter with their words, written in 1935:

> The Harvard system is the product of the experience of a group, headed by Pickering and Miss Cannon, who have looked at a greater number of different stellar spectra than any other group; from this standpoint alone it must be recognised as having a maximum representativeness... As far as the fainter stars are concerned, the original Draper classification, augmented where possible by the later suggested prefixes g and d, will probably represent the spectra with adequate accuracy, until the photographic process has been greatly expedited... The question of revision or replacement of the Draper system concerns only the brighter stars, where high dispersion brings out important additional spectral detail (83).

References

1. Pickering, E.C., *Harvard College Observ. Ann.*, **26** (Part I), 1 (1891).
2. Pickering, E.C., *Harvard College Observ. Ann.*, **27**, 1 (1890) (The Draper Catalogue of Stellar Spectra).
3. Pickering, E.C. and Fleming, W.P., *Harvard Observ. Ann.*, **26** (Part II), 1 (1897).
4. Pickering, E.C., *Henry Draper Memorial, fourth annual report* (1890).
5. Maury, A.C. and Pickering, E.C., *Harvard Observ. Ann.*, **28** (Part I), 1 (1897).
6. Abney, W. de W., *Mon. Not. R. Astron. Soc.*, **37**, 278 (1877).
7. Scheiner, J., *Vierteljahrschrift der Astron. Gesell.*, **33**, 66 (1898).
8. Dunér, N.C., *Vierteljahrschrift der Astron. Gesell.*, **32**, 165 (1897).
9. Dunér, N.C., *Vierteljahrschrift der Astron. Gesell.*, **34**, 233 (1899).
10. Hertzsprung, E., *Zeitschr für wissenschaft. Photographie*, **3**, 429 (1905).
11. Curtiss, R.H., *Handbuch der Astrophysik*, **5**, (Chapter 1) Springer-Verlag, Berlin (1932).

12. Pickering, E.C., *Astron. Nachrichten*, **127**, 1 (1891).
13. Pickering, E.C., *Harvard Circ.*, **12** (1896). See also: *Astrophys. J.*, **5**, 92 (1896).
14. Maury, A.C., *Harvard Ann.*, **84**, 207 (1933).
15. Pickering, E.C., *Harvard Circulars*, **16** (1897); ibid **18** (1897) ibid **55** (1901).
16. Balmer, J.J., *Ann. der Physik und Chemie*, **25**, 80 (1885).
17. Kayser, H., *Astrophys. J.*, **5**, 95 (1896).
18. Pickering, E.C., *Harvard Circ.*, **16** (1897).
19. Fowler, A., *Mon. Not. R. Astron. Soc.*, **73**, 62 (1912).
20. Bohr, N., *Phil. Mag.*, **26** (ser. 6) 1 (1913).
21. Paschen, L.C.F., *Ann. der Physik*, **50**, 901 (1916).
22. Fowler, A., *Phil. Trans. R. Soc.*, **214**, 255 (1914).
23. Lockyer, Sir N., Chisholm-Batten, Captn. and Pedler, A. *Phil. Trans. R. Soc.*, **197**, 151 (1901) (see page 202).
24. Lockyer, Sir N. and Baxandall, F.E., *Proc. R. Soc.*, **74**, 546. (1905). See also *Mon. Not. R. Astron. Soc.*, **65**, [24] (1905).
25. Cannon, A.J. and Pickering, E.C., *Harvard Observ. Annals*, **28** (part II), 131 (1901).
26. Lockyer, Sir N., *Inorganic Evolution*, (1900).
27. Morgan, W.W., *Astrophys. J.*, **77**, 330 (1933).
28. Morgan, W.W., Keenan, P.C. and Kellman, E. *An Atlas of Stellar Spectra with an Outline of Spectral Classification.* University of Chicago Press (1943).
29. Jaschek, M. and Jaschek, C., *Publ. Astron. Soc. Pacific*, **72**, 500 (1960).
30. Cannon, A.J., *Harvard Observ. Annals*, **56**, 65 (no. 4) (1912).
31. Campbell, W.W., *Astrophys. J.*, **10**, 241 (1899).
32. Cannon, A.J., *Harvard Observ. Annals*, **56**, 115 (no. 5) (1912).
33. Pickering, E.C., *Harvard Observ. Annals*, **56**, 1 (no. 1) (1912).
34. Schlesinger, F., *Astrophys. J.*, **33**, 260 (1911).
35. Devorkin, D.H., *Isis*, **72**, 29 (1981).
36. Schlesinger, F., *Astrophys. J.*, **38**, 301 (1913).
37. *Transactions of the International Astron. Union*, **1**, 95 (1922).
38. Fleming, W.P., *Harvard Observ. Annals.*, **56**, 165 (no. 6) (1912).
39. Pickering, E.C., *Harvard Observ. Annals*, **56**, 41 (no. 3) (1912).
40. von Konkoly, N., *Beobachtungen Astrophys. Observ. in O'Gyalla*, **8** (part 2), 1 (1887).
41. Secchi, A., *Comptes Rendus*, **68**, 361 (1869).
42. Fleming, W.P., *Science*, **8**, 455 (1898).
43. Fowler, A., *Mon. Not. R. Astron. Soc.*, **64**, [16] (1904).
44. Pickering, E.C., *Harvard Observatory Circular*, **145** (1908).
45. Pickering, E.C., *Harvard Observatory Circular*, **9** (1896).
46. Rufus, W.C., *Publ. Astron. Observ. Uni. Michigan*, **2**, 103 (1916).
47. Cannon, A.J., *Harvard College Observ. Annals*, **98**, 1 (1923). See also: Shapley, H., *Harvard Bulletin*, **778** (1922).
48. Espin, T.E., *Mon. Not. R. Astron. Soc.*, **51**, 11 (1891).
49. Cannon, A.J., *Harvard Observ. Annals*, **76**, 19 (no. 3) (1916).
50. Wright, W.H., *Publ. Lick Observatory*, **13**, 193 (1918).
51. Cannon, A.J. and Pickering, E.C., *Harvard Observ. Annals*, **91**, 1 (1918).
52. Shapley, H., *Harvard Observ. Circular*, **278** (1925).

53. Bobrovnikoff, N.T., *Publ. Amer. Astron. Soc.*, **8**, 209 (1936).
54. Merrill, P.W., *Publ. Astron. Soc. Pacific*, **51**, 356 (1939).
55. Cannon, A.J. and Pickering, E.C., *Harvard Observ. Annals*, **92**, 1 (1918).
56. Cannon, A.J. and Pickering, E.C., *Harvard Observ. Annals*, **93**, 1 (1919).
57. Cannon, A.J. and Pickering, E.C., *Harvard Observ. Annals*, **94**, 1 (1919).
58. Cannon, A.J. and Pickering, E.C., *Harvard Observ. Annals*, **95**, 1 (1920).
59. Cannon, A.J. and Pickering, E.C., *Harvard Observ. Annals*, **96**, 1 (1921).
60. Cannon, A.J. and Pickering, E.C., *Harvard Observ. Annals*, **97**, 1 (1922).
61. Cannon, A.J. and Pickering, E.C., *Harvard Observ. Annals*, **98**, 1 (1923).
62. Cannon, A.J. and Pickering, E.C., *Harvard Observ. Annals*, **99**, 1 (1924).
63. Shapley, H., *Harvard Observatory Bulletin*, **805** (1924).
64. Newall, H.F., *Mon. Not. Roy. Astron. Soc.*, **80**, 429 (1920).
65. Espin, T.E., *Mon. Not. Roy. Astron. Soc.*, **72**, 546 (1912).
66. Wright, W.H., *Mon. Not. Roy. Astron. Soc.*, **72**, 548 (1912).
67. Shapley, H., *Harvard Observ. Bulletin*, **778**, (1922).
68. Merrill, P.W., *Publ. Astron. Soc. Pacific*, **35**, 217 (1923).
69. Merrill, P.W., *Astrophys. J.*, **65**, 23 (1927).
70. Cannon, A.J., *Harvard Observ. Annals*, **100**, 1 (1925–36).
71. Cannon, A.J., *Harvard Observ. Annals*, **105**, 1 (part 1) (1937).
72. Cannon, A.J., *Harvard Observ. Annals*, **112**, 1 (1949).
73. Shapley, H. and Cannon, A.J., *Harvard Observ. Circular*, **226** (1921).
74. Shapley, H. and Cannon, A.J., *Harvard Observ. Circular*, **229** (1922).
75. Shapley, H. and Cannon, A.J., *Harvard Observ. Circular*, **239** (1922) and Shapley, H., *Harvard Observ. Circular*, **240** (1922).
76. Shapley, H., *Harvard Observ. Circular*, **248** (1923).
77. Shapley, H. and Cannon, A.J., *Harvard Observ. Circular*, **245** (1923).
78. Shapley, H. and Howarth, Helen E., *Harvard Observ. Circular*, **285** (1925).
79. Shapley, H., *Harvard Observ. Bulletin*, **792** (1923).
80. Shapley, H. and Curtis, H.D., The Scale of the Universe. *Bull. Nat. Research Council*, **2**, 182 (1921).
81. Lindblad, B., *Arkiv för Matematik, Astron. och Fysik*, **19A**, No. 21 (1925).
82. Hubble, E., *Astrophys. J.*, **63**, 236 (1926).
83. Russell, H.N., Payne-Gaposchkin, C.H. and Menzel, D.H., *Astrophys. J.*, **81**, 108 (1935).

6 The Doppler effect

6.1 Early history of the Doppler effect An important chapter in the history of astronomical spectroscopy opened on 25 May 1842. On this day Christian Doppler (1803–53), the professor of mathematics at the University of Prague (then part of Austria), delivered a lecture to the Royal Bohemian Scientific Society entitled 'Concerning the coloured light of double stars and of some other heavenly bodies' (1). By analogy both with sound and waves in the sea, Doppler maintained that light waves undergo a change in frequency of oscillation, and hence of colour, either when the luminous source or the observer is in motion relative to the aether (whose existence was at that time supposed necessary for the transport of light waves). He gave formulae for the frequency change $\Delta\nu$ when either the source or observer were in motion, and these amounted to a statement of the now familiar equation $\Delta\nu/\nu_0 = V/c$. Here V is the relative speed in the line of sight, c is the speed of light, and ν_0 is the light wave's frequency for sources at rest.

Doppler then made two incorrect assumptions: first, that the radiation from stars was largely confined to the visual region of the spectrum, and secondly that the space motions of the stars were frequently a significant fraction of the speed of light. As a consequence, stars normally appearing white are seen instead as strongly coloured, either violet or red, depending on their approach towards or recession from the earth. Moreover, the brightness of fast-moving stars also depends on their velocity of recession or approach. If this latter parameter is large enough in either direction, then a star would become fainter or even invisible, when the light is shifted entirely out of the visible part of the spectrum. A white star, according to Doppler, would become invisible at about 136 000 km/s in either direction, while a star which appears red when observed at rest, disappears if its radial speed exceeds 12 000 km/s. In the case of white stars, a perceptible change of colour should be detected at only 240 km/s.

From these false premises Doppler went on to apply his theory to explain the observation that visual binary stars were generally composed of two stars of complementary colours, or at least of a bright white and a fainter coloured star, due to the different motions of the two components in their orbit. In support of his theory he cited evidence for colour changes in binary stars due to their orbital motion. Nevertheless, the evidence for this

Fig. 6.1 Christian Doppler.

was false as he compared the unreliable descriptions of colour of different observers at different times. Moreover, Doppler believed the phenomena of the supernovae of 1572 and 1604 were none other than binary stars whose variable velocities happened to shift their radiation into the visible spectrum. The supposed eccentric orbits explained the rapid rise and slow decline. The periodic colour and brightness changes of Mira were also supposed to be due to the motions of a binary star.

Doppler's hypothesis immediately started a considerable controversy as to its reality. For example, B. Sestini (1816–90), a contemporary of Secchi's both at the Collegio Romano and at Georgetown College, studied the colours of 400 double stars and believed he had found some weak evidence in support of Doppler (2). In Holland, C.H. Buys-Ballot (1817–90) was able to demonstrate Doppler's principle for sound from the change in the pitch of wind instruments played on passing trains (3). Then the great German astronomer H.C. Vogel proved the same thing years later when he found the pitch of the steam whistle on locomotives changed when they passed the observer (4), an experiment since much quoted in physics textbooks. However Buys-Ballot was one of several who contested the applicability of Doppler's principle to light, on the grounds that the ultraviolet or infrared regions should be shifted into the visible spectrum and hence no colour change would then result.

6.2 Fizeau and Mach and the concept of line displacements

H. Fizeau appears to have been ignorant of Doppler's paper when he presented his own version of the theory to the Société Philomatique in Paris in December 1848. Unfortunately the account of this lecture was not published until 1870 and therefore remained practically unknown (5). Fizeau's interpretation of the effect rested on the wavelength shift of spectral lines and not on colour changes, as the following passage shows:

> ... each ray... will take the place of the ray which possessed this same wavelength when the luminous body was at rest; all the rays will thus replace each other in such a way that the lines will no longer be in the same places, but are all displaced towards the red or towards the violet, according to the direction of motion of the luminous body. The colours on the other hand... will suffer no displacement... It is noted that this result only depends on the velocity of the luminous body and not at all on its distance. Such observations could thus lead to data on the intrinsic velocities of the most distant stars... (5).

Considering the infancy of stellar spectroscopy in 1848 (or rather its virtual non-existence since Fraunhofer's observations in 1823), Fizeau's conclusions showed considerable foresight, and were a substantial advance over Doppler's work. In France the principle is known as the Doppler-Fizeau effect, evidently for good reason.

In Vienna in 1860 Ernst Mach (1838–1916), ignorant of Fizeau's lecture, came to substantially the same conclusions when he showed how stellar line-of-sight velocities could be measured:

> The image of the star is dispersed by means of a prism into a spectrum in which two kinds of dark lines are to be seen, one originating from our atmosphere, the other from the star; the latter must now, on a colour change of the star, alter their position and the velocity of the star is then determined from this shift (6).

Since these words were written before the rebirth of stellar spectroscopy in 1862, they also represent remarkable insight.

6.3 First attempts to observe Doppler shifts by Secchi and Huggins

Father Secchi also mentioned the possibility of measuring stellar radial (i.e. in the line-of-sight) velocities in 1866 (7). By this time he could speak with the sure knowledge that the vast majority of stars either have zero or small (as compared to the speed of light) velocities, as the line positions in stars of the same spectral type generally showed close wavelength agreement. The hypothesis of Doppler on stellar colours was therefore already implicitly disproven early on in the 1860s (if not even earlier in Fraunhofer's time, before Doppler had announced his principle). However, Secchi's 1866 comment added little new, as no attempt to measure Doppler line shifts was reported. Nevertheless his paper dated 2 March 1868 to the Parisian Académie des Sciences gave a much fuller account of the possibility of stellar radial velocity work (8). He correctly estimated that a recession of 304 km/s would shift the D_2 line of sodium to the wavelength normally occupied by D_1. Next he compared the position of the F line (Hβ) in his type I stars by lining up the F-line in Sirius with the cross-wires of his spectroscope, and then observing the spectra of other stars of this type. He concluded: '... for the stars of the type of Sirius, there are no appreciable displacements using my measuring apparatus'. For late-type stars he used the magnesium b lines as well, but the results were still negative. 'There is not one whose motion is 5 to 6 times that of the earth in its orbit'.

A dispute now developed between Secchi on the one hand and William Huggins in London on the other over the reality of the Doppler shifts. Secchi appears to have been the first to report an attempt to measure stellar Doppler shifts, but Huggins was the first to claim a positive result. On 23 April 1868 Huggins sent a paper to the Royal Society (9), which was read on 14 May, about 10 weeks later than Secchi's contribution. Huggins had also observed the Hβ line of Sirius, but he claimed to see a measurable shift of the stellar line relative to that of his Geissler tube. After correcting for the earth's orbital motion, he found 29.4 mile/s (47 km/s) recession for the star. Although wrong in amount and sign, the result represents the first claim to have detected stellar radial motion (see Chapter 4). Huggins also maintained that he and Professor Miller had been fully aware of the possibility of using the Doppler method during their early stellar work in 1863 (10). Also included in the 1868 paper is a letter from James Clerk Maxwell (1831–79) in which the theory of the classical Doppler effect using Doppler lines is again expounded, but possibly with greater mathematical clarity than hitherto. Huggins' observations of radial velocities were resumed after his new 15-inch telescope had been installed in 1871. In all the results for thirty stars were given (11), though for some no more than the sign and not the amount of the shift could be estimated. The velocity of Sirius was now reduced to between 18 and 22 mile/s.

In 1869, after Huggins' first positive result had been published, Secchi again tackled the velocity of Sirius. This time he found that:

> the line f of hydrogen, projected in the spectroscope's field, was not found to correspond with the middle of the dark band of Sirius, but, on the contrary, tended towards the less refrangible side of the line. The displacement of the middle is about equal to the width of the lines D′ and D″ of sodium; this could imply a very rapid motion of the star. Mr Huggins, for his part, has arrived at the same conclusion (12).

If Secchi in this article now appears to concede both the reality of a detectable shift in Sirius and the priority of its discovery to Huggins, this feeling of generosity towards his English colleague did not last long. It was replaced by renewed doubts as to the reality of Doppler velocity shifts for any stars. Secchi now claimed that he had made his first albeit negative observations as early as 1863, and he rightly emphasised the large discrepancies between the results of Huggins and those of W.H.M. Christie (1845–1922) who participated in a radial velocity program begun at Greenwich in 1874 (13). To these retorts Huggins replied with the claim that Christie's measurements 'agree in a very striking manner with the observations that I had made on the same stars' (14). He quoted the comparison originally made by Christie, also in 1876 (15). Unfortunately for Huggins and the Greenwich observers, the truth is that results of visual observation continued to be disappointing. Vogel later estimated the typical probable error to be ± 22 km/s for the Greenwich results, and it is likely that Huggins' visual measurements were no better.

6.4 Visual Doppler shift programs of Maunder and Christie (Greenwich) and Seabroke (Rugby)

The Greenwich program just mentioned was initiated under Airy's term as Astronomer Royal (16) and was carried out mainly by E.W. Maunder until 1890 with some assistance from Christie until he (Christie) became Astronomer Royal in 1881. Each year long lists of visual observations of the radial velocities of the brightest stars were presented, based mainly on the Hβ or magnesium b lines (17). The typical probable error was claimed to be $\pm$ 10 mile/s (18), which (if Vogel is to be trusted) was quite a bit smaller than the truth. The results for individual stars often showed wide discrepancies. For example, in paper XIII in the Greenwich series the six values for Sirius ranged from +12 to −79 km/s! (19). Perhaps the only surprising thing about this program is that it should have been continued so long when useful results were obviously not forthcoming. With the introduction of photographic methods at Potsdam, the visual observations could no longer even look respectable, and the program was halted.

G.M. Seabroke (1838–1918) also undertook radial velocity measurements of bright stars in the years from 1877 to 1889 at the Temple Observatory attached to the Rugby School in England. He used a 31 cm Newtonian reflector and observed mainly the displacement of Hβ by comparison with the same line from a Geissler tube (20). His results were unfortunately even more unreliable than Maunder's. His first paper gave eight values obtained for Procyon ranging from + 112 to − 72 km/s. His third paper also showed evidence for a large systematic error, as nearly all the velocities given for 48 stars were large and negative. Such results are not indicative of carelessness at the telescope; they are instructive in showing the extreme difficulty of procuring reliable stellar velocities, and that advances in instrumental techniques would be essential before progress could be made.

6.5 The Doppler effect and solar rotation Significant advances in the technique of Doppler shift measurements were made by Vogel and his colleagues at Potsdam from 1887.

However first we go back some years to the time that Vogel and Lohse spent at the Bothkamp Observatory (see Chapter 4). They also attempted visual radial velocity measurements as early as 1871 with the 28 cm refractor (21), but the results were no improvement over those already obtained by Huggins (for example they found Sirius to be receding at 71 km/s, Procyon at 99 km/s). Far more important is Vogel's application of the Doppler principle in 1871 to the problem of detecting solar rotation, using the fact that the difference in velocity of the sun's east and west limbs should equal twice the equatorial speed of rotation. With a new so-called reversion spectroscope designed by J.C.F. Zöllner (1834–82) in Leipzig (22) (this instrument divides the light into two side-by-side spectra dispersed in opposite directions so as to double the relative Doppler shift between the two limbs), Vogel and Lohse were able to 'observe clearly the very small shift corresponding to the relatively small motion of a point on the solar equator' (21). Their result was only slightly larger than that predicted using sunspots to determine the solar rotation rate. Moreover, it confirmed a similar but qualitative estimate by Secchi the previous year (23). The significance of these solar results was considerable, as they represented the first empirical confirmation that the Doppler effect actually works for light in the same way as for sound.

In the 1870s and 1880s several other observers now took up the challenge of demonstrating the validity of the optical Doppler effect using solar rotation. The work of the American astronomer C.A. Young (1834–1908) (24), using a Rutherfurd grating spectroscope, gave quantitative results in reasonable agreement with sunspot data, although the spectroscopic rotation rate was still slightly faster. Success breeds success, and a spate of increasingly reliable solar rotation observations now followed, from Thollon and Cornu in France, from Crew in the United States and from Dunér in Sweden.

L. Thollon (1829–87) used nearby terrestrial atmospheric (telluric) lines as a stationary comparison (25) and observed the nickel line midway between D_1 and D_2, a technique reminiscent of Mach's suggestion (6). A. Cornu (1841–1902) used a spectroscope with an oscillating lens allowing him to flip quickly from one limb to the other and thereby achieve higher accuracy. His equatorial Doppler shift was within 3% of the value expected from sunspots (26). H. Crew (1859–1953) studied the solar limb up to 45° heliocentric latitude and concluded, though incorrectly, that the sun rotates as a solid body (27).

The application of visual spectroscopy to the solar rotation problem reached its pinnacle with the exhaustive study from 1887–9 by N.C. Dunér at Uppsala in Sweden (28). His spectroscope employed a high quality Brashear grating from Pittsburg, Pennsylvania. With a probable error of only 0.013 km/s, he studied the solar rotation rate as a function of heliocentric latitude to 74.8°, also using the telluric lines as a zero velocity reference. At the equator he found a speed of 1.98 km/s, but this decreased to only 0.34 km/s at the near polar latitude of 74.8°. When converted to angular velocities the results showed differential solar rotation, the angular velocity at 74.8° being nearly six times smaller than at the equator. The agreement with the less extensive sunspot data was excellent, and hence the theory of Doppler and Fizeau for light was finally fully vindicated. We should also mention that Vogel's photographically measured velocities for the planet Venus, in which he compared the results with computations from the known orbit, further confirmed the validity of the Doppler shift's interpretation beyond all doubt (29).

As it happens, the early spectroscopic work on solar rotation was not the first application of the Doppler effect to solar physics. Lockyer's observations of emission lines in the solar chromospheric spectrum in 1869 reported variations in the structure and wavelengths of the $H\alpha$ and $H\beta$ lines (30). This he correctly interpreted as 'extreme rates of movement in the chromosphere'. He estimated the vertical motions to have velocities up to 40 mile/s and the horizontal motions up to 120 mile/s. Lockyer's was thus undoubtedly one of the first reliable observations of Doppler displacements in spectral lines. However it could not be regarded as a demonstration of the validity of Doppler's principle, as corroborating evidence for rapid motions in the chromospheric prominences was not available.

6.6 Visual radial velocity measurements by Keeler at Lick

If visual radial velocity measurements on stars had made depressing progress for over two decades, then at least the work of James Keeler using the giant Lick Observatory 36-inch refractor is a story with a difference. Keeler was the only astronomer to obtain reliable Doppler shifts for stars using a visual spectroscope (31). As part of his classical study of bright-lined nebulae, of which the main purpose was to measure accurately the wavelength of the 5007 Å line with the hope of identifying it (he was unable to do so, but showed the line cannot be due to magnesium oxide as Lockyer believed), he also tested his ability to measure accurate wavelengths of spectral lines by observing the Doppler shifts of Arcturus, Aldeberan and Betelgeuse. These observations were made in the winter of 1890–1 during Keeler's first period at Lick (he briefly returned from Allegheny Observatory to take up the Lick directorship in 1898, but died in 1900 aged 42). The spectroscope employed a Rowland grating with 570 line/mm. The mean probable error of a single velocity determination of

Fig. 6.2 James Keeler.

these bright stars was only 1.8 km/s, well over an order of magnitude better than any other visual observer. As Vogel wrote in 1900:

> With the Lick refractor, which exceeds the Potsdam 11-inch refractor some eight times in light gathering power, it has therefore been possible to determine the motions of the brighter stars, by direct observation, with about the same accuracy as with the Potsdam refractor by the spectrographic method (32).

It is in this classic paper that Keeler also measured the radial velocity of 14 nebulae, so continuing the pioneering work that had been initiated by Huggins in 1868 (33). Keeler's objects were mainly planetary nebulae, although the diffuse nebula in Orion was also observed. For this purpose the Hβ emission line was used, being the brightest identified line in the visible spectrum. Keeler concluded that the motions of nebulae were generally of the same order of magnitude as those of stars, a necessary condition if an evolutionary link between the two was to be invoked. Huggins had reached the same conclusion in 1874 (34), albeit from observations of lower accuracy. Keeler's average probable error of the mean result for a nebula was $\pm$ 3.2 km/s (32), only slightly inferior to his results for the brightest cooler stars. Although nebulae are diffuse and have low surface brightness, they concentrate their emitted light mainly into a few intrinsically sharp lines, which facilitates the accurate measurement of Doppler shifts.

6.7 Photographic radial velocity work by Vogel and Scheiner at Potsdam

The introduction of photography into radial velocity determinations was one of the most significant advances in this chapter of astronomy. The work of Vogel and Scheiner at Potsdam has already been outlined (Chapter 4). The almost immediate reduction in the error bars by nearly an order of magnitude was later eloquently described by Vogel himself (32):

> When I made the first attempt in 1887, with the assistance of Professor Scheiner, to record photographically the displacements of the lines in stellar spectra, and then to measure them as accurately as possible on the spectrograms, it very soon appeared that this constituted a very marked advance in the determination of these motions, which are so significant in stellar astronomy. The accuracy of the observations was increased more than eightfold with the apparatus constructed in 1888; the probable error, which in the

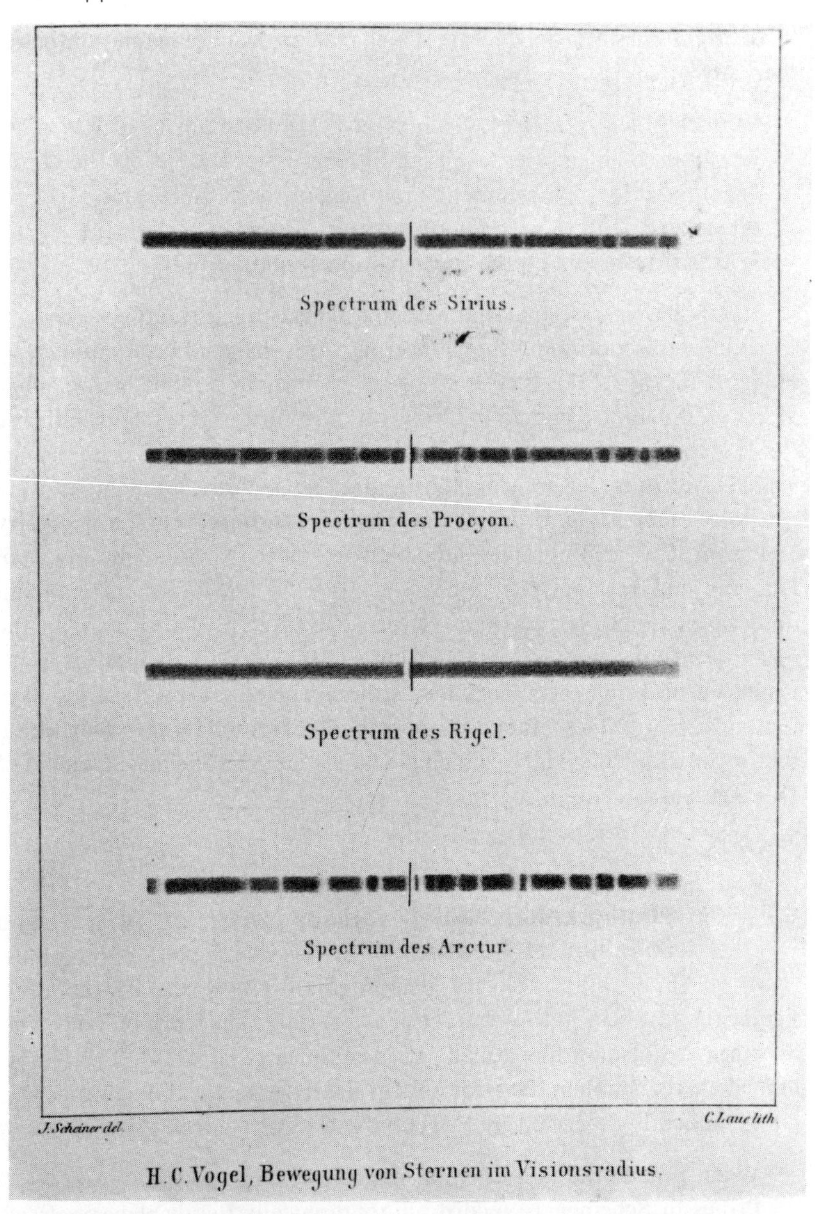

Fig. 6.3 H.C. Vogel's spectrograms showing the motion of stars in the line of sight.

Greenwich observations averaged ± 22 km for an evening, being brought down in the Potsdam observations to an average of ± 2.6 km. We may therefore fairly say that the determination of the motions in the line of sight thus first received a substantial basis in the spectrographic method, and thereby the widest prospects were opened for a period of new investigations and discoveries.

For these pioneering observations by Vogel and Scheiner during the years 1888 to 1890, a light-weight slit spectrograph was used on the 30 cm Potsdam refractor (Fig. 4.17). The spectrograph employed two Rutherfurd compound prisms, and the spectra in the neighbourhood of Hγ were recorded on glass plates held in brass holders. A very thin hydrogen Geissler tube was mounted inside the telescope 40 cm from the slit, and widening of the spectrum was achieved by trailing the telescope rather than by means of a cylindrical lens. Vogel measured the plates with a travelling microscope, either using the stellar and comparison Hγ lines alone, or sometimes using the numerous metallic lines near Hγ in a standard solar spectrum as a secondary reference. The techniques employed are described in detail in Vogel's classic paper of 1892, in which radial velocities for fifty-one stars were presented (35).

After the initial photographic measurements at Potsdam had been completed in 1891, the radial velocity program there languished for some years. Evidently Vogel was well aware that the main limits to accuracy came from flexure and temperature fluctuations, both of which effects limited him to short exposures (mostly of 1 hour or less). His efforts in the later 1890s were therefore directed towards the construction of new spectrographs (36), as well as considerable time being devoted to the erection of the 80 + 50 cm double refractor.

6.8 Radial velocity work of Belopolsky at Pulkova.

Meanwhile other observers were inspired by Vogel's success using photography, among them A.A. Belopolsky (1854–1934) at the Pulkova Observatory near St Petersburg. In 1891 he equipped the 76 cm refractor with a spectrograph practically a replica of Vogel's instrument (37), with which he could reach fourth magnitude stars in an hour. One of Belopolsky's first successes was the detection of radial velocity variations in the pulsating variable star δ Cephei (38), which he incorrectly assumed to be due to the star's binary nature, and his orbital elements, apart from the period, were therefore spurious. A similar analysis was later carried out for another Cepheid variable η Aquilae (39). The fact that minimum brightness and the zero of velocity did not coincide, as would be expected for a binary star, was remarked on in this paper, with the suggestion that an alternative explanation other than eclipses for the light variations would have to be sought.

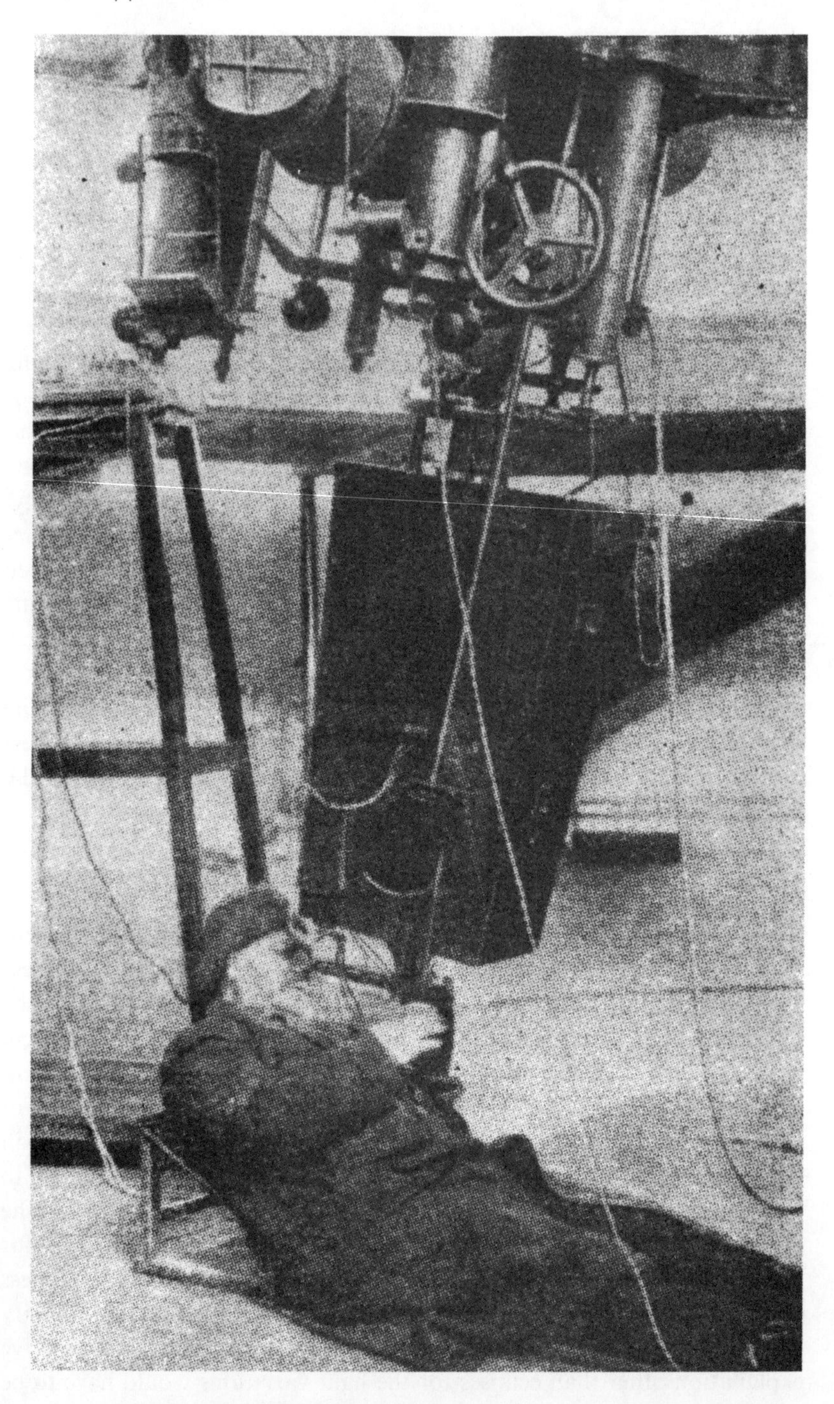

Fig. 6.4 A.A. Belopolsky at the spectrograph of the Pulkova refractor.

Belopolsky gave most of his attention to variable velocity stars over the next few years. However he is also remembered for an ingenious attempt to demonstrate in the laboratory the validity of the optical Doppler effect using the multiple reflections of sunlight produced by mirrors mounted on two counter-rotating wheels (40). His results, although not very precise, were mainly in accordance with predictions.

In 1905 Belopolsky reviewed his stellar radial velocity techniques (41) and included an interesting discussion on the systematic errors arising from guiding on the visual image yet photographically recording the blue-to-ultraviolet spectrum. Because the ultraviolet stellar image may not always be centred on the slit, due to atmospheric refraction, systematic errors of several km/s were found to be possible in extreme cases, as Belopolsky showed from a series of Arcturus spectrograms with deliberate misguiding. The problem must have been particularly acute for Belopolsky, as Hγ was the limit of his violet vision. At this time Belopolsky was using an iron arc for his comparison spectrum. The observed wavelengths of stellar lines (not necessarily iron) were then found using an interpolation formula first applied by the Potsdam astronomer J. Hartmann (1865–1936) (42) and these wavelengths were then compared with the observed solar wavelengths on the scale of the famous Johns Hopkins University spectroscopist, Henry A. Rowland (1848–1901), whose preliminary table of solar spectrum wavelengths (43) was the standard for the first quarter of the twentieth century. Although Vogel had been the first to experiment with iron arcs, Belopolsky had clearly further exploited their practical advantages (44).

6.9 Radial velocity programs in the United States, France and Britain in the 1890s

Belopolsky was only one of several observers who initiated stellar radial velocity programs in the 1890s as a result of the lead shown by the Potsdam astronomers; James Keeler was another. After Keeler left Lick Observatory in 1891 on his appointment to a professorship at the Allegheny Observatory, Pittsburgh, he undertook a stellar spectrographic program with the 13-inch Allegheny refractor (see, for example, 45). His well-known discovery of the rotation of Saturn's rings using the Doppler effect dates from his Allegheny years (46). He was able to show that the rings rotate differentially and not as a solid body. They must therefore be composed of numerous small bodies, or meteorites.

The existence of differential rotation in the rings was confirmed shortly afterwards by H.A. Deslandres (1853–1948) at the Paris Observatory (47). It is fair to say that French stellar spectroscopy had been languishing ever

Fig. 6.5 H.F. Newall.

since the time of Wolf and Rayet in the 1860s, and the Paris Observatory thus resolved to establish a spectroscopic service in 1890 to be headed by Deslandres. The main instrument was the 120 cm reflector, the largest telescope to be employed for stellar spectroscopy at that time. In the years up to 1897 about 200 stellar spectra were secured for a radial velocity program, but substantial results were not forthcoming. Deslandres moved to the new observatory at Meudon in 1897 and the spectrographic program at Paris was then taken over by M. Hamy (1861–1936). Among the results Deslandres accomplished in this time was the demonstration of the rotation of Jupiter using the change in Doppler shift of the spectral lines across the disc (48). As an example of his stellar work, Deslandres confirmed the high velocity of ζ Herculis (-70 km/s; 49) which had earlier been found by Belopolsky in the same year (50). Deslandres used a hydrogen Geissler tube as a comparison and determined velocities from the displacement of Hγ and four or five nearby iron lines.

Photographic radial velocity work was taken up in England at about this time by H.F. Newall. Newall's father was an engineer whose company manufactured wire rope and cables; he was also an amateur astronomer who had a 25-inch refractor by T. Cooke and Sons of York erected in his garden about 1870 (51). This instrument (Fig. 6.6) was donated by Newall's father to the University of Cambridge in 1890, whereupon H.F. Newall resigned his Cavendish assistantship to supervise without pay the telescope's installation. The financial support of Miss C.A. Bruce in the United States enabled him to have a one-prism spectrograph built (52), and

Fig. 6.6 The Newall telescope in Gateshead. The observer is R.S. Newall.

in 1896 he started a spectrographic stellar radial velocity program. The results for seven stars, derived using an iron spark as a comparison, were published the following year (53). As with the results of other observers, the measurements were affected by both flexure and temperature problems, but Newall improved the accuracy when a four-prism spectrograph was installed in 1889. Newall described his reduction procedures with this instrument in 1903 (54). He obtained a precision corresponding to a probable error of 1.04 km/s per plate, as found from 14 plates of the F supergiant α Persei, or of 0.68 km/s from 6 plates of Arcturus.

E.A. Milne later gave a glowing account of how some of the problems were tackled in establishing this program (55):

> All these [problems], Newall with his own unaided instrumental good sense and sound principles of optical design, and his consummate skill with his hands, triumphantly overcame. He deserves the greatest possible credit for his pioneer work in this field. Without Newall's devotion Great Britain would have lagged behind the work then being done by Vogel at Potsdam and Campbell at Lick.

After the radial velocity program had been established, Newall was later appointed to a professorship at Cambridge in 1909, and in 1913 he became director of the Solar Physics Observatory after this institution was transferred to Cambridge from Kensington, following Sir Norman Lockyer's retirement.

Meanwhile the pace of spectrographic work was quickening, especially in the United States. For example, when the Ohio State University in Columbus established an observatory from the funds donated by Mr Emerson McMillan in 1896, the main activity of the first astronomer appointed, Henry C. Lord (1866–1925), was in stellar radial velocity measurements with the $12\frac{1}{2}$-inch Brashear telescope (56). The spectrograph incorporated interchangeable dispersing elements (either one or two prisms, or a grating; 57) and with this instrument Lord was able to attain precisions of around ± 2 km/s (probable error) from one plate. Lord's main publication on radial velocities came in 1905 with results for thirty-one stars (58).

6.10 W. W. Campbell If Vogel was the first to have made substantial progress in the realm of precision stellar radial velocities, there is no doubt about who was his successor as leader in the field. William Wallace Campbell (1862–1938) was born in Ohio in 1862 and first trained to be an engineer at the University of Michigan. However, by 1891 when he joined the Lick Observatory, Campbell had already developed a reputation in the computation of cometary orbits. The Lick post had become available by the departure of Keeler for Allegheny, and Campbell therefore had the opportunity of using Keeler's visual spectroscope on the 36-inch refractor. His spectroscopic interests soon became quite diverse, encompassing the spectra of comets (e.g. he observed Swift's comet of 1892; 59), novae (i.e. Nova Aurigae; 60), nebulae (especially the Orion nebula; 61) and Wolf–Rayet stars. His study of the wavelengths of

Fig. 6.7 W.W. Campbell at the Lick 36-inch refractor in 1893. The spectroscope is Keeler's visual instrument made by Brashear.

emission lines in thirty-one stars of this last type was one of the early classic papers on the subject (62). At this time Campbell was also gaining his first experience in spectrum photography, using a light-weight spectrograph with a wooden frame assembled by the Lick carpenter (61). His interest in radial motions was also kindled in these early years, as is shown by a careful discussion on the reduction of radial velocity measures to the solar frame of reference (63), and also by an account of changes in the motions of the line-

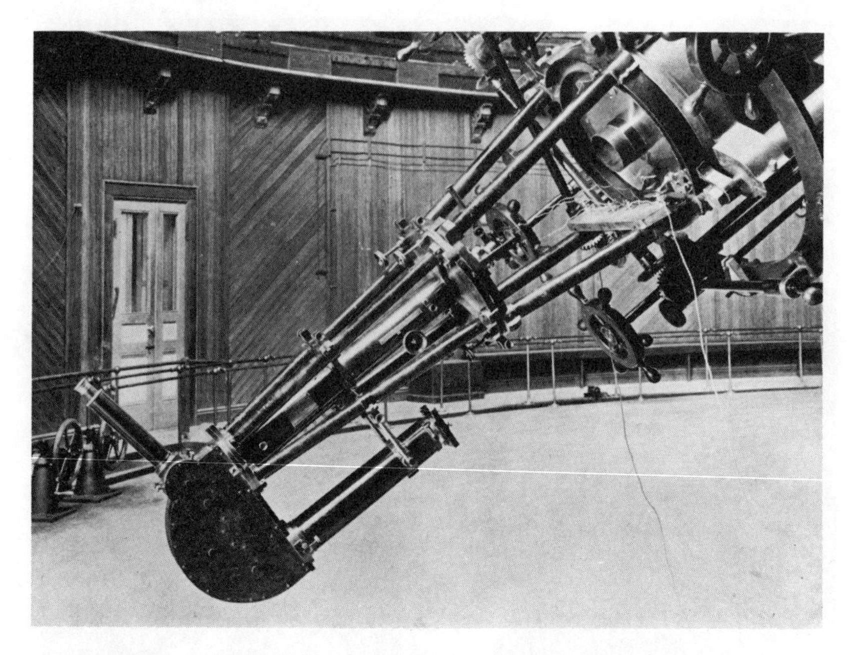

Fig. 6.8 The original Mills spectrograph on the Lick refractor in 1896.

emitting material in Nova Aurigae from visual observations of the principal nebular line (64).

About this time the observatory received a donation from a local businessman, Mr D.O. Mills, for the construction of a spectrograph specifically for stellar radial velocity work with the 36-inch refractor. Mills had been one of the trustees for the will of James Lick, whose bequest of \$700 000 had founded the observatory in 1887. Campbell designed the new spectrograph and the J.A. Brashear Co. built the optical components. Known later as 'the original Mills spectrograph', it was brought into operation in 1896, and as Campbell wrote in his important book *Stellar Motions* (65) the

> probable errors were reduced at once from 2.6 km [/s; referring to Vogel's work at Potsdam] to half a km per second for the brighter stars containing good lines, and down to nearly a quarter of a km. for bright stars containing the best quality of lines. What was still more important, systematic errors appeared to be of minute or vanishing size. One photograph with this instrument gives greater accuracy, depending upon the character of spectrum, than could be obtained from ten to fifty spectrograms made with its predecessors.

Fig. 6.9 The new Mills spectrograph at Lick.

The original Mills spectrograph (Fig. 6.8) employed three dense flint glass prisms and used an iron spark comparison. The collimator focal length was 724 mm, near the maximum considered practicable to avoid flexure, and the spectra were dispersed to 12.5 Å/mm at Hγ (66). Campbell exploited the greatly improved precision for radial velocity determinations that this instrument afforded and he initiated an ambitious program to measure the

Doppler shifts of all the northern stars brighter than apparent visual magnitude 5.51.

In 1897 William Hammond Wright joined the Lick staff, and a particularly fruitful collaboration with Campbell in the radial velocity program dates from this time. One by-product of their work was the discovery of thirty-one new variable velocity stars (mainly spectroscopic binaries) by 1900, adding greatly to the mere handful known hitherto (67). Campbell and Wright put much effort into reducing the errors in the measurement of the minute Doppler shifts of spectral lines (a radial velocity of 1km/s corresponded to a shift of only about 1.2μm of the lines on the plate!). One of their precautions was the thermal insulation of the spectrograph with a woollen blanket to slow the thermal contraction of the instrument when the night air temperature dropped during an exposure (68).

In 1902 a 'new Mills spectrograph' (Fig. 6.9) was constructed with an improved support system to eliminate further the flexure of the optical components. Since the telescope and spectrograph slowly change their orientation in following a star throughout a long exposure, the flexure of the spectrograph can also change, leading to a movement of the spectral image on the plate which blurs the spectrum. The new Mills spectrograph also had more efficient optics. With the original spectrograph only one per cent of the starlight entering the telescope reached the plate; the prism-train's transmission of 25 per cent (68) and the 36-inch telescope objective's transmission of 51 per cent (69), were partly responsible for this inefficient performance. The new instrument superceded the original one in May 1903, and thereafter the bulk of the northern radial velocity program was undertaken with it.

6.11 The D.O. Mills expedition to Chile Meanwhile the year 1900 brought the untimely death of the Lick director, James Keeler, who had returned from Allegheny only 2 years before. Campbell was appointed his successor and he assumed the directorship in January 1901. In this role he was able to channel a major part of the observatory's research effort into determining stellar radial velocities. Even before the original Mills spectrograph had been completed, Campbell was planning his velocity program on a scale that would encompass all the bright stars in both hemispheres.

It was in 1894 that Mr Campbell first described to his colleagues his provisional plans for an expedition to the southern hemisphere to

Fig. 6.10 The D.O. Mills reflecting telescope and 3-prism spectrograph in Chile.

secure radial velocity observations of the brighter stars. The time for organising such an expedition seemed to have come in November, 1900. With the approval and endorsements of Dr Benjamin I. Wheeler, President of the University, the subject was brought to the attention of Mr D.O. Mills, the donor of the Mills spectrograph. Mr Mills most generously offered to provide funds for contributing the instruments, for defraying the traveling,

> erecting, and maintaining expenses, and for the salaries of astronomers engaged in the work (68).

Thus did the D.O. Mills expedition to Chile come into being. In the scale of activities planned, in the breadth of his vision, in the determination to carry a long and arduous program through to its successful conclusion in both hemispheres, Campbell was surely the equal of his great east coast contemporary, E.C. Pickering at Harvard. What Pickering achieved for spectral classification, Campbell now complemented with radial velocities, at this time the two great branches of stellar astrophysics. Like Pickering, Campbell was fortunate to have the continuing financial support of a generous benefactor who was to provide for the running of the Santiago station until 1911. Mr Mills died in 1910, but his son, Ogden Mills, then funded the operation of the southern observatory in full until 1917 and in part from 1917–22.

Work on equipping the expedition began in earnest in 1901. It was decided to build a $36\frac{1}{4}$ inch cassegrain reflector using a parabolic mirror already in the observatory's possession. This however 'broke into a multitude of pieces' during the central hole perforation and a new disc was ordered from the St Gobain works in Paris in June 1901. A year later the Brashear Co. in Pittsburgh had figured the primary and the telescope tube and mounting had meanwhile been completed in December 1901. Unfortunately the mirror continued to put the expedition in jeopardy, as it had to be returned to Pittsburgh to correct figuring errors and was not finally finished until February 1903. The expedition sailed at once under the leadership of W.H. Wright without even having time to test the telescope. The aim was to press ahead and complete within two years the radial velocity program to $m_v = 5.51$ in the remaining quarter of the sky from about $-30°$ to the south pole.

The speed with which the new observatory was established was impressive. Site-testing in Chile took one month and Cerro San Cristobal, 2.3 km from central Santiago, altitude 840 m, was selected. Building began on 27 May 1903, at the start of winter. Observing commenced on 11 September. From first setting foot in Chile to first light through the telescope had taken 148 days, in spite of a major strike and bloody riot in progress on arrival in the port of Valparaiso, when the valuable telescope optics had to be rowed ashore in a small dinghy by Wright himself.

The telescope had a $36\frac{9}{16}$-inch (92.9cm) f/5.7 primary mirror, and a 3-prism spectrograph was built in 1901 for the f/18.2 cassegrain focus, giving a dispersion of 10.3 Å/mm at Hγ. An iron comparison spectrum was employed. Later, in 1906, lower dispersion two- and one-prism spectro-

Fig. 6.11 The San Cristobal Observatory of the D.O. Mills expedition, near Santiago, Chile.

graphs were also installed. Details of the equipment are given by W.H. Wright (69) and by Campbell and J.H. Moore (68). Wright stayed in Chile until March 1906 by which time 899 spectrograms were secured, thereby completing the initial aim of the project. The cost of the operation to this time had been $26 075, including the telescope, dome, building and all equipment, as well as travelling and freight expenses, the rent of land and the astronomers' salaries. The low cost of the telescope itself ($5500) 'is in striking contrast with the eighteen-fold greater cost of the 36-inch refractor' on Mt Hamilton (70).

Reduction of the plates to obtain radial velocities was undertaken by Wright in Santiago using the methods employed by Campbell, namely use of Hartmann's interpolation formula and of Rowland's wavelengths. The accuracy of the results was the same as achieved at Mt Hamilton, with probable errors of about ± 0.5km/s for stars with 'good and fairly numerous lines' and never worse than ± 3km/s for early-type stars with more diffuse lines. The results for 150 southern stars were published in 1911 based on the first $2\frac{1}{2}$ years of observations (71). The publications had been delayed about 3 years through lack of funds.

6.12 Campbell, Wright and Moore at Lick

Wright's departure from Chile in 1906 did not spell the end of his radial velocity work nor of the expedition. A 5 year grant for continuing Chile operations was donated by Mr Mills in that same year. The Santiago station, originally planned as a temporary observatory, thus had a new lease of life. In the event it continued until 1929, and the six leaders who followed Wright include several prestigious names in the subsequent development of stellar radial velocity work, especially Joseph H. Moore (1878–1949) and Ralph E. Wilson (1886–1960), who were in Chile from 1909–13 and from 1913–18 respectively. Wright's immediate successor, however, was Heber D. Curtis (1872–1942). Curiously Campbell himself during nearly three decades of Chile operations never found an opportunity to visit the southern observatory in his time as Lick director. By June 1926, 10 310 spectrograms had been obtained from Chile (68).

In 1928 Campbell and Moore finally published the long-awaited catalogue of radial velocities of stars brighter than visual magnitude 5.51. The catalogue embraced 2771 stars, and was complete down to the limiting magnitude for all but 69 objects. Of the total, 351 had variable radial velocities, and a further 81 were possibly variable. Systemic (i.e. centre of mass) radial velocities were obtained for 57 spectroscopic binaries. About 60 per cent of the plates for the catalogue (some 15 000) were secured at Mt Hamilton up to the year 1927 on the giant Lick refractor. In all thirty-one observers (counting only those who contributed more than 100 plates) were involved on this telescope with the program, of which Moore (2422 plates), Wright (1956 plates) and Campbell (1100 plates) were the three principal contributors. The Mt Hamilton plates were measured by fifty-eight different workers, both astronomers and assistants. The statistics are of interest, as they point to Campbell's ambitions of devoting extensive resources to the relatively narrow goals of a single long-term program on a massive scale.

Like Campbell, Moore came from Ohio and he joined the Lick Observatory staff in 1903 after several years at the Johns Hopkins University where he came under the influence of Simon Newcomb, Henry A. Rowland and R.W. Wood (1868–1955). The last two were the leading spectroscopists in the United States and undoubtedly influenced Moore to accept the Lick position. From the start he was involved with the radial velocity program, and as Campbell's time became increasingly taken up by administrative duties from about 1922,* Moore took over the major part of the work to complete the great catalogue. Later he succeeded Wright in 1942 for a 3 year term as Lick director. Between them, Campbell, Wright and Moore each spent nearly the whole of their professional careers at Lick Observatory, contributing together a grand total of 131 years' service.

* Campbell became president of the newly formed International Astronomical Union (1922–5), of the American Astronomical Society (also 1922–5), of the University of California (1923–30) and of the National Academy of Sciences in Washington (1931–5).

Fig. 6.12 Four radial velocity pioneers from Lick: J.E. Keeler (top left), W.W. Campbell (top right), J.H. Moore (bottom left), W.H. Wright (bottom right).

Campbell was afflicted by failing health and blindness in his final years (this led him to take his own life in June 1938 to spare his family the burden of caring for him). Paul Merrill's comments say much about his character:

> Campbell was not one of those blown about by every idle wind of astronomical doctrine; no one ever called him a weather vane. And the implication that he was not especially receptive to new tenors of thought is true. His method was to push ahead in his own established line of research rather than to experiment with trial balloons in the hope of reaching other levels. He has been criticised on this score, but, if criticism be limited to those who have accomplished as much as he, not many are entitled to be heard (72).

6.13 Campbell's analysis of solar motion The data on stellar radial velocities that came out of the Doppler programs at Mt Hamilton and Santiago were Campbell's main contribution to astronomy. The use he made of the radial velocity data to analyse stellar motions statistically was also of some importance, since a value for the sun's motion relative to the field stars could be obtained. Campbell's was not the first such analysis, in particular R. Kövesligethy (1862–1934) in Budapest had used about seventy stars in the Greenwich visual observations in 1886 (73) and H. Homann in Berlin analysed data by Huggins and by Seabroke in the same year (74). The results from visually determined velocities were generally unreliable. In 1893 P. Kempf at Potsdam (75) and A.D. Risteen in the United States (76) both used Vogel's photographic results and found solar velocities relative to the stars of 13.0 and 17.5 km/s respectively, but the directions found for the sun's motion differed widely. Campbell's 1901 determination of solar motion can be regarded as the first reliable result (77). He used 280 stars observed at Lick up to the end of 1900 and found the sun to be moving at 19.89 ± 1.52 km/s relative to the stars, and he also solved for the direction, or solar apex. Campbell periodically improved this result over the years as more radial velocities became available. 1047 stars were used in 1911 (78), 2034 in 1925 (79), and 2149 in 1928 (68). The improvements were not merely in star numbers, but also refinements in the methods of data handling, such as the elimination of high-velocity stars, the grouping together of stars with common motions, the determination of solar motion relative to stars of different spectral types, and an analysis of the so-called K-term velocity, an apparent expansion of the system of B stars of about 5 km/s, since ascribed to systematic errors in the observed velocities.

6.14 New radial velocity programs established early in the twentieth century In the first three decades of the twentieth century many astronomers took up the challenge of stellar radial velocity work, which became one of the most popular and fruitful areas of research in astrophysics. Apart from the seven observatories already mentioned,* radial velocities were also determined in North America at Detroit (University of Michigan), Lowell, Mt Wilson, Dominion (Ottawa), Dominion Astrophysical (Victoria) and Yerkes Observatories, in Europe at the observatories at Bonn, Vienna, in the Crimea at Simeis, and in South Africa at the Cape. The extensive list speaks for itself; Vogel and Campbell had initiated a wave of radial velocity enthusiasm that swept through many of the world's major observatories during the first decades of the century. Some of the highlights of this activity are mentioned here.

In Bonn, F. Küstner (1856–1936) used a 3-prism spectrograph on the 30cm Repsold–Steinhert refractor from 1903 (80), observing brighter F to M stars with a probable error of ± 0.64 km/s. The work was later extended to fainter stars (limit of sixth photographic magnitude) in a list of 227 late-type stars whose radial velocities were measured to within a probable error of 1.38 km/s (81). Küstner also measured the solar parallax (defined as the ratio of the earth's equatorial radius to the semi-major axis of its orbit around the sun), by determining the earth's orbital speed from eighteen spectrograms of Arcturus taken at different times of year. He found a speed of 29.617 km/s. This in turn gave the distance the earth travelled around the sun in a year and hence a solar parallax of 8.844 ± 0.017arc sec (82).

At the Cape Observatory, Sir David Gill had commenced a radial velocity program in 1903 which was continued by Sydney S. Hough (1870–1923) and Joseph Lunt (1866–1940) using a 4-prism spectrograph on the 24-inch refractor (83). Lunt published several radial velocity papers from 1918, the first being for sixty southern stars (84). After Lunt's retirement in 1926, H. Spencer Jones (1890–1960) succeeded Hough as His Majesty's Astronomer at the Cape but shortly afterwards halted the Doppler program to allow trigonometric parallax work to be carried out on the telescope (85). The results of the entire program with radial velocity data for 434 stars were published by Spencer Jones at that time, mainly from the work of Lunt (86). They were generally regarded as of high quality.

* That is Allegheny, Emerson McMillan, Cambridge, Lick, Paris, Potsdam and Pulkova.

6.15 The Mt Wilson radial velocity program Mt Wilson Observatory was founded by Hale in 1904 and the 60-inch came into service 4 years later. An extensive radial velocity program was undertaken, first with this telescope and later (from 1918) also with the 100-inch, by Walter S. Adams (1876–1956) with the assistance until 1914 of A. Kohlschütter (1883–1969) and then with Alfred H. Joy (1882–1973). The first publication, by Adams and Kohlschütter, discussed the radial velocity results obtained for 100 trigonometric parallax stars (87) from 16 or 36 Å/mm single-prism cassegrain spectrograms in the blue region with an iron comparison. In this paper the authors pointed out that of the stars with high radial velocity (more than 50 km/s), those with negative values exceed those with positive values by more than three to one in number. Two very high-velocity objects were found.* This program continued to be directed towards astrometrically selected stars after Kohlschütter's return to Europe in 1914. The radial velocities for 500 fainter stars, including many selected for very large or very small proper motions, were published by Adams in 1915 (88). The probable errors were around ± 1 km/s. After the 100-inch telescope had been brought into use, Adams collaborated with Joy in 1923 to publish a major catalogue of radial velocities of 1013 stars of types F to M (89). They used the one-prism spectrograph and a camera giving 36 Å/mm dispersion. The mean probable errors were ± 1.35 km/s. Ten to fifteen stellar lines were used on each plate, mainly of iron, and the wavelengths were taken from the solar values of H.A. Rowland, even though these were no longer considered the best available (see section 6.20).

This program continued with the results of a further 741 stars in 1929, many of which were later-type dwarfs with large proper motion, down to about eleventh photographic magnitude (90). By 1935 the declared aim was to observe all the later-type stars in the Boss General Catalogue of proper motions (91) fainter than visual apparent magnitude 5.5, the limiting magnitude of the Lick program. This task was completed only after the war when Ralph E. Wilson had taken over the main responsibility for the radial velocity program at Mt Wilson. Wilson and Joy were thus able to publish results for 2111 stars in 1950 (92) which also included data on many not in the catalogue but known to have annual proper motions in excess of 0.1 arc sec.

* They were Lalande 1966 with $V_R = -325$ km/s, and Lalande 15290, $V_R = -242$ km/s.

6.16 High-velocity stars and the discovery of galactic rotation The large body of radial velocity data accummulated at Mt Wilson was particularly suited to statistical studies of stellar motions in the Galaxy. As an example, we will mention the analysis of the motions of the high-velocity stars. The Mt Wilson observers themselves were the first to draw important conclusions concerning high-velocity objects and this work eventually led to the discovery of galactic rotation. The observations of high-velocity stars thus brought about one of the most striking results from the entire radial velocity program. The existence of a few stars with unusually large motions had been established for some time. Campbell, for example, had published in 1901 a list of seven with radial velocities exceeding 76 km/s (93). The discovery by Adams and Kohlschütter that most high-velocity objects in the northern hemisphere were approaching the sun has already been mentioned (87). This was an important clue that resulted in a more general statement by the Carnegie Institution astronomer Benjamin Boss (1880–1970) in a lecture to the American Astronomical Society in 1918 (94). Boss calculated the total space velocities of stars with measured radial velocities, proper motions and parallaxes.* He found that the directions of motions of those stars with total velocities over 75 km/s, were all confined to galactic longitudes between 140° and 340°.† In other words, the velocities' directions were practically excluded from roughly one half of the celestial sphere.

This result was quickly confirmed by Adams and Joy (95) and also by the Swedish astronomer Gustaf Strömberg (1882–1962) (96). Strömberg had joined the Mt Wilson staff in 1917, and the analysis of stellar motions was his principal interest during 29 years at the observatory. His paper in 1924 on stellar space velocities found about a hundred stars moving at more than 100 km/s, all of them into galactic longitude directions between 143° and 334°, implying that these objects had an asymmetric velocity distribution relative to the sun and to the vast majority of other nearby field stars (96). On average, the sun was moving at about 300 km/s relative to the high-velocity stars, and as Strömberg pointed out in a second paper, a possible interpretation is that these stars defined a system which is at rest through which the sun and most other so-called low-velocity stars are moving (97).

In Holland Jan H. Oort (b. 1900) undertook a thorough study of the

* The latter two parameters give the velocity transverse to the line of sight. When combined with the radial velocity, which is in the line of sight, the total velocity can be found.

† Old galactic coordinates.

high-velocity stars and published a catalogue of 233 which were estimated to have space velocities in excess of 62 km/s (98). His results were in most points similar to Strömberg's in establishing the nature of the asymmetry (see however (99) for some disagreements). He went further than Strömberg in interpretation by suggesting that the sun and local low-velocity stars were actually participating in a general galactic rotation. The high-velocity stars were moving more slowly about the Galaxy, and hence being overtaken, giving rise to the asymmetry in the distribution of their directions of motion. Finally the system of globular clusters and field RR Lyrae stars (a type of variable star found in the field as well as in globular clusters) represented a galactic 'subsystem' that does not partake at all in the galactic rotation. These latter thus appear to have the largest velocities relative to the sun.

Oort's suggestion was studied critically by B. Lindblad of the Uppsala Observatory in Sweden. In Lindblad's model the Galaxy was 'divided up into a series of 'subsystems' having rotational symmetry around one and the same axis, with different speeds of rotation at the same distance from this axis and consequently having different degrees of flattening'. The subsystems of globular clusters and of the high-velocity field stars 'show, on account of their low speed of rotation, a strong asymmetrical drift in velocity nearly at right angles to the radius vector of the big system, when the velocities of their members are measured from a star, like our sun...' (100). This was the first concise statement of galactic rotation, and a major triumph for the observers of Doppler shifts, especially those from Mt Wilson, who had provided the initial stimulus.

6.17 J. S. Plaskett at the Dominion Astrophysical Observatory, Victoria Canadian astronomers under the direction of J. S. Plaskett (1865–1941) at the Dominion Astrophysical Observatory at Victoria, British Columbia, using the 72-inch telescope completed in 1918, cooperated with Mt Wilson in the program of measuring radial velocities of Boss proper motion stars. Plaskett is generally regarded as the founding father of Canadian stellar astronomy. He came to Victoria as the inaugural director from the Dominion Observatory in Ottawa where he had also established a stellar spectroscopy program. At Victoria a list of 720 stars from about magnitude 5 to 8 was on their program, the plan being that they should observe objects with even minutes of right ascension, while Mt Wilson would concentrate on the odd

Fig. 6.13 J.S. Plaskett.

minutes. The Victoria cassegrain spectrograph employed one, two or three prisms (101), giving dispersions between 50 and 7 Å/mm at Hγ. By 1921 Plaskett and his collaborators were able to publish results for 537 stars with apparently constant velocities and give data for a further 57 whose velocities were variable or suspected variable. The probable errors quoted show how the precision of the results deteriorated for early-type broad-lined stars. For F to M stars the probable error of the velocity from one plate ranged from ± 0.2 to 2.5 km/s; for sharper-lined B and A stars, between ± 2.5 and ± 10 km/s (102). W.E. Harper (1878–1940) at Victoria also undertook a separate program of velocities for stars with measured trigonometric parallaxes, extending the work of Adams and Kohlschütter at Mt Wilson. Results for 125 stars were published in 1923 (103). Also noteworthy from the early years at Victoria was J.S. Plaskett's special study of the absorption-line O stars. He published radial velocities for 80 of them in 1924, as part of a wider investigation into this spectral type. The probable error for one plate was ± 4 km/s (104).

Fig. 6.14 The 72-inch D.A.O. telescope at Victoria, British Columbia with spectrograph attached, in 1919.

6.18 Frost at Yerkes Observatory The final radial velocity program of the early twentieth century that will be included in this survey is that initiated at the Yerkes Observatory in Wisconsin about 1901 by Edwin B. Frost (1866–1935), for a time with the assistance of W.S. Adams before this latter went to Mt Wilson in 1904. For many people Frost was the great contemporary of Campbell in the world of radial velocities, and these two astronomers between them ensured the supremacy of the United States in the measurement of stellar Doppler shifts from the beginning of the twentieth century. Their lives had a number of interesting

Fig. 6.15 Edwin B. Frost.

though coincidental parallels; for example, both became directors of their respective observatories in their thirty-ninth years, both devoted the major part of their professional careers to measuring radial velocities, and they had available to them the world's two largest refracting telescopes. Both astronomers were eventually afflicted by blindness; in Frost's case his eyesight was always poor, but this developed into total blindness by 1915 resulting in the premature termination of his work at the telescope. However he proved that a sightless astronomer can still undertake research, direct an observatory and edit the *Astrophysical Journal*. The last of these tasks he undertook for over 30 years from 1902.

The Yerkes 40-inch refractor had been completed in 1897 and the following year Hale had invited Keeler to spend a 5-year period away from Allegheny to use the new facilities in Wisconsin. Although Keeler at first accepted, he was immediately called away to take up the Lick directorship, thus leaving the way open for Frost, then a professor at Dartmouth College Observatory, to fill the post instead. Although initially only a temporary appointment, Hale's subsequent departure for Mt Wilson left the Yerkes directorship vacant, and Frost now filled this post in 1905 and held it for the next 28 years. At this time radial velocity work was already part of his background, as he had spent two years in Germany (1890–92) including one with Vogel in Potsdam. Frost's fluent German enabled him to translate Scheiner's book *Spectralanalyse der Gestirne*, which appeared in English in 1904 and considerably enhanced his reputation. Early on at Yerkes he designed a new three-prism spectrograph (Fig. 6.16) which was funded with a $2300 donation in 1899 by Miss C.A. Bruce, who was then giving financial support to several observatories in the United States and overseas (105).

The radial velocities of B stars became Frost's special interest. His paper with Adams on the radial velocities of twenty bright helium stars with spectra 'of the Orion type' (106) showed generally small velocities for these objects relative to the sun, but even smaller ones relative to the local standard of rest, as deduced from Campbell's solar motion. Later the great Dutch astronomer J.C. Kapteyn (1851–1922) collaborated with Frost to determine the solar motion relative to sixty-one distant B stars near the assumed apex or antapex directions (107). Their result for the apex stars (the direction towards which the sun is moving) differed by 10 km/s from those near the antapex, indicating an apparent expansion of the B stars of about 5 km/s away from the sun. This unexpected result could also be attributed to a systematic error of about this amount in all B star velocities, a view Frost was reluctant to accept. Campbell termed the systematic excess of radial velocity the K-term (108), and found it to be present in stars of all spectral types, though it was largest for the early B and M stars. Shifts of non-Doppler origin in line wavelengths, such as those due

Fig. 6.16 The Bruce spectrograph on the Yerkes 40-inch refractor.

Fig. 6.17 Yerkes measuring machine for stellar spectra, about 1904.

to pressure effects or to gravitation, were for a time postulated as the cause of the K-term. This troublesome problem lingered on for many years, and was only solved after standard wavelengths and standard velocity stars had been carefully selected to eliminate a persistent systematic error – largely as a result of an extensive program at Victoria by R.M. Petrie (1906–66) and J.A. Pearce (1893–1988) (109), which in turn had continued the pre-war B star program of Plaskett and Pearce in Canada (110).

Meanwhile the results of a quarter century's work on B stars at Yerkes were collected into a paper by Frost, S.B. Barrett and Otto Struve (1897–1963) in 1926 (111). Radial velocities for 368 stars were given, from 2431 plates mainly at 30Å/mm. Forty-three per cent of these stars were spectroscopic binaries. The relative imprecision of B star velocities is

Fig. 6.18 J.A. Pearce (left) and R.M. Petrie (right).

shown by the quoted probable error of ± 9 km/s from one Yerkes plate.

Frost is also remembered for his promotion of inter-observatory cooperation in radial velocities. He sent a circular letter in 1902 to Belopolsky, Campbell, Deslandres, Gill, Lord, Newall and Vogel soliciting their cooperation in a program to observe twenty bright stars, nearly all of spectral type F, G or K, to serve as radial velocity standards. All these observers expressed support for the program. Frost and Adams presented results for 13 of these stars in 1903, observed at Yerkes (112), and several other observers also contributed, including Belopolsky at Pulkova (113) and Newall at Cambridge (114). V.M. Slipher (1875–1969) also joined this program and published data for ten of Frost's standard stars in 1905 (115) using the 3-prism Brashear spectrograph on the 24-inch Clark refractor at the Lowell Observatory at Flagstaff, Arizona (see (116) for description of instrumentation).

6.19 The International Astronomical Union and radial velocity programs The foundation of the International Astronomical Union (IAU) in Brussels in 1919 naturally resulted in much closer cooperation and awareness in all branches of astronomy. The effects of this were particularly strong in radial velocity research which reached its zenith in the late twenties to thirties. Four persistent themes recurred during the six inter-war General Assemblies of the IAU in the reports from Commission 30 for radial velocities. These were the need for standard wavelengths of selected lines as a function of spectral type, the need for standard velocity stars of different spectral types, the possibility of inter-observatory collaboration to avoid duplication of effort, and the need for a general catalogue of radial velocities that would apply weights and corrections to the determinations from different observatories to give the best mean values for each star. Astronomers could no longer hope to undertake vast Doppler shift programs in isolation from their peers in the way that Campbell's pioneering program for bright stars had been conceived.

6.20 Standard wavelengths and standard stars H.A. Rowland's solar wavelengths, published at the end of the nineteenth century (43), were known to have small errors which were nevertheless very bothersome for the precise measurement of Doppler shifts, considering that 1 km/s corresponds to only 14 mÅ at Hγ. Work at Mt Wilson by Charles E. St John (1857–1935) (117) showed that errors of this order sometimes occurred, and a major revision of the Rowland scale was undertaken over more than a decade by St John and his Mt Wilson colleagues. The result was the 'Revision of Rowland's table of solar spectrum wavelengths' (118). The new work covered the wavelength region from 2975 to 10 219 Å and gave the measured wavelengths to the nearest milli-Ångström unit.

The problems of standard velocity stars and standard rest wavelengths of selected lines in stellar spectra are intertwined. The revised Rowland scale of 1928 enabled accurate Doppler shifts to be measured for F, G and K stars, and the method could readily be checked by observing bodies in the solar system with accurately known orbits. The problems of early-type stars were, however, far more intractable. At the second IAU General Assembly in Cambridge, England in 1925, a subcommittee was established to recommend the choice of standard radial velocity stars. The committee, consisting of Frost, Moore and Spencer Jones, reported 3 years later at Leiden, listing twenty-eight bright stars of type A0 and later, with their best radial velocities taken from the data obtained at thirteen observatories. The Lick velocities from Mt Hamilton and Santiago were chosen to define the standard system to which the results from eleven other observatories were corrected (119). This list provided a means of ensuring self consistency of results from different astronomers using different equipment and reduction techniques, but of course it was unable to overcome systematic errors for early-type stars relative to solar-type ones, when these errors were inherent in the Lick scale. J.A. Pearce extended the list to O and B stars at the fourth General Assembly (Cambridge, Massachusetts, 1932) (120) and the list of F to M standard stars was extended to fainter objects at the 1935 Paris meeting (121).

Only after the Second World War, thanks to an extensive Dominion Astrophysical Observatory program by the Scottish-born Canadian astronomer Robert M. Petrie, were the early-type stars successfully tied into the solar types. The method was the empirical one of using visual binaries with negligible orbital motion or stars in moving clusters with parallel space motions, provided such systems contain both solar and earlier spectral type stars. Petrie's first paper established wavelength standards for F4 to F8 spectral types (122), followed by a two-step calibration, first to tie A stars to the solar scale (123), and second to bring the B stars into line with those of type A (124). On the basis of these velocities, line lists with standard wavelengths as a function of spectral type were presented which would bring the velocities for all types onto a uniform scale. For the B stars Petrie's list comprised thirty-five standard stars with a final list of only eighteen spectral features from the K line to Hβ that could be relied on for error-free results. The residuals for some lines however depended on luminosity and dispersion, while other lines were only suitable for earlier or later B stars. Average quality results for B stars gave probable errors per 30 Å/mm plate of about ± 3.7 km/s. Petrie concluded that the systematic errors or residuals 'are not significantly different from zero and the velocity system is in agreement with that for the later spectral types' (124). The mysterious K-term that had complicated the analysis of the motions of B stars for over four decades was thus finally conquered and eliminated. In a sense the complexity of the Victoria system was the hallmark of a highly refined system developed over many years. The basic principle of the Doppler effect appears deceptively simple; its actual implementation to achieve consistent and error-free results for any spectral type was a matter of painstaking care in observational and data-reduction procedures.

6.21 Radial velocity catalogues

The popularity of radial velocity work led to a continuing growth in the number of stars for which this type of observation was available. By 1920 over 2000 stars had had radial velocity determinations undertaken at fifteen different observatories. The number was growing at around 250 new stars every year with about ten active observatories at any one time (figures which were maintained, or at times exceeded, up to the Second World War). The need for a general catalogue of stellar radial velocities to bring together this fast-growing body of data was first recognised by the Dutch astronomer J.G.E.G. Voûte (1879–1963). He spent 6 years at the Cape Observatory from 1913 and was able to compile a catalogue of 2071 stars (including nebulae and clusters) from the resources of that institution's library (125). In 1928, when Voûte was director of the Bosscha Observatory in Indonesia, a second edition appeared with 4032 objects (126). K stars (1029 of them) accounted for more than a quarter of the catalogue. Of the variable velocity stars, 317 were listed as spectroscopic binaries while 21 were pulsating stars (Cepheids). These compilations undoubtedly fulfilled a need, but they made no attempt to process the mean velocities listed, by applying observatory corrections or weights.

Meanwhile there was a general consensus in favour of producing a general catalogue that assigned corrections and weights so as to combine the results from all observatories. The project was first discussed by Campbell at the 1925 IAU General Assembly in Cambridge, England. At the 1928 meeting in Leiden the subcommittee of Frost, Moore and Spencer Jones had devised just such a correction and weighting scheme for the standard radial velocity stars. About the end of 1928 Moore at Lick Observatory agreed to undertake a general catalogue, which would incorporate the recently published Lick results for bright stars (68), and adopt the system of corrections and weights (with minor refinements) already used for standard stars. The weights depended mainly on the number of plates, but also on spectral type and dispersion. Cambridge and Columbus results were given half weight, and the pioneer Potsdam observations by Vogel and Scheiner were excluded.

Moore's important catalogue was published by the Lick Observatory 4 years after the work of compilation had commenced, under the title *A general catalogue of the radial velocities of stars, nebulae and clusters* (127). It contained data for 6739 stars which had been published up to the end of 1931, and 6354 of these had mean radial velocities assigned to them. 1320 stars had variable velocities. There were nearly 2000 K stars in the catalogue. The compilation brought together results from nineteen observatories, and to each was assigned a systematic correction depending on spectral type to bring its values to the Lick system. The magnitude of these corrections was mainly less than 2 km/s, except for F.C.P. Henroteau's (1889–1951) results from the Dominion Observatory in Ottawa, obtained during 1920–22, for which a + 9.4 km/s correction was applied (128). The origin of this large error remains unknown. It has even been suggested that Henroteau altered the spectrograph slit between photographing the comparison and stellar spectra, though such a glaring blunder hardly seems plausible (129). Moore's great compendium was not only a valuable reference work which brought together the diverse results of a vast international program inside one cover, it also marked one of the earliest occasions for which extensive astrophysical data from different observers have been processed to a uniform scale.

6.22 Radial velocity programs in the 1930s: David Dunlap and McDonald Observatories By 1948 the number of stars with radial velocities stood at about 12 000, nearly double that in Moore's general catalogue of 1932. Some 3000 stars had been added since 1938, showing that the war years had little effect on this work, which was mainly carried out in North America. However, the 3 years 1935–8 probably represented the peak in productivity; 2500 stars had new radial velocities determined in that period, some as faint as twelfth photographic magnitude. The most active observatories from the mid-1930s were Mt Wilson, Lick, Dominion Astrophysical and Yerkes. In addition, the David Dunlap Observatory of the University of Toronto was opened in 1935. R.K. Young (1886–1977) was the first director and initially he used a one-prism cassegrain Hilger spectrograph on the new Grubb Parsons 74-inch telescope. One of the first radial velocity programs was the observation of 874 stars in one of Kapteyn's selected areas by Young in the years 1935–9. The photographic magnitude limit was 7.59 (130). Another program encompassed all northern stars brighter than $m_{pg} = 8.0$ not already observed for radial velocity.

The other major institution that embarked on radial velocity work in the 1930s was the McDonald Observatory in Texas, with its new 82-inch Warner and Swasey reflector, operated in conjunction with the Yerkes astronomers from Chicago. Otto Struve was the inaugural director, and his wide-ranging interests in early-type stars were pursued at McDonald from the outset. Radial velocity results, especially by Struve, Carl Seyfert (1911–60) and D.M. Popper (b. 1913) were published during the war years, though this work was never a dominant feature of the very broad interests in stellar spectroscopy of the McDonald and Yerkes astronomers. The 82-inch at this time was equipped with a 2-prism cassegrain Brashear spectrograph with high ultraviolet transmission optics. The dispersion was 40 Å/mm at the K line and 20 Å/mm at 3250 Å, the ultraviolet limit. A prism coudé spectrograph for high dispersion work (3 Å/mm at Hγ) was also available. The radial velocity interests of the McDonald observers were diverse, but the observing programs were generally smaller and the goals more selective than the large general programs at either Lick or Mt

Wilson. For example, Seyfert and Popper observed 118 faint (m_{pg} from 9 to 11) B stars to study galactic rotation (131), Popper studied 158 high proper motion stars (132) and an investigation of the radial velocities of Pleiades cluster members was undertaken by B. Smith and Struve (133).

6.23 Radial velocity work in the Soviet Union and in the southern hemisphere, 1930–50

If progress in establishing two major new observatories on the North American continent had been made just prior to the war, then elsewhere the story was different. Both Pulkova and Simeis Observatories in the Soviet Union (the latter had been equipped with a 40-inch Grubb reflector) were destroyed by the Germans and their radial velocity programs terminated. After the war the Simeis Observatory was reconstituted at the newly built Crimean Astrophysical Observatory and a 50-inch reflector was installed there in 1950 (Fig. 6.19), which enabled V.A. Albitzky (1891–1952) and G.A. Schajn (1892–1956) to reembark on radial velocity observations. Moreover, stellar radial velocity work since 1929 was almost entirely confined to northern hemisphere observatories. The Lick Observatory southern station in Santiago was sold to the Catholic University of Chile in 1929 (134), and the Cape Observatory terminated its radial velocity program in 1926 to make

Fig. 6.19 The 1.2 m Grubb reflector at the Simeis Observatory in the Crimea.

time available on the 24-inch refractor for trigonometric parallax work (85). In the south, only the newly established Boyden Station of Harvard Observatory, moved by Shapley from Peru to Mazelspoort near Bloemfontein, South Africa in 1927, was undertaking stellar spectroscopy of any kind.

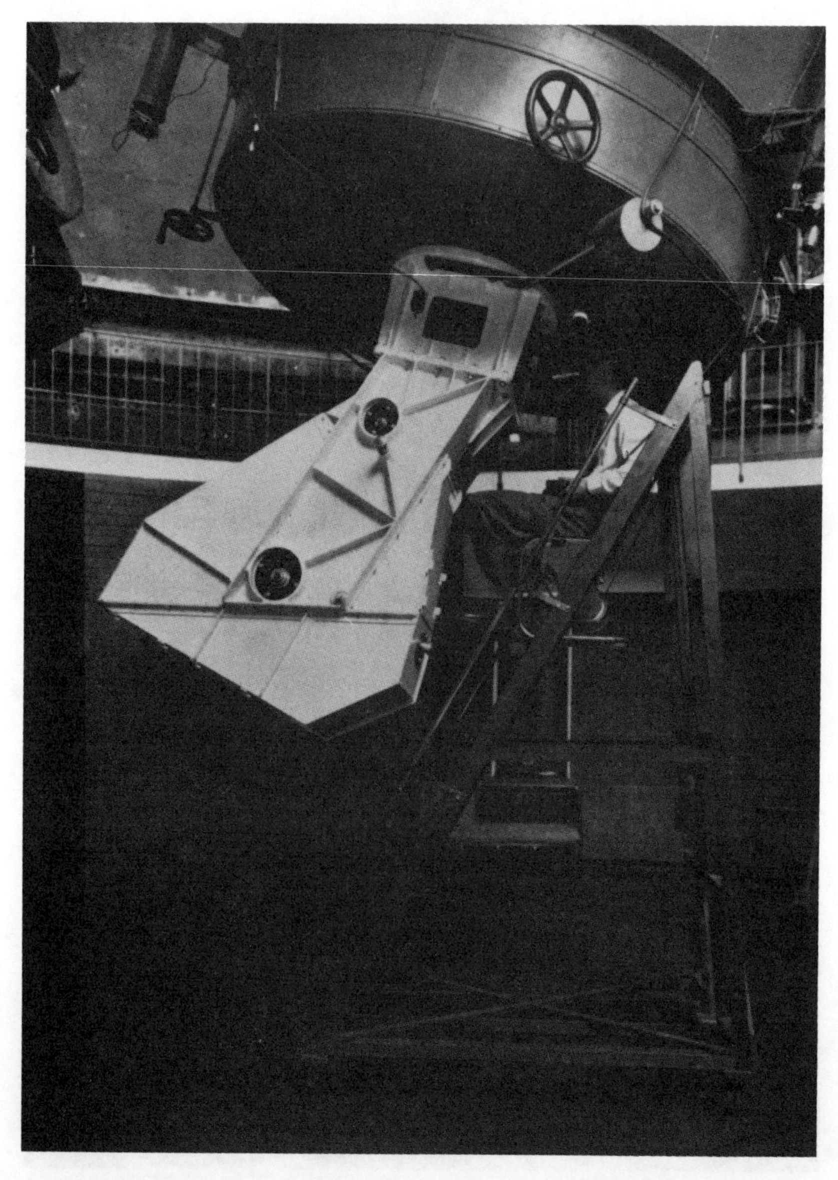

Fig. 6.20 The cassegrain spectrograph attached to the 74-inch Radcliffe reflector at Pretoria.

The result was a long fallow period in which the southern sky was comparatively neglected in radial velocity work from 1929 until the Radcliffe Observatory's 74-inch Grubb Parsons reflector became operational at Pretoria in 1948. This last project, which involved transferring the site of the observatory from Oxford and the ordering of a new telescope, which was installed in 1938, would undoubtedly have been in production much earlier had it not been for the long war-time delay before the mirror was installed (135). Once the 2-prism cassegrain spectrograph had been commissioned in 1951, A.D. Thackeray (1910–78) and his colleagues initiated a program to measure radial velocities of 147 southern B stars (136) from which they were able to demonstrate the effects of galactic rotation by plotting radial velocity against galactic longitude. The B star program continued with observations of 130 fainter, more distant objects (137). More than two decades of southern neglect had ended.

6.24 The Wilson General Catalogue

The immediate post-war years saw the completion of several massive programs at North American observatories, especially those at Mt Wilson on nearly 2500 fainter F to M stars by Wilson and Joy (92), at Lick on F to M stars with apparent photographic magnitudes in the interval 8.5 to 8.6 (138), at David Dunlap on the program in the Kapteyn selected areas (130); another big program at Victoria by Petrie and Pearce on fainter stars of types B5 or earlier was partially complete (139). As a result of this large body of new results reaching the journals after the war, the need for a new version of Moore's general catalogue was expressed at the Zürich IAU General Assembly in 1948. Ralph E. Wilson of Mt Wilson Observatory agreed to take on the task of compilation. The Wilson general catalogue of stellar radial velocities (140) was to become the standard reference from the time of its publication in 1953. A radial velocity conference was held in Pasadena in June 1951 to mark the completion of Wilson's work on the catalogue, and this also coincided with his retirement from the Mt Wilson staff. Wilson himself introduced his future catalogue at this meeting (141). Data for 15 107 stars were to be included in the work. He remarked that:

> Of the twenty-four observatories which have at one time or another made determinations of radial velocities, five only are continuing this line of work, and but two of these are at present engaged on extensive programs. It seems likely, therefore, that a compilation of the data now available should be useful for a good many years.

Fig. 6.21 Ralph E. Wilson.

Ten of the contributing observatories provided 20 637 determinations, and these accounted for nearly 99 per cent of all stars catalogued. The table Wilson gave showing the output of the ten leaders is of interest, especially

Table 6.1. *Source of radial velocities in Wilson's general catalogue*

Observatory	Number of stars	Observatory	Number of stars
Mt Wilson	7434	Simeis	841
Lick	4797	Cape	519
DAO, Victoria	3272	Bonn	253
David Dunlap	1791	Ottawa	174
Yerkes–McDonald	1448	Michigan	108

as Mt Wilson now appears in first place ahead of Lick, mainly due to the productive output of Wilson himself and of A.H. Joy. North American observatories accounted for about 92 per cent of the total, and the three on the Pacific coast for three-quarters of the world output. As with the earlier Moore catalogue, all results were reduced to the Lick system and Moore's system of weights was assigned in determining the mean. Wilson gave each star's mean radial velocity a quality from 'a' to 'e' depending on the probable error of the mean. About 10 per cent of the stars had quality 'a' (p.e. < 1.0 km/s) and 46 per cent were of quality 'b' ($1.0 <$ p.e. < 2.0 km/s). Quality 'e' (1.7 per cent) had probable errors greater than 10 km/s and were considered valueless.

6.25 Changing trends in radial velocity research from the 1950s R.E. Wilson himself was aware that, even as his *chef d'oeuvre* was in press, the popularity of very large radial velocity programs was declining, as more challenging (or perhaps less painstaking?) problems presented themselves in stellar astrophysics. Struve and Zebergs echoed the same thoughts in 1962 when they wrote: 'Now, much of the glamour of such work has faded; indeed, many astronomers now consider it a necessary but tiresome occupation' (142). Certainly the big programs of the major observatories with the emphasis on observing large numbers of stars, are of the past. More observatories now undertake radial velocity work as an incidental but routine part of other programs, than before. In 1972 H.A. Abt and E.S. Biggs compiled a *Bibliography of Stellar Radial Velocities* (143). This was not a catalogue of processed data as Moore's and Wilson's had been, but simply a list of published material. About 44 000 references are given to 25 000 stars. In two decades the number of new stars with measured radial velocities had thus grown by 500 a year, a greater rate than at any other time with the exception of a few years in the mid-1930s. However, as Abt and Biggs point out, the annual production of all astronomical publications increased about tenfold between the inter-war years and 1972. But radial velocity output increased in this interval by no more than twofold, a marked decline in its relative importance compared with other lines of research.

It is also interesting that, after the Second World War, the domination of the United States in radial velocity work was declining in favour of the old countries of the Commonwealth: Canada, South Africa and Australia. Petrie's O and B star program at Victoria continued throughout the 1950s (139) while J.F. Heard (1907–76) at David Dunlap was completing a radial velocity program for 1042 G to M stars (144). Observers at the Cape re-entered radial velocity observations after a break of a quarter century and led by D.S. Evans (b. 1916) they produced lists of *Fundamental data for southern stars* using the Radcliffe reflector (145), while Radcliffe's own concurrent but separate program under Thackeray has already been mentioned (136, 137). Meanwhile a second Grubb Parsons 74-inch telescope, erected at Mt Stromlo Observatory in Canberra, Australia, became operational in 1955, and contributed to a new era for stellar spectroscopy on that continent. The Australian radial velocity program had been initiated 2 years earlier using a 3-prism cassegrain spectrograph (dispersion 35 Å/mm at Hγ) on the 30-inch Reynolds reflector at Mt Stromlo by H. Gollnow (b. 1911) and W. Buscombe (b. 1918). The work was then continued on the Newtonian and coudé spectrographs of the 74-inch. The first two radial velocity projects were on members of the Scorpio-

Centaurus association (146), and on fundamental proper motion stars in the N30 or FK3 catalogues (147). By 1964 it is fair to remark that the principal radial velocity observatories were Dominion Astrophysical, the Cape, Radcliffe and Mt Stromlo, although the Commission 30 report to the IAU General Assembly for that year (148) noted eleven observatories to some degree active in the field.

6.26 Empirical confirmation of Doppler and gravitational shifts Two fundamental investigations during the 1950s had an important bearing on the theory of wavelength shifts of spectral lines. The first concerned the proportionality of $\Delta\lambda$ with λ itself. The use of the neutral hydrogen 21 cm emission line to determine radial velocities of galactic clouds by C.A. Muller and J.H. Oort (149) provided a technique for testing the proportionality over a very long wavelength baseline. The test was first applied by A.E. Lilley (b. 1928) and E.F. McClain in the United States when they observed the radio galaxy Cygnus A (150). The 21 cm recessional velocity was 16 700 km/s, in close agreement with the optical value. The significance of these results is that $\Delta\lambda/\lambda$ varies by no more than 3×10^{-9} for every 1000 Å of baseline (151).

The other development concerned the gravitational redshift, predicted by Einstein's general theory of relativity, which should also increase the measured wavelengths of spectral lines if the photons are emitted from massive stars. This mechanism had for some time been proposed as a possible explanation of the K-term found by Frost and Campbell for B stars. Petrie's careful calibration of the wavelengths of selected lines in B stars (124) effectively reduced this anomaly to near zero, so gravitational redshifts no longer had to be conjectured, at least for normal dwarf and giant stars. The same was not the case for white dwarfs, which have far higher gravitational fields at their surfaces. Daniel Popper used the 60 and 100-inch telescopes at Mt Wilson to measure a gravitational redshift of $\Delta\lambda/\lambda = 7 \times 10^{-5}$ for the white dwarf 40 Eri B, whose radial velocity was known from 40 Eri A, the bright K dwarf primary of the system (152). This result was in close agreement with the predictions of Einstein's theory.

6.27 Objective prism radial velocities The 1950s finally saw come to fruition a method of measuring Doppler shifts which, in concept, is nearly as old as stellar radial velocity work itself. The idea of using an objective prism to determine radial velocities dates back to 1887 when Edward C. Pickering proposed passing starlight through an absorbing cell of gas held in the light path of an objective prism telescope (153). He stated that:

> experiments are in progress with hyponitric fumes and other substances... The stellar spectra will then be traversed by lines resulting from the absorption of the media thus interposed, and, after their wave lengths are once determined, they serve as a precise standard to which the stellar lines may be referred... If, then, satisfactory results are obtained in the preceding investigation, the motion of the stars can probably be determined with a high degree of precision.

The advantages of objective prisms over slit spectrographs to measure radial velocities were clear enough. They could give high light throughput, thus allowing measurements on fainter stars, and also permit simultaneous exposures of several spectra. Unfortunately, Pickering's early experiments mark the start of 60 years of frustrating attempts to follow up a number of proposals to use the objective prism method. The literature on the subject is therefore extensive (see Table 6.2 which summarises some of the important contributions).

Table 6.2 *Radial velocities using objective prisms: a bibliography*

Year	Author(s)	Ref.	Remarks
1887	Pickering, E.C.	153	Use of hyponitric fumes to give reference absorption lines
1891	Pickering, E.C.	154	First suggestion of length method; also proposes stellar images as reference marks
1895	Orbinsky, A.	155	Proposes differential method
1895	Frost, E.B.	156	Criticises Orbinsky's paper
1896	Deslandres, H.	157	Exposure of comparison spectrum using slit and collimator
1896	Maunder, E.W.	158	Follows up Pickering's idea of using stellar images as reference marks
1896	Keeler, J.E.	159	Criticises papers by Deslandres and by Maunder
1896	Hale, G.E. and Wadsworth, F.L.O.	160	First discussion of reversion method using objective prism
1896	Pickering, E.C.	161	Reversion method described and first tests reported
1906	Comstock, G.C.	162	Reversion method using split direct vision prism
1906	de Lisle Stewart	163	Reversion method using two prisms; similar to Comstock
1906	Pickering, E.C.	164	
1910	Wood, R.W.	165	Proposes use of $NdCl_3$ solution as absorbing medium giving line at 4273Å

Table 6.2 (*Contd*).

Year	Author(s)	Ref.	Remarks
1910	Pickering, E.C.	166	First tests reported using $NdCl_3$ cells
1913	Schwarzschild, K.	167	Observes α CrB using $NdCl_3$ solution
1914	Plaskett, J.S.	168	Length method tested; no useful results achieved
1914	Hamy, M.	169	Further proposal using direct stellar images as reference marks
1918	Graham, T.S.H.	170	Harvard trials using $NdCl_3$
1919	Schwarzschild, K.	171	Tests of reversion method, both differential and using $NdCl_3$
1923	Wilson, H.C.	172	Harvard trials using $NdCl_3$ for Praesepe stars
1931	Millman, P.M.	173	Further Harvard trials using $NdCl_3$
1931	Millman, P.M.	174	Review of results in literature to date
1937	Bok, B.J. and McCuskey, S.W.	175	Harvard trials with $NdCl_3$ continue
1937	Cherry, B.	176	Southern hemisphere radial velocities using $NdCl_3$
1938	McCuskey, S.W.	177	
1938	Edwards, D.L. and Barber, D.R.	178	Use of interstellar lines as reference for nova spectrum
1947	Fehrenbach, C.	179	Criticises use of $NdCl_3$ in absorption method
1947	Fehrenbach, C.	180	Reviews and develops reversion method
1948	Treanor, P.J.	181	Reviews available method of objective prism radial velocities
1950	Pearce, J.A.	182	Pearce as Comm. 30 president reports abandonment of Bok's efforts to obtain objective prism radial velocities
1954	Schalén, C.	183	Experiments with reversion techniques at Uppsala Observatory
1955	Panaitov, L.A.	184	Experiments with reversion technique at Pulkova Observatory
1955	Fehrenbach, C.	185	Review of Fehrenbach techniques and results using reversion method
1966	Fehrenbach, C.	186	Review of reversion and other methods; summary of programs at Haute Provence and in South Africa
1967	Fehrenbach, C.	187	
1974	Fehrenbach, C. and Duflot, M.	188	Radial velocities for 1832 stars towards LMC

The basic problem of using the objective prism for radial velocity work is to have a reference position on the plate against which Doppler-shifted stellar lines can be measured. Four different solutions have been proposed. Firstly there is the absorption line method in which lines of an absorbing medium are superposed on the spectrum. The medium could be a gas or liquid solution held in the light path, or the terrestrial atmosphere or even the interstellar medium. Secondly there is the spectrum length technique, which seeks to exploit the stretching of the spectrum of a receding star; that is, the differential shift between two lines is measured. Thirdly come attempts to either record laboratory comparison spectra alongside stellar spectra on objective prism plates, or to have the undispersed stellar images themselves as reference marks from which to measure line shifts in stellar spectra. The last solution to be proposed was the reversion method, in which two spectra of opposing dispersions are recorded sequentially on the same plate by rotating the prism between exposures, and the distance between the same line in each spectrum is measured. It is remarkable that the first three methods all originated with Pickering, who was also the first to experiment with the reversion technique using an objective prism. If he could have applied the objective prism to both stellar classification and radial velocity work, it would have been a considerable triumph. In this goal he was not successful, but the fact that from 1890 to 1920 such well-known names as Frost, Keeler, Deslandres, Maunder, Hale, Schwarzschild and Plaskett all published papers on objective prism radial velocities shows that the method was regarded as potentially rewarding.

When R.W. Wood proposed in 1910 using a weak solution of neodymium chloride ($NdCl_3$) as the absorbing medium, the absorption method received considerable attention. Pickering at Harvard (166) and Schwarzschild at Potsdam (167) both experimented with neodymium chloride, which gives a fairly sharp absorption line near 4273 Å. Pickering claimed to produce results with probable errors of about ± 10 km/s. A long series of tests followed at Harvard between the wars. The last of these investigations was from Bart Bok (1906–83) and Sidney McCuskey (1907–79) using 95 Å/mm spectra on the 16-inch Metcalf telescope (175). They presented results for 200 stars with probable errors in most cases between 10 and 13 km/s as part of McCuskey's doctoral dissertation under Bok's supervision. Only spectral types earlier than G0 could be measured, as the neodymium line became blended with a stellar iron line for cooler stars. The truth is that objective prism radial velocities using neodymium chloride were never very successful. Large systematic errors also plagued the results, and when Bok abandoned the Harvard program after about 2 years of effort, this also marked the end of nearly three decades of tests at Harvard with neodymium solutions and the closing of an interesting but frustrating chapter in radial velocity history. Indeed Fehrenbach later showed that the neodymium method was never likely to give reliable results because of changes in the apparent wavelength of the blue absorption feature (179).

Although Pickering was the first to suggest the length method of obtaining Doppler shifts, the technique became associated with the name of Artémie Orbinsky at the University of Odessa (155). Clearly the differential shift between two lines, giving a stretching or contraction of the spectrum length, is a second order effect, and will thus be much harder to measure with precision, as pointed out by Frost (156). J. S. Plaskett experimented with the method while at the Dominion Observatory in Ottawa, but he produced no useful results (168). Nor have proposals to include a laboratory comparison spectrum on objective prism plates ever come to fruition. Deslandres had suggested a slit and collimator in front of a small part of the prism for this purpose (157), but Keeler correctly pointed out that the low resolution of the comparison that would result if only a small area of prism were employed would vitiate this idea (159).

Soon after Pickering's suggestion of exposing undispersed stellar images on objective prism spectrograms to act as reference points for measuring Doppler shifts (154), Maunder at Greenwich followed up the idea. In one version of Maunder's proposal he even envisaged two astrographs of which only one had an objective prism, and on a single mount so as to feed the starlight to a common focal plane (158). Spectra and direct stellar images were thus to be superimposed on one plate, the latter being the reference marks from which to measure Doppler displacements in the spectra. This concept was too unwieldy to be put into practice, though Maunder's arrangement bears a striking resemblance to that used visually by Fraunhofer (189), in which two rigidly mounted telescopes, one with an objective prism, were used to measure the deviations and hence wavelengths for absorption lines in bright stars.

By far the most successful objective prism technique is the reversion method. The idea was first proposed by G.E. Hale and F.L.O. Wadsworth in 1896 (160) and independently by Pickering later in the same year (161). The method is really a photographic adaptation of Zöllner's reversion spectroscope of 1869 (190). Two stellar spectra are recorded side by side after rotating the prism 180° between exposures. The displacements between a given line in the two spectra are thus twice those of a slit spectrograph, and the method as first applied by Pickering (161) was planned to work differentially using a standard velocity star. K. Schwarzschild used the reversion method in 1913 (171) and obtained results with probable errors of about ± 7km/s from one plate (see Millman's analysis of Schwarzschild's data (174)), which is better than other objective prism radial velocities. After the Second War the reversion method has been pioneered by C. Fehrenbach (b.1914) at the Observatoire de Haute-Provence in France (180). Fehrenbach used a special direct-vision compound prism to eliminate field distortion. His method is otherwise essentially that of Schwarzschild with two reversed spectra side by side on the same plate.

Fehrenbach has reviewed his method at regular intervals. In 1955 he claimed probable errors for late-type stars of between 2.5 and 6.8 km/s. In the 1960s the instruments used were 40cm objective prisms both at Haute Provence and in South Africa, until the latter telescope was installed in Chile in 1968. With a dispersion of only 119 Å/mm at Hγ, Fehrenbach could reach stars as faint as $m_{pg} = 11.5$. One of the most impressive results from this program, which has been conducted with Mme Duflot from the Observatoire de Marseille, is extensive lists of radial velocities of many stars in the direction of the Large Magellanic Cloud (188). The high velocities of Cloud members allow them to be readily distinguished from galactic foreground objects. Without doubt, Fehrenbach and his colleagues have demonstrated the versatility of objective prisms in obtaining large numbers of faint star radial velocities where high precision is unnecessary.

6.28 Photoelectric radial velocities In spite of advances such as the objective prism technique, the photographic determination of velocities is nevertheless still exceedingly laborious work. As well as numerous possible sources of systematic error, the low quantum efficiency of the photographic plate as a detector is one factor preventing the precision inherent in the incoming photons from being fully utilised. A revolutionary photoelectric technique was first proposed by the English astronomer Peter Fellgett (b.1922) in 1953 (191), with the aim of using the relatively high quantum efficiency of photoelectric tubes, as well as simultaneously measuring the Doppler shift of many more lines than normally considered practicable on spectrographic plates. The first experiments were undertaken by Horace Babcock (b.1912) at Mt Wilson and Palomar Observatories about 1955 (192). Babcock's instrument was primarily conceived as a stellar magnetometer to measure the Zeeman splitting of spectral lines arising from a stellar magnetic field. However, its application to radial velocity spectrometry was explicitly stated. Not until 1967 did Roger Griffin (b.1935) at the University of Cambridge Observatories have a successful instrument in operation for determining radial velocities photoelectrically (193). Griffin has been the main pioneer of photoelectric radial velocities since that time.

The principle of the Griffin instrument is to focus a 145 mm length of the blue stellar spectrum (4369 to 4827 Å) on to a mask at the coudé focus of the Cambridge 90 cm reflector. The mask contains 240 fine slots which accurately match the position of absorption lines in the K2 giant star, Arcturus, and is movable along its length by an accurate micrometer screw.

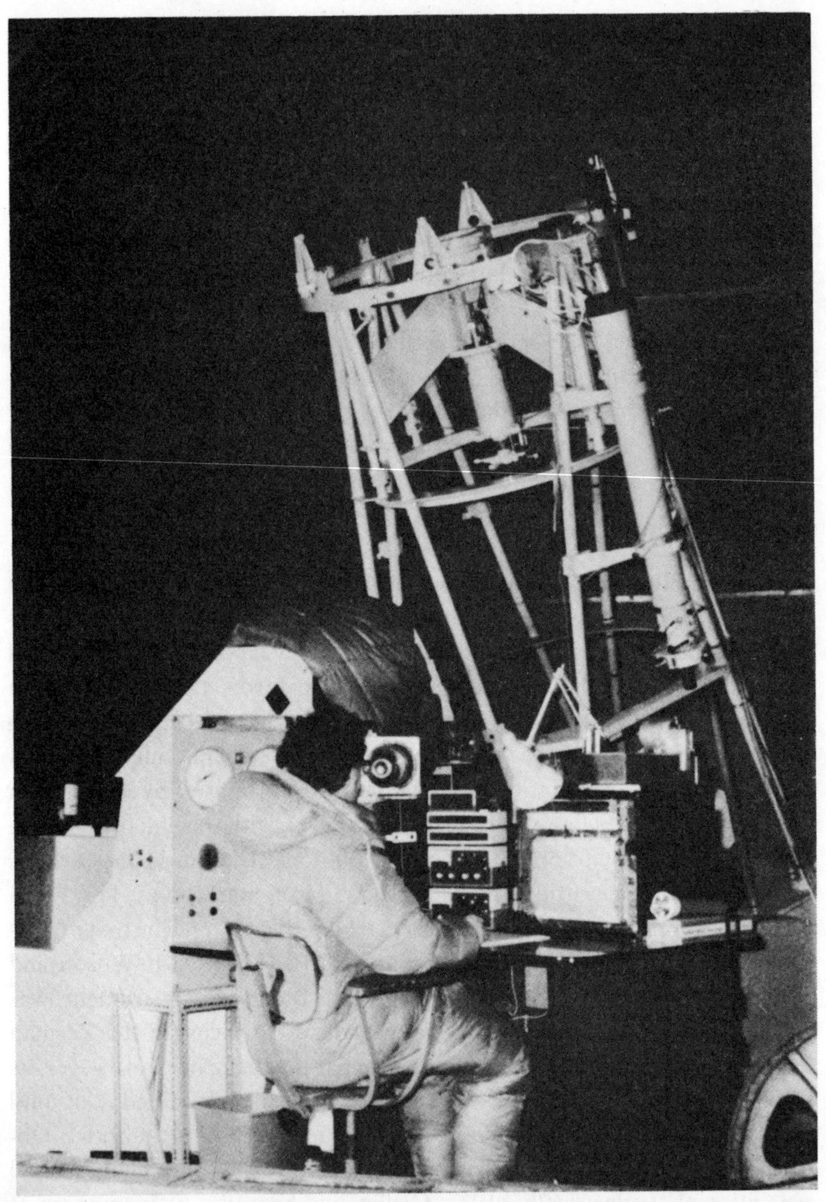

Fig. 6.22 R.F. Griffin using the radial velocity spectrometer at the Cambridge Observatories.

When precisely in register with a Doppler-shifted spectrum of a late-type star, whose spectrum therefore resembles Arcturus, the amount of starlight passing through the mask slots will be a minimum. A large 150 mm aperture f/1 Fabry lens collects the light transmitted by the mask and delivers it to a photomultiplier tube. The position of minimum transmission during the mask's scan, when the stellar absorption lines are focussed onto their respective mask slots, is a measure of the differential radial velocity between a program star and Arcturus. The device necessarily operates differentially, but with the advantage that no wavelength standards are required for spectral lines. The initial results gave standard deviations of about $\pm$ 1km/s from one photoelectric observation at fifth visual magnitude, and about twice this error at the limiting magnitude of nine (193). An inherent precision of 0.1km/s was from the outset believed possible using the mask principle. Even near ninth magnitude, six to eight stars per hour could be observed, a huge speed increase over photographic methods.

By 1969 Griffin had reduced the error for a single observation for a seventh magnitude K star to only $\pm$ 0.64 km/s (194), smaller than the error bars for nearly all seventh magnitude K stars in the Wilson general catalogue. The spectrometer has been used to obtain high precision radial velocities for 528 faint late-type field stars (195) and also for an extensive program on spectroscopic binary orbits.

The remarkable power of these new techniques produced a complete turn-around in interest in radial velocity research work in the 1970s. At least seven other observatories built radial velocity spectrometers in this decade (see (196) and (197) for references). After languishing for three decades since the war years, radial velocity determinations are once again becoming a more popular activity. For over seven decades the basic techniques devised by Vogel and Campbell with slit spectrographs were never fundamentally altered, only perfected; in this sense, the new methods using either photoelectric or solid state detectors represent a radical departure.

6.29 Stellar rotation

So far we have outlined the development of radial velocity measurements on single stars. However, the Doppler effect has other important applications in stellar spectroscopy: the study of stellar rotation, of spectroscopic binary stars, of stellar pulsation, of stellar mass loss phenomena in extended atmospheres or shells, and in the study of turbulence in stellar atmospheres. We conclude this review with an account of the development of ideas concerning stellar rotation.

The history of stellar rotation theory applied to stellar spectra began in March 1877 when Captain (later Sir) William Abney, the well-known English pioneer in scientific photography, published a note with the Royal Astronomical Society (198) on the expected effect stellar rotation could have on spectral lines. He concluded:

Fig. 6.23 Sir William Abney.

> There would be a total broadening of the line consisting of a sort of double penumbra and a black nucleus. It seems that no absorption line can be as sharp and distinct as it is when seen in the solar spectrum. More than this, rotation might account for the disappearance of some of the finer lines of the spectrum.

Abney's hypothesis was based purely on a consideration of the opposing Doppler shifts of a spectral line in the light from receding and advancing stellar limbs. Of course in stellar sources only the integrated light from the whole disc of a star is observed, thus giving shallow broad lines in the spectra of fast-rotating stars.

H.C. Vogel was by this time in Potsdam, and he immediately denounced Abney's paper, attacking it on several points simultaneously (199). First, a fast-rotating star must have all the lines broadened. Stars with broad lines were especially those of Secchi type I, but here only the hydrogen lines were broad and not those due to metals. Contrary to Abney's proposal, spectrum photography would not help resolve the question, as apparent line widths in stellar spectrograms depend on exposure time, plate sensitivity and the development process. Stars such as

Vega and Altair indeed have broad Hβ lines, which would correspond to equatorial speeds of 330 km/s and 185 km/s respectively, which seem totally implausible when compared with the sun's 2 km/s. Finally Vogel believed it doubtful that fast-rotating stars would even give measurable line broadening, as most of the light comes from the central part of the apparent stellar disc where the Doppler shifts are small, whereas the fast-moving edges contribute relatively little.

Twenty-one years later Vogel had good reason to regret his premature negation of Abney's ideas. In the spectrum of α Aquilae (Altair) he now found all the lines to be broad, including the weaker metallic ones. He had no choice but to retract his earlier statements. He wrote: 'I should like to cancel the above statement [referring to his 1877 paper] as applied to α Aquilae. Thus a field is opened up for the explanation of the broadening of lines by rotation, and I consider it quite probable that it applies for the stars α Aquilae, β and δ Leonis, and perhaps also for β Cassiopeiae' (200).

6.30 Rotation in binary stars

The next major advance came in 1909. Frank Schlesinger, who was then director of the Allegheny Observatory in Pittsburg, had observed a series of sixty spectra of the eclipsing Algol-type binary δ Librae. At primary eclipse, when the fainter star (which never contributes significantly to the spectrum) passes in front of the hotter more luminous object, the radial velocity curve for the primary showed anomalously large residuals of nearly 3 km/s from the smooth curve. The eclipses occurred every 56 hours as the two stars move about each other in their orbit. At the same time, Schlesinger reasoned, they would be rotating in such a way that each star maintains the same face towards the other. Schlesinger was able to interpret the residuals as proof of rotation of the primary:

> In general we obtain light from the whole disk and the observed velocity is equal to that of the center of the star. Just before and just after light minimum, however, this is not the case; before minimum the bright star is moving away from us and part of its disk is hidden by the dark star. The part that remains visible has on the whole an additional motion away from us on account of rotation; the observed velocity will therefore be greater than the orbital. On the other hand just after minimum the circumstances are reversed so that the observed velocity is less than the orbital (201).

Later Schlesinger found the same effect in another eclipsing binary, λ Tauri, and the existence of stellar rotation was, at least for certain binary stars, firmly established (202).

Meanwhile other binary stars were studied, and detailed radial velocity curve analyses by Richard A. Rossiter (1886–1977) for β Lyrae and by Dean B. McLaughlin (1901–65) for Algol were undertaken in 1924 at the University of Michigan (203, 204). Both stars showed larger rotational effects than δ Librae. For Algol McLaughlin predicted from the orbital data a $v_e \sin i$ value (i.e., the equatorial speed of rotation times a projection factor for the inclination of the rotation axis to the line of sight) of 27.3 km/s, though admittedly the observations on line broadening suggested about twice this value.

Walter Adams and Alfred Joy at Mt Wilson also studied the short-period (8 hour) contact binary star W Urase Majoris in 1919 and stated 'the unusual character of spectral lines is due partly to the rapid change in velocity during even our shortest exposures but mainly to the rotational effect in each star, which may cause a difference in velocity in the line of sight of as much as 240 km/sec between the two limbs of the star' (205). Then in 1929 the earlier observations in the literature on rotation in binaries were carefully analysed by G.A. Shajn at the Simeis Observatory in the Crimea, and Otto Struve at Yerkes Observatory, Wisconsin, in a collaborative program. They were able to model quantitatively the rotational line broadening for several stars and show that displaced asymmetric lines should occur during a partial eclipse of the primary (206).

6.31 Rotation in single stars and the correlation with spectral type For single stars, however, progress in the analysis of rotation was not so rapid. Struve in his PhD thesis of 1923 states: 'The wide lines in stellar spectra, chiefly of type A, could possibly be explained by rotation' (207), but not until the early 1930s did firmer confirmation come of the earlier conclusions of Abney and Vogel. Struve, with his Yerkes colleagues Christian T. Elvey (1899–1972) and Miss Christine Westgate, tackled the problem in the years 1930–4. First Elvey calculated profiles of the ionised magnesium 4481 Å line for fifty-nine stars of type O, B, A and F to determine rotational velocities from the observed line profiles. The average equatorial velocity (v_e) was 60 km/s for these stars, though the values of $v_e \sin i$ ranged up to 200 km/s for λ Orionis (O8) and δ Ursae Majoris (A2). He concluded 'that a greater number of stars than suspected are in rapid rotation' (208).

Further evidence in support of the rotation hypothesis to explain the broad shallow lines in certain stellar spectra was assembled by Struve (209) that same year. In particular, the line broadening increased with wavelength, as required by the Doppler effect theory, and for binary stars the observed line widths correlated with those expected from the orbital elements assuming co-rotating stars (i.e. stars always showing the same face to each other). Struve was the first to show that rapid rotation of single stars was confined to the earlier spectral types, v_e being sometimes as high as 250 km/s for B and A stars, but for G, K and M stars rotational line broadening was not found at all. Together with the Belgian astronomer Pol Swings (1906–83), who was visiting Yerkes Observatory at that time, Struve showed that the emission line B stars (type Be) such as γ Cas were the most rapidly rotating of all stars. The emission lines originate in an equatorial shell or ring of gas, and the profiles were in accordance with rotational Doppler broadening but not with any known pressure broadening mechanism such as the Stark effect due to charged particles (210).

Meanwhile Christine Westgate at Yerkes continued the program initiated by Struve and Elvey when she studied the distribution of velocities within different spectral classes. By studying the rotational velocities of as many as 800 B, A and F stars she demonstrated that B and A stars have the highest equatorial velocities, whereas rotation ceases to be found in the middle F spectral types (211). The same conclusions were reached almost simultaneously by G. Shajn (212). He found that of single stars with spectral types between O and F2, as many as 52 per cent have detectably diffuse lines ascribable to rotation, with many near the rotational limit to their stability. The theory of line profile calculations for rotating stars was further developed during this active research period in the early 1930s by J. A. Carroll (1899–1974) who was then professor at the University of Aberdeen, Scotland. His calculations included limb darkening and he showed how both $v_e \sin i$ and the original line profile (if there were no rotation) can simultaneously be deduced by Fourier analysis (213).

Undoubtedly the early 1930s represented the golden age of stellar rotation, when the observational evidence for the phenomenon in single stars was firmly established by Struve and his colleagues. After the war, interest in the subject was renewed, mainly by a series of papers by Arne Slettebak (b.1925), who was based initially at Yerkes and later at the Perkins Observatory of the Ohio State University. His first contribution in 1949 analysed the 4026Å line of neutral helium for 123 stars of types O to B5, including all the brighter northern B stars. Among O, normal B and emission-line B stars Slettebak found objects with $v_e \sin i$ in excess of 400 km/s. But the B stars as a class were easily the fastest rotators, their mean rotational velocities exceeding those of normal B stars by 150 km/s (214).

Later Slettebak turned his attention to 179 stars of types B8 to A2 (215). The mean equatorial velocities determined from the ionised magnesium line at 4481 Å for main sequence stars in his sample was found to be around 200 km/s, reaching a maximum of 225 km/s at type B9V. Especially interesting is Slettebak's discovery of special groups of stars with small rotational velocities. Of sixteen peculiar A stars observed, nearly all had $v_e \sin i$ values less than 50 km/s. The peculiar manganese A stars had particularly sharp lines. Six metallic line (Am) stars were also slow rotators, comparable to the Ap star average. Even sharper lines were found for supergiants where 'axial rotation, if any exists among the supergiants, must be extremely small', whereas giants had values intermediate between supergiants and dwarfs. The conclusion that evolved stars have lower rotational velocities than dwarfs was then extended to a study of B2–B5 stars (216).

Allan Sandage (b.1926) at Mt Wilson and Palomar Observatories used Slettebak's data and showed the lower rotational velocities found above the main sequence were consistent with the current theories of stellar evolution, given that an evolving star must conserve angular momentum as it expands during its post-main-sequence evolution to become a giant, or, in some cases, a supergiant (217). These ideas invoking simple angular momentum conservation are also consistent with the data for A3 and G0 stars obtained by Slettebak (218). The slow rotation of dwarfs later than F5 was confirmed, whereas moderate rotation persisted in giants to somewhat later types, presumably because their progenitors were earlier fast rotating A or B stars on the main sequence. Two extensive surveys at Lick on rotational velocities reached similar conclusions. Su-Shu Huang (1915–77) derived $v_e \sin i$ for 1550 stars using the plates in the Lick archives, including those taken in Chile for the radial velocity program (219). His distribution of mean rotational velocities for spectral types from O to F9 was therefore the most extensive statistical sample analysed. G. Herbig (b. 1920) and J.F. Spalding at Lick studied the later-type stars F0 to K5. Using 11 Å/mm. spectra for 656 stars from the new 'Mills' spectrograph, they found moderate rotation in giants only up to type G0, and that F giants as a class rotated much faster than F dwarfs (220).

The cause of the slow rotation for dwarfs cooler than about F5 has been a topic for considerable speculation. Struve in 1955 pointed out that most of the angular momentum in the solar system resides in the planets. The sun's equatorial velocity would not be 2 km/s but 60 km/s if all this angular momentum were to be found in the sun. Hence Struve suggested the possibility of undetected planetary systems being a normal feature for the cooler dwarf stars (221). A more generally accepted interpretation is, however, the idea of angular momentum loss in a stellar wind from the star's surface, with the angular momentum being transferred outwards by magnetic braking. Such a process was first proposed by the French theoretical astrophysicist, E. Schatzman (b. 1920) (222). In this model magnetic braking only occurs in stars with convective envelopes which have solar-like flares giving rise to the necessary mass loss rates of around 10^{-15}–10^{-13} solar masses annually. Convection does not occur in the envelopes of stars hotter than F5 which is also the approximate point at which rotation ceases on the main sequence.

The body of published data on stellar rotational velocity determinations has never been nearly so extensive as for radial velocities. A general catalogue by Boyarchuk and Kopylov in 1964 (223) listed results for 2558 stars. The great majority of these came from determinations at Yerkes Observatory in the 1930s or from Lick in the 1950s. A second general catalogue by Uesugi and Fukuda in 1970 contained nearly 4000 stars (224).

References

1. Doppler, C., *Abhandlungen der königlichen Böhm. Gesellschaft der Wissenschaften*, **2** (5), 465, Jahrgang 1841/2, (1843). Also reprinted separately by Borrosch and André, Prague (1842).
2. Sestini, B., *Astron. J.*, **1**, 88 (1850).
3. Buys-Ballot, C.H., *Poggendorf's Ann.*, **66**, 321 (1845).
4. Vogel, H.C., *Poggendorf's Ann.*, **158**, 287 (1876).
5. Fizeau, H., *Ann. de chimie et physique*, **19** (sér 4), 211 (1870).
6. Mach, E., *Sitzungsberichte der königl. Akad. der Wissenschaften in Wien*, **41**, 543 (1860); see also *Poggendorfs Ann.*, **121**, 58 (1861).
7. Secchi, A., *Comptes Rendus*, **63**, 621 (1866).
8. Secchi, A., *Comptes Rendus*, **66**, 398 (1868).
9. Huggins, W., *Phil. Trans. R. Soc.*, **158** (II), 529 (1869).
10. Huggins, W. and Miller, W.A., *Phil. Trans. R. Soc.*, **154** (II), 413 (1864).
11. Huggins, W., *Proc. R. Soc.*, **20**, 379 (1872).
12. Secchi, A., *Comptes Rendus*, **68**, 358 (1869).
13. Secchi, A., *Comptes Rendus*, **82**, 761 and ibid **83**, 117 (1876).
14. Huggins, W., *Comptes Rendus*, **82**, 1291 (1876).
15. Christie, W.H.M., *Mon. Not. Roy. Astron. Soc.*, **36**, 313 (1876).
16. Airy, G.B., *Mon. Not. Roy. Astron. Soc.*, **36**, 27 (1876).
17. Fourteen papers in *Mon. Not. Roy. Astron. Soc.* from volume **36** (1876) to volume **51** (1891) were issued by the Astronomers Royal.
18. Airy, G.B. *Mon. Not. Roy. Astron. Soc.*, **37**, 22 (Paper II) (1877).
19. Christie, W.H.M., *Mon. Not. Roy. Astron. Soc.*, **50**, 111 (Paper XIII) (1890).
20. Seabroke, G.M., *Mon. Not. Roy. Astron. Soc.*, **39**, 450 (1879); ibid **47**, 93 (1887); ibid **50**, 72 (1890).
21. Vogel, H.C., *Astron. Nachrichten*, **78**, 241 (1871); ibid **82**, 291 (1873).

22. Zöllner, J.C.F., *Astron. Nachrichten*, **74**, 305 (1869).
23. Secchi, A., *Nuovo Cimento*, **3**(2), 217 (1870).
24. Young, C.A., *Amer. J. of Science*, **12**(3), 320 (1876).
25. Thollon, L., *Comptes Rendus*, **88**, 169 (1879); ibid **91**, 368 (1880).
26. Cornu, A., *Comptes Rendus*, **98**, 169 (1884).
27. Crew, H., *Amer. J. of Science*, **35**(3), 151 (1888); ibid **38**(3), 204 (1889).
28. Dunér, N.C., 'Recherches sur la Rotation du Soleil'. *Société Royale des Sciences d'Upsal*, 14 Feb. (1891).
29. Vogel, H.C., *Mon. Not. Roy. Astron. Soc.*, **52**, 87 (1891).
30. Lockyer, J.N., *Proc. R. Soc.*, **17**, 415 (1869); ibid **18**, 74 (1869).
31. Keeler, J.E., *Publ. Lick Observ.*, **3**, 161 (1894).
32. Vogel, H.C., *Astrophys. J.*, **11**, 373 (1900).
33. Huggins, W., *Phil. Trans. R. Soc.*, **158** (II), 529 (1868).
34. Huggins, W., *Proc. R. Soc.*, **22**, 251 (1874).
35. Vogel, H.C., *Publ. Astrophys. Observ. Potsdam*, **7**, 1 (1892). See also: Vogel, H.C., *Mon. Not. Roy. Astron. Soc.*, **52**, 87 (1891) for an earlier progress report.
36. Vogel, H.C., *Astrophys. J.*, **11**, 393 (1900).
37. Belopolsky, A.A., *Astrophys. J.*, **1**, 366 (1895).
38. Belopolsky, A.A., *Astrophys. J.*, **1**, 160 (1895); *Astron. Nachrichten*, **140**, 17 (1896).
39. Belopolsky, A.A., *Astrophys. J.*, **6**, 393 (1897).
40. Belopolsky, A.A., *Astrophys. J.*, **13**, 15 (1901).
41. Belopolsky, A.A., *Astrophys. J.*, **21**, 55 (1905).
42. Hartmann, J., *Publ. Astrophys. Observ. Potsdam*, **12**, 1 (appendix) (1898).
43. Rowland, H.A., *Astrophys. J.*, a series of 19 papers in volumes **1** to **6** (1895–97).
44. Belopolsky, A.A., *Astrophys. J.*, **19**, 85 (1904).
45. Keeler, J.E., *Astron. and Astrophys.*, **12**, 40 (1893).
46. Keeler, J.E., *Astrophys. J.*, **1**, 416 (1895).
47. Deslandres, H.A., *Comptes Rendus*, **120**, 1155 (1895).
48. Deslandres, H.A., *Comptes Rendus,* **120**, 417 (1895).
49. Deslandres, H.A., *Comptes Rendus,* **119**, 1252 (1894).
50. Belopolsky, A.A., *Astron. and Astrophys.,* **13**, 130 (1894).
51. Report of the Council: *Mon. Not. Roy. Astron. Soc.,* **30**, 112 (1870).
52. Newall, H.F., *Mon. Not. Roy. Astron. Soc.,* **56**, 98 (1896).
53. Newall, H.F., *Mon. Not. Roy. Astron. Soc.,* **57**, 567 (1897).
54. Newall, H.F., *Mon. Not. Roy. Astron. Soc.,* **63**, 296 (1903).
55. Milne, E.A., *Nature,* **153**, 455 (1944).
56. Lord, H.C., *Astrophys. J.,* **6**, 424 (1897); ibid **8**, 65 (1898).
57. Lord, H.C., *Astrophys. J.,* **4**, 50 (1896).
58. Lord, H.C., *Astrophys. J.,* **21**, 297 (1905).
59. Campbell, W.W., *Astron. and Astrophys.,* **11**, 523, 698 (1892).
60. Campbell, W.W., *Astron. and Astrophys.,* **11**, 529 (1892), plus seven further papers in volumes **11**, **12** and **13** (1892–4).
61. Campbell, W.W., *Astron. and Astrophys.,* **13**, 384 (1894).
62. Campbell, W.W., *Astron. and Astrophys.,* **13**, 448 (1894).
63. Campbell, W.W., *Astron. and Astrophys.,* **11**, 319 (1892).
64. Campbell, W.W., *Astron. and Astrophys.,* **11**, 881 (1892).

65. Campbell, W.W., *Stellar Motions,* p.45 (1910).
66. Campbell, W.W., *Astrophys. J.,* **8**, 123 (1898).
67. Moore, J.H., *Astrophys. J.,* **89**, 143 (1938).
68. Campbell, W.W. and Moore, J.H., *Publ. Lick Observ.,* **16**, 1 (1928).
69. Wright, W.H., *Publ. Lick Observ.,* **9**, (Part 3), 23 (1907).
70. Campbell, W.W., *Publ. Lick Observ.,* **9**, (Part 1), 3 (1907).
71. Wright, W.H., Palmer, H.K., Albrecht, S., and Campbell, W.W., *Publ. Lick Obs.,* **9**, (Part 4), 71 (1911).
72. Merrill, P.W., *Mon. Not. Roy. Astron. Soc.,* **99**, 317 (1939).
73. Kövesligethy, R., *Astron. Nachrichten,* **114**, 327 (1886).
74. Homann, H., *Astron. Nachrichten,* **114**, 25 (1886).
75. Kempf, P., (reported by H. C. Vogel) *Astron. Nachrichten,* **132**, 81 (1893).
76. Risteen, A.D., *Astron. J.,* **13**, 74 (1893).
77. Campbell, W.W., *Astrophys. J.,* **13**, 80 (1901).
78. Campbell, W.W., *Lick Observ. Bull.,* **6** (No. 196), 125 (1911).
79. Campbell, W.W., *Observ.,* **48**, 274 (1925).
80. Küstner, F., *Astron. Nachrichten,* **166**, 177 (1904) and *Astrophys. J.,* **27**, 301 (1908).
81. Küstner, F., *Astron. Nachrichten,* **198**, 409 (1914).
82. Küstner, F., *Astron. Nachrichten,* **169**, 241 (1905).
83. Hough, S.S., *Ann. Cape Observ.,* **10** (Part 1), (1911).
84. Lunt, J., *Astrophys. J.,* **47**, 201 (1918).
85. Spencer Jones, H., *Transactions I.A.U.,* **3**, 254 (1928).
86. Spencer Jones, H., *Ann. Cape Observ.*, **10** (Part 8), (1928).
87. Adams, W.S. and Kohlschütter, A., *Astrophys. J.,* **39**, 341 (1914).
88. Adams, W.S., *Astrophys. J.,* **42**, 172 (1915).
89. Adams, W.S. and Joy, A.H., *Astrophys. J.*, **57**, 149 (1923).
90. Adams, W.S., Joy, A.H., Sanford, R.F., Strömberg, G., *Astrophys. J.,* **70**, 207 (1929).
91. Boss, B., *General Catalogue of 33342 stars for the Epoch 1950,* Carnegie Institution, Washington D. C. (1937).
92. Wilson, R.E. and Joy, A.H., *Astrophys. J.,* **111**, 221 (1950). See also ibid **115**, 157 (1952) for an extension of this program.
93. Campbell, W.W., *Astrophys. J.,* **13**, 98 (1901).
94. Boss, B., *Popular Astron.,* **26**, 686 (1918).
95. Adams, W.S. and Joy, A.H., *Astrophys. J.,* **49**, 179 (1919).
96. Strömberg, G., *Astrophys. J.*, **59**, 228 (1924).
97. Strömberg, G., *Astrophys. J.*, **61**, 363 (1925).
98. Oort, J.H., *Publ. Kapteyn Astron. Observ. Groningen*, No. 40 (1926).
99. Oort, J.H., *Observatory*, **49**, 302 (1927).
100. Lindblad, B., *Arkiv för Matematik, Astron. och Fysik*, **19A** (No. 21), 1 (1925). See also: *Mon. Not. Roy. Astron. Soc.*, **87**, 553 (1927).
101. Plaskett, J.S., *Astrophys, J.*, **49**, 209 (1919).
102. Plaskett, J.S., Harper, W.E., Young, R.K., Plaskett, H.H., *Publ. Dom. Astrophys. Observ. Victoria*, **2**, 3 (1921).
103. Harper, W.E., *Publ. Dom. Astrophys. Observ. Victoria*, **2**, 189 (1923).
104. Plaskett, J.S., *Publ. Dom. Astrophys. Observ. Victoria*, **2**, 287 (1924).
105. Frost, E.B., *Astrophys. J.*, **15**, 1 (1902). See also: Frost, E.B. and Adams, W.S., *Publ. Yerkes Observ.*, **2**, 143 (1904).

106. Frost, E.B. and Adams, W.S., *Publ. Yerkes Observ.*, **2**, 143 (1904).
107. Kapteyn, J.C. and Frost, E.B., *Astrophys. J.*, **32**, 83 (1910).
108. Campbell, W.W., *Lick Observ. Bulletin*, **6** (No. 195), 101 (1911) and ibid **6** (No. 196), 125 (1911).
109. Petrie, R.M. and Pearce, J.A., *Publ. Dominion Astrophys. Observ. Victoria*, **12**, 1 (1961).
110. Plaskett, J.S. and Pearce, J.A., *Publ. Dominion Astrophys. Observ. Victoria*, **5**, 1 (1930).
111. Frost, E.B., Barrett, S.B. and Struve, O., *Astrophys. J.*, **64**, 1 (1926).
112. Frost, E.B. and Adams, W.S., *Astrophys. J.*, **18**, 237 (1903).
113. Belopolsky, A.A., *Astrophys. J.*, **19**, 85 (1904).
114. Newall, H.F., *Mon. Not. Roy. Astron. Soc.*, **63**, 296 (1903).
115. Slipher, V.M., *Astrophys. J.*, **22**, 318 (1905).
116. Slipher, V.M., *Astrophys. J.*, **20**, 1 (1904).
117. St. John, C.E. and Ware, L.M., *Astrophys. J.*, **44**, 15 (1916).
118. St. John, C.E., Moore, C.E., Ware, L.M., Adams, E.F. and Babcock, H.D., 'Revision of Rowland's Preliminary Table of Solar Spectrum Wavelengths', *Carnegie Institution of Washington Publication No. 396* (1928).
119. Frost, E.B., Moore, J.H. and Spencer Jones, H., *Trans. I.A.U.*, **3**, 171 (1928).
120. Pearce, J.A., *Trans. I.A.U.*, **4**, 181 (1932).
121. Pearce, J.A., *Trans. I.A.U.*, **5**, 191 (1935).
122. Petrie, R.M., *J. Roy. Astron. Soc. Canada*, **40**, 325 (1946).
123. Petrie, R.M., *J. Roy. Astron. Soc. Canada*, **41**, 311 (1947); ibid **42**, 213 (1948).
124. Petrie, R.M., *Publ. Dominion Astrophys. Observ. Victoria*, **9**, 297 (1953).
125. Voûte, J., 'First Catalogue of Radial Velocities' *Natuurkundig Tijdschrift voor Ned-Indie Deel,* **80** (2), 91. Visser and Co., Weltevreden (1920). See Palmer. M., *Astron. J.*, **35**, 99 (1923) for corrigenda to the Voûte catalogue.
126. Voûte, J., 'Second Catalogue of Radial Velocities'. *Ann. v.d. Bosscha Sterrenwacht, Lembang, Java*, **3** (1928).
127. Moore, J.H., 'A General Catalogue of the Radial Velocities of Stars, Nebulae and Clusters'. *Publ. Lick Observ.*, **18**, 1 (1932).
128. Henroteau, F. and Henderson, J.P., *Publ. Dominion Observ. Ottawa*, **5**, 1 (1920); Henroteau, F., ibid **5**, 45 (1921); Henroteau, F., ibid **5**, 331 (1922).
129. Struve, O., and Zebergs, V., *Astronomy of the Twentieth Century*. Macmillan and Co., London and New York. p. 73 (1962).
130. Young, R.K., *Publ. David Dunlap Observ.*, **1**, 69 (1939): ibid **1**, 249 (1942).
131. Seyfert, C.K. and Popper, D.M., *Astrophys. J.*, **93**, 461 (1941).
132. Popper, D.M., *Astrophys. J.*, **95**, 307 (1942); ibid **98**, 209 (1943).
133. Smith, B. and Struve, O., *Astrophys. J.*, **100**, 360 (1944).
134. Sterken, C. and Vogt, N., *E.S.O. Messenger*, **28**, 12 (1982). See also Stone, R.P.S., *Sky and Telescope*, **63**, 446 (1982).
135. Thackeray A.D., *Mon. Not. Roy. Astron. Soc.*, **112**, 318 (1952).
136. Feast, M.W., Thackeray, A.D. and Wesselink, A.J., *Mem. R. Astron. Soc.*, **67**, 51 (1955).

137. Feast, M.W., Thackeray, A.D. and Wesselink, A.J., *Mem. R. Astron. Soc.*, **68**, 1 (1957).
138. Moore, J.H. and Paddock, G.F., *Astrophys. J.*, **112**, 48 (1950).
139. Petrie, R.M. and Pearce, J.A., *Publ. Dominion Astrophys. Observ.*, Victoria, **12**, (No. 1) (1962).
140. Wilson, R.E., 'General Catalogue of Stellar Radial Velocities', *Carnegie Institution of Washington*, Publ. 601 (1953).
141. Wilson, R.E., *Publ. Astron. Soc. Pacific*, **63**, 223 (1951).
142. Struve, O. and Zebergs, V., *Astronomy of the Twentieth Century*, Macmillan and Co., London and New York. p. 64 (1962).
143. Abt, H.A. and Biggs, E.S., *A Bibliography of Stellar Radial Velocities*, Kitt Peak National Observatory (1972).
144. Heard J.F., *Publ. David Dunlap Observ.*, **2**, 105 (1956).
145. Evans, D.S., Menzies, A., Stoy, R.H., *Mon. Not. Roy. Astron. Soc.*, **117**, 534 (1957). Also Evans. Menzies, Stoy and Wayman, P.A., *Roy. Obs. Bull.*, no. 48 (1961).
146. Buscombe, W. and Morris, P.M., *Mon. Not. Roy. Astron. Soc.*, **121**, 263 (1960).
147. Buscombe, W. and Morris, P.M., *Mon. Not Roy. Astron. Soc.*, **118**, 609 (1958).
148. Fehrenbach, C., *Trans. Int. Astron. Union*, **12A**, 495 (1965).
149. Muller, C.A. and Oort, J.H., *Nature*, **168**, 357 (1951).
150. Lilley, A.E. and McClain, E.F., *Astrophys. J.*, **123**, 172 (1956).
151. Minkowski, R. and Wilson, O.C., *Astrophys. J.*, **123**, 373 (1956).
152. Popper, D.M., *Astrophys. J.*, **120**, 316 (1954).
153. Pickering, E.C., *H.D. Memorial, 1st Ann. Report*, p. 9 (1887).
154. Pickering, E.C., *Harvard Ann.*, **26**, 1 (1891).
155. Orbinsky, A., *Astron. Nachrichten*, **138**, 9 (1895).
156. Frost, E.B., *Astrophys. J.*, **2**, 235 (1895).
157. Deslandres, H., *Astron. Nachrichten*, **139**, 241 (1896). See also *Observ.*, **19**, 49 (1896).
158. Maunder, E.W., *Observ.*, **19**, 84 (1896).
159. Keeler, J.E., *Astrophys. J.*, **3**, 311 (1896).
160. Hale, G.E. and Wadsworth, F.L.O., *Astrophys. J.*, **4**, 54 (1896).
161. Pickering, E.C., *Harvard Circ.*, **13** (1896). See also *Astron. Nach.*, **142**, 106 (1897).
162. Comstock, G.C., *Astrophys. J.*, **23**, 148 (1906).
163. Comstock, G.C., *Astrophys. J.*, **23**, 396 (1906).
164. Pickering, E.C., *Astrophys. J.*, **23**, 255 (1906). See also *Harvard Circ.*, **10** (1906).
165. Wood, R.W., *Astrophys. J.*, **31**, 460 (1910).
166. Pickering, E.C., *Harvard Circ.*, **154** (1910). See also *Astrophys. J.*, **31**, 372 (1910).
167. Schwarzschild, K., *Astron. Nachrichten*, **194**, 241 (1913).
168. Plaskett, J.S., *Publ. Dom. Observ. Ottawa*, **1**, 171 (1914).
169. Hamy, M., *Comptes Rendus*, **158**, 81 (1914).
170. Graham, T.S.H., *J. Roy. Astron. Soc. Canada*, **12**, 129 (1918).
171. Schwarzschild, K., *Publ. Astrophys. Observ. Potsdam*, **23**, 1 (no. 69) (1913).
172. Wilson, H.C., *Popular Astron.*, **31**, 93 (1923).
173. Millman, P.M., *Harvard Circ.*, **357** (1931).

174. Millman, P.M., *J. Roy. Astron. Soc. Canada*, **25**, 281 (1931).
175. Bok, B.J. and McCuskey, S.W., *Harvard Ann.*, **105**, 327 (1937).
176. Cherry, B., *Harvard Ann.*, **105**, 355 (1937).
177. McCuskey, S.W., *Popular Astron.*, **46**, 2 (1938).
178. Edwards, D.L. and Barber, D.R., *Mon. Not. Roy. Astron. Soc.*, **98**, 42 (1938).
179. Fehrenbach, C., *Ann. d'Astrophys.*, **10**, 257 (1947).
180. Fehrenbach, C., *Ann. d'Astrophys.*, **10**, 306 (1947) and ibid **11**, 35 (1948).
181. Treanor, P.J., *Mon. Not. Roy. Astron. Soc.*, **108**, 189 (1948).
182. Pearce, J.A., *Trans, I.A.U.*, **7**, 309 (1950).
183. Schalén, C., *Arkiv f. Astron.*, **1**, 545 (1954).
184. Panaitov, L.A., *Soviet Astron. J.*, **32**, 305 (1955).
185. Fehrenbach, C., *J. des Observateurs*, **38**, 165 (1955).
186. Fehrenbach, C., *Adv. in Astron. Astrophys.*, **4**, 1 (1966).
187. Fehrenbach, C., *I.A.U. Symposium*, **30**, 65 (1967).
188. Fehrenbach, C. and Duflot, M., *Astron. Astrophys. Suppl. Ser.*, **13**, 173 (1974).
189. Fraunhofer, J., *Gilbert's Ann.*, **74**, 337 (1823).
190. Zöllner J.C.F., *Astron. Nachrichten*, **74**, 305 (1869).
191. Fellgett, P.B., *Oplica Acta*, **2**, 9 (1953).
192. Babcock's work is reported by Bowen, I.S., *Annual Rep. of the Director of Mt Wilson and Palomar Observatories 1954/55*, p. 27 (1955).
193. Griffin, R.F., *Astrophys. J.*, **148**, 465 (1967). See also: *I.A.U. Symp.*, **30**, 3 (1967).
194. Griffin, R.F., *Mon. Not. Roy. Astron. Soc.*, **145**, 163 (1969) and ibid **148**, 211 (1970).
195. Griffin, R.F., *Mon. Not. Roy. Asyron. Soc.*, **155**, 1 (1971).
196. Griffin, R.F., *Transactions I.A.U.*, **16A** (part 2), 157 (1976).
197. Jones, D.H.P., *Mon. Not. Roy. Astron. Soc.* **152**, 231 (1971).
198. Abney, W. de W., *Mon. Not. Roy. Astron. Soc.*, **37**, 278 (1877).
199. Vogel, H.C., *Astron. Nachrichten*, **90**, 71 (1877).
200. Vogel, H.C., *Sitzungsberichte der königl. Preuss. Akad. der Wiss. zu Berlin*, p. 721 (1898); English version in *Astrophys. J.*, **9**, 1 (1899).
201. Schlesinger, F.J., *Publ. Allegheny Obs.*, **1**, 123 (1909).
202. Schlesinger, F.J., *Publ. Allegheny Obs.*, **3**, 23 (1913).
203. Rossiter, R.A., *Astrophys. J.*, **60**, 15 (1924).
204. McLaughlin, D.B., *Astrophys. J.*, **60**, 22 (1924).
205. Adams, W.S. and Joy, A.H., *Astrophys. J.*, **49**, 190 (1919).
206. Shajn, G. and Struve, O., *Mon. Not. Roy. Astron. Soc.*, **89**, 222 (1929).
207. Struve, O., Uni. of Chicago Abstracts of Theses, *Sci. Ser.*, **2**, 60 (1923).
208. Elvey, C.T., *Astrophys. J.*, **71**, 221 (1930).
209. Struve, O., *Astrophys. J.*, **72**, 1 (1930).
210. Struve, O. and Swings, P., *Astrophys. J.*, **75**, 161 (1932).
211. Westgate, C., *Astrophys. J.*, **77**, 141 (1933); **78**, 46 (1933); **79**, 357 (1934).
212. Shajn, G., *Zs. für Astrophys.*, **6**, 176 (1933).
213. Carroll, J.A., *Mon. Not. Roy. Astron. Soc.*, **93**, 478 (1933).
214. Slettebak, A., *Astrophys. J.*, **110**, 498 (1949).
215. Slettebak, A., *Astrophys. J.*, **119**, 146 (1953).
216. Slettebak, A. and Howard, R.F., *Astrophys. J.*, **121**, 102 (1955).

217. Sandage, A.R., *Astrophys. J.*, **122**, 263 (1955).
218. Slettebak, A., *Astrophys. J.*, **121**, 653 (1955).
219. Huang, S-S., *Astrophys. J.*, **118**, 285 (1953).
220. Herbig, G.H. and Spalding, J.F., *Astrophys. J.*, **121**, 118 (1955).
221. Struve, O., *Sky and Telescope*, **15**, 17 (1955).
222. Schatzman, E., *Ann. d'Astrophys.*, **25**, 18 (1962).
223. Boyarchuk, A.A. and Kopylov, I.M., *Izvestia Crimean Astrophys. Obs.*, **31**, 44 (1964).
224. Uesugi, A. and Fukuda, I., *Mem. Fac. Sci. Kyoto Univ., Ser. Phys., Astrophys., Geophys., Chem.*, **33**, 205 (1970).

7 The interpretation of stellar spectra and the birth of astrophysics

7.1 Some early theories of stellar evolution

At the end of the nineteenth century the two main branches of stellar spectroscopy were spectral classification and radial velocity measurements. The latter department was still in its relative infancy, but classification, thanks mainly to the energy of Pickering at Harvard, was a major activity. Classification had become closely related to theories of stellar evolution and these two aspects could hardly be disentangled in for example the classification devised by Lockyer (1), which involved first rising then falling temperatures of stars over their life cycle, and which had some theoretical support from the work of Lane and Ritter on the gravitational collapse of gaseous spheres.

The classification schemes used by Miss Maury and Miss Cannon were also implicitly evolutionary theories, but with the direction of evolution being from the 'earlier' to 'later' spectral types. Rival theories to Lockyer's were then proposed for stellar evolution in the early years of the century, with the Harvard spectral types as their basis, notably by Sir Arthur Schuster (1851–1934) in Great Britain (2) and by George Ellery Hale (1868–1938) in the United States (3). Schuster's scheme involved gravitational collapse and cooling of gaseous masses on the so-called Helmholtz-Kelvin timescale. Although the Harvard classification was generally believed to be a temperature sequence with stars evolving from the hot Orion-type stars to the cooler solar types, nevertheless Schuster maintained that composition changes at the surface as the evolution proceeded accounted for the differences in the line spectrum from star to star.

Hale's scheme was similar, and based on the quite correct belief that star formation originated in gaseous nebulae which contract under their self-gravitation. However he cautioned against the idea that colour differences are necessarily to be interpreted as meaning 'bluer' stars are hotter than, say, yellow or red ones. Hale believed the sun is yellow because of the large number of lines absorbing light in the short-wavelength part of the spectrum, and that the sun would be bluish-white like Sirius if the metallic lines could be removed.

7.2 Hertzsprung's analysis of the Maury c-type stars The early theories of stellar evolution were plausible enough in the context of the known laws of physics at that time. Developments that took place from 1905 would, in due course, make these simple evolutionary schemes obsolete. The first major step forward took place in Denmark. Ejnar Hertzsprung in 1905 was working as an amateur astronomer at the Urania and University Observatories in Copenhagen. His first professional appointment at Göttingen was still 4 years in the future, and his early training had been in chemical engineering. Despite this limited astronomical background Hertzsprung was able to familiarise himself sufficiently with the recent literature and make an important breakthrough.

His work at this time concerned an analysis of the proper motions of those stars with unusually narrow lines designated by Miss Maury as types c or ac (Chapter 5). He realised that even without knowing the distances to individual stars, from which their absolute magnitudes (that is luminosities) could be deduced, at least a statistical analysis could be applied because the nearer stars have, on average, the larger proper motions. Knowing the actual proper motion, Hertzsprung calculated how bright a star would appear if it could be placed at a distance to bring its proper motion to 1 arc sec per year. This 'corrected apparent magnitude' is a direct statistical measure of luminosity. Similarly, he used as a statistical distance indicator the proper motion a star would have if it were at a distance to bring its apparent magnitude to zero.

Hertzsprung found the A and Orion-type stars were all of a uniformly high luminosity, or small proper motion. But he drew his most important conclusions from the later spectral types. He wrote:

> ... we can say that the annual proper motion, when reduced to zero apparent magnitude, of the average c type star amounts to only about one hundredth of a second of arc. With the relatively large errors in the small values, a possible dependence on spectral type cannot yet be recognised. In other words, the c stars are at least as luminous as the Orion stars... For the stars contained in Annie J. Cannon's Catalogue which have narrow sharp lines, I could also only find small proper motions. This result confirms the assumption of Antonia C. Maury that the c-stars show some intrinsic characteristic (4).

Fig. 7.1 Ejnar Hertzsprung.

Hertzsprung then went on to refer explicitly to the luminosities of the 'collateral series' of c-stars: 'The hypothesis of the collateral series is suitable for explaining the principal observation that among the stars brighter than apparent magnitude 5, the c and Orion stars shine the brightest, and among the remaining stars not the red but the yellow ones are the faintest'.*

In a second paper 2 years later, Hertzsprung further developed these ideas, and was able to conclude:

> We can now accept that the bright red stars (α Bootis, α Tauri, α Orionis etc.) are rare per unit volume of space, and those which belong to the normal solar series form by far the greatest number. The bright red stage is therefore relatively quickly traversed, or the stars which are in that phase belong to a collateral series

(meaning that their evolution is entirely separate from the more numerous fainter stars) (5). These statements constitute the first beginnings of the concept of the colour-magnitude diagram which was to become the basis from which theories of stellar evolution could be built. The sharp dichotomy in later-type stars between those of high and low luminosity, and also the marked differences between their respective space densities, were now empirical facts which stellar evolution theorists would have to

* There are very few red dwarfs of types K and M brighter than fifth magnitude, so the least luminous stars appeared to be yellow dwarfs of solar spectral type.

Fig. 7.2 Hertzsprung (right) and Schwarzschild (left) in professorial gowns in Göttingen, 1909.

incorporate into their theories. Stellar spectroscopic classification alone was no longer an adequate basis from which to explore the evolution of stars.

Unfortunately Hertzsprung's two pioneering articles were at first largely ignored. They were almost the first astronomical publications of a young and unknown scientist without fixed employment and in a journal rarely used by astronomers. Sir Arthur Eddington many years later wrote to Hertzsprung: 'One of the sins of your youth was to publish important papers in inaccessible places' (6). However, Hertzsprung's fortunes were soon to change. In 1908 he visited Karl Schwarzschild in Göttingen. Within a few months he was appointed to a professorship there (a rare achievement for an amateur astronomer who had never attended a lecture in astronomy!), and when Schwarzschild that same year was promoted to the Potsdam directorship, Hertzsprung followed him to that observatory. He thus benefited from close collaboration with one of the leading astronomers at that time. Unfortunately their work together did not last long;

by 1916 Schwarzschild was dead as a result of illness contracted while on military service. After the war Hertzsprung found a position at the Leiden Observatory in Holland, where he stayed until his retirement in 1935. There is little doubt that his immense reputation during his Leiden years owes much to the early support and friendship from Schwarzschild.

7.3 Monck's analysis of proper motion and luminosity

In 1905 Hertzsprung was an amateur astronomer who within a decade was to be known as one of the leading astronomers in Europe. The Irish amateur William Monck (1839–1915) was less fortunate, yet he went nearly as far as Hertzsprung in his analysis of proper motions which preceded Hertzsprung's work by a decade. Monck was an official of the High Court of Ireland, and formerly a philosophy professor at Trinity College, Dublin. In a series of papers mainly in the principal American astronomical journals (the first was in 1892 (7)) he showed that stars of different spectral types have differences in proper motion, and hence also in luminosity. In 1895 Monck wrote:

> I suspect, moreover, that two distinct classes of stars are at present ranked as Capellan, one being dull and near us, and the other bright and remote like the Sirians. Capella itself, perhaps, occupies an intermediate position. α Centauri and Procyon may stand as types of the near and dull Capellan, with large proper motion, while Canopus is a remarkable instance of a bright and distant one, with small proper motions, assuming that there is no doubt as to its spectrum (8).

Unlike Hertzsprung, Monck did not find any spectral peculiarities in the high luminosity stars. However, there is no doubt that this concise statement on luminosities anticipated Hertzsprung's results by a decade, and deserved far more recognition than it received.

7.4 Russell's work on luminosity and spectral type, and his relationship to Hertzsprung

Hertzsprung's famous contemporary in the study of stellar luminosities was Henry Norris Russell (1877–1957) from the Princeton University Observatory. His work on stellar luminosities was one of the earlier contributions of this illustrious astronomer who dominated astrophysics for half a century. Although he was not an observer, Russell was always a very practical person; one of his main interests was the interpretation of data. His line of attack on the

Fig. 7.3 Henry Norris Russell.

luminosity question was through trigonometric parallaxes; some of these he had measured himself during a program initiated when he visited Cambridge, England, from late 1902 till 1905. In a paper written in 1910 he discussed the relationship between luminosity and spectrum and, like Hertzsprung (though apparently oblivious of Hertzsprung's work), he concluded that the brightest stars are found among all spectral types, whereas the faintest were only represented by later spectral types, being progressively fainter the later the spectrum (9). In August that year Russell presented these ideas to the Astronomical and Astrophysical Society of America* meeting at Harvard (10). He described his results as follows: '. . . the redder stars from type G onward, fall into two groups: one remote, of small proper motion and great luminosity, the other near us, of large proper motion and small luminosity'. Russell then linked these two groups into an evolutionary sequence running from the luminous red stars, through the hotter stars of types B and A, followed by a cooling and further gravitational contraction to progressively fainter stars, a scheme which he acknowledged was similar to one proposed by Lockyer. As it happened Schwarzschild also attended the Harvard meeting and evidently told Russell of the earlier work of his young colleague Hertzsprung along similar lines. A reference to 'Herzsprung' [sic] is thus found in the published

* Later known as the American Astronomical Society.

account of the Harvard meeting. Schwarzschild had arranged that Hertzsprung would send Russell copies of his (EH's) papers which Russell gratefully acknowledged in September of that year (Fig. 7.4).

The relationship between Hertzsprung and Russell, as well as the development of the independent research on the same topic, is interesting. They did not meet until July 1913 at the International Solar Union meeting in Bonn. Meanwhile, in 1911, Hertzsprung published colour-magnitude

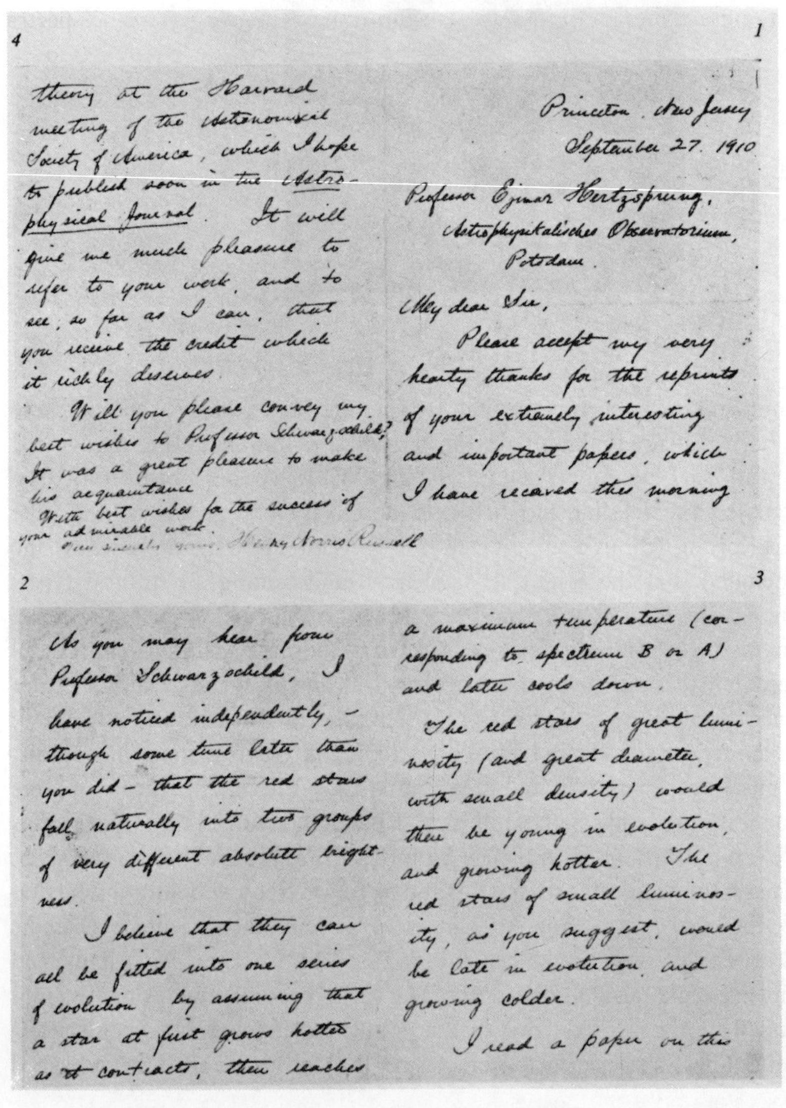

1

Princeton, New Jersey
September 27. 1910

Professor Ejnar Hertzsprung,
Astrophysikalisches Observatorium,
Potsdam.

My dear Sir,

Please accept my very hearty thanks for the reprints of your extremely interesting and important papers, which I have received this morning.

2

As you may hear from Professor Schwarzschild, I have noticed independently,— though some time later than you did— that the red stars fall naturally into two groups of very different absolute brightness.

I believe that they can all be fitted into one series of evolution by assuming that a star at first grows hotter as it contracts, then reaches

3

a maximum temperature (corresponding to spectrum B or A) and later cools down.

The red stars of great luminosity (and great diameter, with small density) would then be young in evolution, and growing hotter. The red stars of small luminosity, as you suggest, would be late in evolution and growing colder.

I read a paper on this

4

theory at the Harvard meeting of the Astronomical Society of America, which I hope to publish soon in the *Astrophysical Journal*. It will give me much pleasure to refer to your work, and to see, so far as I can, that you receive the credit which it richly deserves.

Will you please convey my best wishes to Professor Schwarzschild? It was a great pleasure to make his acquaintance.

With best wishes for the success of your admirable work.

Very sincerely yours, Henry Norris Russell

Fig. 7.4 Russell's letter to Hertzsprung, September 1910.

diagrams for stars in the Pleiades and Hyades clusters, showing in both cases a 'main sequence' (his term) of hot stars, as well as for the Hyades, four luminous red stars, thus presenting a similar dichotomy for later spectral types as already observed in the field (11). These diagrams, the first unpublished version of which Hertzsprung had brought to Göttingen to show Schwarzschild as early as 1908, employed either 'effective wavelengths' or 'colour indices' to represent the stellar temperatures, which were plotted on the vertical axis or ordinate. Effective wavelengths were obtained from the apparent separation of the two first-order spectra on very low dispersion objective grating exposures, while the colour indices were the apparent magnitude differences between photographic (blue, ultraviolet) and visual (yellow) magnitudes. In either case, Hertzsprung clearly recognised that his parameters measured stellar temperature.

Meanwhile Russell appears at first to have been unaware of this new work of Hertzsprung, and presented his own so-called Russell diagram, based on trigonometric parallaxes and Harvard spectral types for field stars, at a Royal Astronomical Society (RAS) meeting in London in June 1913, while on his way to Bonn (12). At the London meeting he explicitly referred to the terms 'giant' and 'dwarf' stars for the two distinct categories for later spectral types, although he erroneously ascribed this terminology to Hertzsprung.* He expanded on these ideas at a meeting of the American Astronomical Society in Atlanta in December that year, and the published account of his paper contains the first Russell diagram to appear in print (13). The Russell diagram employed Harvard spectral types for the temperature-related parameter. It is quite possible that Pickering, who had met Russell at the April 1908 meeting of the Astronomical and Astrophysical Society, had himself suggested that Russell use the Harvard classifications in the interpretation of the stars on Russell's parallax program (14). It is interesting to note that both Hertzsprung and at first Russell (at the London RAS meeting)[†] had plotted absolute magnitude (i.e. luminosity) on the horizontal scale. However, in Atlanta this diagram was now shown rotated clockwise through 90° but not inverted. Astronomers traditionally have ever since plotted these diagrams with the absolute magnitude decreasing towards the top.

Russell acknowledged Hertzsprung's work at this time when discussing the division between dwarf and giant stars. He wrote: 'All I have done in this diagram is to use more extensive observational material' (15). In any

* In December 1908 Schwarzschild referred to some stars as 'Giganten' when discussing Hertzsprung's work in a lecture in Berlin (12).

† Reference (12) contains no diagrams; but these were presented at the RAS meeting and described by Russell in the published text.

event, Hertzsprung's colour-magnitude diagram for clusters continued to be relatively neglected, while the diagram of the Princeton astronomer became widely known as the 'Russell diagram'. This designation stuck for two decades, until Hertzsprung's earlier work was once again rescued and properly acknowledged, this time by a young fellow Dane, Bengt Strömgren, who first used the title 'Hertzsprung–Russell diagram' in a lecture at a meeting of the Astronomische Gesellschaft in Göttingen in August, 1933 (16). Hertzsprung's modesty was unaffected by this belated recognition. He is quoted as saying: 'Why not call it the colour-magnitude diagram? Then we would all know what it is about' (17).

7.5 Adams and Kohlschütter's work on luminosity effects in stellar spectra

Although the wide range in luminosities for late-type stars had been discovered independently by Monck, Hertzsprung and Russell, the only one of these three to emphasise spectral differences arising from luminosity was Hertzsprung. These ideas were pursued in 1914 by the Mt Wilson Observatory astronomers Walter S. Adams and Arnold Kohlschütter. Their classic paper (18) made no reference however to any of the related work already accomplished elsewhere, and it is a fair assumption they were not aware of Hertzsprung's conclusions concerning the Maury c-type stars.

In the years to follow, Walter Adams built up a reputation as one of the leading stellar spectroscopists in the United States. He was assistant director at Mt Wilson in 1913 at the time of his work on spectroscopic parallaxes, the term used for the method of parallax or distance determinations using luminosity effects in stellar spectra. Later, in 1923, he was to become director of Mt Wilson in succession to Hale, a post he retained for 23 years. During this time his interests in stellar spectra were exceptionally wide, but the spectroscopic parallax program will undoubtedly remain his best known achievement. His collaborator, Arnold Kohlschütter, came from Germany. Kohlschütter graduated in Göttingen in 1907 under Karl Schwarzschild before spending the years 1911–14 at Mt Wilson. Returning to Europe in 1914 he was interned in Gibraltar on the outbreak of war and was not released until 1918. His subsequent career was at Potsdam, where he cooperated with F. Becker on the Potsdam southern spectral survey, and then from 1927, in Bonn, where he became director of the university observatory.

The classic paper by these two astronomers in 1914 (18) rediscovered the relationship between intrinsic luminosity and proper motion and parallax already pointed out by Monck, Hertzsprung and Russell. They

Fig. 7.5 Walter Adams.

went further by noting spectral differences between the high and low luminosity stars. The former were found to be weak in the violet and ultraviolet light, as well as having stronger hydrogen lines and abnormal strengths (weak or strong) for various metallic lines when compared to the lower luminosity group. The metallic lines formed the basis for the main part of their investigation. The strengths of certain lines exhibited especially large luminosity effects, and they therefore estimated the ratios of the strengths of these luminosity-sensitive lines to other nearby lines insensitive to luminosity. Their method involved no theoretical treatment for the cause of these luminosity effects, but was a purely empirical technique which was adopted for the measurement of stellar absolute magnitudes from the line strength ratios. Of course the line ratios had to be calibrated against luminosity, using trigonometric parallax and proper motion data to measure distances.

Adams and Kohlschütter then went on to measure absolute magnitudes for 162 late spectral type (F8 to K9) stars with an accuracy they estimated to be about 1.5 magnitudes. Since late type dwarfs and giants were estimated to differ on average by about nine magnitudes in their absolute magnitudes (i.e. the giants are 4000 times more luminous), clearly the two

Fig. 7.6 Arnold Kohlschütter.

types of star could be readily distinguished. Knowing the absolute magnitudes by this method, clearly the spectroscopic parallaxes or distances at once followed. These were accurate to no more than a factor of two, not high precision by any means, but an amazing advance forwards considering the difficulty in measuring distances, especially for higher luminosity stars, by any alternative method. With this research a new quantitative dimension to stellar spectroscopy was born; for the first time astronomers had advanced beyond the restrictive confines of pure classification to spectral analysis.

7.6 New developments in atomic physics and their influence on astrophysics To follow the development of astrophysics in the years following the First World War, we have to go back to the end of the nineteenth century and mention the main advances in laboratory physics that led to the 'Golden Age' of physical astronomy. Four important discoveries can be singled out which were milestones in the progress of atomic physics.

First, in 1897, came the discovery of the electron by Professor J.J. Thomson (1856–1940) at the Cavendish Laboratory in Cambridge (19), which resulted from his work investigating cathode rays. The concept of the electron (the term was introduced by Dr Johnstone Stoney) as a fundamental particle forming a part of the atoms and molecules of nature dates from this time, as does the idea of ionisation, which Thomson himself demonstrated when he found that X-rays could render gases conducting.

At the turn of the century the German physicist Max Planck (1855–1947) in Berlin derived theoretically the spectral energy distribution of radiation from a black body, corresponding to conditions of thermodynamic equilibrium (20) and this was followed in 1905 by Einstein's (1879–1955) discovery of the quantisation of light (21). These discoveries were soon applied to the theory of stellar atmospheres by Schuster and Schwarzschild and they also paved the way for the first quantitative estimates of stellar temperatures by Wilsing and Scheiner.

Next came the important advances in the theory of atomic structure from the New Zealander, Ernest Rutherford (1871–1937) in 1911 and by Niels Bohr in Denmark in 1913. Rutherford in Manchester had proposed a model for the atom comprising a small massive nucleus of positive charge surrounded by a relatively large volume occupied by the orbiting negatively charged electrons (22). In Copenhagen, Bohr developed this atomic model further. In his hydrogen atom the electron energies were quantised into discrete levels known as stationary states (23). This new model was based both on Rutherford's atom and the quantum theory. Not only did it predict the emission or absorption of radiation by atomic hydrogen in discrete quanta of precise energy or wavelength, but the only allowed wavelengths included those already given by the well-known Balmer formula (24).

Bohr's atomic theory had an immediate application to astronomy. This was the explanation of the anomalous lines of the Pickering series found by Pickering in 1896 in ζ Puppis and other O-type stars. Although these lines had been attributed at first to hydrogen (see Chapter 5), Bohr himself was able to assign them to ionised helium (25), whose atomic structure closely resembles that of hydrogen apart from having twice the nuclear charge and four times the mass.

7.7 The first stellar temperatures measured by Wilsing and Scheiner

It is fair to say that the five physicists* just mentioned, more than any others, paved the way for the rapid advances in astrophysics after the Great War. However, one further ingredient went into the melting pot from which astrophysics was cast. This was the first measurement of stellar temperatures, made principally by the German astronomers J. Wilsing and J. Scheiner at Potsdam and also by the Frenchman, Charles Nordmann (1881–1940) in Paris. Wilsing and Scheiner first attacked this problem from 1905. Their method was to

* Thomson, Planck, Rutherford, Einstein and Bohr.

measure the intensities in five spectral pass-bands from 4480 to 6380 Å by visual comparison with the spectrum of a lamp which was in turn calibrated with the radiation from an oven, assumed to be a black body or Planckian source (26). The logarithmic fluxes in the pass-bands were corrected for atmospheric extinction, as well as for wavelength-dependent absorption and chromatic aberration in the optics of the 80cm refractor. These values were then fitted to Planck curves to obtain stellar temperatures. In effect these were colour temperatures, as the slope of the stellar flux curve with wavelength was the parameter being used. One hundred and nine stars of all spectral types were so studied, and the run of temperature with spectral type, which had been suspected for some time, was now quantitatively demonstrated. The hottest star was the B star λ Orionis at 12 800 K. The coolest were two M stars (μ Gem, κ Ser) at 2800 K.

The Wilsing–Scheiner temperatures were criticised by C.G. Abbot (1872–1973) of the Smithsonian Astrophysical Observatory, both because of the short wavelength base of less than 2000 Å, and because of the possible effect of line blocking on the data (27). The latter problem arises whenever absorption lines lower the flux measured in a pass-band; to obtain fits to Planck curves only the continuous flux, avoiding the absorption lines, should be measured.

Also at this time Charles Nordmann was working at the Paris Observatory on the problem of determining stellar colours. His photometer used broad-band blue and red filters made from coloured liquid solutions, with which he visually estimated blue–red colour indices (28). The photometer was calibrated using the radiation from ovens of various temperatures and an electric arc. The effective wavelengths of the filters were estimated to be 6300 and 4600 Å for a 6000 K source.

Nordmann used the Planck curve to convert his colour indices to temperatures for fourteen stars including the sun (29). His results showed the general trend that earlier spectral types were hotter, in agreement with the correlation of spectral type with colour that was already well known. His temperatures for the hotter stars contained substantial systematic errors, the values being generally too high by 2000 or 3000 K. For cooler stars his results were mainly better and for the sun he obtained a temperature of 5320 K.

7.8 Photographically determined stellar energy distributions

The basic techniques for photographic photometry were laid down as early as 1899 by Karl Schwarzschild at the time when he was working as an assistant at the Kuffner Observatory in

Vienna (30). Adolf Hnatek (1876–1960) in Vienna, was the first to apply these techniques to the problem of determining stellar energy distributions (31). This brought benefits as well as complications, the latter being due to the non-linear response of the photographic emulsion to light, and also due to the blue-sensitive plates he used, thus limiting observations to a small baseline (3980 to 4980 Å) where the line-blocking problem already referred to is more severe than in the visual (or yellow–red) region of the spectrum. However, Hnatek exposed calibration plates using the same emulsion batch and development as those carrying the spectra, much as would be done today. Unfortunately he chose to reduce his results for seven stars relative to Altair (α Aql), whose temperature of 7100 K was adopted as a standard from Wilsing and Scheiner's work. Since Altair is in reality over 1000 K hotter than this, the error was reproduced in all his stars, particularly for those of early spectral type.

In Germany Hans Rosenberg (1879–1940) at Tübingen had more success with photography. He went to Göttingen in 1907–9 and obtained spectra on Agfa plates of all the bright northern stars to third magnitude with the aim of deducing energy distributions using the Schwarzschild techniques of photographic photometry (32). He presented his data for seventy stars in 1914 based on Planck radiation curves between 4000 and 5000 Å. His observations were calibrated using C.G. Abbot's absolute intensity data for the sun (33). When compared with spectral types, the Rosenberg temperatures show considerable scatter, but if the average temperature for each type is taken, then a clearly defined trend emerges. Rosenberg's values for B stars are about 2000 K hotter than currently accepted values. This systematic error slowly declined towards later types where quite reliable temperatures were obtained. Rosenberg himself noted large differences between his own and the Potsdam temperatures, especially for the B stars, where differences of 6000 K, and sometimes even over 10 000 K appeared (the Potsdam values were cooler).

Unfortunately, Wilsing had little patience for the results of his colleague from Tübingen, and he published his criticisms of Rosenberg in a full length article in the *Astronomische Nachrichten* (34). The main points he singled out were an unjustified correction Rosenberg had applied for diffraction in the spectrograph slit, his short wavelength baseline of only 1000 Å, and his neglect of absorption lines. The last error was certainly serious, as Rosenberg derived a solar temperature of 4950 K from Abbot's data, which he then employed as a standard source to derive the stellar temperatures. In reality the bluer colour of the hotter stars is only partly due to their greater amounts of blue continuous radiation, but also to the reduction in line blocking at the shorter wavelengths. Rosenberg neglected this effect so would have over-estimated the temperatures of early-type stars.

7.9 Further visual spectrophotometry at Potsdam

Meanwhile Scheiner's death in 1913 after a long illness interrupted the spectral energy distribution program with Wilsing at Potsdam. However, Wilsing continued this work with the help of W.H.J. Münch (1879–1969) who had joined the Potsdam staff as an assistant in 1905. From 1908 to 1913 a further ninety stars were observed, this time in ten pass-bands from 4510 to 6420 Å selected for being relatively line-free (35). In addition, the previous list of 109 stars was re-reduced with the new pass-bands so as to put all 199 stars on a uniform scale.

The Potsdam temperatures for most B stars still lay between 10 000 K and 12 000 K. The difference with Rosenberg's results was considerable, and led another German astronomer, A. Brill (1885–1949), to re-analyse carefully both the Potsdam and Tübingen data (36). Good agreement was demonstrated for the temperatures of stars later than A5, but the wide divergence for the B star results remained. In the 1920s the weight of evidence consistently supported temperatures at B0 of 20 000 K or more, and this implied the presence of a large systematic error of unknown origin in the Potsdam results, while the error in Rosenberg's data was considerably smaller and of the opposite sign (see for example the discussion by W. Becker (37)).

The systematic errors in the stellar temperature scale, however, were hardly a handicap for the next major advance in astrophysics, which concerned ionisation theory. The important aspect that all temperature determinations up to about 1919 confirmed, was the systematic trend with spectral type. This made it seem highly probable that the temperature was the principal parameter that determined the type, and not, for example, the composition.

7.10 Saha and the theory of ionisation

In the first two decades of the twentieth century three principal ingredients went into the mix from which the theory of ionisation was born. These ingredients, as we have seen, were the concept of the Hertzsprung–Russell diagram, the development of atomic quantum theory and the first measurements of stellar temperatures. The person who fitted these three clues together in the interpretation of stellar line spectra on the basis of ionisation in stellar atmospheres was the Indian astronomer Megh Nad Saha (1893–1956).

Fig. 7.7 Megh Nad Saha.

However, he neither invented the concept of ionisation, nor even was he the first to apply ionisation equilibrium theory to astronomy. Professor (later Sir) Arthur Eddington (1882–1944) actually gave a very clear account of the ionisation process in 1917, involving the successive loss of outer orbital electrons as the temperature of a gas increases (38). This work of course explicitly referred to the Bohr atomic theory, but the equilibrium ionisation state as a function of temperature was treated only approximately, and Eddington's discussion was confined to the physical conditions in stellar interiors.

In 1919 Saha travelled to London to visit Alfred Fowler's laboratory at Imperial College, and he then went on to see H.W. Nernst (1864–1941), the renowned German physical chemist who was to receive the Nobel Prize for chemistry in 1920. Nernst had been working on the thermodynamic theory of predicting the equilibrium state of chemical reactions, and the fact that Saha described ionisation as 'a sort of chemical reaction, in which we have to substitute ionisation for chemical decomposition' (39) points to the close influence of Nernst, as Saha readily acknowledged.

Saha would very probably also have met John Eggert (b. 1891), a pupil of Nernst, in Berlin. Eggert calculated the equilibrium state for eight-times ionised iron in stellar interiors (40), using the Nernst dissociation theory. His formula for the dissociation constant for multiply-ionised iron involved a main temperature term of the form $K \propto \exp(-W/kT)$, in which W is the energy required to remove all eight electrons from a neutral iron atom. Eggert's expression was practically identical in form to the equation presented by Saha in 1920 (39) and now known simply as the Saha equation.

Saha initially applied his equation to the problem of ionisation in the solar chromosphere. In particular, he calculated the percentage ionisation of calcium as a function of temperature and pressure in the reversing layer. This layer was believed to produce the bright-lined flash spectrum that had first been seen by C.A. Young in 1870 during a total solar eclipse. Saha assumed the reversing layer to be at 6000 K but with a pressure decreasing outwards from 10 atmospheres to only 10^{-12} atmospheres. The degree of ionisation of calcium was only 2 per cent in the deeper levels, but because of the falling pressure, which also entered into the ionisation equilibrium equation, the ionisation was nearly complete (no neutral atoms remaining) by the point where a pressure of 10^{-4} atmospheres was reached. The great strength of the H and K lines of ionised calcium in the outer solar reversing layer, or chromosphere, could thus be explained, whereas the 4227 Å neutral calcium line was strongest in the deeper layers. This result was in close accord with Lockyer's theory of normal and enhanced lines, supposed to arise from the atoms of elements and proto-elements respectively (see Chapter 4). The terminology was now different and the results were quantitative, but the concept of two distinct spectra arising from the dissociation of the primary material was fully Lockyer's and pre-dated the discovery of the electron (41).

7.11 Saha's analysis of the sequence of Harvard spectral types

Saha's work on stellar spectra in 1921 was his major triumph. He employed the method of 'marginal appearances' based on the first appearance and subsequent disappearance of prominent spectral lines as one progresses through the Harvard sequence of spectral types (42). In spite of the very limited data he had available for ionisation potentials (the energy required to remove an electron from an atom) for the different elements, and no more than order of magnitude guesses on stellar pressures (he took values of the order 1 to 0.1 atmospheres), Saha traced the temperatures at which the H and K lines, the Balmer hydrogen lines, the neutral helium lines, and so on, are marginally present. Thus, for example, he deduced a temperature of 4000 K for Mc stars, since calcium here just begins to be ionised rendering the H and K lines marginally visible. At the other end of the scale, for Harvard Oc stars, double ionisation of calcium leads to the disappearance of the singly ionised calcium lines, and the temperature derived was 20 000 K.

On this basis, Saha deduced a temperature scale for all the spectral types from Oa to Mc and compared his results with the colour temperatures of Wilsing, Scheiner and Münch (35). He found a good correlation between colour and ionisation temperatures, thus supporting the theoretical basis for ionisation changes alone accounting for the spectral type sequence. He concluded his work on stellar spectra as follows:

> It will be admitted from what has gone before that the temperature plays the leading role in determining the nature of the stellar spectrum. Too much importance must not be attached to the figures given, for the theory is only a first attempt for quantitatively estimating the physical processes taking place at high temperature. We have practically no laboratory data to guide us, but the stellar spectra may be regarded as unfolding to us, in an unbroken sequence, the physical processes succeeding each other as the temperature is continually varied from 3000° K to 40,000° K (42).

Saha's temperatures did however average more than 7000 K hotter at type B0 than the Potsdam results. This excess gradually decreased but never vanished as one went to later types. At Cambridge E.A. Milne (1896–1950) gave a critical review of Saha's work that same year (43), and he cited the ionisation temperatures as 'fairly conclusive' evidence for errors in the Potsdam colour temperature scale for hot stars (see section 7.9).

Russell's retrospective and eloquent remarks delivered in the George Darwin lecture of 1935 to the Royal Astronomical Society on this era of stellar spectroscopy are worth quoting (44):

> The first, and still the greatest, triumph of the new conceptions was the interpretation of the sequence of stellar spectra. The systematic study of thousands of spectra at Harvard had shown that the long recognized and very different types of spectra, like those of Vega, the Sun and Betelgeuse, were connected in imperceptible steps by a continuous series of intergrades, but always in the same fashion, so that they could be arranged in a linear sequence – as is recognized in the familiar Draper Classification, with its decimal divisions. Lockyer had explained the disappearance of the metallic arc lines by dissociation. Saha, in 1921, defined this more precisely as ionization of the atoms and pointed out that the appearance of the

> lines of neutral helium only at temperatures high enough to ionize the metals completely was explicable by the high energy of excitation required to get helium atoms into states capable of absorbing the observable lines. He also showed that the degree of ionization might be used to calculate the temperatures of stellar atmospheres, provided that the pressures could be estimated, and suggested that the absolute magnitude effects discovered by Adams and Kohlschütter were probably due to differences in atmospheric density. The later developments of this creative idea have been... numerous and varied... (44).

7.12 Fowler and Milne and the method of line strength maxima The Saha ionisation theory was put on a more secure basis by Professor Ralph Fowler (1889–1944) working with Milne at Cambridge (45). The main development was the explicit inclusion of excitation effects into the calibration which Saha had mentioned but not calculated. Fowler and Milne showed that lines of any subordinate series (i.e. arising from excited levels) should have a maximum somewhere in the spectral sequence due to the combined effects of excitation and ionisation. Transitions from the ground state however only showed a maximum in the case of ions, due to the effects of first and second ionisation at different temperatures (as Saha had himself indicated). Finally, lines arising from the ground state of neutral atoms were always stronger at a cooler temperature, and there was no maximum.

The combined theories of excitation, which obeyed the well-known Boltzmann equation, and of ionisation thus elegantly explained the maxima of the Balmer series around type A0 and of the neutral helium lines for early B stars. The method of maximum strengths has advantages over the less well-defined concept of marginal appearances used by Saha. Fowler and Milne recognised this, and exploited it for fifteen different spectral features to obtain temperatures between spectral types B2 (16 100 K) and K5 (3900 to 4420 K, depending on the lines used). Moreover, the only mutually consistent partial pressure of the electrons was around 10^{-4} atmospheres, lower than hitherto considered, and about 1000 times less than Saha had adopted. Fowler and Milne actually stated that the general agreement between their temperatures and Saha's was fortuitous, partly for this reason, and also because of Saha's lack of explicit calculations for lines arising from excited levels. They wrote: 'It is not clear how he [Saha] obtains a definite temperature for the marginal appearance of the Balmer

Fig. 7.8 E.A. Milne.

lines (4500°, Mb stars) or of the He arc lines (12 000°, A0 stars) without a calculation of the number of atoms in these higher quantum states' (45).

The method of line maxima was next extended to stars of earlier spectral type when Fowler and Milne applied it to the O and early B stars (46). This paper came after a new classification scheme for O stars by H.H. Plaskett (son of J.S. Plaskett) (1893–1980) at Victoria, British Columbia (47). The essence of the Plaskett classification was to place the O stars on a decimal classification based on absorption line strengths only, whereas the old Harvard scheme was based mainly on emission line character (see Chapter 8). Fowler and Milne used as many as 16 different maxima from the spectra of eight mainly lighter elements. The hottest stars of type O5 were found to have ionisation temperatures of 35 000 K; at O9 to B0 this was 26 500 K; at B1, 19 000 K; at B2, 16 500 K.

7.13 Ionisation theory and luminosity effects in stellar spectra

We have already seen how Hertzsprung's work on the Maury c-type stars showed that the spectral peculiarities of these stars we now know as supergiants are associated with very high luminosities, and how this in turn led to the concept of the Hertzsprung–Russell diagram. Saha's work showed that the degree of ionisation of an element depended on the pressure term, or rather the partial pressure of free electrons (written P_e), that forms the link between the large luminosity of the supergiants and their peculiar spectral features, the unusually narrow yet strong lines of certain 'enhanced' spectra.

Saha himself had suggested that the spectroscopic absolute magnitudes of Adams and Kohlschütter might owe their theoretical interpretation to ionisation differences (48). This idea was succinctly expressed again in a short note by the Princeton astronomer J.Q. Stewart (1894–1972) when he wrote: '... a given degree of ionisation should be reached at a lower temperature in stars of low density than of high density, and giant stars should be redder than dwarf stars of the same spectral type' (49); and, later: 'For elements of easier ionisation than the average, the enhanced [i.e. ionised] lines should be stronger in giants than in dwarfs of the same spectral type' (49). Stewart was thus expressing the fact that spectral type does not simply reflect temperature, but more precisely, the degree of ionisation, and thus depends both on density and temperature.

In fact, observations were already available to demonstrate just the effect that Stewart had predicted. They were discussed by F.H. Seares (1873–1964) at Mt Wilson (50) and by Hertzsprung at Leiden (51) in 1922, based mainly on the 1919 colour temperatures of the Potsdam astronomers. Seares commented on the data he had assembled by stating: 'It will be noted that T for a G0 giant is nearly 500° lower than that for a dwarf of the same spectrum' (50). This important result then tied in very nicely with ionisation theory, provided the high luminosity stars could be identified with objects of low density and pressure.

That giants and supergiants do indeed have lower densities and pressures was emphasised by A. Pannekoek (1873–1960) at Amsterdam (52). He argued that giant stars must be much larger than dwarfs because their higher luminosities must be due to a greater radiating surface area. The greater size implies, in turn, a lower gravitational force at their surfaces. The atmospheres are thus less compressed by gravity and hence also less dense than for dwarf stars. Such reasoning was generally current at this time. A necessary additional requirement is that dwarfs and giants have comparable masses, as the supposed lower gravity of giants could otherwise be invalidated by a higher mass offsetting the larger radius. This finding that the masses were indeed roughly comparable (50) thus reinforced Pannekoek's argument.

Pannekoek used this simple reasoning to show that the spectral differences used at Mt Wilson to determine spectroscopic parallaxes empirically, had their explanation in the pressure term entering into Saha's ionisation theory. Fowler and Milne's second paper on ionisation in 1924 put these conclusions into a neat quantitative statement (46). The sun and Capella have about the same spectral types and hence degrees of ionisation.

By applying the Saha equation and adopting a solar temperature of 6000 K they deduced that the forty times lower gravity of Capella implies a temperature for this giant star of 5000 K, in rough accord with its colour, and substantially cooler than the sun.

Later on, Milne developed the theory further in an analysis of the spectral differences between dwarfs and giants (53). He found that both neutral and ionised lines should strengthen in the spectra of the lower gravity giants relative to dwarfs. His conclusions were somewhat hampered by the lack of knowledge about the source of continuous opacity in stellar atmospheres, though under the assumption that the opacity varies in proportion to the electron pressure, many of the observed facts were reproduced. However, the true significance of this finding was not apparent for another decade, until after the discovery of the opacity due to the H^- ion in the atmospheres of the middle- to late-type stars.

7.14 Cecilia Payne and the empirical confirmation of ionisation theory The extensive theoretical predictions of the Saha theory concerning line behaviour and its expected dependence on temperature and pressure in stellar atmospheres were quickly put to the observational test. The most detailed comparisons with stellar spectra were made at Harvard, especially by Cecilia Payne (1900–79) and by D.H. Menzel (1901–76). Menzel's contribution was to test the Fowler–Milne method of line intensity maxima for various elements of different ionisation potential in giant stars (54). The agreement was satisfactory except for the coolest stars, where the observed maxima occurred at somewhat lower temperatures than was predicted, probably because of insufficient knowledge concerning the differences in continuous opacity from star to star.

Miss Payne's contribution, on the other hand, was far more wide-ranging. She tackled the whole field of interpreting the observed line strengths in objective prism spectra using the new ionisation theory. In the process of pulling together the evidence from all the spectral types, she left few stones unturned. Cecilia Payne had been an undergraduate student of Milne's at Cambridge University, where she also came under the influence of Eddington. She then moved to the United States in 1923, and her doctoral dissertation was under Shapley at Harvard College Observatory where she spent the whole of her astronomical career. Her PhD was the first in astronomy awarded at Harvard and was published as a monograph entitled *Stellar Atmospheres* (55). This influential book was firstly a review of the entire field of laboratory and theoretical data of relevance to the interpretation of line strengths in stellar spectra. Its principal contri-

Fig. 7.9 Cecilia Payne in 1924.

bution was the application of the Saha–Fowler–Milne ionisation theory to the wealth of observational objective prism data that had been accumulated at Harvard by Pickering. In practice she mainly used high quality plates taken at Arequipa with the Boyden telescope (both 40 and 19 Å/mm dispersions), together with some 45 Å/mm spectra taken with the 11-inch Draper telescope at Harvard (56).

In her analysis, Cecilia Payne assigned a scale of line intensities to a list of 134 different lines from 18 different elements, including hydrogen and helium. She derived a scale of ionisation temperatures from the line-strength maxima using electron pressures around 10^{-4} atmospheres (57). These were in close agreement with the ionisation temperatures of Fowler and Milne, and confirmed the view that the Wilsing - Scheiner colour temperatures were probably too cool for the hottest stars. She also discussed the effects of gravity on stellar spectra in some detail, showing in addition that the deeper sharper lines of the Maury c-type stars could be interpreted on the basis of very low pressures, as Pannekoek had already established.

Cecilia Payne's most important original contribution in her thesis was probably the emphasis placed on the uniformity of stellar composition from star to star. In so doing she confirmed quantitatively a general feeling that had been raised by Russell at the Atlanta meeting of the Astronomical and Astrophysical Society of America a decade earlier: that temperature, not composition, plays the dominant role in fixing a star's spectral type (see also (58)). Her work was based on the ionisation temperatures at which different lines make their 'marginal appearances' in stellar spectra, a concept first used by Saha (42). For a marginal appearance a line is just visible and this 'must then depend upon the number of suitable atoms above the photosphere' (59).

Miss Payne thus went on to deduce relative logarithmic abundances of the eighteen different elements studied, based on the assumption of uniformity among the stars. Differences in oscillator strength from line to line and in continuous opacity from star to star and with wavelength were all neglected. The simplifications were thus drastic, but results were still forthcoming. She remarked that the 'preponderance of the lighter elements in stellar atmospheres is a striking aspect of the results, and recalls the similar feature that is conspicuous in analyses of the crust of the earth' (59). The agreement between stellar and terrestrial compositions was excellent except in the cases of hydrogen and helium. Relative to silicon these two elements had stellar logarithmic abundances* of 6.2 and 3.5 respectively, much higher than on earth. She commented: 'Although hydrogen and helium are manifestly very abundant in stellar atmospheres, the actual values derived from the estimates of marginal appearances are regarded as spurious' (59). The Harvard astronomer Charles A. Whitney later wrote:

> This was, in fact, one of the earliest clues to the now-accepted high abundance of hydrogen and helium in the universe. In writing that the apparent abundances are almost certainly not real, she was not the last astronomer to mistrust spectroscopic results that contradicted current conceptions (60).

The close agreement for the remaining elements between the compositions of the earth and the stars corroborated the original premise of abundance uniformity throughout the spectral type sequence. The differences between the spectra of the stars first noted by Fraunhofer were thus satisfactorily explained just over 100 years later on the basis of ionisation theory and a sequence of stellar temperatures. Cecilia Payne correctly concluded that: 'the uniformity of composition of stellar atmospheres appears to be an established fact' (59).

Cecilia Payne's *Stellar Atmospheres* received a cautious acceptance when Otto Struve reviewed it in the *Astrophysical Journal* in 1926 (61). His later description of the classic thesis was less reserved: 'It is undoubtedly the most brilliant PhD thesis ever written in astronomy' (62).

* A logarithmic abundance of, for example, 6 corresponds to a factor of 10^6, or a million times the number of silicon atoms.

7.15 The Russell–Adams–Moore analysis of the solar spectrum What Cecilia Payne had done for the stars, Henry Norris Russell and his colleagues now did for the sun. The eminent Princeton astronomer collaborated with Walter Adams at Mt Wilson and with Charlotte Moore (b.1898) who, although based at Princeton, had spent the years 1925–8 as a guest of Adams'.

In the closing years of the nineteenth century Henry Rowland (1848–1901) had published his well-known 'Preliminary tables of solar spectrum wavelengths' (63). Not only were wavelengths tabulated, but also element identifications (thirty-six elements in the sun were identified) and solar line intensities. These latter were based on an arbitrary scale from -3 for the faintest lines just recognisable, to 1000 for the K line. The two D lines had intensities of 20 and 30, while zero was assigned to a weak but readily identifiable line. The wealth of data in Rowland's solar intensities was exploited by Russell, Adams and Moore in 1928 (64). Their method was to consider the solar lines in each multiplet* separately. Russell had shown that the relative number of atoms N in the lower level could be predicted by an approximate method based on the atomic quantum theory (65). Within each multiplet the lines are of different Rowland intensities, R. The calibration of these intensities in terms of the relative number of atoms producing the line, as determined from the quantum numbers, was carried out to find the changes in log N for a given change in R. Putting together the results from 1288 solar lines in 228 different multiplets gave a relationship between log N and R. This relationship was slightly wavelength-dependent, but in the green part of the spectrum (around 5000 Å) for example, a change from $R = -3$ to $+40$ represented a change in log N of about 6, or a million-fold increase in the number of absorbing atoms. They concluded their calibration with the words: 'The most obvious result of the present investigation is to emphasise the enormous differences in the numbers of atoms which are involved in the formation of the stronger and weaker Fraunhofer lines' (64).

The Russell–Adams–Moore analysis was a significant advance over Miss Payne's work, because it did not just use lines of marginal appearance but also those showing saturation. In addition, their wavelength-dependent term in the calibration in effect handled the changes in the solar continuous opacity with wavelength. The results were a primitive forerunner to the present day curve of growth (see section 7.18) as a tool for spectrum analysis.

* The terms between which the transition occurs, as specified by the electron quantum numbers, are the same for all the lines in a multiplet.

7.16 Russell and Adams on stellar composition Russell and Adams immediately applied their calibration to a study of the atomic populations (i.e. values of N) in the atmospheres of seven stars* using coudé spectrograms obtained on the Mt Wilson 100-inch telescope (66). The spectral types ranged from A to M, but only two of the seven were dwarfs. Several hundred lines for each star had their Rowland intensities estimated and they thence deduced a parameter Y for each of fourteen spectra in each star. Y was defined as $\log(N_*/N_\odot)$, where N_* is the number of atoms producing the line in the atmosphere of the star being analysed. Y was tabulated as a function of excitation energy, as there are fewer atoms in the higher energy levels.

One of the principal findings of Russell and Adams was that the supergiant (type c) stars (α Ori, α Sco) have a quantity of metallic vapour 'something like a hundred times that in the sun' while a dwarf such as Sirius had nearly 100 times less (than the sun). On the basis that the ground state (zero excitation energy) populations approximate the total numbers of any given atom or ion, they were able to deduce abundances relative to the sun for all fourteen atoms and ions studied. Their analysis was based on the atmospheric model of Schuster (67) and of Schwarzschild (68), in which the photosphere was considered as the source of a continuous black body spectrum, while the overlying reversing layer contained metallic vapours that scattered the photospheric light at discrete wavelengths to produce dark spectral lines. The abundances were thus expressed as the total amount of metallic vapour above unit area of photosphere. Relative to the sun these were generally 100 times solar for the cool supergiants, but only 0.05 times solar for the A-type dwarf, Sirius. They attempted to account for these results on the basis of the known differences in temperature and surface gravity of the stars, but unfortunately with little success as the incorrect assumption was made that the reversing layers were mainly supported by radiation pressure and not gas pressure.

Since iron, titanium and scandium were all represented in both neutral and singly ionised states, Adams and Russell were able to use the Saha equation to solve for the electron pressure, also expressed relative to the solar value. The electron pressures ranged from 10^{-8} of the solar value for the coolest stars, to 200 times that in the sun for Sirius, the hottest star. They regarded such a wide range in electron pressures as spurious and tentatively ascribed it to departures from the condition of thermodynamic equilibrium, in which the Saha ionisation law is valid. In fact the wide variation, although somewhat in error, corresponds far better to the actual

* These were: α Ori, α Sco, α Boo, α Cyg, α Per, α CMi, α CMa.

conditions now known to prevail in stellar atmospheres than does a constant pressure for all types, as Fowler and Milne had assumed in their first paper (45).

The results were a landmark for two reasons. Firstly, the advantages of differential analysis were exploited for the first time (though Milne had earlier suggested the use of this technique (53)), and secondly, the electron pressure was found to be a parameter which could be interpreted quantitatively to differentiate giants from dwarfs, thus putting Adams and Kohlschütter's empirical spectroscopic parallaxes (18) on a firm theoretical footing.

7.17 Unsöld and Russell on the composition of the sun At about the same time as Russell and Adams were attempting their first analyses of stellar spectra, Albrecht Unsöld (b.1905) in Germany (at the time he was a doctoral student in Munich) undertook an analysis of the solar spectrum, also based on the Schuster–Schwarzschild model with the lines formed by scattering (69). By analysis of the sodium D-line profiles, he gave support to the concept of the transfer of energy up through the solar atmosphere by radiation rather than convection, an idea which Schwarzschild had first proposed in 1906 (70). Unsöld went on to compare the profiles of the solar H and K lines (measured earlier by Schwarzschild) with the inverse square dependence of the scattering coefficient with distance from the line centre, which was expected if the wings of the lines were due to radiation damping. In addition to this he showed the H and K line strengths were consistent with the prediction that they depend on the square root of the calcium abundance.

Unsöld next applied the Saha equation to lines of calcium and strontium, which were observed in both neutral and ionised states, to deduce a solar electron pressure of about 10^{-6} atmospheres. This in turn enabled him to calculate degrees of ionisation and hence total abundances of various elements observed in only one ionisation state, such as sodium which he predicted to be mainly ionised, in spite of the great strength of the D lines due to neutral atoms. His relative solar abundances for sodium, aluminium, calcium, strontium and barium were in reasonable agreement with the results found by Miss Payne for stars, which thus strengthened belief in a common chemistry throughout the stellar universe.

Unsöld's pioneering results on solar composition were soon eclipsed by Russell's classic contribution to the same subject (71). Russell extended the analysis from five elements to a total of fifty-six as well as to six diatomic molecules. A few heavier elements had only one line of doubtful origin ascribed to them, but for the most part the identifications from many lines were well established, being based on the *Revision of Rowland's Table of Solar Spectrum Wavelengths* by St John and his colleagues at Mt Wilson (72). As before, he applied his calibration of the Rowland intensities and then used the Saha equation for five elements with both neutral and ionised spectra. Russell thus deduced an electron pressure of 3.1×10^{-6} atmospheres, close to Unsöld's value. It was then relatively simple to find the total abundances of all the remaining elements in much the same way as Unsöld had done. It is the relative values of these abundances that specify the composition of the sun, and Russell's was the first extensive analysis over such a wide number of elements. He found that the metals from sodium to zinc are far more common than the rest, with a peak in abundance within this group at iron, an important result for future theories of element formation. Some of the lighter elements were, however, more abundant still, especially hydrogen and oxygen, and to a lesser extent carbon, nitrogen, magnesium and silicon. A general tendency for even atomic number elements to be more abundant by about ten times than immediately adjacent ones of odd atomic number was also noted, with the remark that 'the abundance of an element is probably a function of yet unknown properties of the structure of the atomic nucleus' (71).

One of the most important conclusions reached by Russell was the high abundance of hydrogen. The value was quite uncertain since it came from only one saturated Balmer line, but the result, as it happens, was (fortuitously) close to the modern value (relative to iron Russell found log $N_H/N_{Fe} = 4.3$). He concluded that '. . . the great abundance of hydrogen can hardly be doubted' even though the result 'is almost incredibly great'.

A comparison with Cecilia Payne's results (55) showed the same general trends between the element abundances in the sun and stars. In particular, the predominance of hydrogen, the very low abundance for lithium and the peak for iron and neighbouring elements were all common features of Miss Payne's work for stars and Russell's analysis of the sun.

In 1949 a symposium was sponsored by the International Astronomical Union in Zürich on *The Abundances of the Chemical Elements in the Universe*. Twenty years had now elapsed since Russell's solar analysis, and a much improved result for the sun's composition had been obtained by Unsöld (73). Russell wrote to the conference chairman, Struve, on this occasion:

> It may be pardonable to remark that the reconnaissance by my colleagues and me 20 years ago is a rather remarkable example of an approximate theory applied to empirical estimates of low

accuracy. The relative abundance of the commonest and the rarest elements was overestimated by a factor of the order of 10 and the absolute abundance of the former by about the same factor. Owing to ignorance of the correct damping factor, the values given for the elements lighter than sodium were based on scanty uncalibrated data and extreme extrapolation. Fortune appears to have been remarkably kind to audacity. It is much to be hoped that accurate determinations for all the elements will soon replace this exploratory work. The successful study of the solar atmosphere may be attributed to its being nearly in thermodynamic equilibrium – at least for most radiations. The central parts of H and K and of the MgII pair near λ2800 and still more the He absorption near 10,830 would probably tell a different story (74).

7.18 The first curve of growth Russell's technique of spectrum analysis differed from what is commonly known as the 'curve of growth' method in two important respects. Firstly, his line strengths were based on estimates of Rowland intensities, and secondly, his calibration of these intensities was valid only for the few lines within each given multiplet, for which the relative numbers of absorbers could be calculated. The concept of the equivalent width of an absorption line was a key step forward. It is a measure of the fraction of flux (or intensity) removed from the continuous spectrum by the absorption over the whole line. H. von Klüber (1901–78) at Potsdam in Germany (75) and Marcel Minnaert (1893–1970) at Utrecht in Holland (76) introduced the concept almost simultaneously in 1927, to determine the strengths of absorption lines in the solar spectrum. For example, von Klüber obtained 4.2 Å for the equivalent width of the solar Hα line; if a rectangular band of width 4.2 Å were completely dark at the same wavelength, then this would absorb the same amount of light as the solar Hα line. Physicists had for some years prior to this used the term 'total absorption' of a line, which was defined in a similar way in terms of the residual intensity integrated over angular frequency (see (77) where the concept is formally introduced), but the use of total absorption never became established in astrophysics, where the visual Rowland intensities were, until the late 1920s, found to be more convenient.

Minnaert is generally regarded as the founder of the curve of growth technique of spectral analysis. He was a Belgian national whose pro-Flemish activism got him into trouble with the authorities. In November 1918 he fled to Holland to avoid serving a 15-year penal sentence in his native country. It was here, at Utrecht, that he continued his recently begun

Fig. 7.10 Marcel Minnaert.

career as a physicist (his early training having been in botany). At Utrecht he became interested in the techniques of photographic spectrophotometry and his outstanding contributions to solar physics were made there during the 1930s.

In its most primitive form the first empirical curve of growth for Fraunhofer lines was obtained by Minnaert and his pupil van Assenbergh (78) in 1929, after they calibrated Rowland's intensity scale in terms of equivalent width, and plotted the logarithm of this latter quantity against log N for fifty-seven blue lines. The number of absorbing atoms, N, was deduced for lines within a multiplet using Russell's technique. A second paper followed in 1930 with his student Gerard F.W. Mulders (b. 1908), this time applying the method to lines in the blue and green regions of the solar spectrum (79) (Fig. 7.11).

Minnaert and Mulders wrote: 'Russell's calibration of the Rowland scale was used in order to find the relation between "the number of effective resonators" N and the total intensity i [equivalent width]. From Russell's curve $R = F(N)$ and our result $i = f(R)$, R was eliminated'. The result was the logarithmic relationship between equivalent width and the number of absorbing atoms, and it is this curve which Minnaert a few years later christened 'the curve of growth' (80). The strongest lines in solar and laboratory spectra were believed to follow a square root dependence of their equivalent width on N, as Unsöld had demonstrated for the H and K

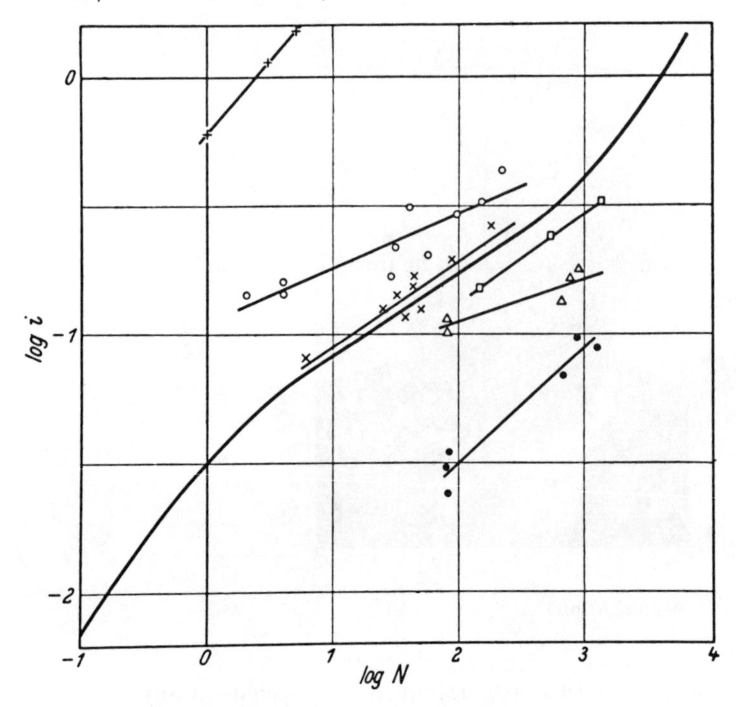

Fig. 7.11 An early empirical solar curve of growth, by Minnaert and Mulders in 1930. The thickly drawn line is the mean curve from the individual multiplets.

lines in the solar spectrum (69). However, Minnaert and Mulders were able to show the square root law to be inapplicable to all lines; intermediate strength ones obeyed a law $i \propto N^{0.31}$, while the weakest possibly approached the $i \propto N$ relation.

At about the same time that Minnaert was working on the empirical solar curve of growth, Wilhelm Schütz in Munich extended the theoretical basis for the curve (81). Assuming that line opacity profiles were a combination of Doppler profiles (due to thermal motion) and Lorentzian profiles due to the damped oscillation of the optically active electrons in atoms, the result is then a so-called Voigt profile (82) which can be integrated over all frequencies to find the 'total absorption' or equivalent width. W. Schütz's theoretical curves for different damping constants showed three distinct regions: weak lines with $W \propto N$, intermediate strength lines where there is almost no increase in the total absorption with N, and strong lines for which $W \propto \sqrt{N}$. His results can be regarded as the first theoretical curve of growth, and he attempted to confirm these findings in the laboratory, though with only partial success, by passing a light beam through calcium vapour and photographing the spectrum.

Minnaert evidently planned to interpret theoretically the solar curve of growth he had constructed. With Mulders, he wrote:

> It was our intention to extend the material to other spectral regions and then to look for a theoretical basis for the curves we observed. Meanwhile the work of Schütz appeared concerning the total absorption of spectral lines... Already the superficial comparison of our curves with the theoretical figures of Schütz gave a surprising similarity (83).

Now, armed with the theory for the curve of growth he was able to analyse further solar equivalent widths. One of the most important results was the conclusion that the damping constants for some of the strongest lines in the solar spectrum were about ten times greater than expected on the basis of the radiative damping of classical electron oscillators (83).

Meanwhile Minnaert and another of his students, C. Slob, calculated their own theoretical curves of growth, similar to those of Schütz, but now more explicitly in an astrophysical instead of laboratory context (84). These theoretical curves formed the basis for the interpretation of the first stellar curve of growth to be plotted, for α Cygni by Minnaert's Dutch colleague, A. Pannekoek.

The whole of this productive era involving Minnaert and his students at Utrecht has been vividly recalled more recently by Minnaert on the occasion of a symposium to celebrate his 70th birthday (85). He starts with a reference to Russell's work using the Rowland intensities and then goes on:

> Notwithstanding our admiration for Russell's wonderful work, I thought it was necessary to get rid of the Rowland scale, to substitute for its estimates the better defined equivalent width and to find how it increases as a function of the number of atoms. Even with our not too perfect spectrograph this could be achieved. We found a curve for the blue region of the spectrum (1929), then another, more or less similar, for the green region (1930). These curves had a curious bend and looked very incomprehensible. However, by a piece of good fortune, just at that moment a paper had been published by Schütz (1930), who had found a similar curve in laboratory spectra. I asked Mulders to study the paper and to see whether it could not be applied in our case; he thought that it looked hopeful. The next days were very exciting; we computed from the morning to the evening and obtained the first theoretical curves of growth for the sun. Our incomprehensible graphs had

found an explanation: apparently the Fraunhofer lines were broadened as well by Doppler effect as by damping; the damping could be found, and proved to be more than ten times the classical damping. This was considerably more than what radiation damping could explain, this could only be collisional damping. We enthusiastically wrote a paper for the *Zeitschrift für Astrophysik*, added a rather accessory calculation about the radiation damping of the green magnesium lines and sent the paper to Grotrian; but two or three weeks later, we noticed that in that accessory part we had made an obvious and serious error. Fortunately the manuscript had not yet gone to the printer and the error could be corrected – it would have spoilt the paper, which now remained one of our finest remembrances (1931). This is the story of what I called the curve of growth (1934), a term where you are free to find a reminiscence from the biological studies of my youth (85).

7.19 The curve of growth applied to interstellar and stellar lines Almost immediately after Schütz had demonstrated the theoretical basis for the shape of the curve of growth, Unsöld, Struve and Elvey applied the method to investigate the strengths of interstellar calcium lines in a number of stellar spectra (86). These lines arise from cool clouds containing calcium ions along the line of sight to a star, and appear as very sharp absorption features especially visible in the spectra of hot and distant stars. The H and K lines of ionised calcium should follow a curve of growth, and to test this Unsöld, Struve and Elvey used Yerkes spectra of three stars to measure the ratios of the equivalent widths of the K to the H line. For all three stars they found ratios near unity, implying the lines were influenced mainly by Doppler effects and hence lay on the intermediate, or flatter, part of the interstellar curve of growth. This was the first application of Minnaert's curve of growth to analyse a spectrum from outside the solar system.

Soon after this work on the interstellar calcium lines, Antonie Pannekoek in Amsterdam constructed the first empirical curve of growth for a star (other than the sun). He measured lines of ionised titanium, iron and chromium in the supergiant, α Cygni (Deneb) (87). Pannekoek's observations came from high quality spectrograms he had himself obtained in 1929 at the Dominion Astrophysical Observatory on the 72-inch (1.8 m) telescope while on a visit as a guest of J.S. Plaskett (Fig. 7.12). The spectra were recorded on a Moll microdensitometer at the Astronomical Institute in Amsterdam. The comparatively new technique of graphical recording of

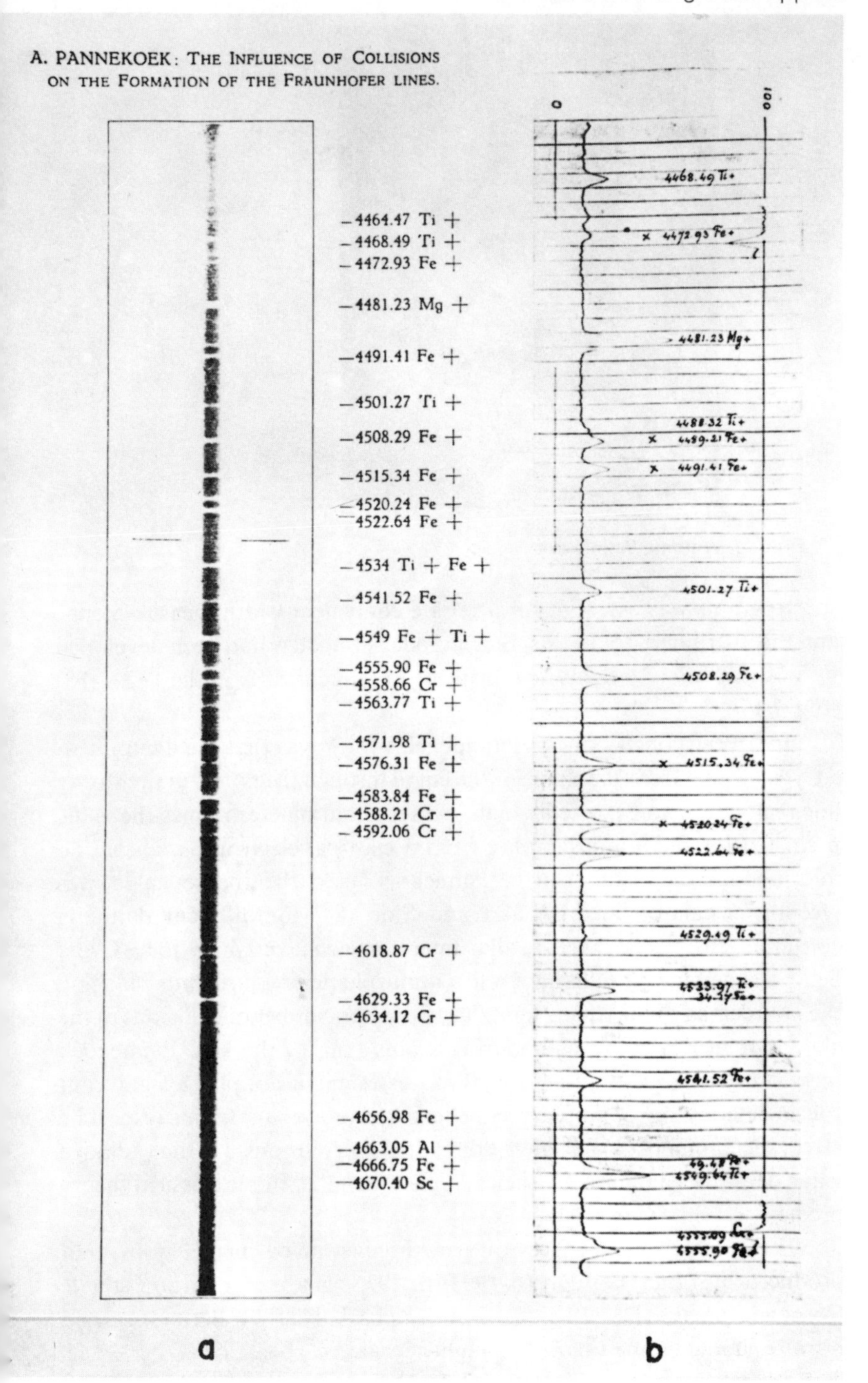

Fig. 7.12 a. Spectrum of α Cygni photographed with the Victoria telescope by A. Pannekoek. b. Part of Pannekoek's microdensitometer tracing of α Cygni.

Fig. 7.13 Antonie Pannekoek.

spectrograms was essential for accurate equivalent width measurements, and the instrument to do this, the microdensitometer, had been developed by W.J.H. Moll at the Physical Institute in Utrecht early in the 1920s (88) (Fig. 7.15).

One of the conclusions of Pannekoek's work was that the damping in α Cygni, found from the shape of the curve through the points in the strong line region, was unexpectedly high, in fact about nineteen times the value predicted for radiation damping from a classical electron oscillator. To interpret his curve of growth, Pannekoek used the theoretical curves recently calculated by Minnaert and Slob (84) for different damping constants. His results were, at that time, somewhat puzzling, for α Cygni being a supergiant should have a low atmospheric pressure, thus reducing the broadening of the strong lines due to a lesser number of collisions of the absorbing atoms with neighbouring atoms than in the sun. Pannekoek therefore concluded that collisional processes might not play a significant role in determining damping in either α Cygni or the sun. In this respect he erred; the damping in the sun is dominated by collisions but the radiative value of a supergiant is nevertheless higher than the simple classical theory predicts.

Of the early stellar curves of growth constructed in the 1930s, that produced by Louis Berman (b. 1903) in 1935 for the hot carbon star R Coronae Borealis (R CrB) was the most remarkable (89). Berman carried out an abundance analysis on this highly peculiar carbon-rich variable star, whose spectral type he classified as cF7p. His spectra came from a variety of

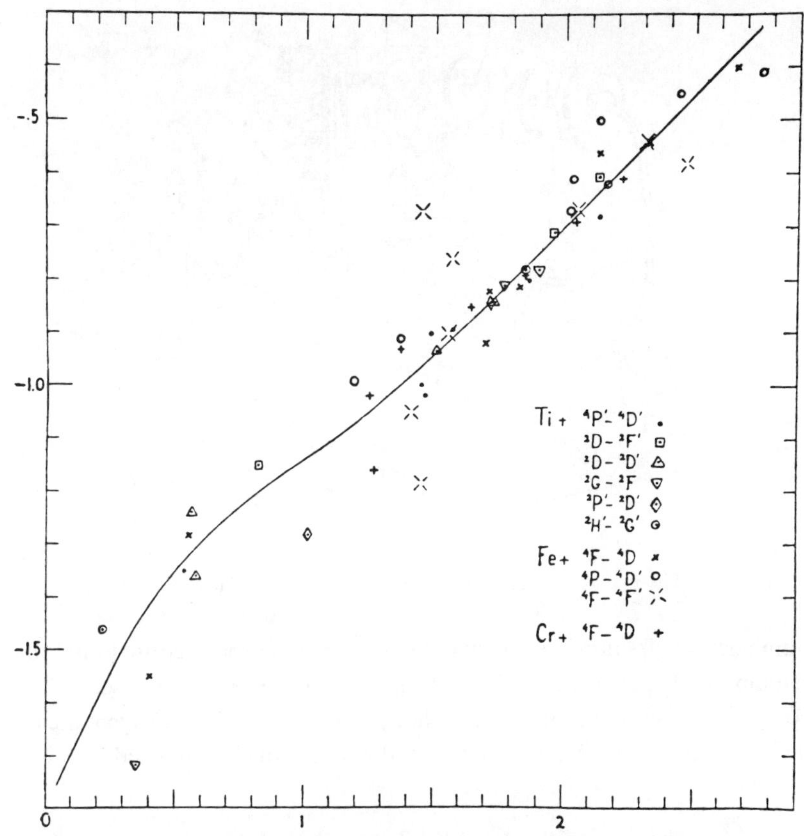

Fig. 3. Aequivalent width-curve for lines of α Cygni. Horiz. *Log C* (concentration); Vert. *Log A* (aequivalent width).

Fig. 7.14 The first stellar curve of growth, constructed by Pannekoek for α Cygni.

sources, but they were mainly from the Mills spectrograph on the giant Lick refractor at a dispersion of 10 Å/mm in the blue. Berman measured equivalent widths from microdensitometer tracings for over 600 lines and hence constructed an empirical curve of growth based on theoretical relative multiplet intensities calculated by Leo Goldberg. He deduced an excitation temperature of 5300 K from the neutral iron curve of growth and then proceeded to measure abundances for twenty-four elements, expressed as the number of atoms above 1 cm^2 of photospheric surface. For an absolute calibration of the zero points of the oscillator strength scale, Berman used the results of Russell's analysis of the solar spectrum in 1929 (71). In effect his abundances can therefore be regarded as differential relative to the sun. The results of this pioneering analysis showed that R CrB contains 69 per cent carbon and only 27 per cent hydrogen (by

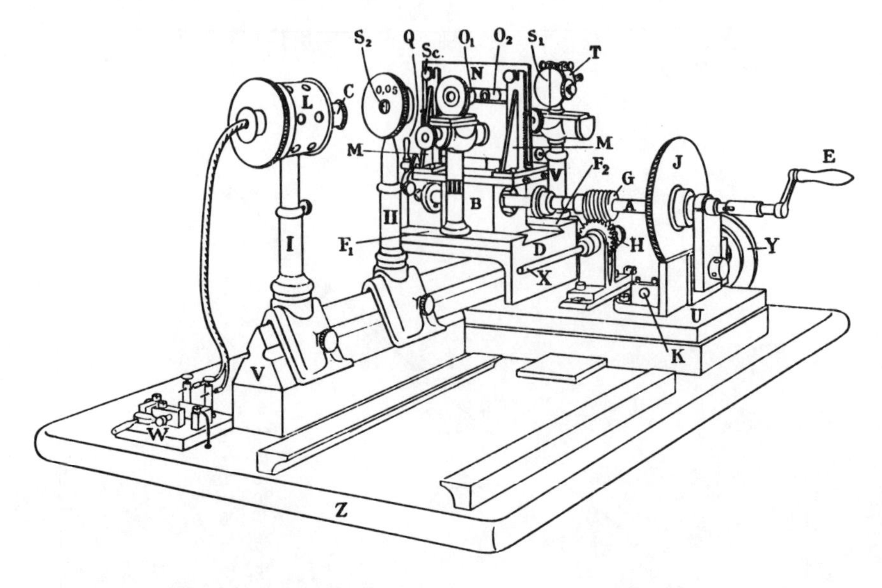

Fig. 7.15 The first registering microdensitometer by W.J.H. Moll.

number), while other elements heavier than carbon contributed the remaining 4 per cent (helium was, of course, not analysed). This composition was recognised as highly peculiar, with hydrogen greatly deficient relative to heavier metals and carbon much enhanced.

7.20 The gradient effect and stellar atmospheric turbulence

The development of the curve of growth concept was almost entirely a European affair, with the main work taking place in Holland and Germany by Minnaert, Mulders, Schütz, Pannekoek and Unsöld. The main foundations were laid over only a 12 month period from July 1930 to June 1931. In the United States the emphasis was slightly different, especially in the analysis of stellar spectra by Otto Struve and his Yerkes colleagues. By 1932 Struve had completed over 2 years of observations on the A0 spectrum variable star, 17 Leporis, using the Bruce spectrograph on the Yerkes 40-inch refractor. He compared the total absorbed energies of lines (proportional to equivalent width) in the spectrum of this star for lines within each multiplet as a function of the theoretical relative numbers of absorbers (called multiplet intensities), as, for example, Russell, Adams and Moore had done for the sun and stars (64,66). The first result was that in 17 Leporis the gradient of the relation between the line intensities and the theoretical multiplet intensities was quite steep, clearly steeper than the square root law then known to be valid for most strong lines in the sun (90). On the other hand the gradients for lines within many multiplets of the F5 supergiant ε Aurigae were generally *less* steep than the square root variation. The same conclusion was reached shortly afterwards by J.A. Hynek (b. 1910), also at Yerkes, but based on more extensive line measurements in the two stars (91).

Two years later, Elvey at Yerkes analysed the ionised iron and titanium spectra of eight stars: α Persei and 41 Cygni both showed smaller gradients for the ionised titanium lines than predicted by the square root law, while three stars (α Cygni, α Lyrae and especially 17 Leporis) all appeared to have steeper gradients for both titanium and iron lines than given by the square root relationship (92). For 17 Leporis the ionised iron lines were practically proportional to the number density of absorbers. Then all these results were brought together, reviewed and further analysed in Struve and Elvey's classic paper on stellar curve of growth gradients of 1934. Several hundred lines were measured in each of six stars using high dispersion blue spectra (10 Å/mm). In the case of α Persei, these came from as many as eight different elements. Composite curves of growth were constructed by horizontally shifting the data points plotted for each multiplet, a procedure which can be checked by observing the emission line intensities from an optically thin gas, as observed in the solar chromospheric flash spectrum.

The results were as follows: the gradient varied from star to star, at one extreme close to equivalent width W being proportional to number density of absorbers N (gradient 45°), while at the other extreme it was nearly zero. In other stars (especially α Per, ε Aur) the gradient was variable, decreasing for the stronger lines. For example, for 17 Leporis the entire curve could be described by $W \propto N$ even for lines as strong as $W = 1$ Å, while for α Persei line saturation led to a flat part once $W = 0.4$ Å was reached.

How could Struve and Elvey explain these gradient differences from star to star? Using the theory of Minnaert and Slob (84) it was clear that the strengths of weaker lines were dominated by their Doppler opacity profiles. Such lines should not continue to grow in proportion to N, but saturate at a certain strength which depends on the spread of characteristic velocities of the absorbing atoms, that is on the photospheric temperature. Using the theoretical curves of Minnaert and Slob, Struve and Elvey were thus able to measure this spread in velocities, called the Doppler velocity v_D, from the shapes of the curves. They then derived the kinetic temperature for each star using $T \propto v_D^2$ as required by the standard kinetic theory of gases. The kinetic temperatures, especially for the supergiants analysed, were much too high to be realistic; for 17 Leporis, 3×10^7 K for ε Aurigae, 2×10^6 K!

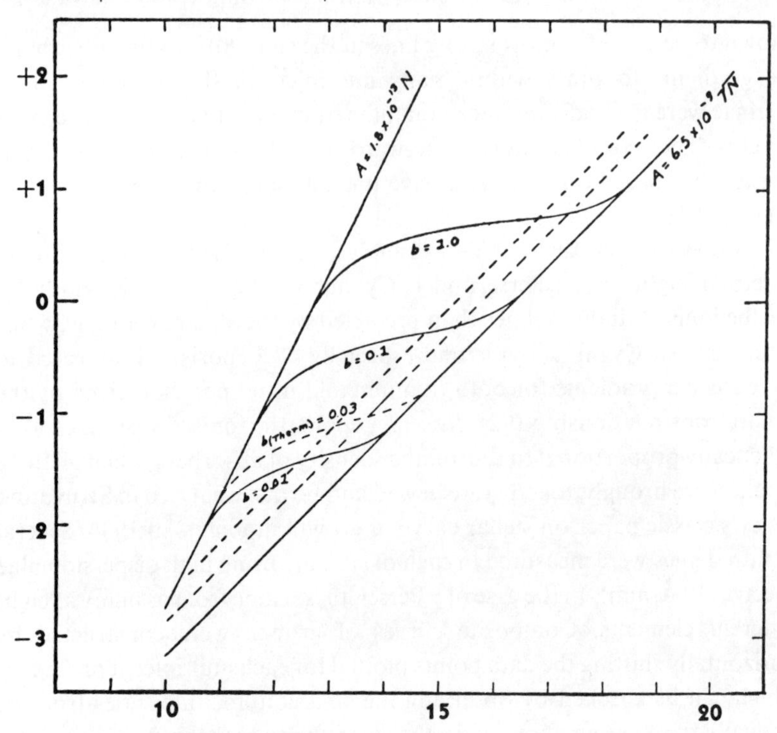

Fig. 7.16 Curves of growth for different degrees of turbulence calculated by Struve.

In order to explain this startling anomaly for stars believed to have temperatures no more than several thousand degrees, Struve and Elvey invoked the following hypothesis: 'The atmospheres of the stars are agitated not only by thermal motion of the individual atoms, but also by currents of a macroscopic character, which we shall refer to as "turbulence" '(93). The turbulent motions in effect considerably increased the overall Doppler velocity and the characteristic turbulent velocity corresponded to as much as 67 km/s for 17 Leporis and 20 km/s for ε Aurigae, but only about 2 km/s or less for the dwarf stars analysed. They concluded: 'We see that the agreement [with theory] is excellent if we assume that the three stars 17 Leporis, ε Aurigae and α Persei have turbulent atmospheres, while in the sun, α Cygni, Vega and Sirius, turbulence is not appreciable' (93).

Although Struve and Elvey are normally credited with the discovery of stellar turbulence, the concept was not entirely new to astrophysics. In 1929 Unsöld had measured a turbulent velocity of 15 km/s for the chromosphere of the sun from the widths of weak absorption lines of ionised calcium (known as H_3 and K_3) (94), and as 1869 Lockyer had reported

Fig. 7.17 Otto Struve at Yerkes Observatory.

extreme rates of movement in the same layer of the sun (95). Nor was turbulence the only explanation for the gradient effect that was proposed. Donald Menzel at Harvard believed a distorted curve of growth could result if photospheric absorption lines were partially filled in by chromospheric emission, which could give the effect of different gradients from star to star (96).

Many years later Struve wrote about this pioneering work on turbulence: 'I had some misgivings about the use of the word "turbulence" in our 1934 article, because I did not wish to imply that the motions necessarily resembled those described by the same term in air. I did, in fact, think of a phenomenon resembling solar prominences, and I have stated so on many occasions' (97).

Following Williams' measurements of helium absorption line equivalent widths in eighty-four O and B stars (98) (see section 8.4.1), Leo Goldberg (1913–87) at Harvard constructed curves of growth for these lines in fifty-seven of Williams' stars (99). The main object of this work was to deduce total Doppler and turbulent velocities for B stars from the level of the flat part of the curve of growth for lines of intermediate strength. For supergiants Goldberg found that the turbulence increased rapidly with earlier spectral type from B8 to O9, but the turbulence in supergiants as a whole was not markedly different from the values for stars of lower luminosity.

The Goldberg survey greatly increased the numbers of stars with determinations of turbulent velocities. The results confirmed the finding of Struve and Elvey, that in general the thermal motions in these stellar atmospheres were insufficient to account for the observed high values of total Doppler velocity. Hence an additional parameter, namely turbulence, had to be introduced. However, quantitatively Goldberg's results may not have been very reliable. Only nine neutral helium lines were used to define the curves of growth. In many stars all these lines were on the flat part of the curve of growth, whereas in practice some weaker lines must be measured for reliable results.

7.21 Subsequent work in solar spectral analysis and the Utrecht Solar Atlas

Minnaert and his colleagues had constructed the first solar curves of growth between 1929 and 1931. Meanwhile C.W. Allen (1904–87) continued this work in Australia using many new determinations of solar equivalent widths to construct a new solar curve of growth (100). One of his conclusions was that if lines from different parts of the spectrum are to be combined into a single curve, then log (W/λ) should be plotted as the ordinate and not log W, where W is the equivalent width and λ the wavelength of the line. This result was soon explained theoretically by Donald Menzel at Harvard (101). He computed a theoretical solar curve of growth and showed that the ratio of the equivalent width to the Doppler width, expressed as log $(W/\Delta\lambda_D)$ gives a unique curve of growth for all elements and wavelengths. The Doppler width is proportional to λ and to the spread in velocities of the individual atoms in the atmosphere, and thus Menzel's ordinate differed from Allen's by only a constant amount.

Menzel's paper was based on the generally used Schuster–Schwarzschild model in which the Fraunhofer lines form by scattering in the solar reversing layer which is over the photosphere, which produces the continuous spectrum.* The line profiles on this model posed a considerable problem, as the centres of the strong lines should be

*This model gives a specific formula for the depression in the continuous intensity as $R = N\alpha_\lambda/(1 + N\alpha_\lambda)$ where N is the number of atoms in the reversing layer over unit area of photosphere and α_λ is the atomic scattering cross-section.

completely black, whereas in practice even the strongest lines are still observed to have some residual intensity in their centres. Because of this problem, Unsöld (102) preferred to compute line profiles on the basis of a semi-empirical formula* first used by Minnaert, in which the maximum depression in the line centre was explicitly included, and he recalculated theoretical curves of growth on this basis.

As a result of Allen's and Menzel's work, the role of the oscillator strength in bringing all the lines even in different multiplets onto a single curve was clearly emphasised. Allen showed that the solar spectrum could be used to measure atomic oscillator strengths for solar lines, while Menzel used an abscissa for the curve of growth written as $\log X_0$ where X_0 is proportional to Nf, the number of absorbing species times the oscillator strength for that particular transition. The oscillator strengths are proportional to the transition probabilities, and can be regarded as the fractional number of effective electron oscillators per atom, for a given transition. The importance of determining oscillator strengths for atomic lines in the laboratory thus became apparent, and the work of Arthur (1876–1957) and Robert King (b. 1908) at Mt Wilson Observatory resulted in the first extensive compilation of electric furnace measurements for over 400 lines of astrophysical interest in the spectra of neutral iron and titanium (103). The random errors of measurement were claimed to be about 10 per cent, at least in the relative oscillator strengths.

The availability of the Australian solar equivalent widths by C.W. Allen and of the Kings' oscillator strengths resulted in several new analyses of the shape of the solar curve of growth. Robert King's analysis at the Massachusetts Institute of Technology in 1937 was the first to make use of both these new sources (104). With oscillator strengths making it possible to predict the relative strengths of lines in different multiplets, the curve of growth method allowed the relative populations of an atom in its different excited energy levels to be measured from the equivalent widths. Since these relative populations depend on temperature (through the Boltzmann equation), a so-called excitation temperature could be derived. King's value of 4400 K for the sun from neutral titanium lines was the first such determination (104) using the laboratory oscillator strengths. At about the same time Menzel, with James Baker (b. 1914) and Leo Goldberg at Harvard also measured solar excitation temperatures of 4350 ± 200 K, from neutral titanium lines, and 4150 ± 50 K from neutral iron; in their case they used Allen's equivalent widths, but with the theoretical relative intensities of the multiplets which Goldberg had calculated (105).

* The Minnaert formula is $1/R = 1/R_c + 1/(N\alpha_\lambda)$. As N becomes large, R approaches the maximum value of R_c.

A monumental work in solar spectroscopy was being undertaken at this time in Utrecht, which culminated in the publication of the *Photometric Atlas of the Solar Spectrum* by Minnaert and his pupils G.F.W. Mulders and Jakob Houtgast (1908–82) (106). The whole work was completed almost on the eve of the outbreak of war, in circumstances which Minnaert later recalled on the occasion of his 70th birthday seminar (85):

> In 1936 Mulders went to the Mt Wilson Observatory and took the plates for our *Photometric Atlas*, while Houtgast developed the modified, home-made instrument, which could be added to the microphotometer and gave direct intensity recordings. All microphotometer curves were obtained by direct photographic recording; we worked mostly in the night, because then the microphotometer was free. You were alone in the building, and in the silence of the darkroom, in the dull red light, you were developing your record. There it emerged, slowly emerged, out of nothingness, and as if by magic there appeared on the paper the profile of the cyanogen band, or of the atmospheric oxygen lines, never earlier observed in their true quantitative shape. The *Atlas* was ready, just before the second world war broke out; and the very last airplane which left Portugal for the United States, the last before the five years long interruption of communications, carried five copies of our *Atlas* for the collegues [sic] over there, our last greeting before the tempest.

The recording of the Utrecht solar atlas had a long-lasting influence on quantitative solar spectral analysis in the post-war years. For the first time the instrumental profile of a recorded spectrum was also measured, and the effect of the spectrograph's resolving power on the observed line profiles could be evaluated, as Minnaert himself discussed in some detail in a lecture he gave before the Royal Astronomical Society in London in 1947 (107).

7.22 K.O. Wright and the solar curve of growth An early product of the careful photometric work on the solar spectrum undertaken by Minnaert and his Utrecht colleagues came with K.O. Wright's (b. 1911) new solar curve of growth in 1944 (Fig. 7.18). Wright, who had been on the Dominion Astrophysical Observatory staff at Victoria, British Columbia since 1936, measured the equivalent widths of about 700 lines in the Utrecht atlas by the laborious process of counting squares on the graphical records to integrate the line profiles. He claimed good agreement with Allen's results and random errors of about 7 per cent,

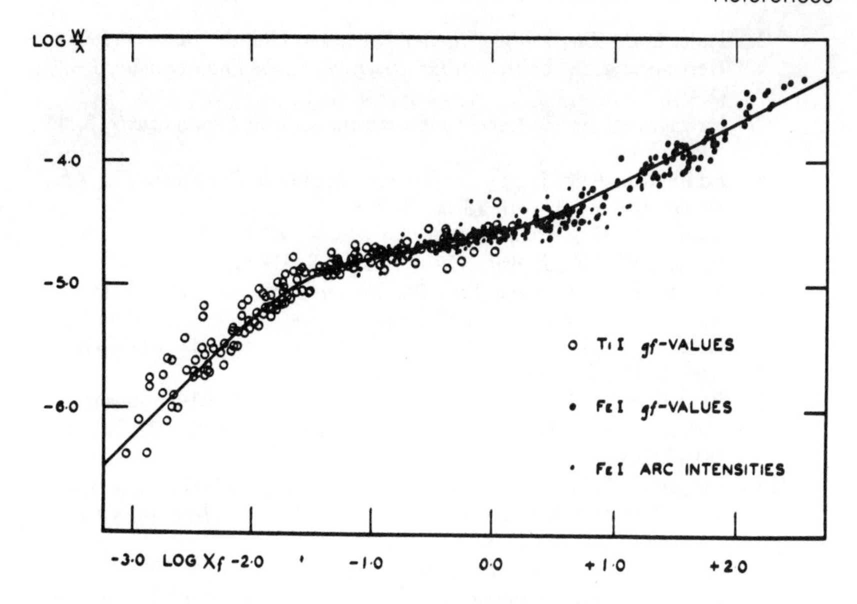

Fig. 7.18 Solar curve of growth by K.O. Wright in 1944.

even for the weakest lines of only a few milliangstroms equivalent width (108).

Wright's solar curve of growth for neutral iron and titanium was easily the best defined curve to date, especially as so many weak lines were included. His curve was constructed using the equivalent widths he had measured from the Utrecht atlas, and supplementing these with those obtained by Allen. The Kings' laboratory data on oscillator strengths (103) was used for the horizontal curve of growth axis. One of the main results of this work was the measurement of solar excitation temperatures from the titanium and iron spectra. Wright obtained 4900 ± 125 K from neutral iron and 4550 ± 125 K from neutral titanium lines, a difference he considered real, though an overall mean value of 4700 K was adopted. According to H.H. Plaskett, the discrepancies in temperature between the team at Harvard (105) and the results of Wright lay in the different oscillator strengths (109). Wright also obtained a turbulent velocity of 0.9 km/s in the solar photosphere, by comparing the empirical data with the theoretical curves of growth that had been most recently computed by Menzel (110).

References

1. Lockyer, J.N., *Inorganic Evolution*, Macmillan & Co. (London) (1900).
2. Schuster, A., *Astrophys. J.*, **17**, 165 (1903).

3. Hale, G.E., *The Study of Stellar Evolution*, Uni. of Chicago (1908).
4. Hertzsprung, E., *Zeitschrift für wissenschaftliche Photographie*, **3**, 429 (1905).
5. Hetzsprung, E., *Zeitschrift für wissenschaftliche Photographie*, **5**, 86 (1907).
6. Eddington, A.S., Letter to Hertzsprung (1925). See Strand, K. Aa., *Dudley Observ. Report*, **13**, 55 (1977).
7. Monck, W.H.S., *Publ. Astron. Soc. Pacific*, **4**, 98 (1892).
8. Monck, W.H.S., *J. Brit. Astron. Assoc.*, **5**, 418 (1895).
9. Russell, H.N., *Carnegie Inst. Washington Publ.*, no. **147**, (1911).
10. Russell, H.N., *Publ. Astron. Astrophys. Soc. America*, **2**, 33 (1915).
11. Hertzsprung, E., *Publ. Astrophys. Observ. Potsdam*, **22**, 1 (1911).
12. Russell, H.N., *Observatory*, **36**, 324 (1913).
13. Russell, H.N., *Publ. Amer. Astron. Soc.*, **3**, 22 (1918). Also published in *Popular Astron.*, **22**, 275 and 331 (1914) and *Nature*, **93**, 227, 252 and 281 (1914).
14. DeVorkin, D.H., *Dudley Observ. Report*, **13**, 61 (1977) *discusses* this point. See also Jones, B.Z. and Boyd, L.G., *The Harvard College Observatory*, Harvard, p. 430 (1971).
15. Russell, H.N., *Observatory*, **36**, 324 (1913).
16. Strömgren, B., *Vierteljahrschrift der Astron. Gesellschaft*, **68**, 306 (1933). See also *Zeitschrift für Astrophys*, **7**, 222 (1933).
17. Quoted by Strand, K. Aa., *Publ. Astron. Soc. Pacific*, **80**, 51 (1968).
18. Adams, W.S. and Kohlschütter, A., *Astrophys. J.*, **40**, 385 (1914).
19. Thomson, J.J., *Phil. Mag.*, **44**, 293 (1897).
20. Planck, M., *Wiedemanns Ann. der Physik*, **4**, 553 (1901).
21. Einstein, A. *Ann. der Physik*, **17**, 132 (1905).
22. Rutherford, E., *Phil. Mag.*, **21** (Ser 6), 669 (1911).
23. Bohr, N., *Phil. Mag.*, (6) **26**, 1 (1913).
24. Balmer, J., *Ann. der Physik und Chemie*, **25**, 80 (1885).
25. Bohr, N., *Nature*, **92**, 231 (1913) and *Phil. Mag.*, (6) **26**, 1 (1913).
26. Wilsing, J. and Scheiner, J., *Publ. Astrophys. Observ. Potsdam*, **19**, 1 (1909).
27. Abbot, C.G., *Astrophys. J.*, **31**, 274 (1910) and ibid **32**, 319 (1910).
28. Nordmann, C., *Bull. Astronomique*, **26**, 5 (1909).
29. Nordmann, C., *Comptes Rendus de l' Académie des Sciences*, **149**, 557 and 1038 (1909).
30. Schwarzschild, K., *Publ. der Kuffnerschen Sternwarte, Vienna*, **5** (1899).
31. Hnatek, A., *Astron. Nachrichten*, **187**, 369 (1911).
32. Rosenberg, H., *Abdhandlungen der Kaiserlichen Leopold. Carol. Deutschen Akademie der Naturforscher, Nova Acta*, **101** (no. 2) (1914).
33. Abbot, C.G., *Astrophys. J.*, **34**, 197 (1911).
34. Wilsing, J., *Astron. Nachrichten*, **204**, 153 (1917).
35. Wilsing, J., Scheiner, J. and Münch, W., *Publ. Astrophys. Observ. Potsdam*, **24**, 1 (1919).
36. Brill, A., *Astron. Nachrichten*, **218**, 209 (1923).
37. Becker, W., *Handbuch der Astrophysik*, Vol. 7, page 459. Springer-Verlag, Berlin (1936).
38. Eddington, A.S., *Mon. Not. Roy. Astron. Soc.*, **77**, 596 (1917).
39. Saha, M.N., *Phil. Mag.*, **40**, 479 (1920).
40. Eggert, J., *Phys. Zeitschrift*, **20**, 570 (1919).

41. Lockyer, J.N., *Proc. R. Soc.*, **44**, 1 (1888).
42. Saha, M.N., *Proc. R. Soc.*, **99A**, 135 (1921).
43. Milne, E.A., *Observatory*, **44**, 261 (1921).
44. Russell, H.N., *Mon. Not. Roy. Astron. Soc.*, **95**, 610 (1935).
45. Fowler, R.H. and Milne, E.A., *Mon. Not. Roy. Astron. Soc.* **83**, 403 (1923).
46. Fowler, R.H. and Milne, E.A., *Mon. Roy. Astron. Soc.* **84**, 499 (1924).
47. Plaskett, H.H., *Publ. Dom. Astrophys. Observ. Victoria*, **1**, 325 (no. 30) (1922).
48. Saha, M.N., *Zeitschrift für Physik*, **6**, 40 (1921).
49. Stewart, J.Q., *Popular Astron.*, **31**, 88 (1923).
50. Seares, F.H., *Astrophys. J.*, **55**, 165 (1922).
51. Hertzsprung, E., *Leiden Annals*, **14**, 1 (1922).
52. Pannekoek, A., *Bull. Astron. Institutes of Netherlands*, **1**, 107 (no. 19), (1922).
53. Milne, E.A., *Mon. Not. Roy. Astron. Soc.*, **89**, 157 (1928).
54. Menzel, D.H., *Harvard Circ.*, **258** (1924).
55. Payne, C.H., *Harvard Coll. Observ. Monographs*, No. 1. 'Stellar Atmospheres' (1925).
56. Payne, C.H., *Harvard Circ.*, **252** (1924).
57. Payne, C.H., See ref. (55), Chapter 9, page 133.
58. Payne, C.H., *Proc. National Acad. of Sci., Washington*, **11**, 192 (1925).
59. Payne, C.H., See ref. (55), Chapter 9, pages 180–9.
60. Whitney, C.A., *Sky and Telescope*, **59**, 212 (1980).
61. Struve, O., *Astrophys. J.*, **64**, 204 (1926).
62. Struve, O. and Zebergs, V., *Astronomy of the 20th Century*, Macmillan Co., New York, p. 220 (1962).
63. Rowland, H.A., A series of papers appearing in *Astrophys. J.*, from vol. **1** (1895) to **6** (1897).
64. Russell, H.N., Adams, W.S., and Moore, C.E., *Astrophys. J.*, **68**, 1 (1928).
65. Russell, H.N., *Proc. Nat. Acad. Sciences*, **11**, 314 (1925).
66. Russell, H.N. and Adams, W.S., *Astrophys. J.*, **68**, 9 (1928).
67. Schuster, A., *Astrophys. J.*, **16**, 320 (1902) and **21**, 1 (1905).
68. Schwarzschild, K., *Sitzungsberichte der königl. Preuss. Akad. der Wissenschaften, Berlin.* p. 1183 (1914).
69. Unsöld, A., *Zeitschrift für Physik*, **44**, 793 (1927); **46**, 765 (1928).
70. Schwarzschild, K., *Nachrichten der königl. Gesellschaft der Wissenschaften, Göttingen, math.-phys. Klasse* p. 1 (1906).
71. Russell, H.N., *Astrophys. J.*, **70**, 11 (1929).
72. St John, C.E., Moore, C.E., Ware, L.M., Adams, E.F. and Babcock, H.D., *Carnegie Inst. Washington, Publ.*, **396** (1928). See also: St John, C.E. *Astrophys. J.*, **70**, 160 (1929).
73. Unsöld, A., *Zeitschrift für Astrophys.*, **24**, 306 (1948).
74. Russell, H.N., quoted by O. Struve, *Trans. IAU*, **7**, 487 (1950).
75. von Klüber, H., *Zeitschrift für Physik*, **44**, 481 (1927).
76. Minnaert, M.G.J., *Zeitschrift für Physik*, **45**, 610 (1927).
77. Ladenburg, R. and Reiche, F., *Ann. der Physik*, **42**, 181 (1913).
78. Minnaert, M. and van Assenbergh, B., *Zeitschrift für Physik*, **53**, 248 (1929).

79. Minnaert, M. and Mulders, G.F.W., *Zeitschrift für Astrophysik*, **1**, 192 (1930).
80. Minnaert, M., *Observatory*, **57**, 328 (1934).
81. Schütz, W., *Zeitschrift für Astrophysik*, **1**, 300 (1930).
82. Voigt, W., *Sitzungsberichte der Königl. Akad. zu München*, p. 602 (1912).
83. Minnaert, M. and Mulders, G.F.W., *Zeitschrift für Astrophysik*, **2**, 165 (1931).
84. Minnaert, M. and Slob, C., *Proc. Acad. of Sci., Amsterdam*, **34**, 542 (1931).
85. Minnaert, M., Fourty [sic] years of solar spectroscopy, in *The Solar Spectrum* p. 3, ed. C. de Jager. D. Reidel (1965).
86. Unsold, A., Struve, O., and Elvey, C.T., *Zeitschrift für Astrophys.*, **1**, 314 (1930).
87. Pannekoek, A., *Proc. Acad. of Sci. Amsterdam*, **34**, 755 (1931).
88. Moll, W.J.H., *Proc. Phys. Soc.*, **33**, 207 (1921).
89. Berman, L., *Astrophys. J.*, **81**, 369 (1935).
90. Struve, O., *Astrophys. J.*, **76**, 85 (1932); see also *Proc. National Acad. of Sci., Washington*, **18**, 585 (1932).
91. Hynek, J.A., *Astrophys. J.*, **78**, 54 (1932).
92. Elvey, C.T., *Astrophys. J.*, **79**, 263 (1934).
93. Struve, O., and Elvey, C.T., *Astrophys. J.*, **79**, 409 (1934).
94. Unsöld, A., *Astrophys. J.*, **69**, 209 (1929).
95. Lockyer, J.N., *Proc. R. Soc.*, **17**, 415 (1869); ibid **18**, 74 (1869).
96. Menzel, D.H., *Popular Astron.*, **47**, 6, 66, 124 (1939).
97. Struve. O., *Astrophys. J.*, **137**, 1306 (1963).
98. Williams, E.G., *Astrophys. J.*, **83**, 279 (1936).
99. Goldberg, L., *Astrophys. J.*, **89**, 623 (1939).
100. Allen, C.W., *Mem. Commonwealth Solar Obs., Canberra*, **1** (no. 5, parts 1 and 2) (1934).
101. Menzel, D.H., *Astrophys. J.*, **84**, 462 (1936).
102. Unsöld, A., 'Physik der Sternatmosphären' Springer-Verlag, Berlin (1938).
103. King, R.B. and King, A.S., *Astrophys. J.*, **82**, 377 (1935); also *Astrophys. J.*, **87**, 24 (1938).
104. King, R.B., *Astrophys. J.*, **87**, 40 (1938).
105. Menzel, D.H., Baker, J.G., and Goldberg, L., *Astrophys. J.*, **87**, 81 (1938).
106. Minnaert, M.G.J., Mulders, G.F.W., and Houtgast, J., *Photometric Atlas of the Solar Spectrum*, Sterrewacht Sonnenborgh, Utrecht (1940).
107. Minnaert, M.G.J., *Mon. Not. Roy. Astron. Soc.*, **107**, 274 (1947).
108. Wright, K.O., *Astrophys. J.*, **99**, 249 (1944).
109. Plaskett, H.H., *Mon. Not. Roy. Astron. Soc.*, **107**, 117 (1947).
110. Menzel, D.H., *Popular Astron.*, **47**, 74 (1939).

8 Spectral classification: from the Henry Draper Catalogue to the MK-system

8.1 The first International Astronomical Union meeting in Rome, May 1922

By the time of the fifth meeting of the International Solar Union in Bonn in 1913 Schlesinger had been able to conclude, as a result of the questionnaire (1) to twenty-eight prominent spectroscopists from the Committee on the Classification of Stellar Spectra, that '... the preference for the Draper classification is nearly unanimous, but... the general feeling among investigators is opposed at the present time of any system as a permanent one' (2).

In practice 1913 represents the point when the Harvard system was universally adopted. Nine years later, at the first meeting of the newly formed International Astronomical Union (IAU) in Rome, in May 1922, the earlier temporary acceptance of this system was formally and unanimously approved as permanent. The chairman of the Spectral Classification Committee was then Walter Adams. By this time six volumes of the HD Catalogue had been published and the classification of nearly a quarter of a million stars had been completed by Miss Cannon 6 years previously. The adoption of the Harvard system was therefore no longer an issue: the Adams Report prescribed that 'the Draper Classification or "Harvard System"... should be the basis on which any further extensions should be built. Classification on other and different systems should be abandoned permanently' (3), although it was conceded that in 'cases of great uncertainty Secchi's types may be employed'. As a further concession to current practice, the terms 'early' and 'late' (meaning spectra from stars which are hotter or cooler) were deemed to be very convenient. F.J.M. Stratton from Cambridge wanted all reference to these words deleted, but Hertzsprung successfully argued in their favour, provided no reference to stellar evolution was to be implied. The overall principles guiding the committee were seen to be the pragmatic acceptance of what had already become a reality.

The Adams committee also recommended an extensive list of additional notations for spectral peculiarities and herein lies much of the historical interest in the committee's report. For example, very luminous stars would be given the prefix 'c' following the practice of Antonia Maury in her classification of 1897 (4), while giants and dwarfs of types F0 or later could be distinguished by the letters 'g' and 'd', based primarily on the greater strength of the 'enhanced' (i.e. ionised) lines such as 4077 Å of strontium in the giants and of neutral lines such as 4227 Å of calcium in the dwarfs. The suffixes 'n' and 's' were chosen for stars with nebulous and sharp lines, a notation first used by Rowland for the solar spectrum, and 'k' was for stars with stationary lines of interstellar origin in their spectra, whether arising from the calcium K-line or otherwise. Emission-line stars were given the suffix 'e' provided these lines are not normally present in that Draper class as they are in types O, P and Q, while a 'p' denoted peculiarities in the preceding symbol. Thus B2pe referred to a peculiar B2 absorption spectrum, B2ep to a star with peculiar emission.

On the notation for individual spectral lines, the committee accepted that only some of the mostly century-old Fraunhofer symbols be retained; only A, a, B, α, D, b, G, H and K were to be preserved because they 'are so well established that it does not seem desirable to abandon them', while 'for the hydrogen lines, the notation Hα, Hβ, etc. should be adopted'.

For late-type stars Miss Cannon's (5) use of the Draper class S was approved, while the classes Ma, b and c were to be relabelled on a decimal system for M0 to M8. Type Md was dropped. The decimal notation in place of letters, however, was considered premature for the subdivisions of the Draper O-type stars, because 'it has not been clearly established that a continuous and unique sequence exists' and the same thoughts prevailed in the classification of the spectra of novae, class Q. In this latter case, seven subtypes with successively broader and stronger emission features were defined from Qa to Qz as a 'basis for discussion' but 'not for immediate use'.

In 1922, with the Henry Draper classification firmly established, the exciting challenge was now how to interpret the sequence of stellar spectral types rather than the relative merits of different classification systems. With tools such as Saha's ionisation theory a whole new era in quantitative stellar spectroscopy was beginning. In the 1920s and 1930s less emphasis was to be placed on a direct evolutionary interpretation of the spectral type sequence, such as Lockyer's scheme had been, and more on the measurement of temperatures and pressures for stars of different spectral types using the new astrophysics.

A few years later Walter Adams nicely summed up the prevailing mood among stellar spectroscopists:

> The first great problem of astronomical spectroscopy, the empirical classification of the spectral lines, is therefore substantially complete, and the way is now open for the second, which is the development of a rational theory of astrophysical spectra, and of stellar spectra in particular. Such a theory, in its fully matured state, should be able to start with a few fundamental data, such as the values of effective temperature and gravity at a star's surface and the general composition of its atmosphere, and predict the intensities and widths of the spectral lines in detail; or conversely, to derive the former from the latter. We are still far from being in a position to do this (6).

As the IAU Committee on Spectral Classification had noted in 1922, there were, however, still areas of the Henry Draper system that needed further attention. One such area, the reclassification of the Harvard O-type stars, was quickly taken up by H.H. Plaskett at Victoria.

8.2 The classification of O stars

Harry Plaskett joined the Dominion Astrophysical Observatory in 1919 where his father (J.S.P.) was the first director and at a time that the new 72-inch (1.8 m) telescope had just been brought into operation. J.S. Plaskett at that time had commenced a major new program to study the O stars which were known to be among the hottest and most massive in the Galaxy.

The Harvard classification of O stars due to Annie Cannon in 1901 (7) had five subclasses Oa to Oe all with emission lines. Only the Oe5 stars, intermediate in type between Oe and B0 had no emission. As discussed in Chapter 5, all of these stars (except those of type Oa) have the Pickering series of ionised helium lines in emission or absorption. Types Oa, b and c are now classified as Wolf – Rayet stars and only types Od, Oe and Oe5 had a clearly present absorption spectrum.

H.H. Plaskett rejected the emission lines from his new scheme and also excluded all the Wolf–Rayet stars. His system (8) reclassified the Od, Oe and Oe5 stars on a decimal scale from O5 to O9, with the ratios of the line strengths of neutral to ionised helium and of ionised helium to Balmer hydrogen lines being the principal criteria, while the ionised lines of magnesium, silicon, carbon and nitrogen were also used. In this way he was able to derive a temperature sequence firmly founded on ionisation theory. He classified 45 stars in this sequence distributed as follows among the five

Fig. 8.1 H.H. Plaskett.

subclasses: O5, 5 stars; O6, 12; O7, 7; O8, 11; O9, 10. The subclasses O0 to O4 were all left vacant, thus allowing the possibility that objects of higher ionisation might be found later.

Soon after this new classification was published Fowler and Milne also applied the Saha theory to the O-type stars (9). Their ionisation temperatures ranged from 22 000 K at type O9 to 30 000 K at O5. Ralph Curtiss in his review article on spectral classification (mostly written in 1926) (10) wrote: 'That the spectral groups Oa to Oe5 stood so long is remarkable in view of the fact that it has been recognised for many years that they do not conform to a physical or well marked descriptive sequence and that the spectra in Class Oe5 share the dark-line characteristics of those in Classes Od and Oe'.

H.H. Plaskett's new classification did not reclassify the Harvard types Oa, b, c at all; he believed these Wolf–Rayet stars may form a side-chain to the sequence O5 to O9. His father did however investigate the excitation levels displayed in the emission lines of Wolf–Rayet and other Harvard O-type spectra (11). He tentatively proposed four groups of decreasing excitation but did not recommend any notation.

At the second IAU meeting in Cambridge, England, in 1925, J.S. Plaskett proposed that his son's O-star classification be formally adopted, as 'This will bring the classification of these stars into conformity and continuity with the rest of the Harvard system' (12). He went on: 'It is suggested that the O-type stars containing broad emission bands... be expressly omitted from the general sequence for the present, as there is no rational basis for their classification...'. His proposal however came into conflict with the view of Cecilia Payne at Harvard. Miss Payne had criticised the Plaskett classification in a recent Harvard Circular article (13). While conceding the new Plaskett scheme an advance over the Harvard types for absorption-line O stars, she argued against adopting any system at the present time because: (*a*) 'The spectra are so different that a classification based on estimates is not easy to apply', (*b*) 'The stars are so few in number and differ so much among themselves that they almost all require special remarks' and (*c*) 'A classification such as that of H.H. Plaskett's is essentially different from the Henry Draper classification, and the two cannot legitimately be used together'.

The result of these objections from Harvard was a letter from J.S. Plaskett to the Commission 29 president, Walter Adams in which he remarked: 'We find ourselves in complete disagreement with Miss Payne's statement so far as it applies to the absorption line O's'(12). The outcome was that the commission made no firm recommendation for changes to the Draper classification at the Cambridge meeting. Formal adoption had to wait another 3 years when the IAU met again in Leiden in 1928. Not only was H.H. Plaskett's classification then approved, but some features of it had even been supported by Cecilia Payne (14). Following P.W. Merrill's suggestion, the notation 'w' was to be used for Wolf–Rayet spectra, and 'e' for normal O stars with lines in emission. The earlier Harvard notation Oa, Ob etc. for Wolf–Rayet stars was retained at this time.

The replacement of the 'e' suffix with an 'f' was later proposed by J.A. Pearce in 1930 (15); an Of star has the ionised helium 4686 Å line in emission as well as prominent emission lines of doubly ionised nitrogen. The Of notation thus avoids confusion with the earlier Harvard Oe spectral type, and this suffix is reserved for O-type stars only; it remains in common usage.

When the new Morgan–Keenan–Kellman (MKK) criteria for spectral classification were published in 1943 (see section 8.8), no significant changes were made to Harry Plaskett's scheme for the O spectral types. The types given by Morgan were 'in very close agreement with those determined by H.H. Plaskett' (16). R.M. Petrie at Victoria measured absorption line intensities and re-applied the Saha equation to these measurements to deduce ionisation temperatures for O stars (17) based especially on the strength of neutral and ionised helium lines. His results were substantially hotter than Fowler and Milne had obtained in 1923: at type B0, $T_{eff} = 28\,600$ K; at O5, $T_{eff} = 36\,300$ K. Petrie also found no luminosity effects in the O star spectra, a result which Anne Underhill later explained, because pressure broadening of strong lines by the Stark effect disappears for these very hot dwarfs (18).

8.3 Spectral classification of nebulae As early as 1864 William Huggins had observed the spectrum of a planetary nebula (NGC 6543 in Draco) and shown that such objects have emission lines typical of a large mass of hot luminous gas (19) (see Chapter 4). This pioneering result was quickly followed by a similar conclusion for the Orion nebula, a very bright diffuse gas cloud of irregular outline. In the early days of visual spectroscopy both Sir Norman Lockyer (20) and W.W. Campbell (21) were prominent for their contributions in measuring the positions of nebular lines, while Wilsing and Scheiner (22) made visual estimates of line intensity ratios. The photographic era, however, led to the discovery of numerous ultraviolet lines, especially the bright double line near 3730Å found by Huggins (23), and to much more accurate measurements of wavelength and intensity than had hitherto been possible.

The work of W.H. Wright at Lick Observatory in 1918 (24) deserves especial mention, because it was more extensive and thorough than any observations made earlier, and it became a standard reference for the next two decades. Wright tabulated a list of wavelengths and intensities of the principal emission lines in the planetary nebulae. Especially noteworthy were the pair at 4959 and 5007 Å (known as N_1 and N_2) and ascribed to the mystery element 'nebulium'; the Balmer lines; the unidentified ultraviolet pair at 3726 and 3729 Å; the line at 4363 Å (also of unknown origin); and in some nebulae the 3869 Å line (unidentified) and that at 4686 Å (recognized as an ionised helium line since 1912 – see Chapter 5). Of the 74 lines listed by Wright, less than half could be identified with any certainty, and the remainder failed to correspond to any known lines in the laboratory. This problem became increasingly acute in the years after Wright's tabulation of nebular lines. Laboratory studies in spectroscopy were being pursued actively, the spectra of more and more elements were being catalogued, and the possibility that 'nebulium' might still be seen in the arc or spark spectra of one of the common elements seemed increasingly remote. The chances of there still being an element undiscovered terrestrially yet cosmically abundant were also unlikely, as all the light elements in the periodic table were accounted for. Such considerations led H.N. Russell to the conclusion that the nebulium lines 'must be due not to atoms of unknown kinds but to atoms of known kinds shining under unfamiliar conditions' which he suggested might be produced in a gas of very low density (25).

With the uncertainty over the identification and physical origin of many of the emission lines in planetary and diffuse nebulae, the problems of devising a classification scheme were especially difficult. In the early Draper Catalogue of Professor Pickering and Mrs Fleming, the letter P had been reserved for planetary nebulae, with no subdivisions. The fact that nebular spectra had been incorporated into a stellar classification scheme reflects the then prevalent belief that planetary nebulae represent a stage in stellar evolution. In both Schuster's and Hale's schemes of evolution, the planetary nebulae were precursors to the O stars, which Miss Cannon in 1901 had placed ahead of B stars in the stellar spectral type sequence. Thus the evolutionary development P, O, B, A, etc, seemed a plausible scenario, which linked the planetary nebulae with O stars, which also in many cases had emission lines, albeit of quite different character and wavelengths.

After the line at 4686 Å had been identified as an ionised helium line (26, 27), Miss Cannon devised the first subdivision of spectral class P into six subtypes Pa to Pf. In Pa 3728 Å (actually two lines) is conspicuous; in Pc 4363, 4959 and 5007 Å are all very strong, while the ionised helium line 4686 Å, and also 3869 Å, become well-marked in Pe (28). She unwittingly placed the nebulae in order of increasing ionisation and excitation temperature (the reverse of the stellar classification, but it seemed natural that nebulae with ionised helium should lead into the O-type stars which also had this feature). This classification was adopted with only a few changes in the Henry Draper catalogue.

In Wright's 1918 study he chose to place nebular spectra showing 4686 Å furthest from the stellar spectral types, and he devised a different classification scheme with these objects as Class I. If the line at 3869 Å was prominent, then the nebula was in Wright's Class II (3869 Å was identified as a forbidden line of doubly ionised neon only in 1933 (29)), while absence of both 4686 and 3869 Å defined Class III. Wright thus reversed the order to one of decreasing temperature, and he preferred to incorporate neither the nebulium lines nor the 3728 Å pair into his classification.

The big breakthrough in understanding nebular spectra came in 1927 from Ira Bowen (1898–1973) at the California Institute of Technology (Caltech) (30). Bowen spent nearly all his working life at Caltech, initially as a physicist working with Professor Robert Millikan. His interests were in ultraviolet laboratory spectra, and his researches in this area led to an understanding of the energy levels and terms of ions of the lighter elements, including oxygen. He showed that the strong nebulium lines corresponded to a 'forbidden' transition between low-lying levels of O^{2+}, which are only observable in large masses of gas at high temperature and very low density. The high temperature ensures enough electrons to excite the ions collisionally, the low density makes subsequent collisional de-excitation improbable. Bowen described how he came to solve the nebulium mystery:

> One night I went down to work and came home about nine o'clock...and started to undress. As I got about half undressed, I

Fig. 8.2 Ira Bowen and his staff at Mt Wilson and Palomar Observatories in 1955. Left to right (seated): W. Baum, F. Zwicky, M. Humason, I. Bowen, J. Greenstein, W. Baade, A. Deutsch; (standing): G. Münch, A. Sandage, E. Pettit, H.W. Babcock, R. Minkowski, S. Nicholson, R. Richardson, D. Osterbrock, O.C. Wilson.

> got to thinking about what happens if atoms get into one of those [metastable] states. Are they stuck there forever? Then it occurred to me, having read this [by Russell (25)], maybe they can jump if undisturbed in a nebula, but we can't see them here [in the laboratory] because of collisions...
>
> Well, I quickly put a reverse on my dressing and went down to the lab. again. Since I had these levels it was very easy to take these differences and check them up... it was a matter of minutes to establish it.
>
> Well, I just went to my table [of energy levels] and there they were within a hundredth of an angstrom. I worked until midnight and I knew I had the answer when I went home that night (31).

Bowen's younger colleague, Olin C. Wilson, described his work thus:

> As a consequence of his own analysis of the ultraviolet spectra of singly ionized oxygen and nitrogen and doubly ionized oxygen, Bowen noticed that the energy differences found in this way gave wavelengths agreeing very closely with those of several of the strongest lines in the spectra of planetary nebulae by W.H. Wright

at Lick Observatory. He reported this result in a three-page note in the *Publications of the Astronomical Society of the Pacific* in 1927. Thus he settled, once and for all, the mystery of the origin of these lines, which had at one time been attributed to the hypothetical element 'nebulium' (32).

In addition to the bright green nebulium lines, Bowen also identified the ultraviolet pair at 3728 Å as forbidden lines of O^+, and two lines near Hα as being forbidden N^+ lines. Several other weaker lines of these same ions, including that at 4363 Å (due to O^{2+}) were also identified.

After this important work by Bowen the physical processes in gaseous nebulae could be understood for the first time; in particular it was now possible to classify them in a temperature sequence with some confidence. Miss Payne regarded the nebulae and O stars as two parallel but separate sequences. She classified the nebular spectra from P1 to P9 in decreasing order of ionisation and excitation (33). The ionised helium line at 4686 Å, was strong at P1; the N1 and N2 lines also decreased from P1 to P7, being absent at P8, while the 3728 Å pair (of forbidden O^+) were weak from P1 to P6, but stronger in the later types P7 to P9. Spectral types P10 to P13 were reserved for reflection nebulae, which give rise to stellar absorption spectra from the scattering of starlight by dust grains. P was thus the only Harvard spectral type that was ever assigned more than ten subdivisions. More importantly, Miss Payne's was the first nebular classification scheme to have a firm astrophysical basis.

A later classification by Thornton Page (b. 1913) at Yerkes Observatory was quite similar. It comprised on eight-point scale with the ionised helium 4686 Å line being the distinguishing feature of classes 1 to 4, and with the strength of 3728 Å of forbidden O^+ relative to 4959 Å of forbidden O^{2+} of classes 5 to 8 (34). Later still Lawrence Aller (b. 1913) at Michigan introduced a ten-point nebular classification in 1956. Class 10 showed the highest excitation, class 1 the least. Aller used the forbidden Ne^{4+} spectrum (lines at 3346 and 3426 Å) as the means of distinguishing nebulae in the hottest classes, where these lines were strong relative to Hβ and the Ne^{2+} line at 3869 Å (35).

However, spectral classification of nebulae, since 1928, has never played a major role as it has done with stars. After 1928 analysis of the spectra became far more important and with the realisation that the classification of nebulae could not be a simple appendage to any scheme for stars, the importance of nebular classification declined. K. Wurm (1899–1975), a contemporary of Bowen's at the Potsdam, and later Hamburg Observatories, summed it up in these words:

> A classification of the [nebular] spectra according to their composition and degree of excitation cannot consequently be carried out, as there are too many variable parameters involved in their production. Next to how hot the central star is, the ionisation of the nebula is also substantially influenced by the gas density in the envelope, by the distance from the star and also finally by the total mass of the nebula. Density fluctuations complicate the [intensity] ratios still further. Each nebular spectrum really requires for its adequate characterisation a special description, and is never the exact duplicate of another (36).

Needless to say, the analysis of nebular spectra made great strides forward in the 1930s. Most noteworthy is the work of Bowen from 1938, when he collaborated with Arthur Wyse (1909–42), at the Lick Observatory. Bowen took to Lick his new invention, the image-slicer. This device greatly improved the efficiency of the spectroscopy of stellar sources when observed with slit spectrographs, and also enabled multiple spectra of extended sources such as nebulae to be obtained in simultaneous exposures. Bowen and Wyse observed three planetary nebulae using an image-slicer on the great Lick refractor, and showed that excitation differences occur not only between nebulae, but also within them (37). For example, NGC 7020 had features of both neutral oxygen and ionised neon (Ne^{4+}) in its spectrum. They analysed the composition of these planetary nebulae, and found them to be similar to the sun. In his last paper before his death while on wartime duties, Wyse extended these observations and published a catalogue of about 270 nebular spectral lines, of which about 160 had identifications. He showed that diffuse nebulae, such as the Orion nebula, also have a solar composition (38).

8.4 The spectroscopy of normal B stars

The two decades of the 1920s and 30s were extraordinarily productive for stellar spectroscopy, and the progress in the study of B stars provides an excellent example of the accelerating pace of stellar astrophysics. The most significant advances were all in North America, especially at Mt Wilson, Victoria and Yerkes. Part of the progress was due to new instruments, especially the installation of a prism spectrograph at the coudé focus of the Mt Wilson 100-inch telescope by Adams in 1925 (39), and also the advent of the recording microdensitometer, an instrument first developed by W.J.H. Moll at Utrecht (40), but soon taken up elsewhere, including at Mt Wilson, Victoria, Michigan and Yerkes.

The advances in normal B star spectroscopy (the discussion of the Be stars with their Balmer emission lines is deferred to Chapter 9) were simultaneously carried out in five closely interrelated areas:

1) the measurement of equivalent widths for lines and their dependence on spectral type;
2) the improvement in the Harvard criteria for B-star spectral classification;
3) the observation of the singlet-triplet anomaly for the neutral helium lines;
4) the confirmation that the Stark effect broadens the strong hydrogen and helium lines; and
5) the study of luminosity effects in B star spectra with the consequent derivation of spectroscopic parallaxes.

The last of these was almost certainly the biggest prize, even though the groundwork had been laid by Miss Maury and by Hertzsprung in the early years of the century.

8.4.1 Line strengths, spectral types and the singlet-triplet anomaly Three papers in the 1930s, all published in the *Astrophysical Journal*, were especially germane to the first of these areas. Otto Struve's 1931 paper, 'A study of the spectra of B stars' was one of the many he subsequently wrote on these stars (41). E.A. Milne described Struve's research at this time:

> [His] output in these years, and indeed in all years since, was and has been enormous. He followed up his studies of axial rotation, in 1931, with a comprehensive standard list of wavelengths in the spectra of B-type stars together with intensity estimates, and applied it to elucidate many points in the theory of high temperature ionization, studying lines of helium (neutral and ionized), and ionized oxygen, nitrogen and silicon (42).

Indeed, during the 1930s, Otto Struve became the undisputed master of B-star spectroscopy. In this rôle he was the natural successor to Edwin Frost. But whereas Frost's reputation rested mainly on B-type radial velocities, Struve's was unmistakably in astrophysics.

Two other papers on the line strengths in B-star spectra were of comparable importance to Struve's. These came in 1936 from Ewan Gwyn Williams (1905–40) and from Paul Rudnick (b.1911). Williams was a talented young astronomer from the Cambridge Solar Physics Observatory who spent 2 years, from 1931, at Mt Wilson. He studied there the prism

Fig. 8.3 E.G. Williams operating a microdensitometer at Mt Wilson.

spectra of eighty-four O and B stars, and measured equivalent widths for many lines as a function of spectral type (43). His results were later described by Anne Underhill as forming 'a landmark in the study of O and B-type spectra' (44).

Rudnick's paper was only slightly less extensive, and similar in its purpose (45). He studied seventy O and B stars from Yerkes spectra, to continue further Struve's earlier work. Perhaps his main misfortune was that Williams' work, although in the same volume of the *Astrophysical Journal,* appeared in print first by a margin of 1 month.

Gwyn Williams improved the Henry Draper spectral subtypes for B stars in a second paper in the same issue of the *Astrophysical Journal* (46). His system was based on line intensity ratios instead of absolute intensities and was designed to give a smooth distribution of star numbers in the different B subtypes, whereas he commented that 'it is well known that the classes B4, B6 and B7 are not used in the Henry Draper Catalogue' (46).

This second paper of Williams, as well as that of Paul Rudnick, reconfirmed the discovery by Struve that the ratio of neutral helium singlet-to-triplet line strengths in the hottest B stars decreases rapidly in going from B2 to the earlier types and class O stars (47). Helium is an example of a two-electron atom where the spectral lines arising from parallel (triplet) and opposed (singlet) electron spins behave quite independently, almost like two elements as, indeed, Lockyer had earlier believed. Transitions between singlet and triplet energy levels are almost never observed. The anomaly of the singlet-to-triplet line intensity ratios was discussed further by Struve in 1933 (48), who subsequently proposed an ingenious explan-

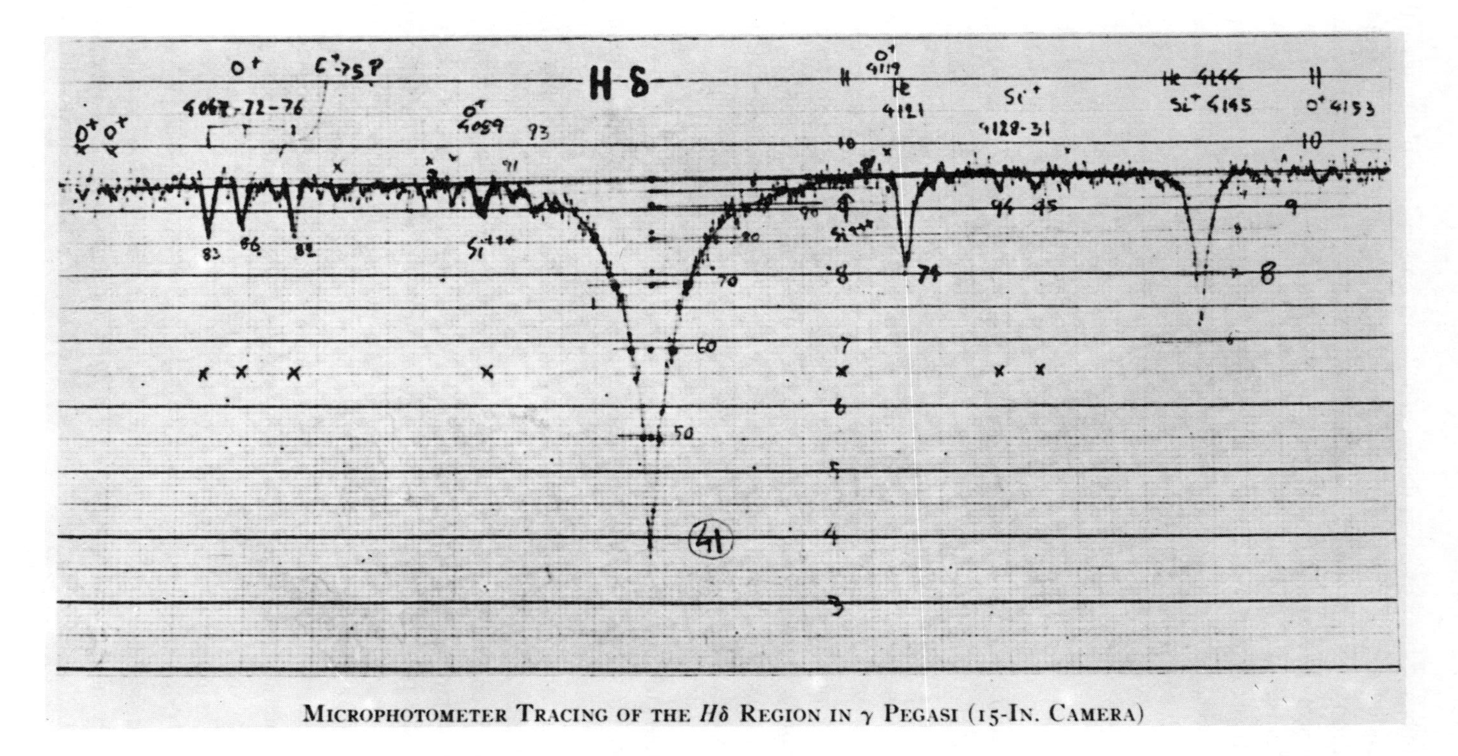

MICROPHOTOMETER TRACING OF THE $H\delta$ REGION IN γ PEGASI (15-IN. CAMERA)

Fig. 8.4 Tracing of the Hδ region in γ Pegasi by E.G. Williams, 1936.

ation based on an overpopulation of helium atoms in the triplet states in the extended atmospheres of very hot early-type B stars resulting from a dilution of the normal Planck black body radiation coming from the lower layers of the star's atmosphere (49). An alternative explanation based on the curve of growth of lines through the effects of line saturation and atmospheric turbulence was not favoured by Rudnick (45), although this was revived briefly by Leo Goldberg in 1939 in his study of the curves of growth for fifty-seven O and B stars (50).

8.4.2 The Stark effect, luminosity criteria and spectroscopic parallaxes The Stark effect is the mechanism operating in B-type and other stars whereby charged particles (mainly protons and electrons) near an absorbing atom in a stellar atmosphere are able to broaden the lines in the spectrum, by spreading the atomic energy levels over a certain range in energy. That such a broadening can occur in the laboratory was announced by the German physicist Johannes Stark (1874–1957): 'One can broaden spectral lines without raising the temperature by increasing the density of the luminous vapour. This type of spectral line broadening does not arise from the conditions of the Doppler effect, but probably through a force which deforms the emitting atom or ion, hence altering the period of the emitted spectral lines' (51).

The English theoretical astrophysicist, R. d'E. Atkinson (1898–1982), appears to have made the earliest reference to the Stark broadening of lines in stellar spectra when he discussed the electric field in a gas containing ions: 'Since it is a chaotic field it will not appear to split up lines, but to broaden them; the hydrogen lines would be broadened by this strength of field [10^4V/cm] to about 4 Å' (52). Russell and Stewart considered the Stark broadening of the Hβ line in the solar spectrum in 1924 (53) but the first major paper studying line widths of hydrogen and helium lines in B stars was that of Struve in 1929 (54). He found that the Stark effect played a major rôle in the broadening of hydrogen and helium lines in these stars. Adams and Joy had earlier firmly established that the more luminous stars have sharper lines, whereas the lines of the less luminous dwarf stars tend to be nebulous in character (55). Struve explained this phenomenon, at least in part, by invoking the Stark effect, which increases in the higher pressure dwarf stars.

At this time the line profiles (or 'contours') of strong lines were poorly known. The use of a recording microdensitometer at Yerkes to study the profiles of the Balmer Hβ and Hγ lines in sixty-four early-type stars provided a major wealth of observational material to confirm Struve's ideas (56). A further study of the hydrogen lines of early-type stars was

Fig. 8.5 Bertil Lindblad.

made by Elvey and Struve, also in 1930 (57). They found far more extensive wings due to the Stark effect in dwarf stars such as α Lyrae than previously recognised. For example, the Balmer lines 3970 Å Hε and 4102 Å Hδ have overlapping wings in this star. In a luminous star such as α Cygni however, this overlap was much less pronounced. In fact, Elvey was able to demonstrate the effect of hydrogen-line wing strength on the colours of stars of differing luminosity as early as 1931, using a photoelectric photometer on the giant Yerkes refractor (58). In his classic study on the spectra of B stars (41) Struve again turned his attention to the neutral helium lines, showing that the Stark effect must be responsible for their widths, and that the ionised helium lines in O stars were also broadened by this effect.

Later, A. Pannekoek and his collaborator S. Verweij made the first theoretical calculations of Balmer line profiles using Stark's theory (59, 60), after applying the statistics of the randomly fluctuating electric fields encountered in a plasma according to the work of the Norwegian physicist Johan Holtsmark (b. 1894) (61). Their results showed that stars with higher gravity produced Balmer lines with more strongly developed wing profiles. Gwyn Williams used these first theoretical line profiles to compare with his Mt Wilson observations (43). He found reasonable agreement with the wings of the hydrogen lines of B stars, but the line centres were predicted to be practically dark, and not at all as observed. Richard

Woolley (1906–86) summarised the position in 1937; 'the Stark effect accounts qualitatively for all phenomena shown by H and He lines in stellar spectra *except* the observed high central intensities of the H lines' (62).

The basis had now been laid for the luminosities, and hence distances, of the normal B stars to be determined spectroscopically. The earliest techniques relied either on line character, or on ratios of line intensities. The line character method was pioneered by Walter Adams and Alfred Joy at Mt Wilson. Two papers treated 544 A stars and then 300 B stars (55). They classified these stars with the suffixes 's' or 'n' for sharp or nebulous lines, and established two separate calibrations of absolute magnitude versus spectral type for the two groups. The sharp-lined stars were more luminous, by 0.8 magnitudes at type A2, 0.6 magnitudes at B5, but falling to zero for B0 and O-type stars.

Donald Edwards (1894–1956) on the other hand, who was at the time chief assistant at the Norman Lockyer Observatory at Sidmouth, Devon, used the method of line intensity ratios of close pairs of helium and hydrogen lines (63). Both these and the Mt Wilson absolute magnitudes had to be calibrated using stars with known distances, obtained from a variety of non-spectroscopic methods. Edwards' line intensities were measured in the pre-microdensitometer era using William Lockyer's graduated density wedge (64). This ingenious device was placed over an objective prism spectrum and adjusted to render a given line invisible when viewed through a microscope. The position of the wedge gave a quick visual estimate of line strength.

Later Edwards also developed the Mt Wilson line character method, but introduced five subclasses ss, s, ns, n and nn, each with its own luminosity calibration (65). In the years up to 1930 he obtained spectroscopic parallaxes for over 200 B stars using one or other of these techniques.

At the same time as Adams, Joy and Edwards were obtaining the first spectroscopic luminosities for B stars, Bertil Lindblad (1895–1965) from Uppsala in Sweden began his studies in the same field. Lindblad made several visits to Mt Wilson from 1921 and his early interest in spectroscopic luminosities and parallaxes thus stemmed from the work of Adams and Joy. A major paper in 1922 contained two important results. Firstly, for hot stars, there is one of the earliest explicit references to the widths of the wings of the Balmer H ζ line depending on the luminosity of the star: 'the extreme limits being represented by very bright stars such as α Cygni and α Leonis with narrow hydrogen lines and the very faint [white dwarf] companion of o_2 Eridani, which has exceedingly wide lines' (66). The other result in the same paper was the discovery of the great luminosity sensitivity of the molecular cyanogen (CN) absorption in cool stars, one of Lindblad's most significant achievements. Both these luminosity criteria were later used for studies in galactic rotation and structure, with great skill by Lindblad himself and by 'the remarkable phalanx of his pupils and successors, among whom we will cite Öhman, Schalén [etc.]...' (67).

A two-dimensional spectral classification for B and A stars was devised by Lindblad in 1925 using the character of the hydrogen lines as the second parameter (68). This scheme used the notation $\sigma\sigma$, $\sigma++$, $\sigma+$, σ, $\sigma-$, ρ, μ, $\chi-$, χ, $\chi+$ as suffixes to the usual Harvard spectral types. The most luminous stars with very narrow nearly invisible hydrogen lines were classed as $\sigma\sigma$, whereas the μ and χ stars have very broad lines characteristic of dwarfs. This notation went through several minor changes in successive years, as given by Lindblad's student Carl Schalén (69), and then by Lindblad and Schalén (70). For example, the last mentioned paper used a scheme with $\tau-$ designating the two classes with the narrowest lines.

Schalén's paper calibrated these luminosity classes using both statistical parallaxes and star clusters, resulting in a two-dimensional table of absolute magnitudes for each spectral type and luminosity class, with an accuracy claimed to be half a magnitude or better. He presented absolute magnitude results for 217 stars, and the agreement with stars also studied in common by Adams and Joy was 'on the whole, remarkably good'. As Schalén noted, the Uppsala classification was essentially different from that used at Mt Wilson. The former relied on the appearance of the wings of the hydrogen lines on Uppsala objective prism spectra, whereas at Mt Wilson the character of the metallic arc lines on slit spectrograms was used for the 'n' and 's' classification. Theoretically the Swedish system was more soundly based, as it isolated the pressure-broadened Stark wings, without allowing rotational broadening to confuse the issue, and also it employed more subclasses; observationally, the Mt Wilson slit spectra were of better resolution and hence superior.

By the late 1930s the problem of measuring electron pressures and hence luminosities of early-type stars was tackled from a slightly different standpoint. It had been known for some years that the number of spectral lines visible in a series depended on electron pressure. Higher pressures result in perturbations of the higher closely spaced energy levels, which thus merge into a continuum (see, for example, (71)). This effect applies in particular to the Balmer series of hydrogen in early-type stellar spectra, and was explained theoretically by Pannekoek in 1938 by invoking the Stark effect (72). D.R. Inglis and Edward Teller (b. 1908) deduced a theoretical relation between electron density N_e and the upper quantum number n_m of the last visible line in alkali and Balmer series, with N_e being proportional to $n_m^{-7.5}$ (73), while Fred Mohler at the National Bureau of Standards confirmed this relationship empirically in the laboratory (74). Mohler expressed the results by the equation: $\log N_e = 23.0 - 7.5 \log n_m$ (N_e in cm^{-3}), which is now generally known as the Inglis–Teller formula. He went on to apply this to stellar spectra. In a supergiant such as α Cygni, the Balmer line from the 30th upper level was the last visible, while for dwarfs this number was in the mid-teens. Mohler thus deduced the individual pressures and densities in five stars.

Using the observed values of n_m as a quick method of finding stellar luminosities for A and B stars now became a possibility. Unsöld and Struve published a joint note from Yerkes Observatory showing how n_m correlated with Williams' luminosities (75). The first thorough exploitation of the method came only after the war, from Gerhard Miczaika (b.1917) in Heidelberg (76). He produced absolute magnitudes for 115 B stars with an accuracy of about ± 0.3 magnitudes.

8.5 Spectral classification programs in the 1920s and 1930s

The publication of the last volume of the Henry Draper Catalogue in 1924 still left plenty of room in this field for other observatories. For example, in the northern sky, the Catalogue was complete only to eighth magnitude, with some stars down to magnitude 9.5 being included; in the south the limit was about a magnitude fainter (77). The HD Extension was planned by Shapley and Miss Cannon to take the completeness limit to about magnitude eleven but only in selected regions of the sky (see Chapter 5).

Meanwhile some very large spectral classification programs were undertaken at other observatories in the 1920s and 1930s. The motivations were either simply to push the statistical sample of classified stars to fainter magnitude limits, or to exploit the method of spectroscopic absolute magnitudes originally devised at Mt Wilson by Adams and Kohlschütter (see Chapter 7). The main centres for classification during these years were Harvard, Mt Wilson, Victoria, Yerkes and McCormick in North America, and Hamburg–Bergedorf, Potsdam, Stockholm and Uppsala in Europe. Of these observatories, Victoria and Yerkes used slit spectra, Mt Wilson both slit and slitless spectra, while the others, including the four mentioned in Europe, employed slitless objective prism cameras. All these programs classified 'on the HD system', though in practice they used somewhat different criteria. Indeed, consistency between institutions was a major concern, and in 1935 the IAU subcommittee on the criteria for classification of stellar spectra, under the chairmanship of Adams, summarised the different criteria in use at five of these observatories (78).

Fig. 8.6 Arnold Schwassmann.

Of the programs mentioned, the huge undertakings at the two German observatories deserve special comment. The origins of these programs go back to the great Dutch astronomer, Jacobus Cornelius Kapteyn of Groningen. He devised his famous *Plan of Selected Areas* in 1906 (79). Kapteyn wrote that his plan 'consists simply in this: For 206 areas, regularly distributed over the sky..., to obtain astronomical data of every kind, for stars down to such faintness as it will be possible to get in a reasonable time' (79). The coordinates of the 206 selected areas were specified. Spectral classifications were recommended as one of eight different types of observation to be carried out in a huge international program.

P.J. van Rhijn (1886–1960), Kapteyn's successor at Groningen, proposed in 1921 that the Hamburg–Bergedorf Observatory undertake the spectral classifications in 115 northern selected areas. Arnold Schwassmann (1870–1964) was put in charge of this project and the objective prism exposures were secured from 1923 to 1933 on the Lippert astrographs at the Bergedorf Observatory. The exposures were up to 4 hours long with the result that the program was complete to photographic magnitude 12.5, and even some stars as faint as magnitude 13.5 were classified (80). The dispersion used for these faint objects was only 400 Å/mm. Schwassmann and his colleagues, A.A. Wachmann and J. Stobbe, classified 173 500

Fig. 8.7 Friedrich Becker.

northern stars in this program on the Henry Draper system, an enormous undertaking comparable in size to the HD Catalogue itself. The catalogue of the Bergedorf Spektraldurchmusterung was published in five volumes (81) from 1935 to 1953. Schwassmann retired in 1934, but he continued working voluntarily on the project through to its completion.

No less interesting is the corresponding Potsdam Spektraldurchmusterung or spectral survey carried out in the remaining ninety-one Kapteyn selected areas of the southern sky (82). The Potsdam Astrophysical Observatory established an observing station near La Paz, Bolivia, in 1926 at an altitude of 3610m.* A 30cm Zeiss astrograph was installed and 180 Å/mm objective prism plates (30 × 30 cm) of all ninety-one southern Kapteyn selected areas were taken from May 1926 to August 1929. The exposures were mainly 3 hours. The whole project, including the observations and subsequent classifications, was under the direction of Friedrich Becker (b. 1900), with the later assistance of Hermann A. Brück (b. 1905), who classified stars in 24 of the fields. The project was somewhat smaller in scope than that at Bergedorf, but still about 68 000 stars were classified in the HD system down to about photographic magnitude 12.

* The La Paz Observatory was dismantled in 1929 at the conclusion of the observing program.

Fig. 8.8 The Potsdam southern station at La Paz, Bolivia, about 1927.

As with the Bergedorf survey, some attempt was made for later-type stars to apply giant or dwarf labels using the CN bands as a luminosity criterion, based on Lindblad's discovery that these bands were much strengthened in giant stars relative to dwarfs (66). In the opinion of C. Fehrenbach, this aspect of both surveys seems to have been largely unsuccessful due to the unreliability of the assigned luminosities (see (67) and see also (83)). At Potsdam the line character for some B to F5 stars was indicated by n, s or c for respectively nebulous, sharp and very sharp lines. Some statistical results of the Potsdam program have been given by Becker (84).

A major concern relating to these large spectral surveys has been the elimination of systematic differences between observers and observatories in the spectral types assigned. For example, Brück showed that the Potsdam types averaged about two-tenths of a spectral class earlier than Harvard types at F0, and about two-tenths later at G5 (85). The situation is somewhat confused as Miss Payne earlier found the principal disagreements to be at types B2 and G2 (86). The most extensive intercomparison of spectral types from five observatories (including Harvard, Bergedorf and Potsdam) was made by Alexander Vyssotsky (1888–1973) at the University of Virginia, whose results for the Potsdam types were consistent with those of Brück (87). Various authors have also considered the precision (or random errors) of these various spectral classifications, including Vyssotsky (87) and Seares and Joyner (88).

This section would not be complete without at least a brief mention of some of the other substantial spectral classification programs of these years. For example, Milton L. Humason (1891–1972) obtained slitless spectra at 110 Å/mm with the Mt Wilson 1.5m telescope of over 4000 stars fainter than eleventh magnitude in the 115 northern Kapteyn selected areas (89). The program was nearly abandoned when he learnt that Schwassmann's program at Bergedorf was reaching to the same magnitude limits. Meanwhile, and also at Mt Wilson, Walter Adams and his colleagues continued a large program of deriving spectral types and spectroscopic absolute magnitudes for the brighter stars, based on a calibration of line intensity ratios (90). The 1935 catalogue of these authors listed results for 4179 stars of type A5 or later and the Hertzsprung–Russell diagram (absolute magnitude v. spectral type) for these produced the most comprehensive such plot for field stars at that time.

8.6 Bertil Lindblad and the spectrophotometry of late-type stars

The early work of B. Lindblad in two-dimensional spectral classification (that is determining parameters for both temperature and luminosity) was already referred to in section 8.4 for early-type stars. Lindblad's discovery of the greatly increased strength of the molecular cyanogen bands in the ultraviolet (λ3883) and blue (λ4216) for G and K stars opened up a new basis for luminosity determinations (66). After his return from Mt Wilson in 1922 Lindblad developed this technique at the Uppsala Observatory using 270 Å/mm objective prism plates with the 15cm Zeiss–Heyde astrograph. From these spectra he defined indices to measure the strength of both the blue CN bands and also of the G band. These indices were calibrated in terms of absolute magnitude using the Mt Wilson results by Adams. The CN index was more sensitive to absolute magnitude (M_v), the G-band index more to spectral type, but together both M_v and the type could be obtained (91). He thus derived in the mid-1920s the absolute magnitudes of about 100 late-type stars to tenth apparent magnitude with an estimated probable error of 0.7 magnitudes. Lindblad's early spectrophotometry involved a fairly crude technique of taking multiple spectra of different exposures to overcome the problem of calibrating the non-linear photographic densities. However, his student Yngve Öhman introduced a new automated method using a densitometer on widened spectra (92).

Lindblad moved to Stockholm Observatory as director in 1927. He continued this work there with Y. Öhman, Erik Stenquist and others. The new Stockholm Observatory (opened 1931) was equipped with a 40cm Zeiss astrograph and a 1.02m Grubb reflector, both of which were used for the spectrophotometric program. His 1934 paper with Stenquist now defined three spectral parameters for cool stars: c (for the CN band), g (G band) and k (spectral flux gradient), which were determined using a self-registering microdensitometer (93). They showed how the cyanogen c-index clearly distinguished dwarfs from giants when plotted against spectral type and this formed the basis for deriving spectroscopic absolute magnitudes. They calibrated the effect, mainly using the spectroscopic parallaxes from Mt Wilson.

Lindblad's early work in spectrophotometry was only the beginning of a considerable program undertaken by his colleagues at Uppsala and Stockholm who over several decades developed the techniques further for studies of galactic structure, using the classification statistics. We only mention here the names of Jöran Ramberg (94), of Tord Elvius (95) and of Gunner Malmquist (96) as being among those who continued such work. Meanwhile the theory of the luminosity effect of the cyanogen molecule was not so obvious. As Lindblad and Stenquist pointed out, this is partly due to the cooler temperatures of late-type giants than dwarfs of the same spectral type favouring the molecule's formation (93). However, that is not the full explanation. A theoretical study by Miss Yvonne Cambresier and L. Rosenfeld in Liège showed that the luminosity effect in part also resulted from the effect of the competing equilibrium of the molecules CH, CO and C_2 on the formation of CN (97).

8.7 Barbier, Chalonge and the Balmer jump

William Huggins was the first observer of the Balmer jump in stellar spectra. On spectrograms of Vega and stars of similar type he noted a sudden fall in the intensity of the 'background' spectrum beyond the end of the series of Balmer lines in the ultraviolet (98). W.H. Wright at Lick observed this phenomenon in 1918 and produced an intensity plot of Vega's spectrum showing the absorption beginning about 3700 Å and extending into the ultraviolet (99). Wright's student at Lick, Ch'ing Sung Yü was the first to study the Balmer jump in detail in 1926 (100) (Fig. 8.9). Yü took spectra of ninety-one late B- and A-type stars on the Crossley reflector. The spectra were carefully calibrated with a spot sensitometer (this put six spots of known relative intensities onto the plate) and the intensity curves were thus plotted using a microphotometer. The limit of the Balmer series of lines was well known to be at 3647Å from laboratory studies. But Yü showed that the

Balmer lines merge together and the strong continuous ultraviolet absorption generally starts to the long wavelength side of this limit. The amount of absorption of ultraviolet light varied from star to star, depending, so he concluded, on the strength of the Balmer lines.

The study of the ultraviolet continuum was taken up again a decade later by Daniel Barbier* (1907–65), Daniel Chalonge (1895–1977) and several of their colleagues. These French astronomers helped to establish in 1931 the Scientific Station on the Jungfraujoch (Switzerland) at an altitude of 3457 m. The high altitude of the site was ideally suited to ultraviolet spectroscopy and a program using a quartz optics objective prism telescope (aperture 8 cm) was commenced in 1934 to observe the region from 3 100 to 4600 Å (105). The early work consisted in measuring the gradients† in the continuous spectra on either side of the Balmer jump (Φ_1 and Φ_2) as well as the size of the discontinuity itself in terms of the logarithmic intensity drop at 3700 Å (106). The spectra were calibrated by means of the light from an artificial star consisting of a hydrogen lamp 600 m from the telescope. They found the Balmer jump D increased from B0 and reached a maximum for stars around A0 or A2. The jump of $D = 0.50$ in α Lyrae was among the largest measured; γ Cassiopeiae (type Be) was unusual and had a negative jump ($D = -0.13$), corresponding to a brighter spectrum below 3700 Å than at longer wavelengths. For two F5 stars studied, D was larger in the supergiant α Persei than in the dwarf Procyon. The shortest wavelength within the jump was hard to determine, owing to overlapping Balmer lines,

* It is Daniel Barbier whom we must thank for one of the most amusing 'true stories' in the whole history of stellar spectroscopy. In 1962 Barbier and Mlle Morguleff reported observing very strong and broad potassium emission in the two infrared lines near 7700 Å of HD 117043, a sixth magnitude G5 dwarf. The coudé spectrograph of the 193 cm telescope at the Observatoire de Haute Provence was used for this observation. Only one spectrum showed the emission (101). A second potassium flare star was found 2 years later – HD 88230 (K7V), once again with very strong broad lines and once again, only on one spectrum (102). A third star, 4 Her, was reported to have a potassium flare by Mme Y. Andrillat in 1965 (103), this time of spectral type B9e. An entirely new and mysterious class of astrophysical phenomenon appeared to have been discovered, but in spite of an extensive search by Mme Andrillat, no further potassium flares were ever found. A likely explanation of the strong potassium lines seen in these stars was given in 1967 by Robert Wing and his associates at the University of California at Berkeley (104). The spectrum of a match struck near the coudé slit produced an effect identical to that reported at Haute Provence. They concluded that this was probably the most plausible explanation. Potassium flare stars have not been heard about since. Unfortunately, Barbier passed away while on an observing run at Haute Provence on 1 April 1965, 2 years before the match explanation was announced. He was therefore unable to share this delightful story with the astronomical world.

† The gradient was defined as $\Phi = d(\log_e I)/d(1/\lambda)$ and the jump by $D = \log_{10}(I_{3700+}/I_{3700-})$.

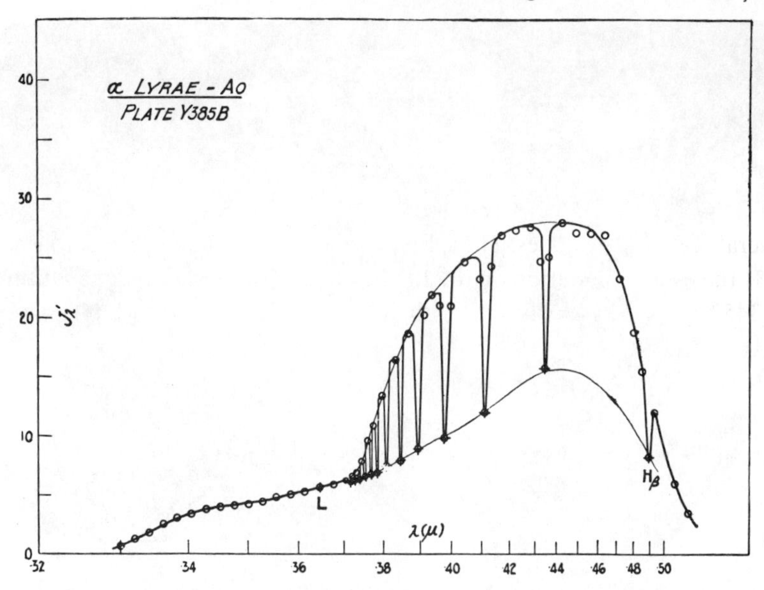

Fig. 8.9 The Balmer jump in Vega observed by Yü at Lick in 1926.

but was generally between 3690 and 3720 Å, which is to the red of the theoretical value of 3650 Å. For some stars, however (such as the supergiants β Orionis and α Cygni), the jump occurred nearer to the theoretical value.

Barbier and Chalonge continued this important work at the Jungfraujoch up to 1939 by which time they had observed 204 stars. By 1938 they were working on the possibility of classifying the early-type stars from the Balmer jump (107). They defined two parameters, the size of the jump D (as before) and the wavelength of the mid-point of the intensity decrease, λ_1. This varied from star to star between 3690 and 3780 Å and, for A0 stars, could be measured to within 1 Å and for earlier or later types, to within 3 or 4 Å. For each spectral type they found a unique relationship between D and λ_1; or conversely, if D and λ_1 are measured, then the Harvard spectral type can be deduced. Moreover, for A0 stars λ_1 correlated well with absolute magnitude, being smaller for the more luminous stars, and hence spectroscopic values of M_v were deduced for twenty-two stars with probable errors of about 0.4 magnitudes. Thus a new two-dimensional spectroscopic classification was born. The detailed results for 204 B, A and F stars were presented in 1941 (108) with values of Φ_1, Φ_2 as well as D and λ_1 for each star. The variation of the Balmer jump was plotted for the entire range from the O stars ($D \simeq 0.03$) to a maximum at A0 ($D \simeq 0.47$) and falling again to about 0.06 for type G0. The Balmer jump for γ Cas was

Fig. 8.10 Daniel Chalonge (left) and Daniel Barbier (right) at the Paris Observatory, about 1937.

found to be variable, reaching its greatest negative value of $D = -0.29$ in September 1937. The two Ap stars α^2 CVn and α And had Balmer jump parameters of normal B5 stars. 'These are the only two cases of stars of normal absolute magnitudes for which we have found such a difference between the spectral class and the collective properties of the continuous spectrum' (108).

After the war Chalonge returned again to the Jungfraujoch and installed there a 25 cm reflector with cassegrain prism spectrograph to resume their ultraviolet spectroscopy. This instrument had been constructed and tested at the Institut d'Astrophysique in Paris, and it was during the war years that he and Barbier, with Mlle R. Canavaggia, perfected the concept of the two-dimensional (λ_1, D) classification by plotting the loci of constant spectral type (for stars of different luminosities) in the (λ_1, D) diagram (109). The first results with the new telescope after its installation on the Jungfraujoch were reported by Chalonge and Mlle Lucienne Divan (110). Here, for clarity, the authors plotted two (λ_1, D) diagrams, for stars A1 and earlier and A2 and later, since for any pair of (λ_1, D) values they recognized an ambiguity exists with two different spectral types being a possible solution. Provided this ambiguity could be resolved (this was usually easily done, from the spectral gradients for example) they then concluded that λ_1 and D actually classified stars far more precisely than the

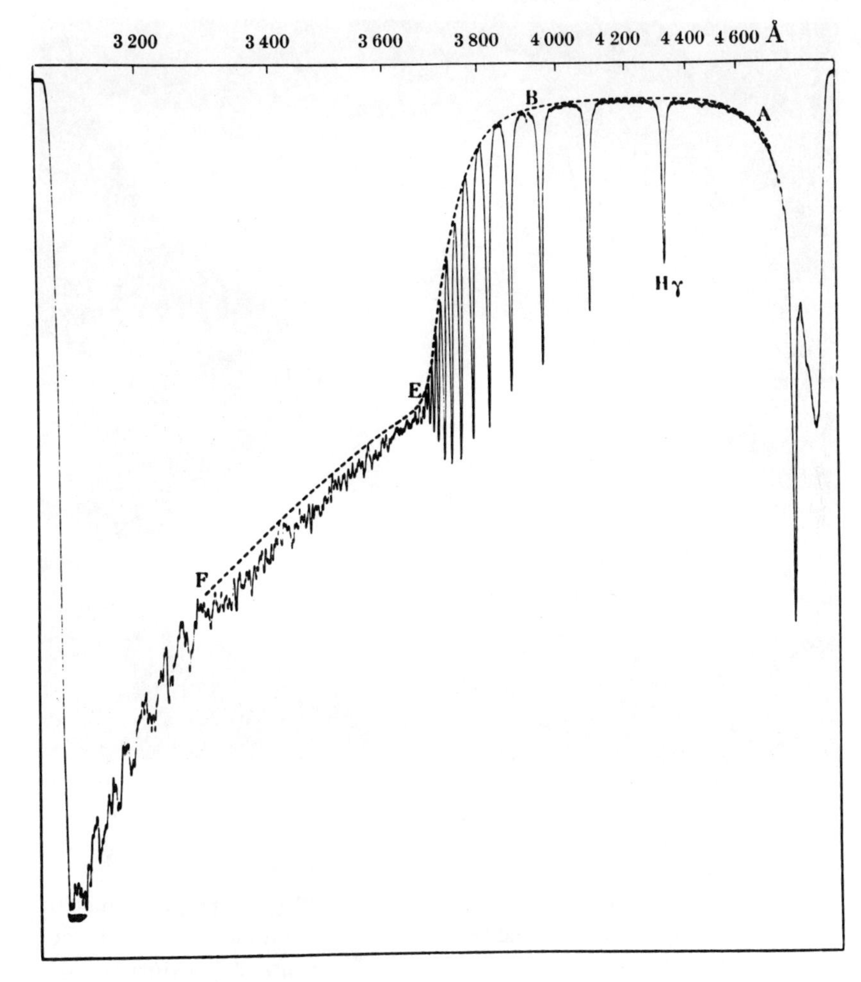

Fig. 8.11 The ultraviolet spectrum of Vega, observed by Barbier and Chalonge on the Jungfraujoch in 1939.

systems based on line strength used hitherto: 'for example between two stars classified B3V the Balmer jump can differ by 0.10. A star is thus better defined by the values λ_1 and D than by the simple statement of its spectral classification' (110).

Meanwhile Chalonge and Mlle Divan realised that the blue spectral gradients between 3800 and 4800 Å, which had always been measured along with λ_1 and D, provided the basis for a three-dimensional stellar classification (111). They showed how normal stars define a surface called Σ in the (D, λ_1, Φ_b) space (Fig. 8.12), in general Φ_b being dependent on the other two parameters. However, peculiar stars could be usefully isolated by their deviation from the Σ surface and in addition, stars near spectral type

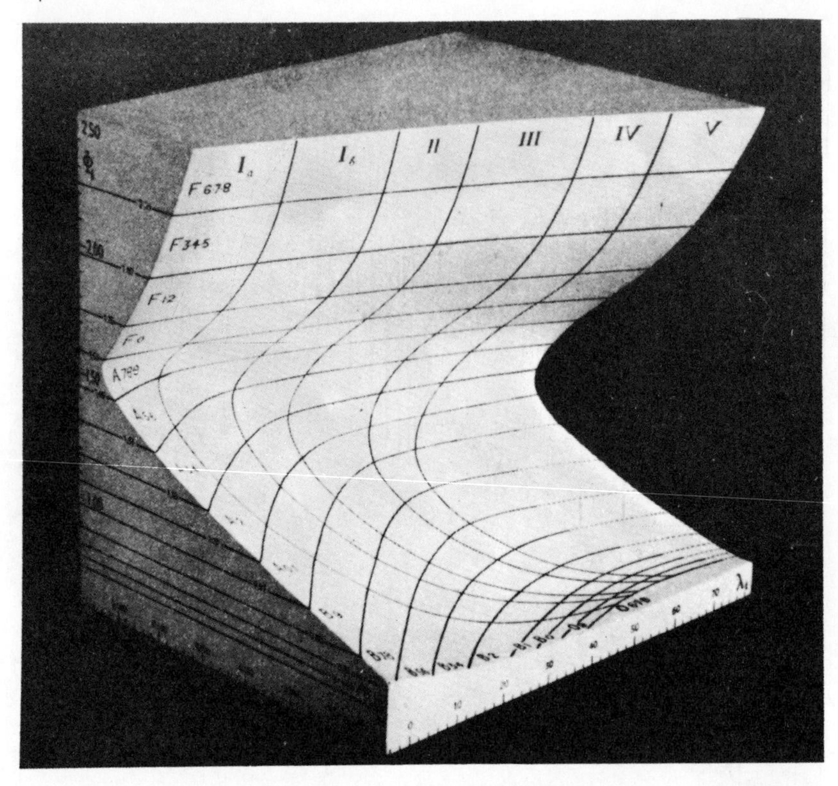

Fig. 8.12 The Barbier–Chalonge–Divan Σ-Surface for analysing ultraviolet stellar spectra.

A0, where D is at its maximum and rather insensitive to temperature, can be classified with better precision in the three-dimensional scheme, provided Φ_b is not affected by reddening due to interstellar dust clouds. Lucienne Divan applied the 3-dimensional classification to subdwarfs. These are cool stars with low metal abundances giving them weak lines. She found they lay below the upper Σ surface with observed Φ_b values corresponding to hotter stars than λ_1 and D would normally indicate (112). Similarly the strong-lined metallic line stars (type Am) were found to be above the Φ surface by J. Berger and his colleagues (113).

This three-dimensional system has been continuously in use at the Institut d'Astrophysique up to the present time, and has become known as the BCD (Barbier–Chalonge–Divan) system. It has been used extensively for studying galactic clusters, stars of peculiar compositions and stellar evolution (for a review of applications up to 1964 see (114) and see also (115) for some more recent comments on chemically abnormal stars). The BCD system has remained a peculiarly French product devised and used almost solely by astronomers at the Institut d'Astrophysique in Paris. More international use of the BCD system and recognition of its merits has been overdue.

Chalonge has been described as 'the creator of stellar astrophysics in France' (116) by J.-C. Pecker. Indeed, he was right, for the science of stellar spectroscopy had fallen behind in France of the 1930s compared with America, Germany and England. H.A. Deslandres had established a stellar spectroscopy 'service' at the Observatoire de Paris in 1890 where at first he, then later, M.T.A. Hamy and P. Salet (1875–1936) measured stellar radial velocities up to about 1931. France in all these years had no Huggins or Secchi, no Vogel or Draper, no Pickering or Campbell. J. Janssen and L. Thollon she did have, but these men were primarily solar astronomers, not stellar.

8.8 To the MKK classification and beyond

8.8.1 The origins of the MKK two-dimensional classification Ralph Curtiss concluded his well-known review on spectral classification just before he died in 1929 with these thoughts (10):

> The classification of the future will undoubtedly be based on physical principles in addition to temperature-ionization and will be expressed numerically in terms of definite parameters... at least three parameters will be necessary. They include a quantity based on temperature, another depending on the abundance of neutral atoms, and a third depending on the abundance of ionized atoms.

When he wrote this the final publication of the one-dimensional system of the Henry Draper catalogue was barely 5 years old. At Yerkes, Otto Struve echoed Curtiss' thoughts with this question: 'If two stars of equal atmospheric temperature and pressure are given, are their spectra necessarily identical?' (48). In Struve's opinion they were not; it was already becoming apparent that the discrepancies between spectral types for the same star from different observatories were not simply due to random errors but to the use of different classification criteria, which resulted in small discrepancies if 'new physical parameters such as abundances of elements etc.' were operative.

Although Curtiss and Struve both hastened to add their qualified support for the HD system, both men were already looking to systems with not two but at least three parameters. In a sense, two-dimensional systems had long been in use, represented either by the spectroscopic absolute

Fig. 8.13 W.W. Morgan.

magnitudes at, for example, Mt Wilson or the c, g and d prefixes introduced by Commission 29 of the IAU in 1922. The former were, however, based on good resolution slit spectra and the latter symbols were never quantitatively defined, which was especially difficult for early-type stars where the distinctions in luminosity are more of degree than of kind.

Struve's paper in the *Astrophysical Journal* in 1933 can be interpreted as meaning that he was not satisfied with the current state of spectral classification, and that moves were underway for an evolution of the Harvard scheme at least to a fully fledged two-dimensional classification. What is more, he seemed to be giving notice that Yerkes was where the new development would originate. He already had the man for the job in W.W. Morgan, whose paper in the previous volume of the *Astrophysical Journal* marked the start of his long interest in the overall classification of stars (117). In this paper Morgan had studied strengths of certain lines seen in low dispersion spectra and their dependence on both HD spectral type and luminosity.

Struve's remarks in 1933 brought a very cool response from Henry Norris Russell and the Harvard astronomers Cecilia Payne-Gaposchkin and Donald Menzel. They insisted that the HD classification criteria 'express the most *conspicuous* features from type to type. It is doubtful whether more outstanding bases for classification could be selected' (118).

They advocated 'a fairly liberal use of the letter 'p' ... to denote characteristics which are common to very few objects'. What is more, 'the mere existence of the Draper classification, applied to hundreds of thousands of stars, is a strong argument for its substantial conservation'. To Struve's question (see p. 283),

> the answer, of course, is 'No', unless the surface gravities, relative atomic abundances, and other factors are all identical, which we have every reason to believe is seldom the case. A complete description of the spectrum of a star requires not only the wavelengths and intensities of all the lines, but also their profiles... But we cannot admit that a mere listing of line intensities *constitutes* sufficient classification; classified material differs from tabulated data as a rogues' gallery differs from an undiscriminating photographic album (118).

Much later Morgan commented on Struve's 1933 paper with the remark:

> That's the only paper Struve wrote in spectral classification. It's an interesting paper, but it violates some established rules.... It just leapfrogs over a certain region of the conceptual development [of spectral classification] that you can't leapfrog over.... Within a year Russell, Donald Menzel and Cecilia Gaposchkin jumped on him with both feet (119).

The irony is that the classifications of Miss Maury and Miss Cannon which early in the century were attacked for their over-complexity, especially by Scheiner at Potsdam (120), now had to be defended by Harvard for their over-simplicity. In truth, Pickering, Annie Cannon and Antonia Maury were by temperament classifiers, more interested in unifying the bewildering diversity displayed by stellar spectra into an ordered system. Struve was an analyser, who relished the challenge of unexpected detail in spectra and giving an astrophysical explanation.

Fortunately Morgan's temperament was for classification, and he had already made progress on a two-dimensional classification system. It was far from 'a mere listing of line intensities'. In 1937 he had estimated surface gravities g for a sample of bright stars, including supergiants, giants, dwarfs and a white dwarf (lying below the main sequence), based on the mass-luminosity relationship, on colour temperatures T and available parallaxes (mainly trigonometric). He plotted the first $\log g$ versus $\log T$ diagram and at once recognised it as a fundamental form of the (Hertzsprung –)Russell diagram (121). '... the main sequence is shown to have the characteristic of approximately constant surface gravity over the whole range from A0 to M' and he went on:'Each successive luminosity group is separated in g from the others'. This was the first time that the near constancy of gravity along the main sequence had been remarked upon. The discovery attracted very little attention at the time, a point that Morgan himself mentioned later (122). Morgan saw that his $\log g - \log T$ diagram formed the basis for a new two-dimensional classification, for a star could be assigned to one of these groups using luminosity (i.e. gravity)–sensitive features in a low dispersion spectrum, without having to assign a calibrated numerical value for the spectroscopic absolute magnitude. 'If a second dimension in the classification is to be introduced, it therefore seems advisable to go back to the actual spectra and to give measures of the value of certain criteria on an arbitrary scale which is defined by type stars' (121).

This is just what he did the following year, 1938. This is the year Morgan introduced five luminosity classes labelled from I to V to cover the range of luminosities from supergiants (I) to main sequence stars (V) for stars of spectral type F4 or later. In addition, the supergiants between F4 and F8 only were divided into Ia and Ib. Class III were normal giants while the subgiants (Class IV) were only found between G5 and K2 (123). The criteria he adopted for the luminosity classification were line intensity ratios, mainly of a neutral to an ionised species, with high weight being accorded to the ratio of 4045 Å (neutral iron) to 4077 Å (ionised strontium). Lindblad's CN bands were not used.

In devising this two-dimensional classification, Morgan put great emphasis on the stellar colours, using photoelectric data obtained by K.F. Bottlinger at the Berlin–Babelsberg Observatory (124). He found for stars unaffected by reddening from interstellar dust, that 'the color indices of the stars of spectral types F4-M0 are uniquely determined by their spectral type and luminosity' (123). In other words, two parameters seemed to be sufficient to describe completely the low dispersion spectrum of a star.

Calibration was therefore unnecessary for the new luminosity classes, though Philip C. Keenan (b. 1908), working with Morgan at Yerkes, realised this could be useful for certain applications of the classified material. He thus calibrated the classes for late-type stars using mainly trigonometric parallaxes, and hence was able to estimate the relative numbers of stars of types G7 to K2 in the different classes in a given volume of space (the luminosity function) (125). He also used Morgan's classes for classifying variable stars of type M (126).

8.8.2 The MKK *Atlas of Stellar Spectra* In 1943 Morgan, Keenan and Edith Kellman published their celebrated *Atlas of Stellar Spectra, with* *an Outline of Spectral Classification* as a special monograph of the *Astrophysical Journal.* The atlas comprised fifty-five prints produced by Miss Kellman displaying Yerkes 40-inch refractor slit spectra at the modest dispersion of 125 Å/mm, considered to be ideal for classifications of high accuracy (16). The prints were in three categories: 23 showed luminosity effects in spectra at different spectral types, 9 displayed the spectra of standard stars of given luminosity class but varying spectral type (e.g. Fig. 8.14), while the remainder illustrated various spectral peculiarities which fell outside the normal two-dimensional classification now proposed.

As the authors said, the atlas was designed for the 'practical stellar astronomer' and the emphasis of the classification was on 'ordinary' stars down to 12th magnitude. The historical basis of their system was summarised thus:

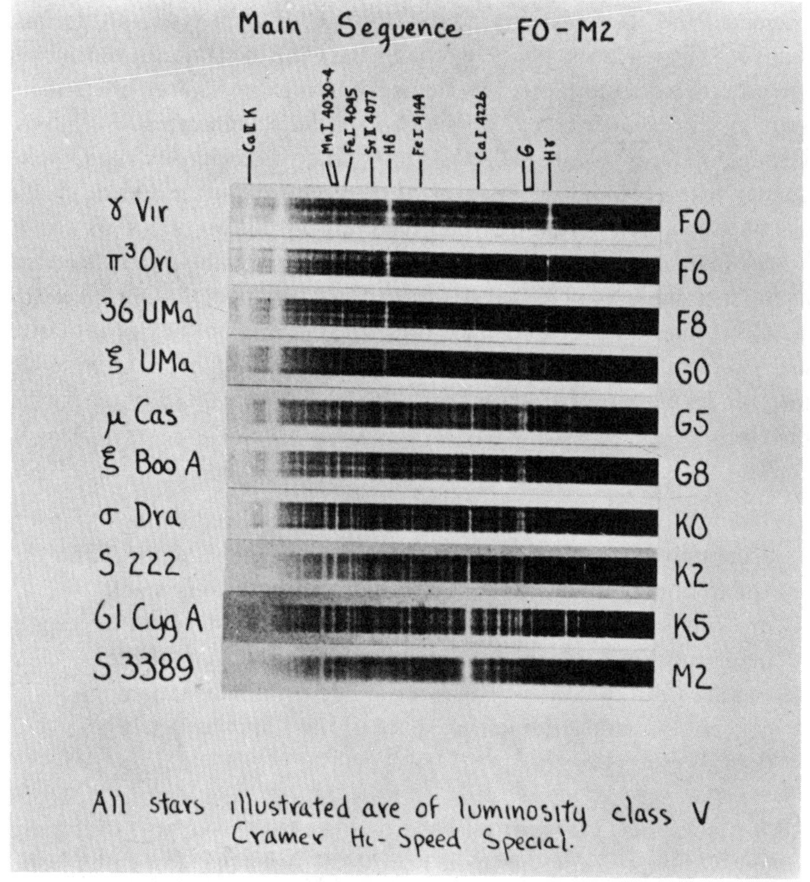

Fig. 8.14 A typical print from the MKK *Atlas of Stellar Spectra.*

> The *Atlas* and the system it defines are to be taken as a sort of adaptation of work published at many observatories over the last 15 years. No claim is made for originality; the system and the criteria are those which have evolved from a great number of investigations... By far the most important are those of the investigators at Harvard and Mt Wilson. The idea of a temperature classification is based on the work of Miss Maury and Miss Cannon at Harvard and of Sir Norman Lockyer. We owe to Adams the first complete investigation of luminosity effects in stellar spectra. If we add to this the work of Lindblad on cyanogen and the wings of the Balmer lines in early-type stars and the investigations of the late E.G. Williams, we have the great majority of the results on which the new classification is based (16).

The outline of spectral classification accompanying the atlas carefully presented the classification criteria for each spectral type and decimal subtype. The spectral types were broadly based on the Harvard system, but the criteria were more precisely defined and a large number of standard stars given for each type. For the O stars the classification was almost identical to that of Harry H. Plaskett (8), and no luminosity classes were defined earlier than O9. M dwarf stars were not included in the classification, nor were the carbon (Harvard R and N) or S stars.

Six luminosity classes (including Ia and Ib supergiants) were described again, in a similar way to that of Morgan's 1938 paper (123). In effect, the standard stars themselves defined these classes, and not any quantitative measurements of line ratios. A preliminary calibration of the absolute magnitudes of these classes was given, although in no way was this a part of the classification *per se*.

The Yerkes astronomers concluded their monograph thus:

> The two-dimensional classification can be used to describe accurately the spectra of the normal stars brighter than eighth apparent magnitude. Since this includes all but a very small percentage of the total number of stars brighter than that limiting magnitude, it is possible to derive from the extension of the classification to fainter objects certain general information concerning the distribution in space of the stars absolutely brighter than the sun (16).

8.8.3 Some commentaries on the MKK system Naturally enough, Struve at Yerkes gave the new MKK Atlas the most enthusiastic welcome:

> In 1913, W.S. Adams invented the method of spectroscopic luminosities by relating the intensity ratios of certain lines to the intrinsic luminosities of the stars. This gave a new impetus to the problem of spectral classification – and its culmination is now found in the Yerkes system of W.W. Morgan, P.C. Keenan, E. Kellman, Nancy Roman, W.P. Bidelman, and their numerous collaborators.

He continued: 'The stars were . . . arranged, not in a linear sequence, but in a rectangular pattern of pigeonholes, like the mail boxes at a post office' (127). Struve goes on to list the principal values of the MKK system. These were its generality, its refinement over many years using uniformly obtained material, the purely empirical basis of its criteria, that all the brighter stars have so far been classified in the new system (a statistically significant sample), its use of low dispersion spectra, and that the system limited itself to stars of normal composition without attempting to tackle the complexities of peculiar stars with abnormal element abundances. Struve continued: '. . . the main purpose of the Yerkes system was to aid in probing the Milky Way by establishing reliable criteria of stellar spectroscopic distances. More or less as a by-product this system is the best we now have for following Secchi's original quest "to see if the composition of the stars is as varied as the stars are innumerable"' (127).

The response from Mt Wilson was not so warm. Alfred Joy reviewed the new work: 'In the opinion of the reviewer this nomenclature [for the luminosity classes] adds little in convenience and detracts attention from the physical picture represented by the Russell diagram. . . . The new system for designating luminosity . . . merely adds new names for groups previously recognised' (128). Morgan himself, much later, remarked on Joy's review: 'The review was not all positive, but it was fine. I think that it was his review that brought it to the attention of people, and this, I think, is the reason why I finally made tenure at the University of Chicago' (129).

The Astronomer Royal, H. Spencer Jones, after a very bland and factual summary, seemed to find a spark of enthusiasm for the new classification:

> The Atlas and the descriptive outline will be most valuable to the practical astrophysicist. The Editors of the *Astrophysical Journal* are to be congratulated upon their enterprise in including such a useful working tool in their series of 'Astrophysical Monographs'. The authors are also to be congratulated for the care with which the investigation has been carried through and for the clearness and excellent quality of the photographic prints (130).

From France, Charles Fehrenbach could hardly wait for more copies of MKK: 'The results published in various reviews have been combined in the excellent *Atlas of Stellar Spectra* which unfortunately has rapidly sold out, so that numerous European observatories have been unable to procure this fundamental work which appeared in 1942. Its reprinting is awaited with impatience' (67).

From Harvard there was complete silence. A scan of the astronomical literature for the decade following publication of the MKK Atlas shows no commentary on the MKK system from Mrs Payne-Gaposchkin or others.

8.8.4 Spectral classification at Yerkes after the MKK When Struve wrote in 1953 (127) that the main purpose of the Yerkes system was to aid in probing the Milky Way, he had the advantage of hindsight of some of the remarkable achievements that his 'protégés' had already achieved in putting the Yerkes spectral classifications to practical use. Still, he was not mistaken! The most significant results were made for high luminosity OB stars by Morgan, Bidelman, Sharpless and Osterbrock and for the high-velocity stars by Nancy Roman, not to mention the revision of the MKK system by Morgan and Johnson.

Classifications of new supergiants as MKK standard stars were made in 1950 by Morgan and Nancy Roman (b. 1925) (131) and Class I was divided at this time into Ia and Ib for spectral type B stars (which in the MKK Atlas had not been done). William P. Bidelman (b. 1918) used the new standards to classify 102 suspected high luminosity (Maury c-type) stars. Most of these were Class I supergiants though thirty turned out to be less luminous than Class III (132). He noted the close association of supergiants with the galactic plane in these results. Later Bidelman used the plates taken for radial velocity work between 1903 and 1929 at the Lick southern station in Chile to classify further southern supergiants (133).

Morgan in 1951 had shown that in blue spectra as little dispersed as 230 Å/mm the hot early-type stars (types O to B3II) can be identified as a 'natural group' in the Hertzsprung–Russell (HR) diagram, which he called the 'OB stars' (134); 918 stars of this type were catalogued by J.J. Nassau (1892–1965) and Morgan, using Warner and Swasey Observatory objective prism plates (135) and for the nearer stars they showed these lay in a plane around the sun tilted at 20° to the overall plane of the Galaxy. This was the local system of stars long ago recognised for brighter naked eye stars as Gould's belt (136).

Now came the real 'breakthrough'. A paper presented by Morgan in December 1951 to a symposium organised by the American Astronomical Society in Cleveland on the HR diagram was entitled 'Some features of galactic structure in the neighbourhood of the Sun'. The authors were Morgan with Stewart Sharpless (b. 1926) and Donald Osterbrock (b. 1924) and the result, based on spectroscopic luminosities for OB stars near regions of ionised hydrogen, was the discovery of

> 'two long narrow belts similar to the spiral arms observed by Baade in the Andromeda nebula [galaxy, M31]. The nearer arm... passes at it nearest point about 300 parsecs distant from the sun in a direction opposite to that of the galactic centre... A second arm can be traced... parallel to the first and is situated at a distance of about 2000 parsecs from it in the anti-center direction. There is some evidence of another arm located at a distance of around 1500 parsecs in the direction toward the galactic center' (137).

This contribution represents the discovery of spiral structure in our Galaxy, certainly one of the great achievements of the Yerkes astronomers, as it showed our own star system to be similar in kind to the countless thousands of other spiral objects then known to be at great distances throughout the universe. Strangely enough, only the abstract of this important paper presented at the Cleveland meeting was ever published. A full length paper was planned but never appeared, as a result of an illness requiring hospitalisation that Morgan went through in 1952.

The definitive study of the spectroscopic luminosities and distances of OB stars in the so-called O-associations (loose aggregates of young stars) by Morgan, Whitford and Code the following year fully confirmed the presence of three spiral arms within about 2500 parsecs of the sun (138).* Calibration of the luminosity classes was of course a key step in deriving absolute magnitudes and hence distances. Although a preliminary calibration had been given in the MKK discussion, a more refined version, which tackled the difficult problem of supergiant distances, was presented by Keenan and Morgan in 1951 as part of the volume commemorating the fiftieth anniversary of Yerkes Observatory (139).

The other important development at Yerkes in these years, which also pertained to galactic structure, came from the analysis of the spectra of giant and dwarf stars of types F5 to K5 by Nancy Roman. The first paper studied ninety-four bright stars no later than G5. She noticed certain inconsistencies in the classifications depending on whether hydrogen or metallic lines were used, and this in turn led to the discovery of what she

* Compare this to the diameter of the Galaxy, about 30 000 parsecs. The spiral structure was confirmed soon afterwards by the Dutch radio astronomers Oort, van de Hulst and Muller observing clouds of cold hydrogen gas. Note that 1 parsec = 3.26 light years.

termed a group of 'weak-lined' stars. Using spectroscopic distances she was able to compute the speed at which each star moves relative to the sun. Slow-moving stars were found in both groups, but a few high-velocity stars ($V > 70$ km/s) were solely in the weak-lined group (140). A more extensive compilation of new spectral classifications for bright F5 to K5 stars followed in 1952; the same two groups were found, and in addition two rarer groups whose spectra were characterised either by abnormally weak CN bands (even though they were giants), or by the absorption in the 4150 Å region (due primarily to CN) being unusually strong (the '4150 stars') (141). The 4150 stars have a large spread of velocity extending to high velocities, as in the more common weak-lined stars, while the weak-CN stars also had high-velocity members, but relatively few low-velocity ones.

At this time the discussion of stellar populations was receiving much attention from many astronomers. W. Baade had shown that the Andromeda galaxy contains two types of star, which he denoted Population I for the bluer objects in the spiral arms and Population II for the on average redder stars of the nucleus. The same distinction of two populations was now being recognised in our Galaxy, not so much from intrinsic stellar colours but from other properties such as chemical composition and kinematics. Both types of star were intermingled to some extent in the solar neighbourhood, and Nancy Roman's classifications showed the correlation of the chemical and dynamical properties. In reality there were relatively few of the high-velocity Population II (or halo) stars in her sample, but even within the younger Population I (or disk) stars she was observing a correlation between their chemical (as shown by their weak or strong lines) and dynamical properties.

By including fainter stars to magnitude 9.5 many more of the fast-moving Population II objects near the sun could be found. Nancy Roman listed 571 such stars in a catalogue (142). All these were provisionally identified as moving at more than 63 km/s. Many had weak lines, indicating low abundances of heavier elements. The weak-lined stars she labelled 'sd' (subdwarf) if of type G0 or earlier. She also used the luminosity class label 'VI' for a few apparently subluminous G and K dwarfs. The addition of Class VI and also the further subdivision of Classes I to V was subsequently suggested by Keenan and Morgan to the IAU (143). In the opinion of Keenan in 1963 however, the question of introducing this last luminosity class for subdwarfs should be 'left open for discussion' (144). This was not the first catalogue of high velocity stars; for example, G. Miczaika published a list of 555 in 1940 (145) and W.W. Campbell had first recognised several of these as early as 1901 (146). Nancy Roman's space velocities were, however, based on spectroscopic distances (via the luminosity class calibration), and therein lies the difference, for Miczaika had used the generally less reliable trigonometric parallaxes.

Finally we mention the paper of Harold L. Johnson (1921–1980) and Morgan in the *Astrophysical Journal* of 1953 (147). This paper is certainly one of the most often quoted in the whole of astrophysics in the ensuing quarter century. No doubt it would be quoted yet more if it had not meanwhile assumed that rare distinction that its contents are so well-known as not to require an explicit reference. Most of this paper deals with defining the UBV system of photoelectric photometry. Morgan had 10 years earlier recognised the value of photometry in achieving a uniform system of spectral classification. According to Morgan, Johnson 'made all of the photoelectric observations and devised the Q Method [of reddening-free indices]. The UBV system was devised by me; and I wrote most of the paper' (148). He now formalised some small changes to the original MKK system of classification, some of which were already mooted earlier. For example, the early B supergiants were divided into Classes Ia and Ib, some of the F giants were downgraded a class from III to IV and G and K giant standards were redefined to give a smoother (B-V) colour versus spectral-type relationship. The type K7V was added between K5V and M0V and the classification extended to the M dwarfs by defining Barnard's star as an M5V standard. They were all cosmetic changes to tidy up the system after a decade's practical use. The resulting MK system* has remained easily the most used spectral classification system until the present time, although further small changes were introduced in 1976 (149) and in 1978 (150). A useful compilation listing all MK spectral classifications up till January 1963 has been published by C. Jaschek and his colleagues at La Plata Observatory (151).

Let us conclude with these words of Morgan's in 1979: 'The MK system has no authority whatever; it has never been adopted as an official system by the International Astronomical Union† – or by any other astronomical organization. Its only authority lies in its usefulness; if it is not useful, it should be abandoned' (152).

8.9 The classification of the carbon stars

8.9.1 Introduction and summary of principal band systems

The classification of the carbon stars of types R and N in the HD catalogue was discussed in Chapter 5. We recall that in this classification the subtypes

* The notation MK (Morgan–Keenan) refers explicitly to the 1953 revision of the original MKK system of 1943.

† The HD system is still the only system on which that honour has been bestowed by the IAU.

Fig. 8.15 F. Ellerman (right), a pioneer in the study of carbon star spectra, at Yerkes Observatory with A.A. Belopolsky (centre) and E.B. Frost in 1899.

R0, 3, 5, 8 and Na, b and later, in the last two volumes, Nc, were recognised. The IAU at its first meeting in Rome in 1922 recommended a change in this notation to N0 and N3 in place of Na and b, while no decimal notation was assigned to replace type Nc for the time being. This decimal notation was adopted in the HD Extension.

The first extensive analysis of the spectra of carbon stars was by George Ellery Hale, Ferdinand Ellerman (1869–1940) and J.A. Parkhurst (1861–1925) at Yerkes in 1903 (153). They studied the wavelengths of numerous absorption lines and bands and emission lines in eight stars and emphasised the many similarities with stars of Secchi's type III (M stars) rather than the differences. Fine prints of the banded spectra of these stars were included in their results (Fig. 8.16).

The Ph.D. thesis of W. Carl Rufus at Michigan represents the other major work in the detailed study of carbon stars prior to the publication of the HD Catalogue (154). He investigated ten R stars and compared these with five of class N. He believed that 'stars of class R form a connecting link between class N and the solar type' and he showed, as Pickering had in 1908 in introducing the R class (155), that the R stars were distinguished from

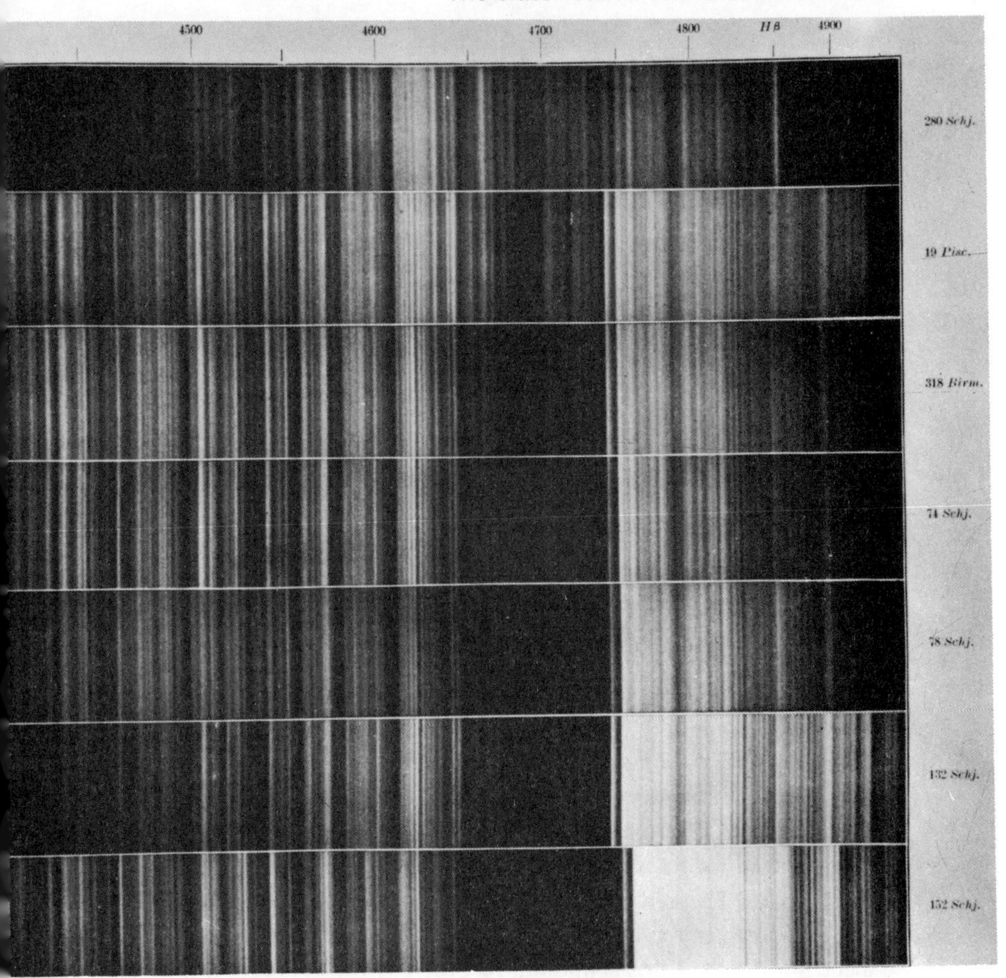

Fig. 8.16 Spectra of carbon stars in the blue, photographed by Hale, Ellerman and Parkhurst at Yerkes in 1903.

the N by the greater general ultraviolet absorption in N stars relative to R, giving R stars a generally less red colour.

Since the strong molecular bands of carbon stars are their characteristic feature, a short summary of the stronger ones could at this point be useful. In the visible spectrum that Secchi observed, the Swan bands of C_2 are the most prominent features. Those with heads at 5636, 5165 and 4737 Å are the strongest. All three were seen by William Swan in emission in candle flames in 1857 (156); he believed their origin to be due to the combustion of carbon and hydrogen. These bands are seen weakly in the solar absorption spectrum and were ascribed by Lockyer to carbon vapour (157).

The cyanogen molecule CN gives rise to ultraviolet and red band

Fig. 8.17 W.C. Rufus.

systems, but the strongest three bands are those with heads at 4216, 3883 and 3590 Å in the ultraviolet. A weaker band with a head at 4606 Å was found by Hale, Ellerman and Parkhurst (153) to be common to all carbon star spectra, but the shorter wavelength region was outside the range of their spectra. These bands were identified as due to cyanogen by H. Kayser and C. Runge in the laboratory (158). They are seen weakly in the solar spectrum and the 4216 Å band was recorded by Rufus in carbon stars (154). Miss Maury had observed strong absorption of two of these bands in many late-type objective prism spectra (4), but the observation of all three of the strong bands in stellar spectra and their explicit identification as being those of the well-known cyanogen spectrum appears to be due to Lindblad in 1922 (66), from spectra he obtained earlier at Lick using a quartz spectrograph on the Crossley reflector.

The CH molecule gives the well-known Fraunhofer solar G band at 4300 Å. It is very strong in carbon stars and was identified by Newall, Baxandall and C.P. Butler (1871–1952) in 1916 (159). Among the weaker molecular bands of other species, Paul Merrill found an unidentified band at 4976 Å in the R star UV Aurigae in 1926; this band was unusual in being degraded towards the red side (unlike the C_2, CN and CH bands) (160). Sanford found another similar band at 4868 Å in 15 N-type stars (161).

They are generally known as the Merrill–Sanford bands. C.D. Shane at Lick added six more of this band type, three of which Sanford had suspected (162). In spite of a detailed study by Andrew McKellar (1919–60) (163) at Victoria, the origin of these bands remained a complete mystery for many years, until Bengt Kleman in Ottawa was able to identify them as due to the triatomic SiC_2 molecule in 1956 (164), from a study of the spectral emission of a silicon-charged carbon furnace.

Finally we mention the discovery by McKellar of a group of bands in the ultraviolet spectrum of the N3 star Y Canum Venaticorum centred around 4050 Å, which he erroneously ascribed to another polyatomic molecule, CH_2 (165). These bands were found in other carbon stars by Swings, McKellar and Rao (166) and they are also seen in the emission spectra of comets. They were finally identified in the laboratory by Clusius and Douglas in 1954 (167) as due to triatomic carbon, C_3.

8.9.2 Classification of carbon stars after the HD Catalogue

Donald Shane's (1895–1983) study of two bright carbon stars 152 Schjellerup (= Y CVn) and 19 Piscium, marks the start of his interest in this spectral class (168). He showed that the 4216 Å CN band was stronger in type R than N but remarked that Rufus had found that the weaker 4606 Å CN band was always stronger in the N stars. Such complications in the band strengths resulted in Shane's major contribution to carbon star classification in 1928 (162). Here he observed 50 carbon stars in the blue spectral region on the Crossley 90 cm reflector at Lick Observatory. The dispersion was mainly only 650 Å/mm with some 120 Å/mm spectrograms also being used. However, even on the low dispersion material he could plot the band strength for the main blue bands of CN, C_2 and CH. The CN bands (but not that at 4606 Å) increased from R0 to R5 then declined to vanish completely by N3. But the prominent 4737 Å Swan band had a maximum at R5 then a minimum at N0 before increasing again for progressively redder N types. The G band had a broad maximum near R3 followed by a steep decline to vanish by N3.

The behaviour of the blue Swan band was remarkable, as was also the extreme red colour of the later types. Shane noticed that the 650 Å/mm spectrum of Y CVn (HD type Nb, a variable brighter than sixth magnitude) could be exposed in only 1 s at 4800 Å but a 5-hour exposure of this same spectrum still revealed no trace of light at 3900 Å or less. Shane also studied the enigmatic Merrill–Sanford bands and found them to be especially prominent only in the very reddest stars; they were absent in type N0.

These observations of band strengths in this large sample of stars

Fig. 8.18 Jason Nassau.

allowed Shane to devise an improved classification, still based on the Harvard R and N system but with more decimal subtypes. He retained the four R subtypes R0, 3, 5 and 8 but gave improved definitions based on band strengths and behaviour of the ultraviolet continuum. For the N stars he adopted types N0, 3, 5, 6, 7, the last three of which were marked by progressively stronger Merrill–Sanford bands and lesser ultraviolet extent of the continuum. The stars had mainly been classified Nc at Harvard.

Some quotes from Shane's 1928 article point to the direction of carbon star classification of the future:

> Despite the progressive redness of the carbon stars, there is considerable doubt that temperature plays as important a role on this branch of the sequence [as it does for spectral types O to M] ... It appears from these considerations [i.e. the variation of band strengths with type] that temperature may not be of prime importance in determining the spectrum of carbon stars... [There is] a very definite indication that a one-dimensional system is valuable only as a rough means of describing the spectra (162).

It thus became clear that the deficiency of the Harvard R, N system was its failure to order the stars in terms of decreasing temperature, a problem recognised but not rectified by Shane. Philip Keenan and W.W. Morgan at Yerkes thus devised an entirely new classification for carbon stars in 1941

(169) at a time when the basis of the MKK classification for types O to M was also taking shape. They decided not to base their temperature criteria on the overall strength of the molecular bands at all, in view of their erratic behaviour, but to employ the intensities of the less prominent atomic absorption lines, especially the NaD lines and the ratios of several violet metallic lines, supplemented where feasible by estimates of the 'continuum' flux gradient in the yellow and red. They used the symbols C0 to C9 for these stars in order of decreasing temperature. As far as C7 the analogues of the same temperatures in the G, K, M sequence from about G4 to M4 could be identified using the same metallic line criteria. The C notation had of course been employed by Pickering and Mrs Fleming for certain early-type stars in the Draper Memorial Catalogue of 1890. But this class was abandoned by 1897 (170) and the symbol had remained unused for spectral classification since then.

Keenan and Morgan used a second dimension in their classification to represent not luminosity (as in MKK), but molecular band strength, which they took to be an indication of carbon abundance. This second parameter was indicated by a subscript after the temperature class and was given by a digit from 1 to 5 based on the strength of the prominent C_2 band at 4737 Å. Thus, on the new C system, Y CVn became $C5_4$ instead of N3, 19 Piscium became $C6_2$ instead of N0. In general Keenan and Morgan found a good correlation between R subtype and C subtype between C0 and C4. The N stars were all in types C5 to C9 but for them there was no correlation between the N and C subtypes. Thus in general they found N stars to be cooler than R, but no stronger statement on the temperature ordering of the N subtypes was possible. The coolest star in a catalogue of fifty-six carbon stars in the Keenan–Morgan paper was WZ Cas, unusual for its strong 6708 Å lithium line, and classified as $C9_1$, whereas Shane had assigned the type N1p.

The Keenan–Morgan classification has remained as the principal system for carbon stars up to the present time. It has been used extensively in a series of carbon star classifications by Yasumasa Yamashita at Victoria and Tokyo (171) and for example by Brian Warner in South Africa (172).

Finally, an innovative program of low dispersion carbon star classifications in the infrared deserves special mention. At Warner and Swasey Observatory (Cleveland, Ohio) Jason Nassau and Attilio Colacevich (1906–53) (visiting from Arcetri Observatory, Florence) observed carbon stars on objective prism plates in the near infrared spectral region from 6880 to 8700 Å, as part of a much larger program on late-type stars. The 61 cm aperture Burrell–Schmidt telescope at the observatory, equipped with a 4° prism, was the instrument used. The telescope had been installed in 1941, and was the first large instrument with Schmidt optics (Fig. 8.19). The dispersion was only 1700 Å/mm, but even on these minute spectra, only 1 mm long, the red molecular bands, mainly of the CN molecule (especially that at 7945 Å), were visible, and the stars could be placed into one of four groups on the Keenan–Morgan system: C1-C2, C3, C4-C6 and C7-C9 (173).

8.9.3 Other proposed carbon star classifications A variety of

Fig. 8.19 The Burrell–Schmidt telescope at Cleveland, Ohio.

other classification systems has at times been proposed for these most difficult spectra. Rupert Wildt was one who pioneered in infrared photography, no doubt because during his brief time at Mt Wilson in 1935 he came in contact with Paul Merrill who in the early 1930s had also done so much to develop this new technique (174). Using a grating spectrograph on the 100-inch telescope to observe Y CVn and U Hydrae, Merrill had shown:

> that a very large part of the extraordinary structure in N-type spectra between λ6910 and λ8780 is to be attributed to absorption by carbon or cyanogen molecules. Many, and perhaps all, of the apparently bright lines in this region are only bits of continuous spectrum emerging through narrow interstices in the complex networks of absorption lines.

Merrill had used new infrared dye-sensitised plates from Eastman Kodak. Wildt however employed Agfa infrared plates for his quite different study (175). Wildt's spectra on the 60-inch telescope were only 2 mm long between 4000 and 9000 Å. At this dispersion (even lower than Nassau had used for carbon stars) the carbon stars showed no absorption features at all, but the spectra could be used to estimate the relative amounts of blue and infrared light, a kind of colour index. He divided sixteen carbon stars into eight different colour classes based on visual inspection, a primitive form of spectral classification. He concluded, as Shane had also done, that the red colour of N stars may not be due simply to cooler temperatures, but to the presence of an unidentified 'absorption process going on in the N atmospheres which reduces the ultra-violet intensity very efficiently'. In this respect his hunch was correct; John G. Phillips and Leo Brewer at Berkeley, University of California, found an ultraviolet 'continuum' opacity associated with the C_3 molecule in 1954, which could account for the extinction of ultra-violet light in N-type carbon stars (176).

In 1950 Yoshio Fujita (b.1908) in Tokyo proposed extending the Keenan–Morgan classification to all late-type stars. Carbon stars would still have the C classification from C0 to C9, but other late-type stars, where molecular oxide bands dominate the spectra at low temperatures, would be given the types Ox0 to Ox9 (Ox for oxygen-rich). If necessary, Fujita suggested that a nitrogen-rich sequence could also be defined (N0 to N9) (177). This proposal was never followed up. The explanation that carbon stars differ from the other late-type stars of the G, K, M sequence in the ratio of carbon-to-oxygen, we owe to Miss Yvonne Cambresier and L. Rosenfeld in Liège (97) who proposed that the carbon stars have an abundance excess of carbon over oxygen in their atmospheres whereas the

G, K, M sequence has oxygen as the more abundant element. Henry Norris Russell (178) came to the same conclusion shortly thereafter. In this context we note here an interesting early reference by Edward Baly (1871–1948), assistant professor of chemistry at University College, London, who found in 1892 that small amounts of oxygen, when introduced into a discharge tube filled with pure carbon monoxide, were completely able to extinguish the characteristic spectrum of Swan bands (179).

Roger Bouigue (b. 1920) at the Observatoire de Toulouse developed the Keenan–Morgan classification further (180). He obtained spectra on the 1.2m Haute Provence telescope for ninety-five carbon stars and measured two parameters: the vibration temperature T from the relative band strengths of the red CN system, and the half-width W of the unresolved pair of sodium D lines (a measure of the intensity of D-line absorption). Bouigue found T to be a more sensitive temperature parameter for the earlier type carbon stars, W better for the later types. The temperatures lay between 1900 and 4000 K. He then defined a composite index C, in terms of T and W, which showed good sensitivity to the changes in the spectrum through the whole range displayed by carbon stars. This index was used as the parameter for the decimal subtype in a one-dimensional classification which closely resembled the temperature subtype of Keenan and Morgan.

Bouigue also claimed to have found two distinct sequences among the hotter carbon stars, distinguished by the strengths of the 6260 Å band of $C^{13}N^{14}$ and of the C_2 Swan bands. The stars with strong 6260 had weak C_2 and were given the nomenclature of subclass I; those with the converse properties, of subclass N.

More recently Harvey B. Richer (b.1944) (181) devised a new carbon star classification, which was claimed to present a better temperature sequence than achieved by Keenan and Morgan, since infrared photometry was found to correlate poorly with the Keenan–Morgan type. Richer preferred not to use the NaD lines because of contamination from the lines from interstellar clouds along the line of sight. Instead, one of the infrared lines of the ionised calcium triplet (at 8662 Å) was the main temperature discriminator. Richer's system was two-dimensional and the first system in which the second dimension had been luminosity. He defined two luminosity classes (I, II) based on the strength of infrared CN bands and of neutral titanium and iron lines between 8300 and 8450 Å. Most of the stars classified were of luminosity class II. His spectral types used the notation C0 to C9 as before. However Yamashita (182) has shown that Richer types correlate very poorly with those he obtained on the Keenan–Morgan system.

8.10 Classification and spectra of S stars The S stars comprise a rare type of red giant that was recognised as a separate spectral class in 1922 by the IAU (3). Mrs Fleming had noted peculiarities in three stars π^1 Gruis, T Cam and R Gem as early as 1912 (183) in her extensive monograph on peculiar spectra. She wrote: 'The spectra of T Camelopardalis and π^1 Gruis resemble each other, and are very peculiar. The absorption band between 4640 and 4750, is present, and almost all of the continuous spectrum between this band and Hγ is cut out by other strong absorption bands'.

The introduction of the S-class, its incorporation into the last two volumes of the HD Catalogue, and the discovery by F.E. Baxandall and P.W. Merrill (184) of zirconium oxide bands in the spectra of these stars as their characteristic feature, are all described in Chapter 5. Although the peculiarities were a Harvard discovery, the early history of the study of S-type spectra was thereafter dominated by the work of Paul Merrill at Mt Wilson. His list of twenty-two S stars in 1922 included sixteen which were long-period variables which generally had the Balmer lines in emission at maximum light as in the more common Md long-period variable stars (185). He also found emission lines of ionised iron in their spectra, an unexpected occurrence as these lines were normally associated with hotter stars. Merrill noted the unusual strength of the lines of two of the heavier elements, namely barium (at 4554 Å) and strontium (at 4607 Å), as a general feature of S stars. Near the beginning of this paper he wrote: 'The general conclusion may be anticipated here by stating that the S stars probably form a third branch of the spectral sequence in addition to the G-K-M and the G-R-N branches' (185) (Fig. 8.20).

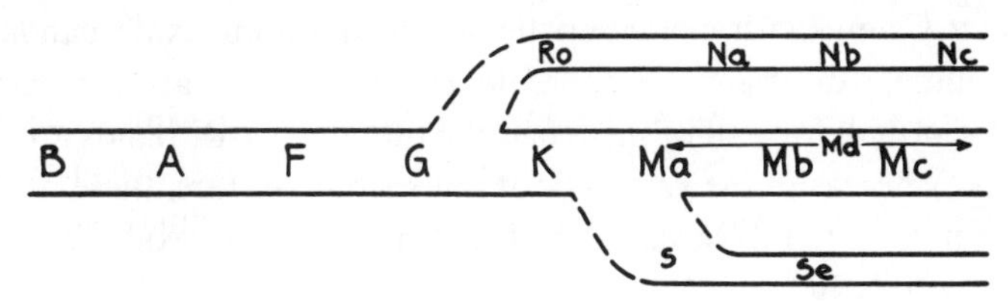

Fig. 8.20 A sketch by Merrill in 1922 to illustrate the trifurcation of late spectral types.

This was the first comprehensive study devoted to the new spectral class and remained as the basis for further investigations, mainly those of Merrill himself. As Alfred Joy noted: 'The complex spectroscopic behaviour of the long-period variable stars was, from first to last, Merrill's foremost astronomical concern and in the advance of our knowledge of these everchanging bodies his contributions over half a century were preeminent'

Fig. 8.21 Paul Merrill.

(186). On the discovery in 1923 that the red-degraded bands, of which the most prominent are at 4619, 4640 and 6474Å, are due to zirconium oxide (184) one of the major problems of these spectra was thus quickly resolved. Merrill continued to explore the details of their complicated spectra. In 1927 he listed thirty-one S stars, twenty of them variables (187). He showed there was no sharp dichotomy between the S and M types. Thus AA Cyg has strong bands of both zirconium oxide and titanium oxide while S UMa has strong zirconium oxide but the titanium bands are absent. The M-type star χ Cygni (with characteristic strong titanium oxide bands) still had zirconium oxide bands of moderate strength. The S stars were certainly rare and unusual, for half of them had the titanium oxide bands absent or very weak, whereas in the vast majority of such cool stars, these bands are the dominant spectral features. The fact that zirconium oxide bands were also seen weakly in some M stars was also pointed out by N.T. Bobrovnikoff (b.1896) using the Perkins Observatory 69-inch reflector (188). He found twenty zirconium oxide bands in β Peg (M2) and ρ Per (M4), and the only plausible conclusion was that an abundance effect, involving the elements titanium and zirconium, rather than the temperature and surface gravity of the stars, was the explanation for the differences in the bands from star to star.

These differences between individual stars led to the first S-star classification by Dorothy N. Davis (b.1913) in 1933 (189). Prior to that year, the stars had been listed simply as S (or Se if they showed Balmer emission lines). Miss Davis proposed five subclasses. Class I were 'very strange' and bands of neither zirconium oxide nor titanium oxide were seen but other S-star features (e.g. strong ionised barium and strontium lines) were present. The classes II to V ranged from strong zirconium oxide and no titanium oxide (class II) to titanium oxide dominating and zirconium oxide only weakly present (class V). This formed the basis for a classification from S1 to S5, the index corresponding to the Roman numeral class, and which she believed to be a temperature sequence. However, it is now clear that abundance effects play a greater role than temperature in determining the index. She classified thirty-two stars on this scheme (including χ Cygni as either M6e-M8e or S5e).

Meanwhile Merrill continued detailed studies of a number of the S stars. For example, for R Andromedae (Davis type S4e) he listed 27 elements found in the atomic line spectrum (190) and he again emphasised that the trifurcation of the cool giant sequences was probably an abundance effect. A detailed study of χ Cygni, with its M- and S-type features, also belongs to this period (191). In 1952 Merrill announced the discovery of four neutral lines of the element technetium, in the spectra of a number of S and M giants, but the S stars R Gem, R And and AA Cyg were outstanding for the strength of these lines, which were weak in the M types (192) (Fig. 8.22). The interest in the stellar technetium lines was intense, since the element is not found to be naturally occurring on earth and no stable isotopes are known; the longest-lived (Tc^{97}) has a half-life of 2.6 million years, much less than the presumed stellar ages. It thus seemed (and has since been confirmed) that this intriguing discovery pointed to a stellar origin for the technetium by some high-energy nuclear process. Though Merrill suspected this, his conclusion at the time was not certain as only one of the three longest-lived isotopes in 1952 had had its half-life measured.

Several elements apart from technetium were now known to be unusually abundant in S stars, including strontium, barium, yttrium and lanthanum, the last being discovered by Keenan from the lanthanum oxide bands in the near infrared (193) while the unusual strength of yttrium oxide bands in several S stars was noted by Merrill (194). The time was now ripe for a new classification scheme that would take the abundance peculiarities of S stars into account. Such a scheme was derived by Keenan at the Perkins Observatory in 1954 (195). He remarked that the Davis classification 'has proved of practical value' but 'the usefulness of Miss Davis' arrangement is limited, however, by the fact that it is a one-dimensional scheme, in which the subclasses are probably sensitive to more than one physical variable'. Keenan firstly defined an index for overall band strength

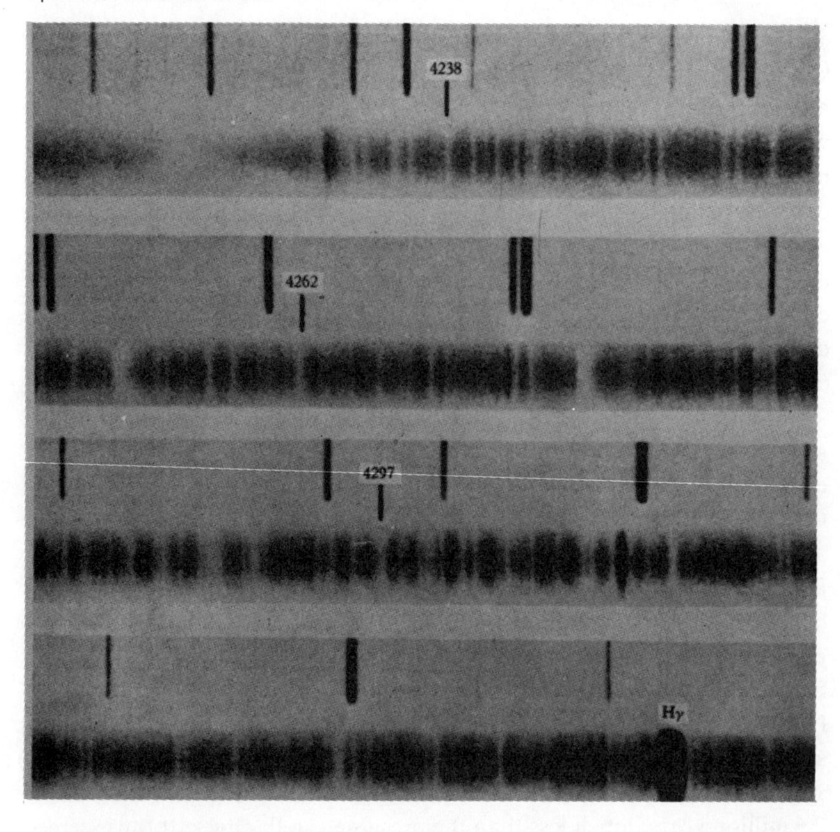

Fig. 8.22 Neutral technetium lines discovered by Merrill in the spectrum of the S star, R Andromedae.

of titanium oxide. Since in M stars the titanium oxide bands strengthen uniformly from M0 to M8, a convenient index was devised as equal to the decimal subtype plus one. A titanium oxide index of unity indicated bands just visible; 9 indicated bands of great strength. On the same scale a zirconium oxide band strength index was also defined. This was of course zero for many M stars with no zirconium oxide bands visible.

Keenan was next able to define both a temperature and an abundance parameter from the titanium oxide and zirconium oxide band-strength indices, each parameter having a possible range from one to nine. The stars with zirconium oxide bands just visible were abundance class 1, while those with zirconium oxide very strong relative to titanium oxide were in abundance class 9. The abundance class thus indicated the relative prominence of the zirconium oxide and titanium oxide bands. Examples of the Keenan classification are AA Cygni with titanium oxide intensity 5, zirconium oxide intensity 4, classification S7,5; S UMa with no titanium oxide bands, zirconium oxide intensity 1.5, classification S1.5,9 (at maximum light); and χ Cygni, titanium oxide strength 7, zirconium oxide strength 0 to 2, classification S7,le. In each case the temperature class precedes the abundance class in the Keenan notation.

Keenan also found that the infrared lanthanum oxide bands were a useful temperature parameter, if they were visible. They only appeared in the spectra of the cooler and more zirconium-rich S stars.

A total of 69 S-type stars, including all the known northern objects brighter than eleventh magnitude, were classified by Keenan on his two-dimensional scheme (195). Keenan defined an S star to be one with the zirconium oxide band intensities greater than unity. For weaker but still detectable zirconium oxide bands he assigned the mixed type MS and he classified 14 stars in this category. For example, AA Cam was given type M5S, the normal titanium oxide temperature class being M5. He also listed seven cool stars with neither zirconium oxide nor titanium oxide bands (Davis class S1) but with other S-type features such as an unusually strong ionised barium line at 4554 Å. R Canis Minoris was a typical member of this rare group, which had also been noted by Merrill (187). They tended to have strong CN and C_2 bands, though not as strong as in recognised carbon stars.

The Keenan classification has remained as the main tool, with minor recent modifications, for classifying the S stars. In 1962 Tsuji (196) explained some of the features of the Keenan classification, in particular why the titanium oxide bands generally weakened in the zirconium-oxide-rich stars. He reasoned that if zirconium and other heavier elements were enriched, the oxygen-to-carbon ratio was also simultaneously usually lower in these stars, and the reduced oxygen content and increased carbon monoxide abundance inhibited the formation of oxides such as titanium oxide.

Finally, we note that Keenan proposed no luminosity criteria for the S stars (195). He considered they probably had the absolute magnitudes of giants (III) or bright giants (II), but the evidence was mainly kinematic rather than spectroscopic. As with the carbon stars, it seems certain that no late-type dwarfs are to be found in this group.

8.11 Vanadium oxide and metallic hydrides in the M-type stars

Bands of vanadium oxide (VO) were among those of several oxides reported in M stars by N.T. Bobrovnikoff to the American Astronomical Society in 1936 (197). By 1939 he was able to list twelve metallic oxides seen in stellar spectra (198). The presence of

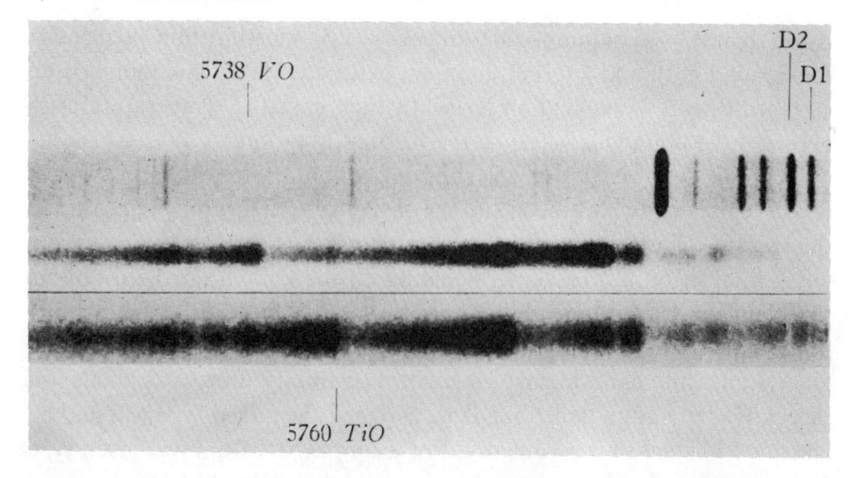

Fig. 8.23 The VO band at 5738 Å in R Leonis (above) is compared with Mira Ceti's spectrum (below). R Leo was photographed by Merrill in 1935.

vanadium oxide bands in the yellow–red region of the spectrum of five M stars was then confirmed by Merrill (194); the band with a head at 5738 Å was the most prominent. All these stars were long-period variables of type M6e or later. He noted: 'The VO bands appear to increase with advancing spectral type ... The fact that the vanadium bands are inconspicuous in the earlier subdivisions of class M indicates a lower energy of dissociation for VO than for TiO'.

Meanwhile Jason J. Nassau and G.B. van Albada, using the Warner and Swasey Observatory Schmidt telescope, found unidentified bands on infrared plates at 7900 Å for stars of type M7e or later, and P.C. Keenan was able to confirm their presence using higher dispersion Perkins reflector spectra. They found groups of bands near both 7400 and 7900 Å (199). Stars such as R Aql (M8e) and U Her (M7) showed the new bands especially strongly. These features were identified by Keenan and Leon Schroeder at the Perkins Observatory in 1951 as being also due to the vanadium oxide molecule (200) and this opened the way for an accurate temperature classification for these cool M stars using the extreme temperature sensitivity of vanadium oxide. As Keenan and Schroeder commented, 'the astrophysical interest of the infrared vanadium oxide bands lies chiefly in the fact that they occur in a fairly clean region of the spectrum'.

This technique was exploited by Nassau and Donald Cameron at Warner and Swasey. In late M stars, the vanadium oxide bands were so strong 'that almost the entire continuum on the short-wavelength side of λ7900 seems to disappear ... This gives the spectrum a unique appearance, definitely different from other types of spectra' (201). They were able to define four subtypes a, b, c and d with increasingly strong infrared vanadium oxide bands. In the last type, the vanadium oxide bands were so strong near 7900 Å, they 'produced the appearance of a sharp break at this wavelength'. The Nassau–Cameron subtypes generally varied with phase for the twelve variables in Cygnus studied. Cameron and Nassau extended this classification the following year, arbitrarily identifying subtypes b, c and d with stars of spectral type M8, 9 and 10 respectively, although this nomenclature was not based on any MK standards (202). They classified several dozen stars from M6 to M10 using this system, including four which near minimum light were of the latest subtype, M10 (W And, R Aur, χ Cyg, S Ori). Although these types represent a 'non-standard' extrapolation of the MK-system, Keenan later defined MK-standards for M giants and supergiants, extending as late as M8 for the standard star RX Boo (144).

The discovery of metallic hydrides in the spectra of M-type dwarf stars was one of the significant achievements of the Swedish astronomer Yngve Öhman when he visited Mt Wilson in 1933 (203). Here he was involved with the spectroscopic program in the red spectral region that Merrill and Humason were pursuing. The following year he was able to announce the discovery of calcium hydride in several M dwarfs, the first time this molecule had been seen in any stellar spectra although its presence in

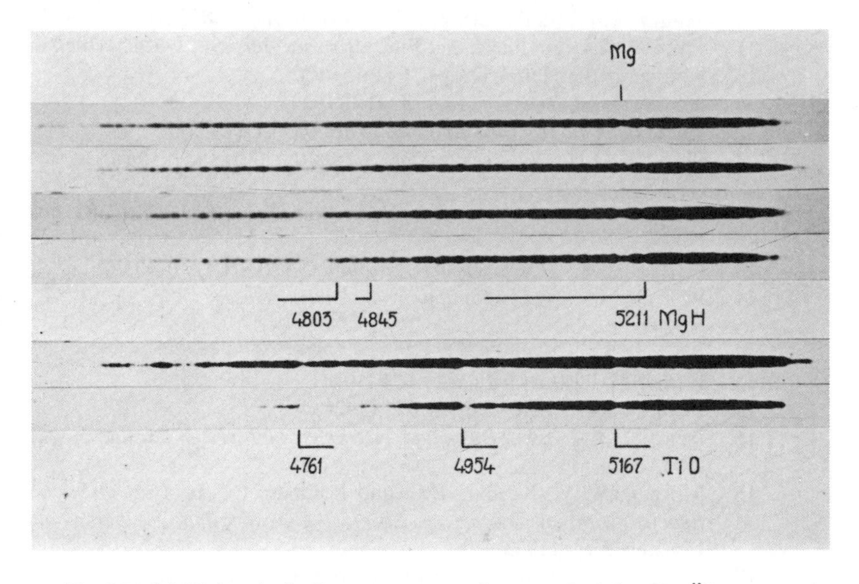

Fig. 8.24 MgH bands in late-type stars photographed by Y. Öhman at the Stockholm Observatory in 1936. Top to bottom: 61 Cygni A (K6V), ηCas (M0V), Cin. 1218 (M0V), 61 Cygni B (M0V) and Boss 1604 (2 spectra) (M3III).

sunspots was already well known (204). It was visible in several bands in the yellow and far red regions, the most conspicuous being at 6389 and 6382 Å. It appeared that M dwarfs from M0 and later displayed the new bands.

In the same paper magnesium hydride absorption was reported in 61 Cygni B in two bands, the stronger being at 5211Å. This molecule was also already known in sunspots (205) and had been identified by Joy in emission in the spectrum of Mira during the unusually faint 1924 maximum of this long-period variable star (206). Öhman was able to show that both these metallic hydrides provided a good luminosity discriminant for late-type stars as the absorptions were very weak or absent in giants. Shortly afterwards, following his return to Stockholm, he found further late K and M dwarfs in which MgH bands were visible (207) (Fig. 8.24). More recently, the earlier observations of this molecule in stellar spectra have been summarised by H. Spinrad and D.B. Wood (208).

References

1. Schlesinger, F., *Astrophys. J.*, **33**, 260 (1911), See also Chapter 5.
2. Schlesinger, F., correspondence to G.E. Hale, 18 Dec. 1912, quoted by D.H. DeVorkin, *Isis*, **72**, 29 (1981).
3. Adams, W.S., *Trans. I.A.U.*, **1**, 95 (1922); also in *Astrophys. J.*, **57**, 65 (1923).
4. Maury, A.C. and Pickering, E.C., *Harvard Ann.*, **28** (Part I), 1 (1897).
5. Cannon, A.J. and Pickering, E.C., *Harvard Ann.*, **98**, 1 (1923) (the 8th volume of the Henry Draper Catalogue).
6. Adams, W.S., *Trans. I.A.U.*, **3**, 162 (1929).
7. Cannon, A.J., *Harvard Ann.*, **28** (Part II), 1 (1901).
8. Plaskett, H.H., *Publ. Dominion Astrophys. Observ. Victoria*, **1**, 325 (1922).
9. Fowler, R.H. and Milne, E.A., *Mon. Not. Roy. Astron. Soc.*, **83**, 403 (1923).
10. Curtiss, R.H., *Handbuch der Astrophys.*, **V** (Part I), 1 (1932).
11. Plaskett, J.S., *Publ. Dominion Astrophys. Observ. Victoria*, **2**, 287 (1924).
12. Plaskett, J.S., *Trans. I.A.U.*, **2**, 117 (1925).
13. Payne, C.H., *Harvard Circ.*, **263** (1924).
14. Payne, C.H., *Harvard Bull.*, **855** (1928).
15. Pearce, J.A., *Publ. Dominion Astrophys. Observ. Victoria*, **5**, 110 (1930).
16. Morgan, W.W., Keenan, P.C. and Kellman, E., *An Atlas of Stellar Spectra with an Outline of Spectral Classification*, University of Chicago Press (1943).
17. Petrie, R.M., *Publ. Dominion Astrophys. Observ. Victoria*, **7**, 321 (1947).
18. Underhill, A.B., *Publ. Dominion Astrophys. Observ. Victoria*, **8**, 385 (1950).

19. Huggins, W. and Miller, W.A., *Phil. Trans. Roy. Soc.*, **154**, 437 (1864).
20. Lockyer, N.J., *Phil. Trans. Roy. Soc.*, **186A**, 73 (1894).
21. Campbell, W.W., *Astron. and Astrophys.*, **13**, 384 and 494 (1894).
22. Wilsing, J. and Scheiner, J., *Astron. Nachrichten*, **159**, 181 (1902); see also *Astrophys. J.*, **16**, 234 (1902).
23. Huggins, W., *Proc. Roy. Soc.*, **33**, 425 (1882).
24. Wright, W.H. *Publ. Lick Observ.*, **13**, 191 (1918).
25. Russell, H.N., Dugan, R.S. and Stewart, J.Q., *Astronomy*, p. 837 (1927).
26. Bohr, N., *Phil. Mag.*, **26** (ser.6), 1 (1913).
27. Fowler, A., *Phil. Trans. Roy. Soc.*, **214** (A), 225 (1914).
28. Cannon, A.J., *Harvard Ann.*, **76**, 19 (1916).
29. Boyce, J.C. Menzel, D.H., Payne, C.H., *Proc. Nat. Academy of Sciences*, **19**, 581 (1933).
30. Bowen, I.S., *Nature*, **120**, 473 (1927). See also Bowen, I.S., *Publ. Astron. Soc. Proc.*, **39**, 295 (1927) and Bowen, I.S., *Astrophys. J.*, **67**, 1 (1928).
31. Bowen, I.S., *American Institute of Physics, Oral History Interview with I.S. Bowen*, p. 11 (1968).
32. Wilson, O.C., *Sky and Telescope*, **45**, 212 (1973).
33. Payne, C.H., *Harvard Bull.*, **855**, 1 (1928).
34. Page, T., *Astrophys. J.*, **96**, 78 (1942).
35. Aller, L.H., *Gaseous Nebulae* publ. J. Wiley and Sons (New York) (1956).
36. Wurm, K., *Handbuch der Physik*, **50**, 138 (1958) (see p. 146).
37. Bowen, I.S. and Wyse, A.B., *Lick Observ. Bull.*, **19** (no. 495), 1 (1939).
38. Wyse, A.B., *Astrophys. J.*, **95**, 346 (1942).
39. For instrumental developments on the Mt Wilson 100-inch coudé see: Dunham, T., *Vistas in Astronomy*, **2**, 1223 (1956).
40. Moll, W.J.H., *Proc. Phys. Soc.*, **33**, 207 (1931).
41. Struve, O., *Astrophys. J.*, **74**, 225 (1931).
42. Milne, E.A., *Mon. Not. Roy. Astron. Soc.*, **104**, 112 (1944).
43. Williams, E.G., *Astrophys. J.*, **83**, 279 (1936).
44. Underhill, A.B., *The Early Type Stars*, Astrophys. and Space Sci. Library, publ. D. Reidel Publ. Co. (1966), p. 15.
45. Rudnick, P., *Astrophys. J.*, **83**, 439 (1936).
46. Williams, E.G., *Astrophys. J.*, **83**, 309 (1936).
47. Struve, O., *Nature*, **122**, 994 (1928). See also ref. (41).
48. Struve, O., *Astrophys. J.*, **78**, 73 (1933).
49. Struve, O., *Astrophys. J.*, **82**, 252 (1935). See also Struve, O. and Wurm, K., *Astrophys. J.*, **88**, 84 (1938).
50. Goldberg, L., *Astrophys. J.*, **89**, 623 (1939).
51. Stark, J., *Annalen der Physik*, **21**, 422 (1906) and *Verhandlungen der Deutschen Physik. Gesellschaft*, **8**, 109 (1906).
52. Atkinson, R.d'E., *Mon. Not. Roy. Astron. Soc.*, **82**, 396 (1922).
53. Russell, H.N. and Stewart, J.Q., *Astrophys. J.*, **59**, 204 (1924).
54. Struve, O., *Astrophys. J.*, **69**, 173 (1929).
55. Adams, W. and Joy, A., *Astrophys. J.*, **56**, 242 (1922) and ibid. **57**, 294 (1923).
56. Elvey, C.T., *Astrophys. J.*, **71**, 191 (1930).

57. Elvey, C.T. and Struve, O., *Astrophys. J.*, **72**, 277 (1930).
58. Elvey, C.T., *Astrophys. J.*, **74**, 298 (1931).
59. Pannekoek, A. and Verweij, S., *Proc. Amsterdam Academy*, **38**, 479 (1935).
60. Verweij, S., *Publ. Astron. Inst.*, University of Amsterdam, no. 5 (1936).
61. Holtsmark, J., *Phys. Zeitschrift*, **20**, 162 (1919).
62. Woolley, R. van der R., *Observatory*, **60**, 239 (1937).
63. Edwards, D.L., *Mon. Not. Roy. Astron. Soc.*, **83**, 47 (1922).
64. Lockyer, W.J.S., *Mon. Not. Roy. Astron. Soc.*, **82**, 226 (1922).
65. Edwards, D.L., *Mon. Not. Roy. Astron. Soc.*, **87**, 364 (1927).
66. Lindblad, B., *Astrophys. J.*, **55**, 85 (1922).
67. Fehrenbach, Ch., *Handbuch der Physik*, **50**, 1 (1958).
68. Lindblad, B., *Nova Acta Regiae Soc. Scient. Upsaliensis* IV:6, no. 5 (1925).
69. Schalén, C., *Arkiv för Mat. Astron. och Fysik*, **19A** No. 33 (1926).
70. Lindblad, B. and Schalén, C., *Arkiv för Mat. Astron. och Fysik*, **20A**, No. 7 (1927).
71. Robertson, H.P. and Dewey, Jane, M., *Phys. Rev.*, **31**, 973 (1928).
72. Pannekoek, A., *Mon. Not. Roy. Astron. Soc.*, **98**, 694 (1938).
73. Inglis, D.R., and Teller, E., *Astrophys. J.*, **90**, 439 (1939).
74. Mohler, F., *Astrophys. J.*, **90**, 429 (1939).
75. Unsöld, A. and Struve, O., *Astrophys. J.*, **91**, 365 (1940).
76. Miczaika, G.R., *Zeitschrift für Astrophys.*, **25**, 268 (1948); ibid **29**, 262 (1951).
77. Shapley, H., *Harvard Circular*, **278** (1925).
78. Adams, W.S., *Trans. I.A.U.*, **5**, 180 (1936).
79. Kapteyn, J.C., *Plan of Selected Areas*, published by Astron. Lab., Groningen (1906).
80. Schwassmann, A., *Vierteljahrschrift der Astron. Gesell.*, **70**, 352 (1935).
81. Schwassmann, A., *Bergedorfer Spektraldurchmusterung*, **1** (1935), **2** (1938), **3** (1947), **4** (1951) and **5** (1953).
82. Becker, F., *Himmelswelt*, **46**, 41 (1936) has given a general review of the Potsdam spectral survey. The survey was published in six parts: Becker, F., *Publ. Astrophys. Observ. zu Potsdam* **27** (parts 1, 2 and 3) (1931). Brück, H.A., ibid **28** (part 4) (1935). Becker, F., ibid **28** (parts 5 and 6) (1938).
83. Hoffleit, D., *Astrophys. J.*, **90**, 621 (1939).
84. Becker, F., *Sitzungsberichte der Preussischen Akad. der Wissenschaften*, p. 174 (1930) and p. 86 (1932).
85. Brück, H.A., *Mon. Not. Roy. Astron. Soc.*, **105**, 206 (1945).
86. Payne, C.H., *Astrophys. J.*, **90**, 321 (1939).
87. Vyssotsky, A.N., *Astrophys. J.*, **93**, 425 (1941).
88. Seares, F.H. and Joyner, M.C., *Astrophys. J.*, **98**, 244 (1943).
89. Humason, M.L., *Astrophys. J.*, **76**, 224 (1932).
90. Adams, W.S., Joy, A.H., Humason, M.L. and Brayton, Ada M., *Astrophys. J.*, **81**, 187 (1935).
91. Lindblad, B., *Nova Acta Reg. Soc. Sci. Upsala* (series IV) **6** (no. 5) (1925) and ibid., *Medd. Astr. Obs., Upsala*, no. 11 (1926), no. 18 (1927), no. 28 (1927).
92. Öhman, Y., *Arkiv för Mat., Astron. och Fysik*, **20A**, (no. 23) (1927).

93. Lindblad, B. and Stenquist, E., *Astronomiska Iakttagelser och Undersökningar å Stockholms Observatorium*, **11** (no. 12) (1934).
94. Ramberg, J., *Stockholms Observ. Ann.*, **13** (no. 9) (1941).
95. Elvius, T., *Stockholms Observ. Ann.*, **16** (no. 5) (1951).
96. Malmquist, G., *Uppsala Observ. Ann.*, **4** (no. 9) (1960).
97. Cambresier, Y. and Rosenfeld, L., *Mon. Not. Roy. Astron. Soc.*, **93**, 710 (1933).
98. Huggins, Sir Wm. and Lady Huggins, *An Atlas of Representative Stellar Spectra*, William Wesley and Son, p. 85 (1899).
99. Wright, W.H. *loc. cit.* See ref. (24), p. 257.
100. Yü, C.S., *Lick Observ. Bull.*, **12** (no. 375), 104 (1926).
101. Barbier, D. and Morguleff, N., *Astrophys. J.*, **136**, 315 (1962).
102. Barbier, D. and Morguleff, N., *Comptes Rendus de l'Académie des Sciences*, **258**, 4209 (1964).
103. Andrillat, Y., *Comptes Rendus de l'Académie des Sciences*, **261**, 321 (1965). See also: Andrillat, Y. and Morguleff, N., *Ann. d'Astrophys.*, **29**, 17 (1966).
104. Wing, R.F., Peimbert, M. and Spinrad, H., *Publ. Astron. Soc. Pacific*, **79**, 351 (1967).
105. Chalonge, D., *Contrib. Inst. d'Astrophys.* (sér. A) no. 97 (1951). Chalonge here gives a general review of the French program on the Jungfraujoch and a list of publications from the program up to 1951.
106. Barbier, D., Chalonge, D. and Vassy, E., *J. de Physique et le Radium*, **6**, 137 (1935). Also Arnulf, A., Barbier, D., Chalonge. D. and Canavaggia, R., *J. des Observateurs*, **19**, 149 (1936).
107. Barbier, D. and Chalonge, D., *Annales d'Astrophys.*, **2**, 254 (1939).
108. Barbier, D. and Chalonge, D., *Annales d'Astrophys.*, **4**, 30 (1941).
109. Barbier, D., Chalonge, D. and Canavaggia, R., *Annales d'Astrophys.*, **10**, 195 (1947).
110. Chalonge, D. and Divan, L., *Annales d'Astrophys*, **15**, 201 (1952).
111. Chalonge, D. and Divan, L., *Comptes Rendus de l'Académie des Sciences*, **327**, 298 (1953). Also Chalonge, D., *Annales d'Astrophys.*, **19**, 258 (1956).
112. Divan, L., *Annales d'Astrophys*, **19**, 287 (1956).
113. Berger, J., Fringant. A.-M. and Menneret, C. *Annales d'Astrophys.*, **19**, 294 (1956).
114. Chalonge, D., *I.A.U. Symposium*, **24**, 77 (1966).
115. Chalonge, D. and Divan, L., *I.A.U. Symposium*, **72**, 143 (1976).
116. Pecker, J.-C., *Quarterly J. Roy. Astron. Soc.*, **21**, 481 (1980).
117. Morgan, W.W., *Astrophys. J.*, **77**, 291 (1933).
118. Russell, H.N., Payne-Gaposchkin, C.H. and Menzel, D.H., *Astrophys. J.*, **81**, 107 (1935).
119. Morgan, W.W., *American Institute of Physics, Oral History Interview with W.W. Morgan* (1978) p. 27.
120. Scheiner, J., *Vierteljahrschrift der Astronomischen Gesellschaft*, **33**, 66 (1898).
121. Morgan, W.W., *Astrophys. J.*, **85**, 380 (1937).
122. Morgan, W.W., *American Institute of Physics, Oral History Interview with W.W. Morgan* (1978), p. 25.
123. Morgan, W.W., *Astrophys. J.*, **87**, 460 (1938).

124. Bottlinger, K.F., *Veröffent. der Universitätssternwarte zu Berlin-Babelsberg*, **3** (part IV) (1923).
125. Keenan, P.C., *Astrophys. J.*, **91**, 506 (1940).
126. Keenan, P.C., *Astrophys. J.*, **95**, 461 (1942).
127. Struve, O., *Sky and Telescope*, **12**, 184 (1953).
128. Joy, A.H., *Astrophys. J.*, **98**, 240 (1943).
129. Morgan, W.W., *American Institute of Physics, Oral History Interview with W.W. Morgan* (1978), p. 98.
130. Spencer Jones, H., *Observatory*, **65**, 127 (1943).
131. Morgan, W.W. and Roman, N.G., *Astrophys. J.*, **112**, 362 (1950).
132. Bidelman, W.P., *Astrophys. J.*, **113**, 304 (1950).
133. Bidelman, W.P., *Publ. Astron. Soc. Pacific*, **66**, 249 (1954).
134. Morgan, W.W., *Publ. Observ. University of Michigan*, **10**, 33 (1951).
135. Nassau, J.J. and Morgan, W.W., *Astrophys. J.*, **113**, 141 (1950).
136. Gould, B.A., *American J. of Science*, **8**, 325 (1874).
137. Morgan, W.W., Sharpless, S. and Osterbrock, D., *Astron. J.*, **57**, 3 (1952).
138. Morgan, W.W., Whitford, A.E. and Code, A.D., *Astrophys. J.*, **118**, 318 (1953).
139. Keenan, P.C. and Morgan, W.W., *Astrophysics: A Topical Symposium*: editor, J.A. Hynek, publ. McGraw-Hill. Chapter 1, p. 12 (1951).
140. Roman, N.G., *Astrophys. J.*, **112**, 554 (1950).
141. Roman, N.G., *Astrophys. J.*, **116**, 122 (1952).
142. Roman, N.G., *Astrophys. J. Suppl.*, **2**, 195 (1955).
143. Keenan, P.C. and Morgan, W.W., *Trans. I.A.U.*, **11A**, 346 (Appendix II) (1962).
144. Keenan, P.C., *Stars and Stellar Systems*, vol. 3: *Basic Astronomical Data*: editor K.A. Strand, publ. University of Chicago Press. Chapter 8, p. 78 (1963).
145. Miczaika, G.R., *Astron. Nachrichten*, **270**, 249 (1940).
146. Campbell, W.W., *Astrophys. J.*, **13**, 98 (1901).
147. Johnson, H.L. and Morgan, W.W., *Astrophys. J.*, **117**, 313 (1953).
148. Morgan, W.W., *American Institute of Physics, Oral History Interview with W.W. Morgan* (1978), p. 17.
149. Keenan, P.C. and McNeil, R.C., *An Atlas of the Spectra of the Cooler Stars*. Ohio State University Press, Columbus (1976).
150. Morgan, W.W., Abt, H.A. and Tapscott, J.W., *Revised MK Spectral Atlas for Stars Earlier than the Sun*. Yerkes and Kitt Peak Observatories (1978).
151. Jaschek, C., Conde, H., de Sierra, A.C., 'Catalogue of stellar spectra classified in the MK system', *Observ. Astron. Universidad Nacional de la Plata, Serie Astron.*, **28** (2) (1964).
152. Morgan, W.W., *Ricerche Astronomiche*, **9**, 59 (I.A.U. Coll **47**) (1979).
153. Hale, G.E., Ellerman, F. and Parkhurst, J.A., *Publ. Yerkes Observ.*, **2**, 251 (1903).
154. Rufus, W.C., *Publ. Observatory University of Michigan*, **2**, 103 (1916).
155. Pickering, E.C., *Harvard Circ.*, **145** (1908).
156. Swan, W., *Edinburgh Trans.*, **21** (ser.3), 411 (1857).

157. Lockyer, J.N., *Proc. Roy. Soc.*, **27**, 308 (1878).
158. Kayser, H. and Runge, C., *Wiedemann's Annalen* **38**, 80 (1889).
159. Newall, H.F., Baxandall, F.E. and Butler, C.P., *Mon. Not. Roy. Astron. Soc.*, **76**, 640 (1916).
160. Merrill, P.W., *Publ. Astron. Soc. Pacific*, **38**, 175 (1926).
161. Sanford, R.F., *Publ. Astron. Soc. Pacific*, **38**, 177 (1926).
162. Shane, C.D., *Lick Observ. Bull.*, **13** (no. 396), 123 (1928).
163. McKellar, A., *J. Roy. Astron. Soc. Canada*, **41**, 147 (1947).
164. Kleman, B., *Astrophys. J.*, **123**, 162 (1956).
165. McKellar, A., *Astrophys. J.*, **108**, 453 (1948).
166. Swings, P., McKellar, A. and Rao, K.N., *Mon. Not. Roy. Astron. Soc.*, **113**, 571 (1953).
167. Clusius, K. and Douglas, A.E., *Canadian J. of Physics*, **32**, 319 (1954).
168. Shane, C.D., *Lick Observ. Bull.*, **10** (no. 329), 79 (1920).
169. Keenan, P.C. and Morgan, W.W., *Astrophys. J.*, **94**, 501 (1941).
170. Pickering, E.C. and Fleming, W., *Harvard Ann.*, **26** (Part II), 1 (1897).
171. Yamashita, Y., *Publ. Dominion Astrophys. Observ.*, **13** (no. 5), 67 (1967); ibid. *Ann. Tokyo Astron. Observ.*, **13** (series 2), 169 (1972) and **15** (series 2), 47 (1975).
172. Warner, B., *Mon. Not. Roy. Astron. Soc.*, **126**, 61 (1963).
173. Nassau, J.J. and Colacevich, A., *Astrophys. J.*, **111**, 199 (1950).
174. Merrill, P.W., *Astrophys. J.*, **79**, 183 (1934).
175. Wildt, R., *Astrophys. J.*, **84**, 303 (1936).
176. Phillips, J.G. and Brewer, L., *Mém. Soc. Roy. des Sciences de Liége* (4e série), **15**, 341 (1955) (6th Liège International Astrophysics Colloquium).
177. Fujita, Y., *Publ. Astron. Soc. Japan*, **1**, 171 (1950).
178. Russell, H.N., *Astrophys. J.*, **79**, 317 (1934).
179. Baly, E.C.C., *Spectroscopy*, p. 444 (1905).
180. Bouigue, R., *Ann. d'Astrophys.*, **17**, 104 (1954).
181. Richer, H.B., *Astrophys. J.*, **167**, 521 (1971).
182. Yamashita, Y., *Ann. Tokyo Astron. Observ.*, **13** (series 2), 169 (1972).
183. Fleming, W., *Harvard Ann.*, **56**, 165 (1912).
184. Merrill, P.W., *Publ. Astron. Soc. Pacific*, **35**, 217 (1923).
185. Merrill, P.W., *Astrophys. J.*, **56**, 456 (1922).
186. Joy, A.H., *Quarterly J. Roy. Astron. Soc.*, **3**, 45 (1962).
187. Merrill, P.W., *Astrophys. J.*, **65**, 23 (1927).
188. Bobrovnikoff, N.T., *Astrophys. J.*, **79**, 483 (1934).
189. Davis, D.N., *Publ. Astron. Soc. Pacific*, **46**, 267 (1934).
190. Merrill, P.W., *Astrophys. J.*, **105**, 360 (1947); ibid **107**, 303 (1948).
191. Merrill, P.W., *Astrophys. J.*, **106**, 274 (1947).
192. Merrill, P.W., *Astrophys. J.*, **116**, 21 (1952). See also Merrill, P.W., *Trans. I.A.U.*, **8**, 832 (1954).
193. Keenan, P.C., *Astrophys. J.*, **107**, 420 (1948).
194. Merrill, P.W., *Publ. Astron. Soc. Pacific*, **51**, 356 (1939).
195. Keenan, P.C., *Astrophys. J.*, **120**, 484 (1954).
196. Tsuji, T., *Publ. Astron. Soc. Japan*, **14**, 222 (1962).
197. Bobrovnikoff, N.T., *Publ. American Astron. Soc.*, **8**, 209 (1936).
198. Bobrovnikoff, N.T., *Astrophys. J.*, **89**, 301 (1939).

199. Nassau, J.J., van Albada, G.B. and Keenan, P.C., *Astrophys. J.*, **109**, 333 (1949).
200. Keenan, P.C. and Schroeder, L.W., *Astrophys. J.*, **115**, 82 (1951).
201. Nassau, J.J. and Cameron, D.M., *Astrophys. J.*, **119**, 175 (1954).
202. Cameron, D.M. and Nassau, J.J., *Astrophys. J.*, **122**, 177 (1955).
203. Öhman, Y., *Astrophys. J.*, **80**, 171 (1934).
204. Olmsted, C.M., *Astrophys. J.*, **27**, 66 (1908).
205. Fowler, A., *Mon. Not. Roy. Astron. Soc.*, **67**, 530 (1907); see also: ibid., *Phil. Trans. Roy. Soc.*, **209A**, 447 (1909).
206. Joy, A.H., *Astrophys. J.*, **63**, 281 (1926); see p. 327.
207. Öhman, Y., *Stockholm Observ. Ann.*, nos. **3** and **8** (1936).
208. Spinrad, H. and Wood, D.B., *Astrophys. J.*, **141**, 109 (1965).

9 Spectroscopy of peculiar stars

9.1 Introduction

The study of peculiar stellar spectra has always played an important role in stellar spectroscopy, ever since emission-line stars such as γ Cas, T CrB and the Wolf–Rayet stars were first recognised during the 1860s as being unusual (see Chapter 4).

Thirteen types of peculiar star are discussed in this chapter. In general, the peculiarities are either the presence of emission lines in the spectra, or the presence of absorption lines of abnormal strength or profile. These peculiarities often aggravate attempts to classify stellar spectra in a two-dimensional scheme such as the MK system, described in the previous chapter. This is especially so if the peculiarities are due to photospheric abundance abnormalities which result in either unusually strong or weak absorption lines.

The list of types of star having peculiar spectra is certainly not exhaustive; emphasis is given to discussing those in which the spectral peculiarities predominate or were the first observed. Other peculiarities, such as intrinsic light variability or those arising for binarity, may also be present in many of these types. However a full discussion of the peculiarities in the spectra of all known types of variable or binary star is omitted for want of space.

9.2 C.S. Beals and the Wolf–Rayet classification

By 1928 the problem of how to classify the absorption-line spectra of O stars had been essentially settled by the IAU Commission 29. The question of how to classify the broad and mainly unidentified emission-line spectra of Wolf–Rayet (WR) stars, still remained in an unsatisfactory state. To summarise the position prior to that date, it is noted that the spectral subdivisions Oa, b and c for WR stars were introduced by Annie Cannon in 1901 (1) (see section 5.8). Class Oa had a bright emission band at 4650 Å as the most conspicuous feature, and this was only later identified as being from doubly ionised carbon. Likewise, Oc stars showed a strong 4634–41 Å feature, since recognised as arising from doubly ionised nitrogen. For class

Ob, the 4686 Å emission line (ionised helium) was especially prominent. All these stars showed Balmer lines and often the ζ Puppis lines (also ionised helium) in emission. This classification, which was part of a larger scheme for the Harvard O stars, was no doubt based to some extent on the visual descriptions and drawings of the emission bands in WR spectra by W.W. Campbell during his early years at Lick in 1894 (2) (Fig. 9.1). Here he observed thirty-one WR stars with Keeler's spectroscope on the Lick refractor; a result of this work was the discovery that the bright blue band in these stars was not in the same position in the spectrum in every star.

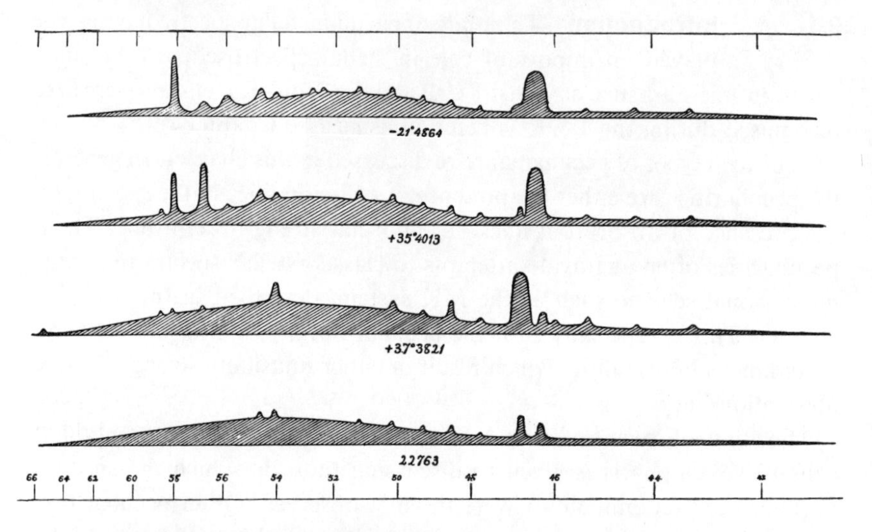

Fig. 9.1 The intensity distribution in visual spectra of Wolf–Rayet stars, drawn by W.W. Campbell at Lick in 1894.

One of the major problems of the Harvard classification was that many of the prominent emission lines on which the subdivisions were based remained unidentified. Carlyle S. Beals (1899–1979) at Victoria made the spectra of Wolf–Rayet stars his special interest after his appointment to the staff there in 1927. One of Beals' earliest papers on this topic concerned the interpretation of the broad emission lines in the Wolf–Rayet spectra, which were often found with a violet absorption feature similar to the line profiles of P Cygni. Beals interpreted this as being due to rapid mass loss of material in an extended transparent envelope around the star (3). The broadening was due to the Doppler effect and the violet absorption resulted from light from the stellar surface being scattered or absorbed in the envelope. The spectra of P Cygni and η Carinae could be explained in a similar way, though the ejection velocities were less than for Wolf–Rayet stars. This model for the P Cygni-type emission lines still stands even if it

Fig. 9.2 Carlyle S. Beals.

cannot explain all the emission-line broadening in the Wolf–Rayet spectra (4). The theory of Beals' is quite similar to a much earlier but seldom-quoted model proposed by the German-born astronomer Jacob K.E. Halm (1866–1944) in Edinburgh in 1904 (5).

In 1928 the IAU Commission 29 for Stellar Spectra had adopted P.W. Merrill's proposal to use the notation w to denote the presence of Wolf–Rayet emission and w! when these features were especially strong, and Cecilia Payne also proposed the same scheme at this time (6). Meanwhile, Beals reviewed the whole field of Wolf–Rayet spectra, including possible classification criteria, in a Dominion Astrophysical Observatory monograph (7), and a classification scheme was at that time put forward with the notation OW5 to BW1 for the WR stars.

A major advance came two years later; Bengt Edlén (b. 1906) working in the Physical Laboratory of Uppsala University used Beals' wavelength tables for WR emission lines (7) to identify many of them with the same features of highly ionised carbon, nitrogen and oxygen that were observed in the laboratory (8). This discovery resolved a very significant puzzle concerning the chemical origin of the lines, which were not observed in other stellar spectra, and which differed in their strengths between the Wolf–Rayet stars. At the 1932 IAU meeting in Cambridge, Massachusetts, a committee was set up to examine further the classification of WR spectra.

C.S. Beals was the natural choice for chairman, while Cecilia Payne and Harry Plaskett were the other members. Miss Payne had herself been working on identifying the WR emission lines and she was able to find ionised carbon and nitrogen lines in several stars at about the same time as Edlén (9).

The line identifications of Edlén were immediately taken up by Beals who wrote to the editor of the *Observatory*:

> A major advance in our understanding of Wolf–Rayet spectra has recently been made by Edlén, who successfully identified a majority of the emission bands of hitherto unknown origin with the highly ionised atoms of oxygen, nitrogen and carbon. The existence of two parallel sequences of Wolf–Rayet spectra .... seems to be established beyond reasonable doubt by the new identifications, which indicate a group of spectra containing bands due to CIII, CIV, OIV, OV and possibly OVI, to the exclusion of nitrogen, and a separate group containing nitrogen in various stages of ionization to the exclusion of oxygen and carbon. [These groups] ... may be referred to as the carbon sequence and the nitrogen sequence respectively (10).

Although Beals was undoubtedly the architect of the scheme of two parallel sequences for the WR stars, it is less well-known that several other astronomers had foreshadowed these ideas over a decade earlier. The concisest statement came from Charles Perrine (1868–1951), the American-born director of the Cordoba Observatory in Argentina. He found that in twenty-two WR stars either the emission was at 4650 Å, or that at 4634–41 Å was present, but never both, and he recommended that this observation be included in any scheme for the classification of Wolf–Rayet stars (11).

The year following Edlén's line identifications a major paper by Beals came out in the Dominion Astrophysical Observatory Publications with the carbon and nitrogen sequences defined and with classification subtypes from 6 to 8 for the carbon stars and 5 to 8 for the nitrogen sequence (12). From the high ionisation in Wolf–Rayet spectra, Beals deduced very high temperatures in the range 50 000 to 100 000 K, much hotter than previously ascribed to any other stellar types. This work on WR spectra remains Beals' most significant achievement. When the IAU met again in 1935 in Paris, the subcommittee for WR spectra reported back to Commission 29 (13). The report was dominated by Beals' ideas. The carbon and nitrogen sequences were again defined in their report, the use of the letter W for the nitrogen sequence and W′ for the carbon sequence was proposed with subtypes

W5 to W8 and W′6 to W′8, while the letter O should be restricted to absorption line O stars.

The report was signed by Beals and Harry Plaskett; the third member, Cecilia Payne, again (as she had in 1925 over Plaskett's classification of O stars) raised objections: 'I feel, however, that as the Sub-Committee was appointed to consider the possibility of classifying the Wolf–Rayet stars, rather than to devise a classification, I am at liberty to state that in my opinion the adoption of a classification is not at present to be desired ' (14). Possibly as a result of this dissenting voice from Harvard, no recommendations on WR spectral classification were adopted in 1935. Instead, the committee was invited to continue its work, ostensibly as the spectra of all southern WR stars had not yet been secured.

Another three years were to elapse before the sixth IAU General Assembly in Stockholm. By this time Mrs Payne-Gaposchkin (she had married Sergei Gaposchkin in 1934) had mellowed and a unanimous report was presented and adopted. The two sequences were described in detail and the notation WC and WN followed by the decimal subtype proposed and accepted (15) (Fig. 9.3). The Beals classification of WR stars has survived with only slight modifications for nearly half a century and remains as a fine tribute to its originator.

In 1942 P. Swings had argued that many Wolf–Rayet stars fell mid-way between the two sequences and should be classified simply as W, not WC or WN (16). Although he correctly identified carbon lines in some WN stars, the need to make the dichotomy between the two sequences less distinct has not been widely accepted.

In 1966 W.A. Hiltner and R.E. Schild (b. 1940) at Yerkes Observatory extended the classification with types WC5 and WN4 and further divided the nitrogen sequence into two further subsequences WN-A and WN-B. The A stars have narrower lines and most exhibit absorption features characteristic of O and B stars; many are binaries. On the other hand the B sequence is characterised by broader lines (17).

9.3 Spectral classification of novae

The spectroscopic study of novae has of necessity been spasmodic with real progress being limited to a handful of stars which have become naked-eye objects on outburst. T Coronae Borealis was the first nova ever to be observed spectroscopically, by Huggins at Tulse Hill (18), by E.J. Stone at Greenwich (19), and by Wolf and Rayet in Paris (20). Nova Coronae reached second magnitude in May 1866 and all observers noted an unusual spectrum of broad emission lines.

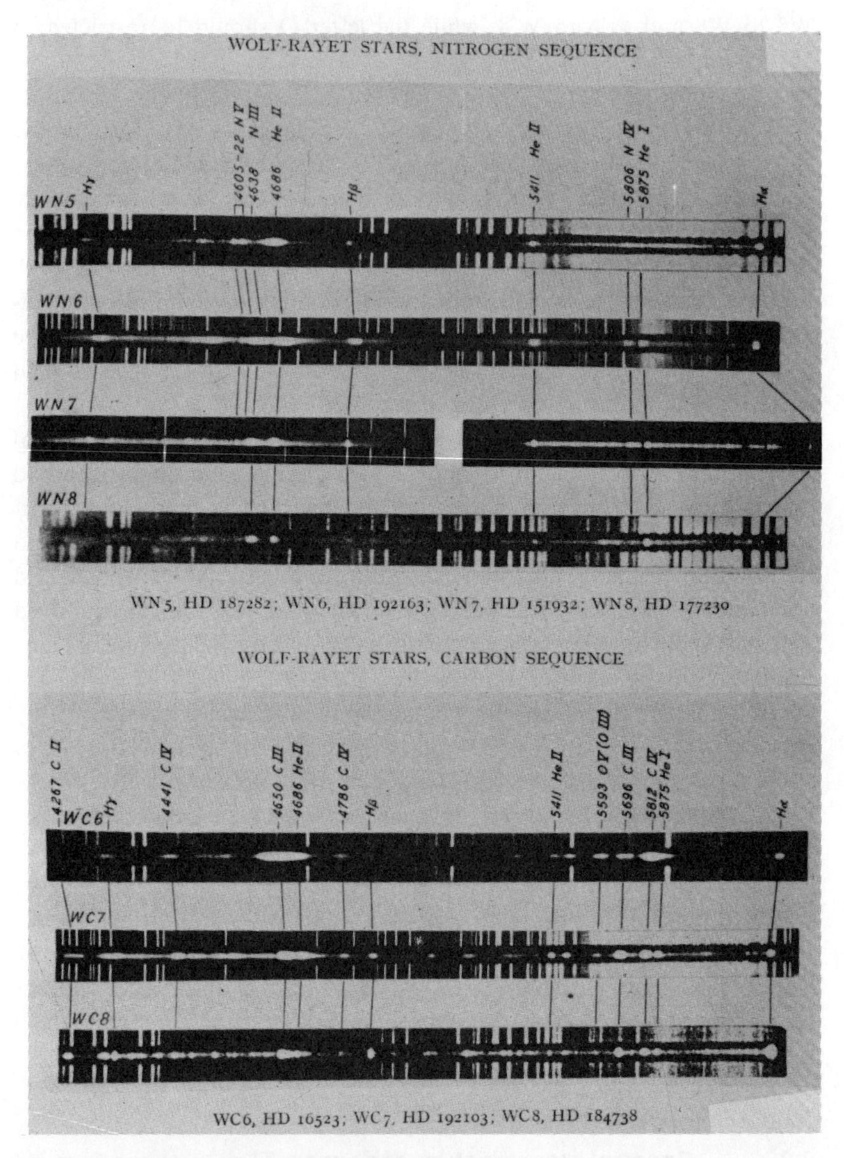

Fig. 9.3 Beals' spectra of WR stars, showing the nitrogen and carbon sequences, presented to the IAU in 1938.

Ten years later Nova Cygni (Q Cygni) reached about third magnitude in November 1876 and Cornu measured the wavelengths of eight emission lines in its spectrum including one at about 5000 Å which developed later, becoming very intense as in a gaseous nebula (21). Vogel found numerous dark lines in this spectrum (22).

Nova Aurigae (T Aur) was the first nova to be well-observed spectroscopically, in 1892, though it was discovered over 6 weeks after its rise to maximum. Several observers recorded the spectrum photographically including W.W. Campbell at Lick (23) and many visual descriptions of the spectrum were published. Huggins found dark absorption lines on the violet edges of the broad emission features, with a velocity difference of 880km/s if displacement was due to the Doppler effect (24). Campbell observed the development of a nebular spectrum about 8 months after the outbreak with unidentified lines of the 'nebulium' doublet near 5000 Å in the green.

From the start of the twentieth century all naked-eye novae have been observed intensively with the spectrograph; Table 9.1 lists some of the more prominent examples including Nova Persei in 1901, and Nova Aquilae in 1918, two of the brightest observed. New naked-eye novae are discovered at the rate of over one a decade but fainter examples are found at about six times this rate. The objective prism plates at Harvard were especially productive in new discoveries and Annie Cannon listed twenty in her 1916 list of emission-line stars (including the supernovae S Andro-

Table 9.1 *Principal naked-eye novae 1866–1965*

Nova	Outburst	m_v(max)	Principal spectroscopic references
T Coronae Borealis	May 1866	2.0	Huggins (18), Huggins and Miller (25)
	Feb. 1946	2.0	Sanford (26), McLaughlin (27)
Q Cygni	Nov. 1876	3.0	Cornu (21), Vogel (22), Lockyer (28)
T Aurigae	Dec. 1891	4.2	Campbell (23), Huggins (24)
RS Ophiuchi	Jun. 1898	4.3	
	Aug. 1933	4.3	Adams and Joy (29)
	Jul. 1958	5.0	
	Oct. 1967	4.8	
GK Persei	Feb. 1901	0.2	McLaughlin (30)
DN Geminorum	Mar 1912	3.5	Stratton (31), Wright (32)
V603 Aquilae	Jun. 1918	−1.1	Wyse (33)
V476 Cygni	Aug. 1920	2.0	Baldwin (34)
RR Pictoris	May 1925	1.2	Spencer Jones (35), Payne-Gaposchkin and Menzel (36)
DQ Herculis	Dec. 1934	1.4	Stratton (37), McLaughlin (38), Stratton and Manning (39).
CP Lacertae	Jun. 1936	2.1	Harper et al. (40), Wyse (41), McLaughlin (42).
V630 Sagittarii	Mar. 1936	4.5	Joy, Adams and Dunham (43)
CP Puppis	Nov. 1942	0.2	Sanford (44), McLaughlin (45), Weaver (46)
DK Lacertae	Jan. 1950	5.4	Larsson-Leander (47)
V446 Herculis	Mar. 1960	3.0	Meinel (48), Saweljewa (49)
V533 Herculis	Jan. 1963	3.0	Friedjung and Smith (50), Doroschenko (51)

medae and Z Centauri (52), see section 9.14). It was here at Harvard that Annie Cannon devised the first attempt at nova spectral classification (52). The Cannon classification had nova spectral types from a to e; types a,b and c differed primarily in the appearance of the nebular spectrum while type d showed no nebular spectrum after 6 months and e was used only for Nova Centauri whose 'spectrum was totally unlike any other new star yet photographed'.

In the original Draper Catalogue published by Pickering with Mrs Fleming in 1890, the class Q had been used for any objects otherwise unclassified by the letters A to P (53). In her classification of 1901, Annie Cannon had also used Q for any peculiar star with bright lines (1). In the Henry Draper Catalogue the Q notation was dropped in favour of 'Pec' which explicitly included the spectra of novae, but without any further subdivisions.

Annie Cannon's a,b,c,d,e classification had attempted to place individual novae with variable spectra into a single class. From the outburst of T Aur and more especially GK Per, the characteristic spectral changes were more carefully catalogued, and the possibility of assigning time-dependent types to characteristic spectral changes arose. Sir Norman Lockyer had listed four principal phases in a nova's spectral development as being (*a*) a continuous spectrum sometimes with a few dark lines; (*b*) broad emission lines or typical nova stage; (*c*) a stage where the bright emission and 4640 Å dominates (now identified as being due to doubly ionised nitrogen); (*d*) the nebular stage with a very weak continuum and emission lines resembling a gaseous nebula spectrum (54). At Cambridge F.J.M. Stratton (1881–1960) identified seven stages each with their spectral characteristics appearing in chronological sequence for a given nova, based especially on a study of Nova Geminorum, 1912 (31) (Fig. 9.4). When the International Astronomical Union met in Rome in 1922, these ideas were extended further with the revival of the Harvard Q classification being proposed by the spectral classification committee under Walter Adams. An evolutionary sequence Qa, Qb, Qc, Qu, Qx, Qy, and Qz was adopted. The choice of letters allowed room for further insertions (in fact Qd was added by the IAU at H.F. Newall's suggestion in 1925 (55)) and the classification Qz5O was reserved for a spectrum intermediate between that of a post-nova and a Wolf–Rayet star (56).

Although the Q classification of 1922 was officially adopted by the IAU its use was never widespread, probably because of the paucity of bright novae to classify spectoscopically, and because of the individual peculiarities and complexities of each new nova which made generalisations difficult to formulate and also less useful.

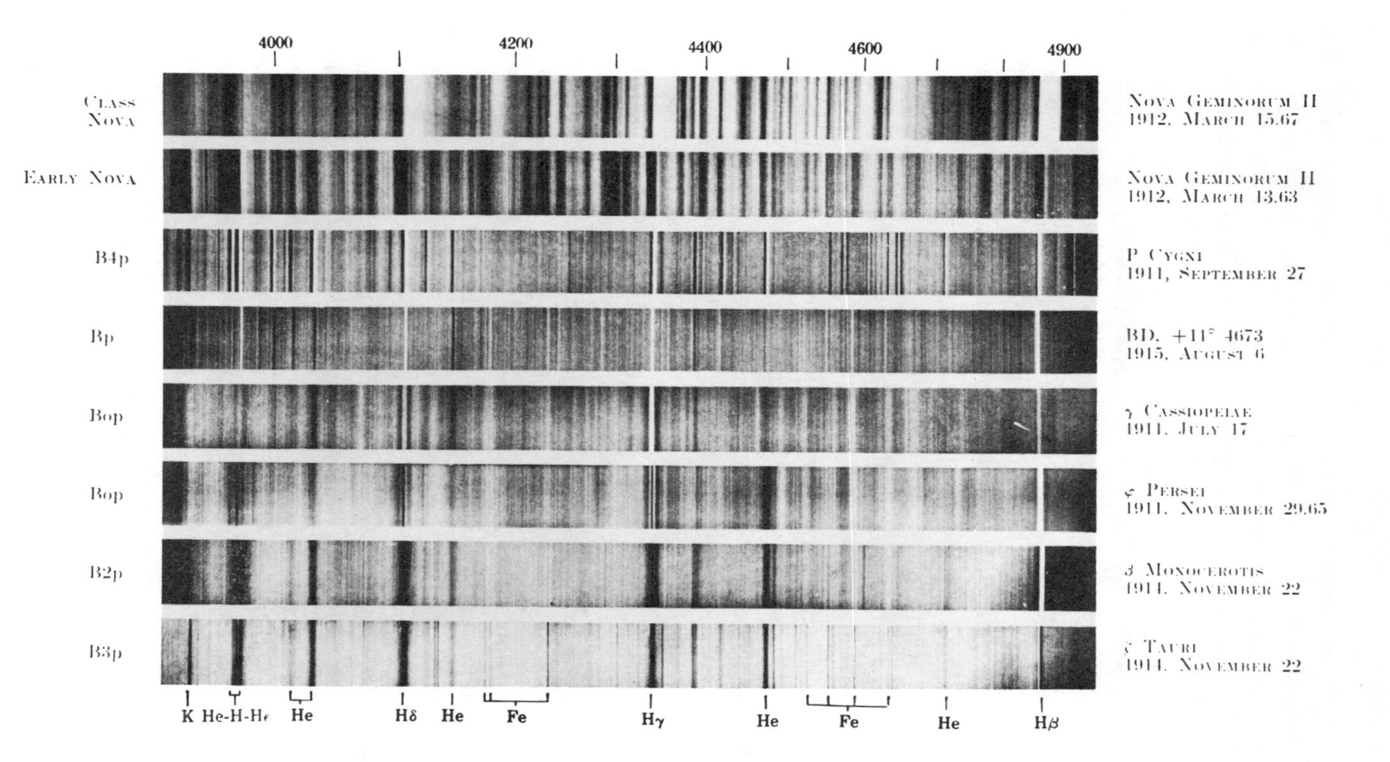

Fig. 9.4 Spectra of Nova Geminorum, P Cygni and various Be and shell stars observed by R.H. Curtiss at Michigan (published 1923).

In 1935 at the Paris IAU meeting, H.N. Russell (57) proposed that a subcommittee on the classification of the spectra of novae be established. This proposal was passed and F.J.M. Stratton became the chairman. The committee was established partly because of the wealth of data from Nova Herculis in 1934. In the first report of this subcommittee Stratton noted that 3200 spectrograms of Nova Herculis had been secured at different observatories within the first 110 days after discovery, a situation he described as 'excessive multiplication'. The wealth of spectrographic material enabled Stratton and W.H. Manning to produce *An Atlas of Nova Herculis 1934* (39) a work which Mrs Payne-Gaposchkin described as 'beautifully illustrated, [it] reproduces almost daily spectra, and furnishes a model for the future discussion of novae...Nova Herculis is perhaps the best observed of all novae' (58).

Dean McLaughlin was one of those who studied the spectral changes of Nova Herculis in great detail, and it was from his analysis that he identified a progression of seven evolutionary stages in this star (38). These were given Greek letters from α (premaximum) to η (nebular stage). He developed this scheme the following year and suggested a somewhat more suitable means of classifying the spectral changes of a nova's development, using the notation in ten steps from Q0 to Q9 (59). Soon after this time (c. 1940) McLaughlin began a thorough study of the spectral changes in the seven bright novae observed spectrographically in the period 1901–36 using plates exposed not only at Michigan but others loaned from Mt Wilson, Lick, Yerkes, Harvard, Lowell and Allegheny. As a result of his detailed intercomparison, he found 'a greater degree of similarity in the spectra of individual novae than has been admitted previously' (60). McLaughlin identified four distinct absorption systems characterised by different times of the first appearance of each system, and different blue-shifted Doppler velocities. They were named the pre-maximum, principal, diffuse enhanced and Orion systems. A subsystem of the Orion absorption, which appeared later, he called the nitrogen absorption (as doubly ionised nitrogen lines at 4097 and 4103 Å were prominent). There was a general trend for higher ejection velocities of the absorbing gases for the later occurring systems. Each of these systems has an associated emission spectrum generally resulting in P Cygni-type profiles, although the premaximum emission was often weak. The nitrogen absorption was associated with the so-called 4640 emission, a very strong broad emission line also due to ionised nitrogen and sometimes 50 to 100 Å wide. Following the '4640 emission' stage came the well-known 'nebular emission' with a spectrum resembling a planetary nebula and no absorption lines, followed by a 'post-nova emission' spectrum where the lines are all relatively narrow and stellar, superficially resembling a Wolf–Rayet star (60).

McLaughlin's terminology for the absorption and emission spectra of novae has been widely used. His descriptions of the various stages were both more detailed than those of earlier observers, and they were more generally applicable to the majority of novae.

Table 9.2 *Classification of spectral stages in novae*

Based on the proposal by McLaughlin to the I.A.U. Symposium on Novae and adapted from Trans. I.A.U. **7**, 305 (1950)

(*a*) *Spectral systems*

Name	Absorption	Emission	Description
Pre-maximum	I	I	Resemble normal stars
Principal	II	II	
Diffuse enhanced	III	III	Broad lines of HI and ionised metals
Orion	IV	IV	Lines of NII, OII, CII, HeI etc.
Nitrogen III (λ4640)	V	V	Strong NIII band at 4640
η Carinae (forbidden Fe II)	—	VI	
Nebular	—	VII	
Wolf–Rayet	—	VIII	

(*b*) *Evolutionary sequence of spectral types*

Type	Stage	Spectral systems present
Q0	Pre-nova	Practically unobserved
Q1	Pre-maximum	Absorption and emission I
Q2	Maximum	Abs. and em. II
Q3	Diffuse enhanced	Abs. and em. II; abs. and em. III (often multiple lines)
Q4	Orion	Abs. and em. II; abs. and em. IV (abs., em. III weakly present)
Q5w	(Fast novae only)	Abs. and em. II, IV, V (abs. and em. III faint or absent)
Q5f	Slow novae only / η Carinae	Em. VI; abs., em. II, IV (abs. and em. III sometimes present)
Q6w	(Fast novae) Transition stage	Em. V, VII
Q6f	(Slow novae) Transition stage	Em. VI, VII
Q7	Nebular	Em. VII only
Q8	Wolf–Rayet	Em. VIII on weak continuum; em. VII present but belongs to nebula
Q9	Final stage	Em. VII from nebula; practically continuous spectrum from star, which may be the same as type Q0

At the seventh IAU General Assembly in Zürich in 1948 the subcommittee on novae of Commission 29 reconvened under Stratton's chairmanship. Although McLaughlin was not present, his ideas on the classification of nova spectra were reported to the subcommittee which closely followed McLaughlin's proposals (61). It used essentially the same terminology and notation as proposed in McLaughlin's earlier publications (59,60). Five distinct absorption line spectra were identified, and eight emission spectra, the first five appearing with the corresponding absorption spectrum. These different spectra sometimes appeared simultaneously during the spectral evolution of a nova's development. Ten distinct stages were identified in an evolutionary sequence from Q0 to Q9, as shown in Table 9.2.

9.4 Emission line B stars: the Be stars

Although the notation Be was only applied to stellar spectra from 1922, at the time of the first General Assembly of the IAU in Rome, the history of this common class of star extends right back to the mid-1860s. The description in W.S. Adam's first report of Commission 29 is more than a definition, and nicely summarises the essential characteristics of Be stars (62).

> The hydrogen series may be thought of as comprised of the normal Class B absorption lines, increasing in strength from Hα towards the violet, each having superposed upon it, in a nearly symmetrical position, one of a series of bright lines which decrease in strength from Hα towards the violet. Frequently in the hydrogen series...one or more of the lines Hβ to Hε will show *both* emission and absorption components, the lines towards the violet showing no emission, and those towards the red no absorption; but in some cases Hα is the only distinct emission line. The emission lines often are double, and in some cases may appear as bright edges to a well defined absorption line. Fainter emission lines (enhanced metallic) may or may not be present.

The brightest of the Be stars is γ Cassiopeiae, and the emission lines in the spectrum were discovered by Angelo Secchi in August 1866 (63). He described the spectrum as having broad emission lines, whereas ordinary type I stars have absorption lines in the same places, due to hydrogen. The emission lines of γ Cas were the second to be discovered in any stellar spectrum, after those of the nova T CrB in May of the same year (18). Secchi also announced the discovery of narrow emission lines in another B star, β Lyrae, but since this object is a spectroscopic binary, it does not belong to the classical Be stars described at the first IAU meeting. The D_3 line of neutral helium was suspected in γ Cas by Vogel in 1872, during his time at Bothkamp. He wrote that the star 'showed a bright line on the boundary of green and blue, another was suspected in the yellow (D_3?), and in the red I believed I observed a dark band' (64). The D_3 line was also probably seen by Secchi, as early as December 1868 (65).

Norman Lockyer observed γ Cas from Kensington from 1888 onwards using objective prism plates. His results were reported to the Royal Society in 1894 (66) and for the clarity of their conclusions they represent Lockyer at his best. He saw other bright lines than those of hydrogen, found no evidence for emission intensity variations and showed that the hydrogen emission lines were double corresponding to a constant velocity difference of 115 mile/s. He found too that the emission lines were superimposed on broader absorption bands while other ill-defined dark lines occurred in the same positions as in normal B stars such as ζ and γ Orionis.

At this time W.W. Campbell also observed γ Cas from Lick: '... my photographs show many striking facts concerning γ Cassiopeiae. The bright hydrogen lines decrease in intensity very rapidly as we go to the violet, and are situated within broad dark hydrogen lines' (67). This proved to be a general property of the 'bright Hβ stars' as Campbell called them. At that time thirty-two were known which Campbell lists, mostly discoveries made from the Harvard objective prism work.

The success of E.C. Pickering's objective prism program at Harvard in finding new early type stars with bright hydrogen lines is shown by Mrs Fleming's posthumously published work *Stars Having Peculiar Spectra* (68). Here she lists ninety-two mainly B or A emission-line stars of which 84 were Harvard discoveries from the late 1880s onwards. Many of the best known bright Be stars, such as ϕ and ψ Persei, α Columbae, Alcyone and Merope (in the Pleiades), μ and δ Centauri and α Arae are to be found listed there.

Miss Maury at Harvard had earlier described in detail the spectra of several of the brighter emission-line B stars, including γ Cas, ϕ Per and π Aqr. For the first of these she wrote: 'This star is one of the most interesting among those of the Orion-type having bright lines. Its most important peculiarity is shown in the double reversal of its hydrogen lines' (69) – her way of describing the narrow central absorption separating the two emission peaks, thus confirming the finding that Lockyer had published earlier of doubled lines.

At Potsdam Julius Scheiner speculated on the origin of the emission lines as early as 1890 in his book *Spectralanalyse der Gestirne*. 'Much more plausible [than the in-falling meteorites hypothesis] seems to be the hypothesis that these stars seem to be surrounded by a very extensive atmosphere, and that the emission spectrum arises from the parts of the atmosphere which stick out over the stellar disk, as seen by us in projection' (70).

In the first half of the present century, one name stood out more than any other in the study of the Be stars – that of Paul Willard Merrill. His interest in the emission line B stars began at Lick about 1912, was continued for a while (1913–16) at Michigan (where Ralph Curtiss was also active in Be star research) and then especially at Mt Wilson where he took up an appointment early in 1919 under Hale. In his early days at Lick, Merrill observed 'nearly all Class B stars north of declination $-40°$ that were known to have bright lines' (71). He recorded the hydrogen line profiles on a microdensitometer and found the double lines to be a general characteristic. For γ Cas the first six Balmer lines were all in emission, as well as D_3 (neutral helium) though this last feature was variable. He placed the thirty-eight classical Be stars observed into four groups which were defined according to the number of Balmer lines in emission, the emission width and its duplicity. Although this classification was not adopted in later work, it served to illustrate the differences among Be stars known at that time. It was this study that Merrill submitted for his PhD thesis from the University of California in 1913.

At Mt Wilson, Merrill again turned to the Be stars, applying his expertise in red-sensitised photographic emulsions to photograph the Hα line on objective-prism plates using the ten-inch Cooke telescope. For this work he collaborated with Milton Humason and Miss Cora Burwell. The dispersion of 440 Å/mm was modest, but as Merrill pointed out, the emission of the red Hα line was always stronger than at Hβ (blue-green) or for shorter wavelength lines, where indeed in many Be stars there was no emission at all (Fig. 9.5). Thus the Harvard blue objective prism plates had failed to detect the emission characteristic of many B stars, and their first paper listed ninety new Be stars with Hα emission in areas of the sky where previously only sixty Be stars were known (72). Variability of the bright lines seemed to be common in these stars. Their spectral types were nearly all B0 to B5. Slit spectra were obtained in the blue for all of them, and almost half had Hβ emission absent or so weak as to be undetectable on a blue objective-prism plate. This work was continued with a second paper in 1932 listing a further 132 newly discovered stars with bright Hα lines. The red colour of many Be stars was noted, a possible result of the absorption of blue light from distant stars by interstellar dust particles (73).

The accumulation of data from the Hα program at Mt Wilson over the next two decades led directly to the Merrill–Burwell catalogue of Be stars.

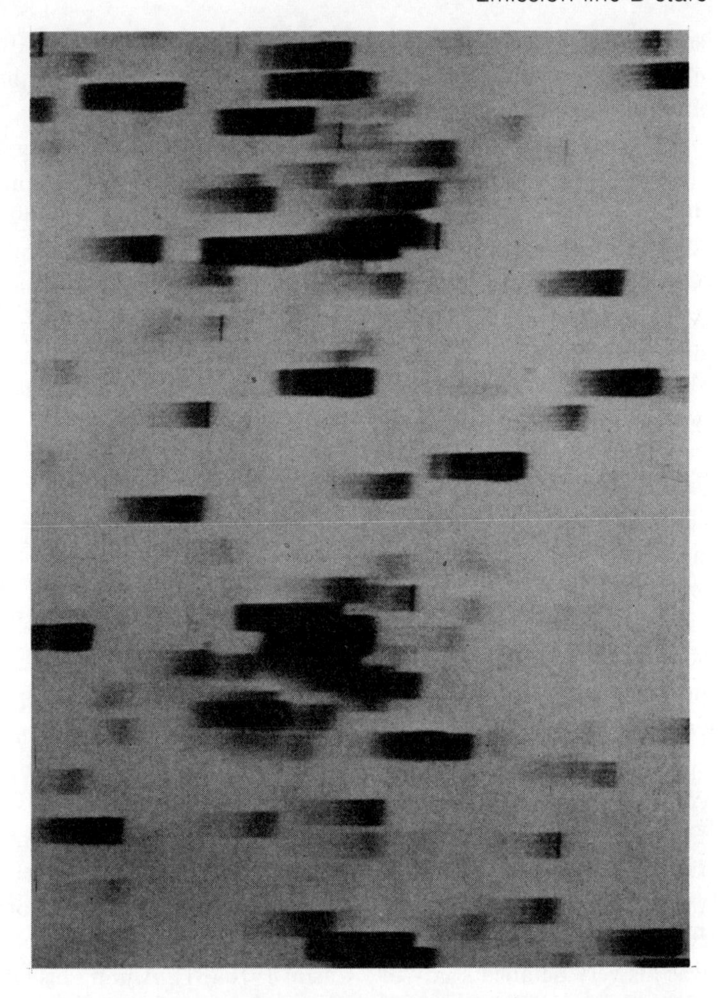

Fig. 9.5 The double cluster in Perseus; part of a typical Mt Wilson objective prism photograph with the 10-inch telescope. The Hα line in emission is seen in a number of Be stars, which were catalogued by Merrill and Miss Burwell in 1949.

This was published in parts in the *Astrophysical Journal*: the first three sections from 1933 to 1949 not only listed over 1000 Be stars but gave comprehensive references to the literature, star by star (74). In section four of the catalogue a further 519 emission-line stars, mostly of uncertain spectral type, were listed (75). This survey work was continued in the 1950s by Karl G. Henize (b. 1926) in the southern hemisphere observing with the 10-inch telescope installed at Bloemfontein, South Africa (76).

On the theoretical side Otto Struve in 1931 sought to explain the origin of the emission lines in Be stars by suggesting they are 'rapidly rotating single stars [which] form lens-shaped bodies which eject matter at the equator, thus forming a nebulous ring which revolves around the star and gives rise to emission lines' (77).

This model is believed to be essentially correct; W.W. Morgan showed that the Be star equatorial rotational velocities must be around 300 to 400 km/s (78) and this was confirmed in a thorough study by Arne Slettebak (79) in which he found Be star rotational velocities sometimes of over 400 km/s and averaging 150 km/s more than for B stars as a whole. Most of the stars rotated at somewhat less than their break-up speed however, contrary to Struve's hypothesis, and it is now clear that a simple equatorial ring for the extended atmosphere is an over-simplification. V. Kourganoff later wrote:

> Everyone knows...the magnificent contribution of Struve who knew how to qualitatively explain the Be profiles by ingeniously combining Rosseland's hypothesis concerning cyclic emission for atmospheres with diluted radiation, with those of Laplace and of Jeans, concerning the ejection of rings by fluid masses in rapid rotation (80).

Following the discovery of so many new Be stars from the objective prism surveys at Harvard and Mt Wilson, many detailed studies of the spectra of individual stars were undertaken. The most outstanding contributions came from Ralph Curtiss (1880–1929) and his successor Dean McLaughlin at Michigan. Curtiss was assistant director at the Detroit Observatory of the University of Michigan when he commenced the Be star program there as soon as the 95 cm ($37\frac{1}{2}$ inch) reflector had been commissioned in May 1911. One of his earliest contributions to the field was his very detailed study of γ Cas in 1916 (81). Among his conclusions was: 'The approximately linear relation between emission line widths and wave-length, found in γ Cassiopeiae, seems consistent with hypotheses of widening due to rotation and convection currents'.

Much of the work at Michigan centered on the discovery and discussion of spectral changes in Be stars, first found by Curtiss in f^1 Cygni (82) and in HR 985 (83). The former showed changes predominantly in the overall emission line intensities, whereas in the latter star the most marked variations were in the relative strength of the violet or red peaks (V/R intensity ratio) of the double hydrogen lines. When Curtiss died in 1929 McLaughlin continued the Be star programs, especially the work on spectral variations. He published a paper under Curtiss' name, detailing spectral V/R intensity variations in HD 20336, 25 Orionis and κ Draconis with periods of several years (84). McLaughlin's own contribution of that year discussed the spectra of forty-five Be stars. Two types of variability were treated, the so-called *E/C* variations (ratio of emission intensity relative to the continuum) and the *V/R* variations. The first type of variability was documented in fifteen of the stars, the second type in twenty-four. Thirty stars showed variations of one type, the other or both. 'It can therefore be concluded that variability of emission is the rule, and constancy the exception, among the Be stars' (85). Although γ Cas was not one of the variable stars listed by McLaughlin, *V/R* variations of a period of about 4 years were found in 1933 in this star too by William J.S. Lockyer (1868–1936), son of Sir Norman, and director of the Norman Lockyer Observatory in Devon (86).

A subclass of the Be star was defined by Paul Merrill in 1949 and known as a 'shell star' (87); the term however had been used widely before then and was introduced by Struve. The salient features of the shell star spectrum were hydrogen lines with narrow dark cores, absorption lines of ionised metals narrow and sharp and neutral helium lines absent or weak and diffuse. As the name implies, the dark narrow lines of these stars arise from an extended layer or 'shell' above the photosphere. Several of the brighter Be stars were listed by Merrill and Burwell in the shell-star category, including γ Cas, ϕ Per, ζ Tau, ψ Per, Pleione, β Mon A and 48 Lib (88). The spectra of many shell stars are variable and Pleione (28 Tauri) represents an excellent example. The hydrogen emission disappeared from this star in 1906 (89); its reappearance was discovered in October 1938 by McLaughlin (90) accompanied by many weak lines of ionised metals throughout the spectrum. In the early 1940s the shell was at its greatest optical thickness and the rich absorption-line spectrum, resembling the supergiant α Cygni, was described by Otto Struve and Pol Swings who found low amplitude radial velocity variations of a period of about 4 months (91). This shell spectrum persisted until 1951 when it was weakening and finally disappeared early in 1952 when the shell was blown off the star. Its disappearance was recorded that year by Merrill (92). In conclusion to this section on Be stars it is noted that three of the greatest names in the field, those of McLaughlin, Struve and Merrill, all contributed to the study of Pleione's spectral changes between 1938 and 1952.

9.5 The peculiar A-type stars: an astrophysical enigma

The great American stellar spectroscopist William W. Morgan (b. 1906) commenced his 1935 treatise on the spectra of the A-type stars with these words:

> There is probably a greater diversity in appearance among the spectra of class A than in any other type. It is well known that lines of singly ionized calcium, europium, manganese, chromium, silicon, and strontium show considerable differences in intensity between certain stars. Many of the most striking dissimilarities cannot be explained by differences in temperature or surface gravity in the stellar atmospheres and form at the present time one of the most puzzling problems in astrophysics (93).

This was written in 1933, soon after Morgan had first recognised the peculiar A stars as a distinct group whose peculiarities correlate with temperature (94). The early 1930s thus represent the turning point in the study of these stars; before then the study of their spectra had been spasmodic and the results lacked organisation. However, there are some interesting papers on peculiar A stars in the pre-1930 period which are worth recalling.

9.5.1 Discovery of peculiar A stars and early progress up to 1930 The work of Annie Cannon at Harvard in her 1901 paper classifying bright southern stars has already been mentioned (1) – (see Chapter 5). Several stars with either abnormally strong lines of ionised silicon or of ionised strontium were found, including ν For, τ^9 Eri and α Dor (all silicon stars) and ξ Phe, θ^1Mic and ι Phe (of the strontium type). Antonia Maury had also classified and remarked on over a dozen peculiar northern stars, including α^2 Canum Venaticorum, for which she wrote: 'It has, however, marked peculiarities. Thus the line K is extremely faint, and the lines 4131.4 and 4128.5 [of SiII] have greater intensity than in any other stars except those of Division c in Group VIII' (95). This star was later to have one of the best known and studied of all the peculiar A-type spectra.

The first study of peculiarities of A stars not originating from Harvard was from Sir Norman Lockyer and Frank E. Baxandall (1868–1929) at the South Kensington Solar Physics Observatory (96). They discussed the spectra of four stars: α Andromedae, θ Aurigae, α Canum Venaticorum and ε Ursae Majoris. For the first of these a number of prominent but unidentified 'strange lines' were noted in addition to strong lines of silicon; strengthened silicon lines were also noted in the next two stars, while the last mentioned had chromium enhanced. One of Baxandall's significant achievements was the identification of the strange lines of α And as being due to ionised manganese (97) – αAnd is now the best-known of the peculiar manganese stars. He also found several ionised europium lines in α

CVn, in particular very strong ones at 4130 and 4205 Å (98) – this was the first rare-earth element to be positively identified in any stellar spectrum.*

At the same time as the work in London was proceeding, Hans Ludendorff (1873–1941) in Potsdam found intensity variations for several lines in α CVn (101). Belopolsky at Pulkovo studied these variations in some detail and found a 5.50 day periodicity; in addition, some of the lines showed radial velocity variations of the same period (102). This was followed up by the pioneering photoelectric photometry of Paul Guthnick (1879–1947) and Richard Prager (1883–1945) at the Berlin–Babelsberg Observatory, using an argon-filled sodium cathode photocell (Fig. 9.6). α CVn was one of several stars placed on their list, and they found small amplitude light variations with the same $5\frac{1}{2}$-day period. A beautiful slightly asymmetric light curve of amplitude 0.051 magnitudes (Fig. 9.7) was the result (103). Ludendorff also noted 'that the majority of the lines are exceptionally weak and narrow' (101). This too is a general property of these stars, later ascribed to slow rotation. Thus as early as 1914, three of the most important properties of peculiar A stars, the spectrum and light variability, the line sharpness and the enhancement of certain elements, had all been established for this star.

Meanwhile the list of peculiar A stars being discovered at Harvard continued to grow. Annie Cannon put them into two separate groups, those with abnormally strong silicon lines at 4128 and 4131 Å, and those in which the strontium lines were strong, at 4078 and 4216 Å. In a sense this was the first peculiar A star classification. In her two papers on the classification of fainter stars in 1912, the discovery of thirty-eight new silicon stars and of twenty-six new strontium stars was announced (104). When the Henry Draper Catalogue was published this number grew yet further; 165 stars with spectral types from B8p to F0p were remarked upon as being peculiar. The great majority were of type A0p or A2p (105).

In spite of this wealth of new data, surprisingly very little work was done to explore further the strange peculiarities of the Ap stars in the period prior to about 1926. One exception was the work of Carl C. Kiess (1887–1967) who studied α CVn during his brief time at the University of Michigan, 1916–17. Kiess was able to add yttrium and seven rare earth elements, principally lanthanum, gadolinium, terbium and dysprosium, to the heavy elements identified in this star (106). His catalogue of spectral lines in α CVn was the first detailed study of the line wavelengths and identifications in any Ap star. He also confirmed Belopolsky's discovery of periodic radial velocity and line intensity variations for some lines only,

* J. Lunt's earlier identification of europium in Arcturus in 1907 (99) has been doubted by W.W. Morgan (100)

Fig. 9.6 Early photoelectric photometer used by Paul Guthnick and Richard Prager on the 30 cm refractor at the Berlin-Babelsberg Observatory in 1914. They detected periodic light variations in α CVn with this apparatus.

especially those of europium and terbium.

The origin of the abnormal strengths of the lines of certain elements in Ap stars remained a complete mystery on which few astronomers ventured to speculate. Shapley in a brief comment did raise the possibility of

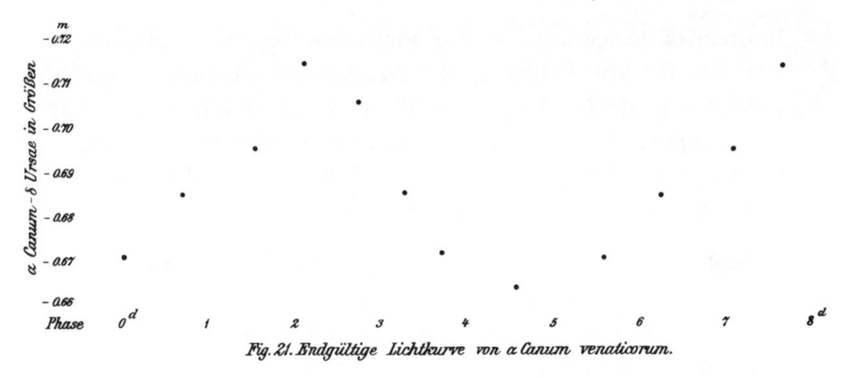

Fig. 9.7 The 'final' light-curve of α Canum Venaticorum by Guthnick and Prager, Berlin, 1914.

abundance abnormalities (107) but such a suggestion was rejected by Miss Payne (his student!) in her doctoral thesis: '...it is not very probable that, if silicon is unevenly distributed in the universe, the irregularity would be revealed in stars at one temperature only' (108).

9.5.2 W.W. Morgan and Ap stars, 1931–5 The years 1931 to 1935 mark a period of greatly increased interest in the peculiar A stars. In 1931 W.W. Morgan had just finished his doctoral thesis at Yerkes Observatory under Struve. The results of his studies into the Ap stars followed in a series of eight papers or notes to the *Astrophysical Journal* during 1931–3 as well as what B. Strömgren has called 'his monumental detailed investigation of A stars' in the Yerkes Observatory publications of 1935 (93).

His achievements were as follows: The peculiar manganese star α Andromedae was found to be one of a class of manganese stars, to which Morgan added another thirteen (109). He showed that the peculiar star BD-18°3789 had strong lines of ionised chromium which varied periodically in antiphase to the europium lines (110). Thirty-four peculiar stars of the strong europium class were catalogued. Whereas the manganese stars had B8, B9 or A0 spectral types, the europium stars were significantly cooler, with A0 to A3 classifications (100). He defined a new class of peculiar star, the chromium stars, with ι Cas and 17 Com being representative members (111). A special study was then made of 73 Draconis, also a chromium star, in which the variable intensity of the europium lines (period 20.7 days) was particularly striking (112). Finally, his most important result of all, was the announcement of a classification for peculiar A stars in which the abundance anomalies correlated closely with ionisation temperature (94). He defined five groups:

> In order of decreasing degree of ionization the groups and type stars are: the Mn II stars (α Andromedae and μ Leporis), stars in which the unidentified line at λ4200 is well marked (θ Aurigae), the Eu II stars (α^2 Canum Venaticorum at maximum of Eu II), the Cr II stars (73 Draconis), and the Sr II stars (γ Equulei). All of these groups overlap the contiguous ones.

The silicon stars (lines at 4128 and 4131 Å strong) were not explicitly included, as they overlapped with the first three groups. In fact the unidentified 4200 Å feature characteristic of Morgan's second group was identified in 1962 by W.P. Bidelman as being a blend of two high excitation ionised silicon lines (113).

Morgan's 1935 treatise (93) studied the line strengths in 125 A stars, thirteen of them in considerable detail. The individual differences from star to star of the same HD spectral type emphasised the inadequacy of current classification schemes for A stars. He concluded: 'From the foregoing discussion it seems safe to conclude that there is some physical factor other than temperature and surface gravity concerned in the production of spectra of the A stars and that the additional factor is probably variable effective abundance in a number of elements observed, if not in all of them.'

9.5.3 Horace Babcock and magnetic fields in the Ap stars It has been known since 1896 that a magnetic field can split spectral lines into a number of components whose separation increases with increasing field strengths. The effect was discovered by the Dutch physicist Pieter Zeeman (1865–1943) after whom it is named (114). Although Hale had shown that magnetic fields were present in sunspots in 1908, and also as a general solar field in 1913, early attempts at measuring stellar fields using the Zeeman effect by W.H. Wright in 1910 (115) and P.W. Merrill in 1913 (116) were unsuccessful. The problem was reviewed again by Minnaert in 1937 (117) who suggested that some rapidly rotating stars seen pole-on may have dipole magnetic fields (assumed to be parallel to the rotation axes) which would give measurable Zeeman splittings. The effect was therefore to be looked for in rapidly rotating pole-on stars which should have sharp lines. At Mt Wilson Horace Babcock assumed that the sharp-lined peculiar A stars were just such objects and from April 1946 he commenced a program to study Zeeman effects in these stars using the coudé spectrograph on the 100-inch telescope.

It can only be regarded as good fortune that he chose to observe the Ap stars in this program, albeit for the wrong reasons, for results showing a strong magnetic field in 78 Virginis (a fifth magnitude chromium-europium

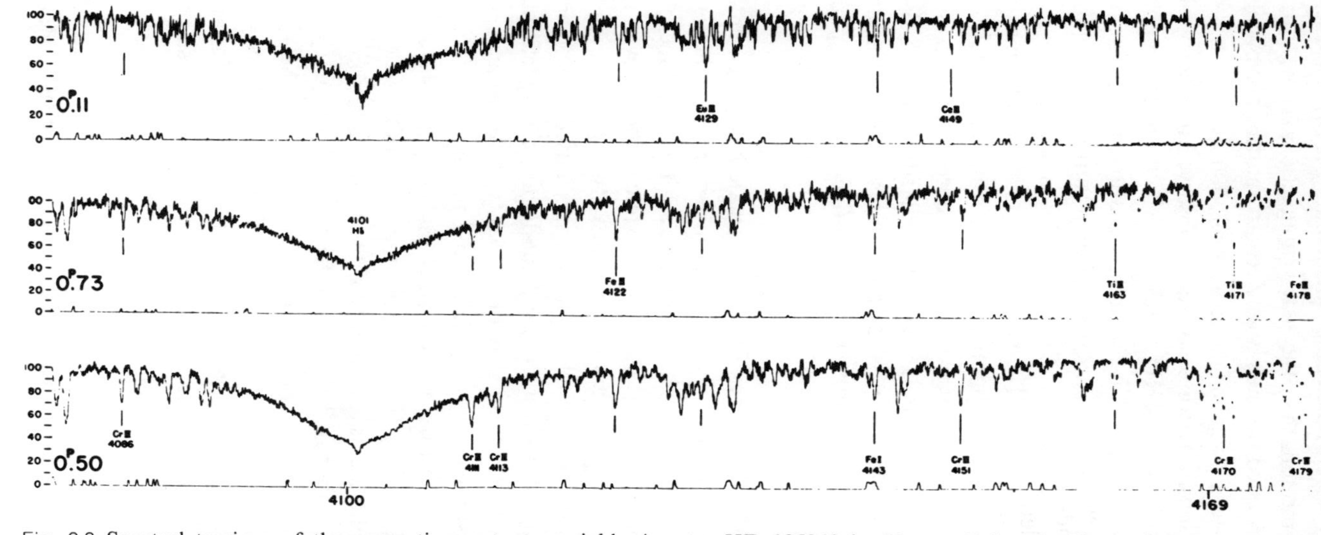

Fig. 9.8 Spectral tracings of the magnetic spectrum variable Ap star HD 125248 by Horace Babcock. Three phases are shown, corresponding to Eu II maximum (top), cross-over (middle) and Cr II maximum (bottom).

star) were immediately forthcoming (118). On the (probably mistaken) assumption that the star's magnetic axis was towards the earth, he interpreted the measured Zeeman splittings (typically of about 0.01 Å) as due to a polar magnetic field of 1500 gauss.

Soon afterwards Babcock observed another Cr-Eu star, BD-18°3789 (also known as HD 125248) in which the chromium and europium lines were known to have periodic variations (110). The results were even more interesting: the dipolar field was yet stronger and varied between + 7000 G and −6200 G at the pole with the same regular 9.3 day period as the line strengths (119). The field was at its maximum positive intensity when the rare earth absorption lines were strongest, while the reversed polarity field was at its strongest when the chromium lines were most prominent (Fig. 9.8). A similar reversing polarity field was also found for α^2 CVn (120); it was also periodic and in phase with the line-strength variations, but for this star the maximum positive field corresponded to the phase of strongest chromium lines and maximum negative field to that of strongest rare earth lines.

The discovery of stellar magnetic fields was, in the opinion of Struve (121), 'the most significant advance in this field' in the previous decade. The interpretation of the results gave a whole new dimension to the study of the anomalies in Ap stars. Meanwhile Babcock was the only observer to undertake this work for nearly two decades, presumably because it required a large telescope and specialised equipment.* In 1958 Babcock compiled his results into a catalogue of magnetic stars (124). The great majority of these were of the Ap type, for which seventy were listed with measurable magnetic fields, and magnetic variability was found to be a general feature of these stars. Babcock here listed the effective magnetic field intensity H_e, being the component of the field in the line of sight on the simplifying assumption that the field is uniform and unidirectional. This field is about 0.3 of the polar field previously quoted, and is derived without any a priori assumption on the orientation of the magnetic dipole. The star with the strongest field was 53 Cam with H_e varying between 3700 and −4350 gauss. Since then Babcock's discovery of a varying magnetic field of up to 34 000 gauss in the eighth magnitude silicon star, HD 215441, represents the highest such field found (125).

9.5.4 The oblique rotator model for magnetic stars The oblique rotator model has been easily the most successful model for explaining

* H. Gollnow later reported stellar magnetic observations in 1962 (122) at Mt Wilson and Palomar; G.W. Preston and D.M. Pyper in 1965 at Lick (123).

the spectroscopic observations of magnetic Ap stars. Babcock never supported this model with any enthusiasm, though in fact he was in 1949 the first to suggest it:

> It is true that I have suggested as a revised working hypothesis that intense magnetic activity may be correlated with rapid stellar rotation, but at this stage an equally good case can probably be made for the alternative hypothesis that the spectrum variables of type A are stars in which the magnetic axis is more or less highly inclined to the axis of rotation and that the period of magnetic and spectral variation is merely the period of rotation of the star (126.)

This model thus conceded that magnetic Ap stars were generally slowly rotating stars seen at any orientation, with the inclined magnetic dipole being somehow 'frozen in' to the rotating star.

This model was independently developed by D.W.N. Stibbs (b. 1919) while at Mt Stromlo Observatory in Australia. Evidently Stibbs was unaware that Babcock had already proposed it. Stibbs observed the light curve for BD-18°3789 and showed how an oblique rotating magnetic dipole could account for the observed variations of the light, spectral lines and magnetic field (127). Babcock however favoured an alternative explanation called 'the magnetic oscillator theory' in which the magnetic stars were seen pole-on and undergoing magnetic oscillations (128). In this theory too the rare earth elements and chromium were supposed to be concentrated in separate atmospheric zones, and a non-uniform surface composition to explain the spectral variations henceforth seemed to be a conclusion hard to refute on any model.

Armin J. Deutsch (1918–69) at Mt Wilson, more than anyone, became the champion of the oblique rotator model. As G.W. Preston has put it: 'Babcock first considered it as a possibility, Stibbs worked out its consequences for a dipole field and Deutsch became its most ardent advocate' (129). Deutsch first studied A-type spectrum variables for his doctoral thesis at Yerkes Observatory in 1946 (130). He continued this work at Mt Wilson from 1951. He showed that the rotationally broadened profiles of the ionised magnesium 4481 Å line agreed closely with the profiles expected from the periods derived from the line intensity variations, the stars with the narrowest lines having the longest periods (131). Such evidence strongly appeared to favour the oblique rotator model. For α^2 CVn Deutsch had the benefit of some impressively thorough spectral observations of varying line strengths to give weight to his assertions: notably those of Struve and Swings in 1943, in which 3107 absorption lines were catalogued (132), and later of G.R. Burbidge (b.1925) and E.M.

Fig. 9.9 Armin Deutsch.

Burbidge (b.1922) (133). He reviewed the evidence for the oblique rotator model in 1956 (134). More recent studies, especially that of Diane Pyper (b.1939) on α^2 CVn (135), have only strengthened the case for this model.

In conclusion, Preston's comments are worth quoting:

> In retrospect most of the observational ingredients of the oblique rotator model were available long ago – periods, line widths, the spectrum variations.... But prior to the discovery of magnetic fields in the Ap stars..., no one was prepared to seriously entertain the notion that a star could possess permanent non-uniformities on its surface (129).

9.5.5 Spectral classification of Ap stars Morgan's 1933 classification (94) of the Ap stars into the groups Mn II, λ4200, Eu II, Cr II and Sr II, with Si II stars forming an additional overlapping class, has remained the basis for future work. However, some refinement of this scheme has proved useful. Thus in 1958 Mercedes and Carlos Jaschek (both b. 1926) at La Plata Observatory, Argentina defined the following five groups: λ4200, manganese, silicon, europium–chromium and strontium (136). They chose not to divide the chromium and europium stars. In addition, they recognised that the λ4200 and silicon groups frequently overlapped giving rise to silicon – λ4200 stars. Similarly two further intermediate groups Si, Eu-Cr and Eu-Cr, Sr could also be identified. They then studied the correlation between the Ap groups and mean photoelectrically-measured stellar colours and concluded that the λ4200 and silicon – λ4200 stars were the bluest (and hence hottest) with a well-defined progression to the strontium stars which were reddest and coolest. This classification has remained easily the most frequently used since. The distinction between the silicon (with the strong 4128 and 4131 Å Si II doublet) and λ4200 groups is of course rather artificial in view of Bidelman's identification of λ4200 being also an ionised silicon feature, though of higher excitation (113). As with Morgan's scheme, it was to be understood that the group names simply referred to the most prominent spectral peculiarities on low dispersion spectra.

K. Osawa at Tokyo attempted to classify all the Ap stars north of $-30°$ declination which appeared in a catalogue compiled by Charles Bertaud in Paris (137). His classifications were made from 60 Å/mm spectra and covered 244 stars (138). Osawa identified eleven attributes of peculiarity and assigned if necessary several such attributes to each star. Thus α^2 CVn received the classification 'SiHgCrEu', the 'Hg' referring to the great strength of the mercury line at 3984 Å.

9.5.6 The manganese stars and other early-type peculiar stars

The manganese stars were first recognised as a class by Morgan in 1931 when he identified thirteen further stars with spectra similar to α Andromedae (109). Lockyer and Baxandall had earlier called attention to unidentified lines at 3944, 3984, 4137, 4206 and 4282 Å in this star (96). All these were soon identified as manganese lines by Baxandall (97) with the exception of that at 3984 Å. This line remained a mystery for over half a century until William P. Bidelman (b. 1918) was able to identify it with ionised mercury (139). This discovery must rank as one of the more interesting of the many surprises peculiar stars have turned up, especially because mercury is virtually unknown in stellar spectra of any other type. The exact wavelength of the line was found to vary by about 0.2 Å from star to star, an effect Bidelman interpreted as due to different proportions of the six different mercury isotopes. The star χ Lupi for example is abnormal in having almost pure Hg^{204}. Nearly all manganese stars also show the 3984 Å mercury feature (an exception is 53 Tau) whereas few other stars do, hence the common appellation 'HgMn stars'.

Some manganese stars also show other unexpected elements; notable are the discoveries of phosphorus in κ Cancri by Bidelman (140) and of gallium in 53 Tauri by Aller and Bidelman (141).

Since the late 1960s it has been clear that the HgMn stars are not simply late B-type analogues of the Ap stars, but represent a class of peculiar objects discrete from the silicon and other peculiar stars. The evidence for this was first summarised by G.W. Preston (129) who cited the lack of magnetic fields in these stars, the lack of known light or spectrum variations, the lack of high rare earth abundances and their relatively high incidence (compared to classical Ap stars) in spectroscopic binaries. They may also on average rotate even slower than the classical Ap stars.

The HgMn stars are thus non-magnetic stars with spectral types mainly B8 or B9. There is a possibility of a link between these and hotter B stars which also show peculiarities. The best known such group are the helium-weak stars which can be best identified by the discordance between their blue colours and their spectral types based on helium line strengths. This discordance between colour and type was first noticed for about a dozen stars in Upper Scorpius by R.F. Garrison (b.1936) at Yerkes Observatory (142) and by P.L. Bernacca at the Asiago Observatory in Italy (143). Their hydrogen-line spectral types lie in the range B3 to B7.

For spectral types of B2 or earlier (to about B0) some stars conversely show the peculiarity of helium-richness. The best known such object is σ Orionis E discovered by Jacques Berger in Paris in 1956. In a very brief note he reported: 'While investigating multiple stars of early spectral type, it has been ascertained that one of the companions of σ Ori (designated E) is a star very rich in helium' (144). Another member of the class to be identified at about this time is HD 135485, described by J.C. Stewart as 'a sharp-lined dwarf showing abnormally strong helium lines' (145). He estimated the helium content to be about ten times normal. In recent years several further members of the class have been recognised, especially by W.A. Hiltner *et al.* (146) and by D.J. MacConnell *et al.* (147). The latter authors give many references to the earlier literature.

9.6 The λ Boötis stars

The λ Boötis stars represent a rather esoteric and little-studied group of weak-lined A stars with only a handful of known members. The peculiarity of the type-star was commented on by Morgan, Keenan and Kellman in their MKK atlas (148).

> The spectral type of λ Boö is near A0, as far as can be determined. The spectral lines, while not unusually broad, are very weak, so that the only features easily visible are a weak K line and the Balmer series of hydrogen. The trigonometric

parallax indicates that the star is probably located below the main sequence.

Arne Slettebak added 29 Cygni and possibly ξ Aurigae to this class of weak-lined A stars in 1951 (149), and π^1 Ori 2 years later (150). He showed that fast rotation, low space motion and subluminosity (i.e. lying below the main sequence) were additional properties.

P.P. Parenago (1906–60) at the Sternberg Astronomical Institute in Leningrad emphasised the position of the stars about 1.5 magnitudes below the main sequence (151) more than the line-weakness. He appears to have been the first to use the label 'λ Boö star'.

The λ Boötis stars have been reviewed by W.L.W. Sargent (b. 1935), who defined them as low velocity early-type A stars *either* with weak lines *or* lying below the normal main sequence (152). However it seems that their under-luminosity is not well established. J.B. Oke (b.1928) at Mt Wilson, from a study of all the known λ Boö stars, found only 2 And to be below the main sequence of the Hyades (153).

The element abundances were found for λ Boötis and for 29 Cygni by the Burbidges; in λ Boö iron was ten times deficient and magnesium a hundred times, thus indicating that the spectral peculiarity arises from an overall heavy element deficiency (154). More recently Slettebak has suggested that HR 1754 (ADS 3910B), which has the same colour as a B6V star, may be a hot member of the λ Boö class (155). This is an especially interesting object as it is the fainter star of a visual binary system. The primary component (HR 1753) is a normal B5 dwarf. W. Sargent analysed the spectra of both stars and showed that the peculiar secondary was deficient in all heavier elements studied, by factors of between ten and a hundred (156). He conjectured that although the λ Boö stars are fairly fast-rotating stars, they may nevertheless be somehow related to the peculiar A stars, in that 'some unknown process' has taken place to change the composition 'at least on the stellar surface'.

Bodo Baschek (b.1935) and Leonard Searle (b.1930) in Australia analysed five λ Boö stars in 1968, mainly from Mt Wilson coudé spectra. They emphasised that this is probably not a homogeneous class of peculiar star. Three stars appeared to have moderate metal deficiencies by about a factor of three below the normal solar values (these were λ Boö, 29 Cyg, π^1 Ori). They found θ Hya and HR 1754 were peculiar stars, but of a different type, while γ Aqr seemed to be normal (157). If this is right, then a definite group of weak-lined A stars exists, but its members are extremely rare, and the measurement of the peculiarities is only marginally significant.

9.7 The metallic-line stars (Am)

9.7.1 Early history to 1960 of Am stars The discovery of several peculiar stars at Harvard by Miss Maury in 1897 (95) and by Miss Cannon in 1901 (1), such as τ UMa and δ Nor respectively, attracted little attention from other observers over the next four decades. The features of these stars as found at Harvard were many fainter metallic lines characteristic of mid-F spectral type but the hydrogen lines and the ionised calcium K lines were typical of A stars. Vogel and Wilsing in 1899 also classified some of these stars, mainly assigning type Ia3 (corresponding to about A3 to F3) but sometimes with the comments 'KsS' indicating an unusually narrow and sharp K line (as for ζ Lyr A), or '*lll*' (as for τ UMa) for very many lines seen in the spectrum apart from those of hydrogen (158).

Explicit reference to the unusual spectra of stars of this type was made again in 1940, without any detailed investigations having been made in the intervening four decades. John Titus (b. 1910) and W.W. Morgan classified A and F stars in the Hyades and found six* with the following features:

> The spectra of these objects are all similar in the characteristic that the metallic arc lines are considerably stronger than normal; they are of the same strength as in a normal F star. From the intensity of the metallic lines in the green and ultraviolet it seems likely that all lines originate in the atmosphere of a single star. They are of the same general type as the fourth-magnitude star 15 Ursae Majoris, whose Draper type is A3p. If the A stars had been classified by the intensity of the metallic arc lines, these objects would be classified as F stars with unusually weak K lines (159).

In addition to the six Hyades cluster stars, which Titus and Morgan considered to be subluminous A dwarfs (based on their K-line spectral types), they also mentioned 15 UMa, α GemB and ζ UMaB as having similar spectra.

These comments are generally regarded as the first to draw attention to this new class of peculiar star. In a sense they added little to what had already been done for other stars at Harvard. What was new in the Yerkes paper was the collecting together of a group of stars into one basket having common peculiarities, rather than the individual idiosyncrasies which Misses Maury and Cannon had earlier described.

The 'metallic-line stars' were designated as such in the MKK Atlas of Morgan, Keenan and Kellman (148). This atlas listed nine stars of the

* The stars were 16 Ori (A2p), 60 Tau (A3p), 63 Tau (Alp), 81 Tau (A5p), HD 28226 (A5p) and HD 30210 (A3p).

metallic-line class. The Titus–Morgan paper was surprisingly not referenced and only one Hyades star (63 Tau) was assigned to the new class. The differences between the spectral types based on the strength of the K-line and most other metallic lines were again emphasised; thus for τ UMa, the K line gave type A3, the metallic lines, F6II.* Nancy Roman with Morgan and Olin J. Eggen (b.1919) from Washburn Observatory, Wisconsin, took up this theme again in 1947 when they listed thirteen metallic-line stars (curiously only four in the Hyades and again the Titus–Morgan paper is not referenced) (160). They found K-line types from A1 to A6 on the MKK system, hydrogen-line types A5 to F2 and metallic-line types A5 to F6 for these stars. Photoelectric colour indices were in definite discord with the K-line types and corresponded to those of F stars; such a finding seemed to remove the need for the metallic-line stars to be subluminous. Two other important properties were noted by these authors: at least six of the thirteen stars were in known spectroscopic binary systems, an unusually high proportion; and the rotational velocities were lower than the average for A stars, as shown by the sharp lines. Morgan and Bidelman had also studied A-star colour indices and concluded '...it seems definite that the type derived from the metallic lines should be used for stars of this peculiar class...' (161), a view which was also supported by Harold F. Weaver at Lick (162).

This conclusion was in hindsight unfortunate for the work that now followed. In 1947 Jesse L. Greenstein (b.1909) at Yerkes Observatory undertook the analysis of five F-type stars of differing luminosities by the relatively new curve-of-growth technique. τ UMa was included. Part of this analysis was to find the ionisation equilibrium for neutral and singly ionised species of iron, chromium and titanium using the Saha equation. Solution of the equation gave an ionisation temperature and electron pressure on the assumption that these elements have the 'same fractional abundance per gram of stellar material in all stars and in the sun' (163). The assumption resulted in too cool a temperature for τ UMa which in turn gave much too low an electron pressure. Greenstein thus believed τ UMa had the normal temperature of an F star (of type F5, the same as Procyon, which was also analysed) but the low gravity of a giant. He also found a high turbulent velocity (4 km/s), typical of high luminosity stars. The implications were discussed in a second paper in 1949 (164). Since τ UMa was believed to have a main-sequence luminosity, a variety of models were discussed involving a low-pressure turbulent envelope to promote the observed high degree of ionisation. However, among the important conclusions of this study was the abnormally low abundance of several

* Note that MKK class II = MK class III for spectral type F6.

lighter elements, especially scandium and calcium.

Greenstein moved to the California Institute of Technology in 1948 and he later analysed there two other metallic-line stars, 8 Com and 15 Vul with Armin Deutsch, Gerhard Miczaika (who was by then at Harvard) and F.A. Franklin (b. 1932) (165). The method of analysis was similar and the results, if somewhat less extreme, also indicated low giant-type gravities for these stars.

The early history of the metallic-line stars was also confused by the claim in 1958 by Horace Babcock to have found weak but detectable magnetic fields in six Am stars (124). The belief in weak fields being the norm for Am stars persisted until 1969, when Peter S. Conti (b.1934) at Lick failed to find them in 16 Orionis, a star for which Babcock thought a periodically reversing field was present (166). Although no detectable fields are now known in any Am stars, the apparently erroneous early result led to further speculation on the Am phenomenon. Thus Erika Böhm-Vitense at Kiel University, Germany, reviewed the whole field of observations of Am stars in 1960. Although the hydrogen-line profiles and the distributions of energy in the continuum were those of normal A and not F stars, the only plausible model she could devise was as follows:

> ...that the atmosphere of a metallic line star is very strongly extended in the outer layers. Hence gas and electron pressures decrease outwards and also the continuous absorption coefficient becomes smaller. Since the line intensities get bigger with decreasing continuous absorption coefficient, the metallic lines appear strengthened...What forces might produce the extension of the atmosphere? ... Magnetic forces remain as the only possibility. In this context it is reassuring that H. Babcock has measured magnetic fields of a few 100 gauss for a series of Am stars (167).

9.7.2 Clarification of some of the metallic-line problems From an historical point of view, the metallic-line stars present an interesting case study, as a surprising number of the deductions made from the observational material were based on unsound premises and thus led down dead-end alleys. Much scientific research of course proceeds in this fashion, but the Am stars had more than their share of false starts. Intrinsically their spectra are a more homogeneous and less complex class than the hotter Ap stars, but early progress was if anything less rapid for the Am type. Meanwhile the number of known metallic-line stars had increased substantially. Harold Weaver (b.1917) found thirty-four in five galactic star clusters and claimed they comprised as many as 35 per cent of all the cluster stars with absolute magnitudes in the narrow range between 1.7 and 3.3 (168). Mercedes and Carlos Jascheck published a list of eighty-five Am stars, mainly from a compilation of Bidelman's, in 1957 (169). Spectroscopic studies confirmed the preliminary findings of Roman, Morgan and Eggen concerning rotation and duplicity. The confirmation of slow rotation was primarily the work of Arne Slettebak at Perkins Observatory in the mid-1950s (150, 170). He found a mean equatorial velocity of 57 km/s, substantially below that of normal dwarf stars of types mid-A to early F. Duplicity was studied by the Jascheks (169) and especially by Helmut Abt (b.1925) at Kitt Peak National Observatory, using spectra obtained at McDonald and Mt Wilson in a very thorough investigation (171, 172). Abt studied twenty-five Am stars and found twenty-two of these had variable radial velocities, indicating orbital motion in a close binary system. Since not all binaries would be detected because of their unfavourable orientation, he wrote 'we can safely conclude that all Am stars are members of spectroscopic binaries' (171).

Several of these papers point the way to substantial progress from the early confusion surrounding these stars. The Jascheks' paper of 1957 studied absolute magnitudes, spectral types and colours (169). They found 'that the M-L [i.e. metallic-line] stars should be considered as intrinsically A-type stars, rather than F-type stars, as had been thought before'. Abt saw that the duplicity in close binaries would be the cause of the slow rotation found by Stettebak, as tidal interactions between the two stars result in slow synchronous rotation with the orbit provided the orbital period was greater than about 2.5 days. This in turn led him to speculate that the slow rotation could be linked to the line strength anomalies: '...if the abnormal spectra denote actual abundance anomalies in the surface layers, it may be that rapid rotation will produce enough meridional circulation to dilute the abnormal surface material through a larger amount of material, producing a more nearly normal composition in the rapidly rotating stars' (172). Abt collaborated with Bidelman to produce an even more concise statement on the duplicity of Am stars: '...all stars in the approximate spectral type range A4-F1 on the main sequence and that are primaries in binaries with periods between approximately 2.5 and 100 days have metallic-line or peculiar spectra' (173).

If one work had to be singled out for going straight to the heart of the metallic-line problem and for producing order where before confusion had reigned, it would be the spectral analysis of the Am star 63 Tauri by Claude Van't Veer-Menneret (174). She completed this study in 1963 at the Institut d'Astrophysique in Paris under the guidance of Daniel Chalonge and Roger Cayrel. She summarised her work thus: 'After numerous fruitless attempts using an abnormal structure, a detailed analysis of the Am star 63 Tauri is undertaken in order to determine the chemical composition of its atmosphere'. Madame Van't Veer showed how the metallic lines distorted the colours of these stars in the UBV system of photometry, making them appear redder and hence vitiating the early attempts to determine their temperatures. But the three-dimensional classification of Barbier, Chalonge and Miss Divan placed 63 Tauri at a temperature and luminosity corresponding to dwarf stars of spectral type A7 to A8,* provided care was taken to allow for the effects of the metallic lines on the λ_1, D and Φ_b indices. She then went on to analyse high dispersion spectra obtained at Mt Wilson by R. and G. Cayrel. The ionisation equilibrium of four elements, using the higher temperature that now was deemed appropriate, gave the surface gravity of a normal dwarf star. Her conclusion ran as follows:

> Are the spectral abnormalities of 63 Tau due to an abnormal atmospheric structure or to a peculiar chemical composition?... We assign to 63 Tauri a higher temperature than that which has been adopted on average for metallic-line stars, but a normal pressure of a dwarf. It seems possible to explain all the spectral anomalies by a peculiar chemical composition.... Calcium is underabundant by a factor 10, iron and chromium overabundant by a factor of 3 to 5, and elements heavier than strontium that we could observe were overabundant by a factor of 10 to 25. This last result is new and deserves to be stressed.... The only physical peculiarity would be then the high turbulence (174).

Shortly after this important paper of Mme Van't Veer, Peter S. Conti at Lick came to similar conclusions from an analysis of five Am stars in the Hyades including 63 Tauri (175). He was unable to explain the observations from models with an abnormal structure, given that the temperatures (from multicolour photometry) were those of the later A stars. For 63 Tauri he derived a temperature $T_{\text{eff}} = 7650\,\text{K}$, very similar to Mme Van't Veer's value. His calcium and scandium abundances for this star were about ten times deficient, while heavier elements including chromium, copper, zinc, strontium, yttrium and barium were all enriched. Iron had only twice the normal solar abundance. Results for the other Am stars studied were similar (175).

9.7.3 Am stars in 1970 In 1970 Conti reviewed the whole field of Am phenomena and proposed the following definition of this class: 'The

* These stars have an effective temperature of 7760 K and a solar surface gravity.

Am phenomenon is present in stars that have an apparent surface underabundance of Ca (and/or Sc) and/or an apparent overabundance of the Fe group and heavier elements' (176). This therefore identified three possible subtypes, with only weak calcium or scandium lines, with only strong lines of heavier elements, or thirdly, with both of these peculiarities. Whether any stars actually populate the group in which weak calcium or scandium lines are the only peculiarity was, Conti noted, debatable; possibly α Geminorum B was such a star. However some Am stars certainly had normal calcium and scandium lines but strong lines of heavier elements (e.g. 17 Hydrae).

Several other abundance analyses had been undertaken by this time and they now all confirmed the general picture of Van't Veer-Menneret and Conti of abnormal line strengths due to abnormal abundances. The only other spectroscopic peculiarities were high turbulence and slow rotation. Frederick H. Chaffee, however, studied turbulence in dwarfs along the main sequence and concluded that high values were in any case normal for late A-type stars (177). It seemed that the slow rotation could give a clue to the cause of the problem, as Abt had earlier suggested. Michel Breger in Texas studied the pulsating δ Scuti-type stars in this same part of the HR diagram; he concluded there was a sharp dichotomy between the pulsating stars and those which showed metallic-line characteristics (178).

In 1970 progress was also made in the theoretical explanation of the Am phenomenon. Conti, in his review, noted that earlier attempts by various people to explain Am abundances by nuclear processes, either in the stars' interiors or on their surfaces, did not look promising. In 1967, Françoise Praderie in Paris had discussed in her doctoral thesis how diffusion of elements in stellar atmospheres may help to explain the Am abundances (179). Her suggestion was that Am stars had normal abundances whereas many heavier elements had gravitationally settled in other stars, including the sun. Georges Michaud (b.1940) in Canada put the diffusion hypothesis the other way round: he proposed that the combined effects of gravitational settling and radiation pressure could make some elements diffuse up, others down, in Ap and Am stars, where the slow rotation produces atmospheres of exceptional stability (180). If this is the explanation for Am stars, it meets many of the observational requirements; rotation sets up circulating meridional currents in a star that ensure uniformity of element abundances by mixing between the observable surface and the deeper stellar interior. Pulsation of the δ Scuti stars would also cause mixing. Only the slowly rotating stars that have been rotationally braked by tidal effects in a close binary can therefore exhibit the effects of diffusive element separation as a metallic-line star.

9.8 White dwarf spectra

9.8.1 Discovery of three white dwarfs The story of white dwarf stars goes back to 1844 when the great German observer Fredrich Wilhelm Bessel (1784–1846) reported on the variability in the proper motions of Sirius and Procyon (181), which he ascribed to the presence of a dark companion for each of these stars in a binary system. This prediction and the subsequent visual sighting of the faint companion to Sirius by the American optician Alvan G. Clark (1832–97), about ten arc sec from the bright star, was one of the triumphs of astrometric astronomy in the mid-nineteenth century (182). In 1896, John M. Schaeberle (1853–1924) at Lick discovered that Procyon too had a very faint close companion (183) of about twelfth magnitude.

The presence of these faint companions was not regarded as being in any way unusual, as visual binaries with much fainter secondaries were known to be relatively common. Another such example was o^2 Eridani (=40 Eri), a K dwarf with a fairly distant double companion, consisting of a faint white and a faint red star. It was the faint white companion (40 Eri B) that attracted Henry Norris Russell's attention in 1910. At this time the Russell diagram was in its early stages of formulation. Russell recognised that all of the intrinsically faint stars were cool and red. The white star 40 Eri B was the only apparent exception which appeared in the lower left-hand corner of the first published Russell diagram. Russell all too quickly dismissed the exception when he addressed the Astronomical and Astrophysical Society of America in December 1913: 'The single apparent exception is the faint double companion to o^2 Eridani, concerning whose parallax and brightness there can be no doubt, but whose spectrum, though apparently of Class A, is rendered very difficult of observation by the proximity of its far brighter primary' (184).

The story of the discovery of the A-type spectrum of this ninth magnitude star that Russell found so puzzling was only retold by Russell in print in the 1940s. In 1944 he noted this 'bit of history' as he described it:

> The first person who knew of the existence of white dwarfs was Mrs Fleming; the next two, an hour or two later, Professor E. C. Pickering and I. With characteristic generosity, Pickering had volunteered to have the spectra of the stars which I had observed for parallax looked up on the Harvard plates. All those of faint absolute magnitude turned out to be of class G or later. Moved with curiosity I asked him about the companion of 40 Eridani. Characteristically, again, he telephoned to Mrs Fleming who reported within an hour or so, that it was of Class A. I saw enough

of the physical implications of this to be puzzled, and expressed some concern. Pickering smiled and said, 'It is just such discrepancies which lead to the increase of our knowledge.' Never was the soundness of his judgment better illustrated (185).*

Hertzsprung was one of those who learnt of the A-type spectrum from Russell when they met in Bonn at the Solar Union meeting of 1913. He commented that the exception was 'very strange' but that the spectral type agreed with his measurement of the colour or 'effective wavelength' (187). Confirmation of the spectral type came from Walter Adams in 1914. The spectrum was of type A0 (188). He calculated an absolute magnitude of 10.3 for the star which confirmed its position in Russell's diagram, about 9 magnitudes (4000 times) fainter than other A0 stars.

In 1915 Adams obtained his first spectrum of Sirius B, the faint companion discovered by Clark. He had attempted this observation for two years with the Mt Wilson 60-inch telescope, and extreme care was necessary to prevent scattered light from the very bright primary from entering the spectrograph. His result: 'The line spectrum of the companion is identical with that of Sirius in all respects so far as can be judged from a close comparison' – in other words, of type A0 (189). This result is what Willem Luyten later described as 'the bombshell' (190). To this point, however, no inkling of consternation had appeared in the literature. Sirius B and 40 Eri B were simply odd quirks of nature. Russell had almost certainly worked out the implications, but chosen not to publish, probably because he considered the observations too few and unreliable.

Another subluminous star was found by Adrian van Maanen (1884–1946), the former student of Kapteyn who was at Mt Wilson from 1912. Apparently by chance he noticed on a pair of Mt Wilson plates a faint high proper motion star moving at over 3 arc sec annually: 'A spectrum of the star of $4\frac{1}{2}$ hours exposure was taken on October 24, 1917, with a small spectroscope at the 80-foot focus of the 60-inch reflector. The scale of the spectrum is 2mm from K to Hβ; according to Mr Adams the spectrum is about F0' (191). Adams and Joy later measured a remarkably high velocity

* The story, which arose from Russell's visit to Harvard in 1910, was also narrated by Russell at the Paris conference on novae and white dwarfs (186) in July 1939. He added:

> I knew enough about it, even in these paleozoic days, to realize at once that there was an extreme inconsistency between what we would then have called 'possible' values of the surface brightness and density. I must have shown that I was not only puzzled but crestfallen, at this exception to what looked like a very pretty rule of stellar statistics; but Pickering smiled upon me, and said: 'It is just these exceptions that lead to an advance in our knowledge', and so the white dwarfs entered the realm of study.

of recession of +238 km/s, in fact, the largest value then known for any star (192). Van Maanen's star was in due course to be recognised as the third white dwarf.

Bertil Lindblad obtained further spectra of 40 Eri B when at Mt Wilson in 1921. His spectral type was B9 and his description of the spectrum was more concise than hitherto:

> The most conspicuous feature is in the very strong widening of the wings of the hydrogen lines connected with a decided weakening of the high-numbered members of the Balmer series. On the spectrograms of low dispersion taken here, Hη is the last hydrogen line to be identified with certainty. ...the photographs show a faint extension of the continuous spectrum much farther into the ultraviolet than is the case for an ordinary A-type star (193).

So what were the 'physical implications' of these spectra that had puzzled Russell in 1910? Luyten's 'bombshell' actually exploded in 1924 when Arthur Stanley Eddington in Cambridge wrote his classic paper on the mass-luminosity relationship (194). He used the term 'white dwarfs' here for the first time. It was well known that the size of a star could be calculated if both its temperature (from the spectral type) and luminosity were known. Eddington found a radius of only 19 600 km for the companion to Sirius. The mass, deduced from the double star orbit, was taken as 0.85 times the mass of the sun. The very small size had to contain a normal stellar mass, which implied a density 53 000 times that of water. 'Such a density', wrote Eddington 'is not absurd, and we should accept it without demur if the evidence were sufficient'. The value was over ten thousand times greater than Eddington's mean density of the sun. R.H. Fowler then gave a theoretical explanation in terms of quantum mechanics in which the white dwarfs were supported not by thermal gas pressure, but by the much greater forces of a degenerate electron gas (195).

From this time on, white dwarfs were a physical reality, with an observational and theoretical basis. The details of white dwarf structure using the theory of a relativistic degenerate electron gas were worked out by S. Chandrasekhar in Cambridge in the early 1930s. He predicted an upper limiting mass of 1.44 solar masses for a helium white dwarf, and also a unique mass–radius relation for these stars (196).

9.8.2 The Einstein redshift Eddington had pointed out in 1924 that Einstein's general theory of relativity predicted a gravitational redshift of the light emanating from a small massive body, such as a white dwarf (194). For Sirius B he predicted a gravitational shift equal to that produced by the Doppler effect for a star receding at 20 km/s, which should provide an excellent test of the high density model. Walter Adams took up this challenge in 1925 using the single prism cassegrain spectrograph on the 100-inch telescope. The problem was to demonstrate an apparent Doppler velocity of the companion 20 km/s in excess of that of the bright star from the difference in spectral line displacements. A small correction was also applied for the relative orbital motion of the two stars. Adams' result was 21 km/s (or 0.32 Å) for the gravitational shift, almost exactly as predicted. 'The results may be considered, therefore, as affording direct evidence from stellar spectra....for the remarkable densities predicted by Eddington for the dwarf stars of early type of spectrum' (197).

At Lick, Joseph Moore repeated these difficult observations by obtaining four spectrograms of Sirius B on the giant refractor. His shift was coincidentally in exact agreement with that of Adams. His spectral type was about A5 (198).

Strangely the test was not applied to 40 Eri B, the other fairly bright white dwarf in a binary, until 1954, when Daniel Popper measured the gravitational shift from three Balmer lines using Mt Wilson spectra. The measured gravitational redshift, expressed as a velocity, was also 21 km/s, compared with 17 km/s expected from general relativity (199).

The problem of detecting gravitational line shifts in single white dwarfs would at first sight seem insurmountable, as the radial velocity must also be known. However, this problem can be tackled statistically, as has been done by Jesse Greenstein and Virginia Trimble (b.1943) from a large sample of shifts (200). Interpreting all these as Doppler shifts gave an apparent positive expansion of 50.4 km/s for the sample of 46 field white dwarfs studied, similar to the old K-term found in early B star radial velocities. They interpreted this apparent expansion as the Einstein effect, from which mean masses could be derived.

9.8.3 New white dwarfs in the 1930s In 1930 the number of known white dwarfs was still only three. An interesting paper by Russell and R. d'E. Atkinson considered two further stars as members of this class, Wolf 489 and the companion to the long-period variable, Mira, namely *o* Ceti B (201). The exceptionally hard to observe companion to Mira had been discovered by Alfred Joy in 1918, and in 1920 he found an unusual emission-line spectrum which was classified B8e (202). Recent observations have confirmed this star to be a white dwarf accreting material from the long-period variable. Wolf 489 is a nearby red star with a large parallax and proper motion, later confirmed as a white dwarf. Russell and Atkinson took the spectrum as type G (presumably from the colour), and using these five stars and also the hot O-type nuclei of planetary nebulae, they found a tentative correlation between their absolute magnitudes and spectra: 'It looks as if this sequence were roughly parallel to the main sequence, probably separated from it by a sparsely populated band, and with considerable scattering within it' (201). The evidence was quite insecure, yet the result was the first indication of the existence of a distinct white dwarf sequence in the HR diagram.

The fourth confirmed white dwarf was van Maanen no. 1166, a faint white star in Perseus with a high proper motion, discovered after P. Th. Oosterhoff at Leiden Observatory recognised that faint blue or white high proper motion stars are most likely nearby white dwarfs (203). Slitless spectra were obtained by Yngve Öhman (1903–88) at the new Stockholm Observatory. It was of type A0, with very broad Balmer hydrogen lines, which could be seen to Hη but no farther. The normal continuous absorption of an A0 star in the ultraviolet was, however, absent. His microphotometric scans were the first to be published for the white dwarf class (204) (Fig. 9.10).

The mid-1930s represent the epoch when white dwarfs began to be discovered in ever increasing numbers. The first results came from Gerard Kuiper (1905–73). Soon after his arrival at Lick Observatory in 1933 (he had then just obtained his doctorate at Leiden under Hertzsprung) he began using the 36-inch Crossley telescope to study the spectra of stars with large parallax. Later he joined the staff at Yerkes, and when the 82-inch had been completed at McDonald Observatory in 1939, he extended the program to increasingly fainter stars, including those with large proper motion of over 0.3 arc sec annually. The first results, in 1934, announced two new white dwarfs, including one (A.C. 70°8247)* of thirteenth magnitude which showed no lines at all, but a continuum whose light distribution resembled that of a B0 star (205). A series of publications now followed with further nearby white stars identified as white dwarfs, and also some comments on the remarkable 'red' white dwarf, Wolf 489, which showed no lines in its spectrum, but was classified K5$\pm$ from its colour (206). This program resulted in a list of thirty-eight stars in 1941 (207), including the tentative inclusion of Procyon B, which Eddington in 1924 had suggested might perhaps be a white dwarf (194). Some of these objects were as faint as fifteenth magnitude, but Kuiper obtained spectra for all but Procyon B. In this paper Kuiper proposed the first white dwarf spectral classification scheme in the form which he had presented to the Paris

* In the *Astrographic Catalogue* 1900.0 Greenwich Section, Vol. I, W.H.M. Christie (1904); the other star was Wolf 1346.

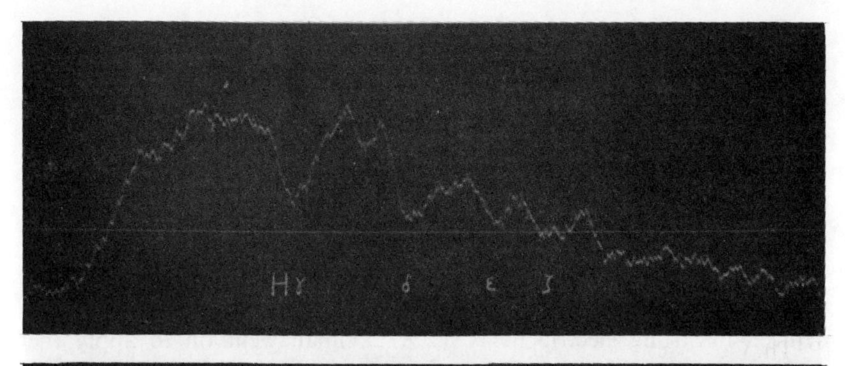

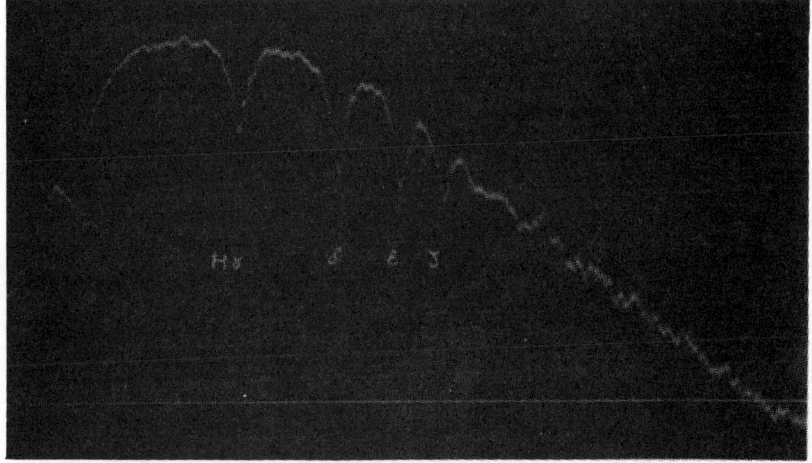

Fig. 9.10 The first microphotometric tracings of white dwarf spectra, by Yngve Öhman, 1931: van Maanen no. 1166 (above) and o^2 Eridani (below).

conference on novae and white dwarfs in 1939. There were seven classes as listed in Table 9.3.

Table 9.3. *Kuiper's spectral classification scheme for white dwarf stars, 1941.*

Kuiper class	Typical features	Typical star	No. of stars known
Con	Continuous spectrum, no lines visible	AC70°8247	7
wA	Broad Balmer lines to Hη only	40 Eri B	22
wAs	Balmer lines narrower	L870–2	4
wA5	—	Sirius B	1
wF	Only H,K seen but less strong than in normal F star	L745–46	1
wF5	H, K stronger. Also line at 3835Å (MgI?)	Ross 640	1
wG	H,K strong; broad blends in UV resemble G star spectrum	van Maanen 2	1

The wA stars with strongly Stark-broadened Balmer lines were easily the most common, while Kuiper noted that the second-commonest group, the Con stars, might have lines not visible at the necessarily low dispersion of his spectrograms.

The very broad lines of the wA stars had been interpreted as being due to the Stark effect by Yngve Öhman in 1935. He wrote that 'The observed effects must evidently be connected with the high pressure existing in the atmospheres of the white dwarfs.... The phenomenon is probably also related to the Stark broadening, as the number of Balmer lines is much reduced in strong electric fields' (208). Öhman went on to apply the theoretical Balmer line profiles of S. Verweij (209) to measure the surface gravities of two white dwarfs. His value for Wolf 1346 was $\log g = 7.0$ (cgs units) or 100 km/s^2, which is 400 times greater than that on the sun's surface and 10 000 times greater than on the earth!

9.8.4 Luyten's white dwarf discoveries and spectral classification scheme By far the most productive program in discovering white dwarfs has been the proper motion work carried out by William J. Luyten (b.1899) at the University of Minnesota. He began observing low luminosity stars spectroscopically at Lick in 1921 (210), and indeed one of the stars then studied later proved to be a white dwarf;* but his association with Harlow Shapley during his time at Harvard (1923–31) led to the inception in 1928 of the Bruce Proper Motion survey, in which the 24-inch Bruce refractor at Harvard's Boyden Station in South Africa was used to obtain second generation plates for proper motion measurements. The plates were taken from 1928 to 1937 and results on high proper motion stars of white colour published from 1940 (211). By 1949 about 100 new white dwarfs had been announced (212) and in 1960, 400 (213). Some of these also arose from a survey of faint blue stars in high galactic latitudes or in dust-obscured regions which had been commenced by M. Humason and F. Zwicky at Mt Wilson (214) and continued by Luyten from 1952 (215), and more recently others have come from Luyten's northern proper motion survey with the Palomar Schmidt telescope, commencing in 1963. Luyten's 1970 compendium listed 2934 'probable white dwarfs' based on his proper motion and faint blue star surveys (216) although many of these stars could be high-velocity 'intermediate' or subdwarfs. Some were as faint as magnitude 21.0. As Luyten himself has remarked: 'White dwarfs are the easiest stars to discover, but the hardest to observe' (217).

These surveys provided a wealth of data for subsequent spectroscopic observations. Luyten himself undertook some of this work using the 82-

* This was Cincinatti 20: 398.

inch McDonald telescope while Struve was still director. In 1950 Luyten wrote to Struve, who was President of the IAU Commission for Stellar Spectra, recommending a new classification for white dwarf spectra (218). First, he noted that white dwarfs were frequently confused with other less extreme subluminous stars that Kuiper had termed 'subdwarfs' in 1939 (219). Luyten wanted this term dropped altogether, and for the true degenerate stars he suggested the capital letter D followed by A, B, C or F. This scheme was presented in detail in 1952 with the classifications of forty-four white dwarfs mainly from McDonald spectra (220). The DA stars showed Balmer lines of hydrogen and were further subdivided from DA-0 to DA-7 depending on whether the lines were very broad or sharper. DB-0 and DB-1 were for the two white dwarfs found with helium lines in their spectra. In the spectrum of that with the stronger lines (L930-80; type DB-0) Morgan reported that the hydrogen lines were probably absent, while the neutral helium lines 'are probably stronger than in any known stellar spectrum' (221). DC was used for stars with no lines visible on low dispersion spectra, while type DF showed the ionised calcium H and K lines but 'very little else' (e.g. van Maanen's star). This scheme did not depart radically from Kuiper's earlier proposal, but to the scheme of his ex-patriot Dutch rival Luyten made no reference.

9.8.5 Classification and analysis of white dwarf spectra, 1957–67 The spectra of twenty-three white dwarfs, taken mainly from Luyten's lists, were discussed by Beverley Lynds (b.1929) at Berkeley in 1957. She used spectra at 430 Å/mm from the Crossley telescope at Lick. Of these stars seventeen were of Luyten type DA, for which she measured the hydrogen line strengths and profiles. Gravities of about $\log g = 7.0$ were found (in agreement with the much earlier work of Öhman, who was not acknowledged!) (222), and also a variation of the Hγ widths with colour index that fitted Verweij's theoretical results quite well.

The work of Jesse Greenstein at the California Institute of Technology in white dwarf spectroscopy also dates from the beginning of this period. Minkowski had in 1938 found two mysterious diffuse shallow bands at 4135 and 4475 Å in AC 70°8247 (223). Greenstein confirmed the existence of these by co-adding several spectra from the 200-inch Palomar telescope (224) and the presence of a third weaker band at 3650 Å was shown. He also found two stars with a broad band at 4670 Å being the only discernible feature in the spectrum. In L879-14 this band was at most 20 per cent deep but about 150 Å wide (224). Both stars have a yellow colour suggesting a cooler temperature. The C_2 molecule was given as a tentative origin for this feature. Greenstein later undertook a large program of photometry and spectroscopy with Olin J. Eggen. From data for 166 white dwarfs, they selected sixty-nine with the best absolute magnitudes and were able to demonstrate the presence of two distinct parallel sequences in the colour–magnitude diagram (225).

Greenstein introduced a new classification scheme in 1958 for white dwarfs (226). In its initial form it recognised nine types DB, λ4135, DC, DA, DAe, DF, λ4670, DG and DK without further subdivision. The λ4135 type was assigned to one star, AC70°8247, while there were just two stars in the λ4670 group. Two DA stars showed sharp emission lines and hence received the e suffix. Van Maanen's star was the only DG star, and Wolf 489 (mainly from its colour, though Greenstein noted that weak H and K lines are just visible) was the sole DK white dwarf. Apart from these details, the Greenstein classification was apparently based on that of Luyten, although this was not explicitly stated. The subdivision of the DA stars according to absorption line character was not attempted apart from the suffixes wk or s for those with weak diffuse Balmer lines, or sharper lines respectively. Two years later some minor modifications to the scheme were made, including the addition of type DO (ionised helium lines present; e.g. Humason–Zwicky star no. 21) (227). Fig. 9.11 shows some white dwarf spectra observed by Greenstein, with his spectral classifications.

Apart from this new classification, Greenstein has obtained the line profiles of Balmer, neutral and ionised helium and H and K lines in a variety of white dwarfs of different types (227). Greenstein's Hγ profiles for twenty-two DA stars were analysed by Volker Weidemann (b. 1924) at the Mt Wilson and Palomar Observatories so as to obtain surface gravities and temperatures by comparing the observations with theoretical profiles from model atmospheres for white dwarfs (228). The average surface gravity was found to be $\log g = 8.0$ while the temperature varied over a wide range from 10 000 to 25 000 K. From an analysis of these stars' positions in the HR diagram, a DA white dwarf has a typical mass of 60 per cent of that of the sun and a typical radius slightly over one hundredth of the solar radius. Weidemann also analysed Greenstein's spectra of Van Maanen's star (229). The iron and calcium lines were studied and fitted to a model with a solar temperature, $\log g = 8.0$ (cgs units) and with large deficiencies of hydrogen and especially of heavier elements.

9.9 The hydrogen-deficient stars As early as 1888 Mrs Fleming found the Hβ line in emission in the southern 4th magnitude star υ Sagittarii (230). This star was one of many found to show emission lines on the Harvard objective prism plates towards the end of the

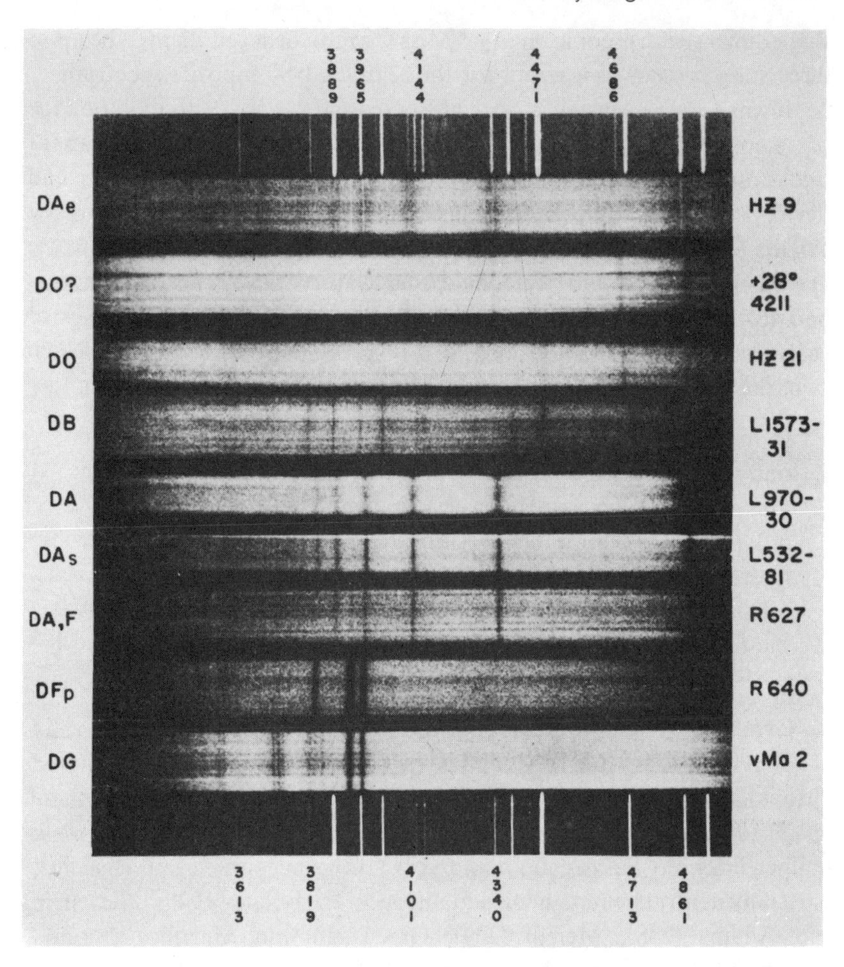

Fig. 9.11 White dwarf spectra observed by Greenstein, 1959.

nineteenth century. W.W. Campbell at Lick found Hα also in emission in 1894 (231) and he discovered radial velocity variations in 1899 (232). υ Sgr continued to receive intermittent attention from various observers over the next four decades. For example, Annie Cannon described changes in the line intensities in the notes to her 1912 catalogue of spectral classifications of the brighter stars (104). The spectral type was given as 'Pec' but in the later HD Catalogue it appeared as B8p. However spectral features of both types A2 (the metallic lines) and B3 (the helium lines) were present, as noted by Miss Cannon, and later in 1928 by J.S. Plaskett (233), who also measured the wavelengths of nearly 3000 blue lines, assigning identifications where possible. Although υSgr was recognised quite early on as a single-lined spectroscopic binary,* Miss Cannon believed there to be up to three stars present to account for the apparently composite spectrum.

It was Jesse Greenstein's work at Yerkes in the early 1940s that took the first steps towards understanding this peculiar spectrum. Greenstein was at that time a post-doctoral fellow under Struve, who in the 1930s had developed Yerkes to be the leading centre for astrophysics, surpassing Mt Wilson for quantitative interpretation of stellar spectra. With access to the new 82-inch telescope at McDonald Observatory, the Yerkes astronomers had nearly the equal of Mt Wilson in telescope power. Struve had suggested that Greenstein work on υ Sagittarii at this time. Later on Greenstein remarked:

> In a certain sense I made a living for five or six years out of that one star and it is still a fascinating, not understood, star. It's the first star in which you could clearly demonstrate an enormous difference in chemical composition from the sun. It had almost no hydrogen. It was made largely of helium, and had much too much nitrogen and neon. It's still a mystery in many ways ... But it was the first star ever analysed that had a different composition, and I started that area of spectroscopy in the late thirties (235).†

Greenstein's first paper on υ Sgr estimated a hydrogen-to-helium ratio of about 0.01, compared to around 10 for normal stars (236). Four hundred ultraviolet lines were catalogued (Fig. 9.12). A second paper extended these results to the visual region (237). The general features were extremely weak Balmer lines, no Balmer jump and very strong lines of certain elements, especially neutral helium, neon and nitrogen. In the later 1940s Greenstein and Adams (238), Merrill (239), Greenstein and Merrill (240) and Bidelman (241) all studied coudé spectra of υ Sgr, finding absorption and emission lines with anomalous velocities or asymmetries at various phases of the binary orbit. Bidelman hypothesised a stream of gas between the primary and invisible secondary to account for large Hα absorption blueshifts seen at certain phases. The concept of mass loss of the outer hydrogen envelope to a companion star may account for the abundance and velocity anomalies in υ Sgr, and in this respect it is not a typical hydrogen-deficient star, although a very interesting object in its own right.

A few other binary stars possibly of similar type to $\boldsymbol{\upsilon}$ Sgr have been discovered, most notably the spectroscopic binary HD 30353 by Bidelman in 1950 (242). He wrote:

* R.E. Wilson gave the first orbit in 1914 (234).

† In making this remark, Greenstein appears to have overlooked Berman's analysis of R CrB in 1935 (see later this section).

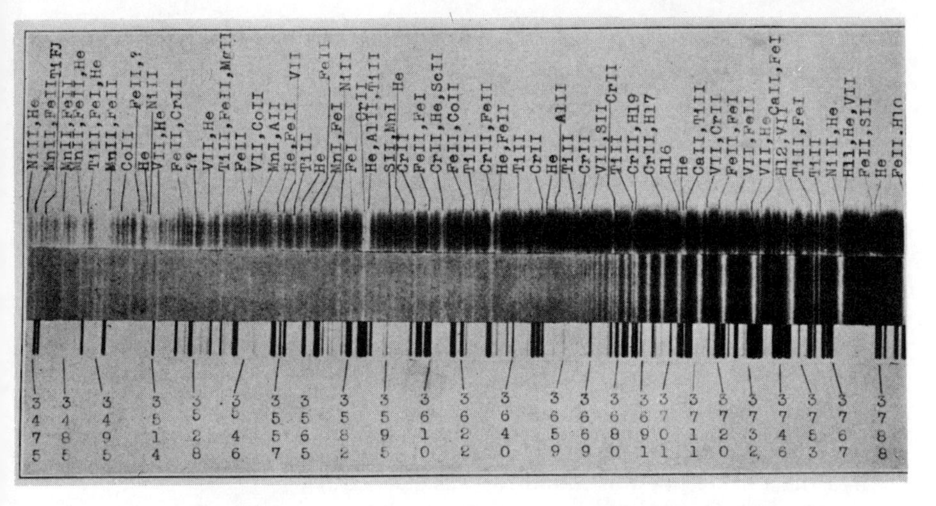

Fig. 9.12 υ Sagittarii (top) is compared to α Cygni (bottom). Note the absence of a Balmer jump in υSgr due to its low hydrogen content. McDonald spectra analysed by Greenstein, 1940.

> The comparative weakness of the lines of neutral iron and of ionized calcium make it necessary to classify HD 30353 as an A-type star, although, properly speaking, it cannot be given a spectral-type at all. The most remarkable feature of its spectrum, however, is the extreme weakness of the lines of hydrogen...a deficiency of hydrogen is strongly indicated ... (242).

A new class of single hydrogen-deficient star showing no hydrogen lines at all was announced by Daniel Popper at Yerkes in 1942 (243). HD 124448 is a ninth magnitude star classified as B2 in the Henry Draper Catalogue. McDonald spectra indicated the complete absence of Balmer lines and the Balmer jump. Popper wrote: 'The abundance of hydrogen appears to be very low in the atmosphere of this star. Besides the strong HeI spectrum, lines of OII and of CII are present' (243).

HD 124448 is an example of a very rare class of hot star showing extreme hydrogen deficiency (the ratio hydrogen-to-helium probably less than 10^{-3}). A few other examples were found in the 1950s and 60s. For example, HD 160641 is an O star discovered to be without hydrogen lines by Bidelman in 1952 (244) and HD 168476 is another B star discovered at Radcliffe Observatory in South Africa by Thackeray and Wesselink to be similar to Popper's star (245). Thackeray argued that the lack of hydrogen results in a low opacity, and hence the spectrum of HD 168476 mimics that of a supergiant (246).

Two other less extreme examples with very weak but not absent Balmer lines were HD 96446 found in 1959 (247) and BD + 10°2179 in 1961 (248). Both these stars are examples of intermediate hydrogen-deficient stars, with the hydrogen-to-helium ratio one to two orders of magnitude below the normal value of 10. Abundance analyses of several stars in the extreme and intermediate groups were undertaken in the 1960s, especially those by P.W. Hill (b. 1937). His observations were mainly Radcliffe 74-inch coudé spectra and he gave abundance results for HD 168476, HD 124448 and BD + 10°2179. Apart from the hydrogen deficiency of these stars, carbon and possibly nitrogen and neon were overabundant while oxygen was deficient (249). About the same time the Mt Stromlo coudé spectrograph in Australia was used to observe HD 96446 and an analysis was undertaken by Anne Cowley (b.1938), L. H. Aller and T. Dunham; only a mild hydrogen deficiency was found for this star (250).

In their hydrogen deficiency and high carbon abundance, the early-type hydrogen-deficient stars bear some resemblance to the late-type carbon-rich stars of the R Coronae Borealis class. R CrB itself and for example RY Sgr are members of this class, and a possible link between such objects and the early-type stars was discussed by Bidelman in 1952 (251). The deficiency of hydrogen in R CrB had been known for many years, since Hans Ludendorff in 1906 reported the Balmer lines to be invisible on spectra obtained at Potsdam (252). The high abundance of carbon on the other hand was demonstrated by Louis Berman in 1935 (253) using mainly Lick and Mt Wilson Observatory spectra. He found 69 per cent of all atoms (by number) to be carbon in this star. Berman's is a very remarkable piece of research, since it represents one of the most extensive applications in the 1930s decade of the new curve of growth theory developed by Minnaert for the sun and first applied by Struve and Elvey to stars.

A possible evolutionary link between the early-type hydrogen-deficient stars, the R CrB variables, and a class of hydrogen-deficient non-variable carbon stars has been proposed by Brian Warner (360). He analysed spectra of five such carbon stars and showed that hydrogen was strongly deficient, and carbon was over-abundant. Most heavy elements however had normal abundances relative to iron.

9.10 The T Tauri variable stars

In 1945 Alfred Joy at Mt Wilson defined a new class of variable star with T Tauri as the type star (254). His original definition was based partly on non-spectroscopic criteria: an irregular light-curve, a spectral type from F5 to

G5, a low luminosity (probably indicating the stars are dwarfs) and an association with bright or dark nebulosity were features in common to all eleven stars he listed. The principal spectroscopic peculiarity that Joy noted was the presence of 'emission lines resembling those of the solar chromosphere, particularly in the great strength of H and K of calcium'. There were initially eleven members in this new variable class. They were generally about ninth magnitude or fainter, which made them difficult objects to observe spectroscopically.

Fig. 9.13 Alfred Joy.

The presence of irregular variables associated with nebulosity was not new (stars of this type had been known, in the case of T Tauri, since 1852). However, their peculiar emission-line spectrum was a feature that Joy found to be common to all these stars. Joy continued:

> The most significant characteristic of the spectra of the T Tauri variables is the low excitation bright-line spectrum which is prominent in most of the stars at certain phases...This bright-line spectrum is quite different from other steller emission spectra, but its resemblance to the solar chromosphere is sufficient to invite a detailed comparison....The stellar emission lines are much broader than those of the chromosphere...yet the general resemblance of the spectra leads to the conclusion that they must have their origin under similar physical conditions (254; Fig. 9.14).

Joy was able to suggest a classification of the emission-line spectra based on their maximum intensity for a given star. Those with the least strong emission lines were assigned emission strength 0 (e.g. UX Tau); those

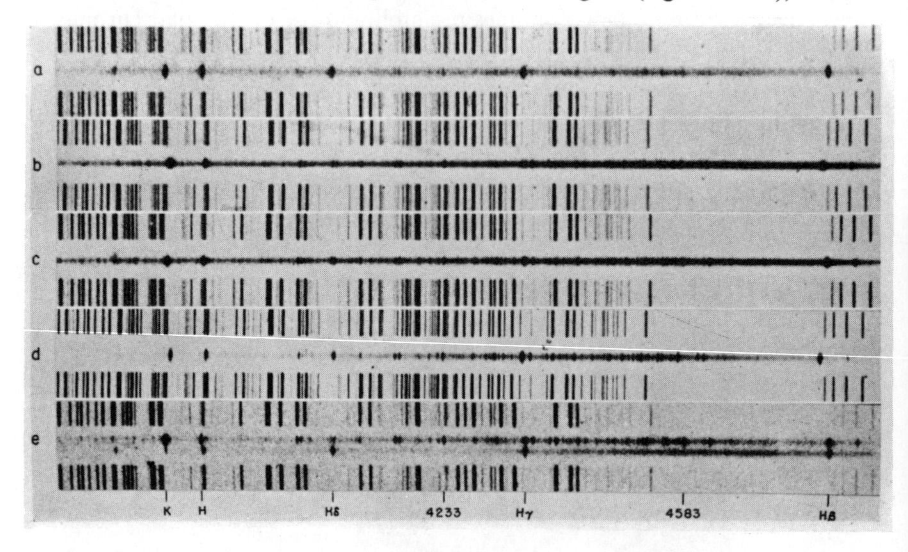

Fig. 9.14 Spectra of T Tauri stars photographed by Joy, 1945.

with the strongest emission received an index 10 (e.g. UZ Tau). More than 160 bright lines were identified in Joy's eleven stars; about half of these were lines of neutral or ionised iron. However the Balmer lines and lines of neutral helium, and for some stars 4686 Å of ionised helium, were also present in emission.

Six of the stars listed by Joy had been studied earlier spectroscopically. Indeed, the presence of emission features had been found in four of the stars between 1915 and 1918, by Adams and Pease for T Tauri (255), by V.M. Slipher for R Mon and R CrA (256) and by Annie Cannon for RU Lup (257). Sanford's study of T Tauri in 1920 gave measurements on eleven bright emission lines (all ionised iron) in this star (258), while Merrill tentatively found emission lines from seven different elements in RU Lup in 1941 (259). Although the spectral similarity of R CrA to T Tau had been noted by Hubble in 1922 (260) and of UY Aur to T Tau by Joy in 1932 (261), none of these earlier investigations linked the different stars into a common group.

A further forty emission-line stars, nearly all of the T Tauri type and located in the Taurus dark cloud, were reported by Joy from Mt Wilson objective prism spectra in 1949 (262). All these stars were faint, mostly apparent magnitude 12 to 15, but the results showed the relatively high frequency of occurrence of T Tauri stars in certain discrete regions associated with nebulosity. In this work Joy noted the presence 'in many stars of an unknown hot source which produces an overlying continuous spectrum, strong in the violet region'. The ubiquity of faint T Tauri variables in certain nebulous regions is well illustrated by their high rate of discovery; thus George Herbig catalogued 126 stars of this type in 1962 (263).

Joy's paper on the T Tauri stars in the Taurus cloud was published soon after his retirement from Mt Wilson in 1948. In George Herbig's words: 'It is difficult to be certain which was Joy's most far-reaching contribution [to astronomy], but the T Tauri stars are his most famous monument' (264). It was Herbig in the 1950s and 1960s at Lick who now made the principal contributions to T Tauri research.

In 1952 Herbig found that T Tauri stars generally have significantly broader absorption lines than normal stars of the same spectral type (265). If interpreted as being due to rotation, then the values of $v \sin i$ were in the range 20–65 km/s.* At first such results were considered anomalous, but subsequently were explained by Herbig (266) on the basis that T Tauri stars are gravitationally contracting pre-main sequence stars. Their rotational velocities on reaching the main sequence at earlier spectral types can be calculated using angular momentum conservation, and these agree with the rotation observed for normal dwarf stars of the corresponding main sequence spectral type.

The line width observations of T Tauri stars were thus consistent with those of normal dwarfs, if the gravitational collapse hypothesis, first proposed (267) by the Soviet astronomer Viktor Ambartsumian (b.1908) in 1947 is accepted. This theory was not immediately accepted in the West; at first the idea that the emission spectrum of a T Tauri star arose from the passage of a normal star through a nebula was put forward by several authors, including Herbig (268), Greenstein (269) and Struve and Rudkjøbing (270). Not until 1954 was Ambartsumian's theory firmly established. The observation of T Tauri stars in very young clusters gave strong support to these ideas – notably from Merle Walker (b. 1926) at Lick (271) for the cluster NGC 2264. The Hertzsprung–Russell diagram showed the later-type stars (including many T Tauri stars) about 2 magnitudes above the main sequence, even though the more massive B stars were already main sequence objects. The overluminous nature of T Tauri stars was consistent with the theoretical pre-main sequence tracks for

* v is the equatorial rotation speed; i the axial inclination to the line of sight.

gravitationally contracting stars calculated by Louis Henyey (1910–70) and his colleagues at Yerkes (272).

Herbig in 1958 redefined a T Tauri spectrum using purely spectroscopic criteria (273). All T Tauri stars show the H and K lines and the hydrogen lines in emission, and the fluorescent neutral iron emission lines at 4063 and 4132 Å are also present. Several forbidden emission lines of ionised sulphur and neutral oxygen are usually but not always seen, while strong lithium absorption at 6708 Å is also characteristic. Herbig thus removed any mention of variability or association with nebulosity from the definition. These however are properties of all T Tauri stars, but they would not be sufficient basis for a definition. Herbig commented in 1962:

> There have been instances where the name ['T Tauri star'] has been applied to any emission-line star in nebulosity, and even to any irregular variable found in a nebula, regardless of spectral characteristics. Such indiscriminate use of the name today can only lead to unnecessary confusion, although it was entirely excusable in the state of knowledge of a decade ago (263).

Two further spectroscopic discoveries in the early history of the study of T Tauri stars deserve mention: the lithium line in T Tauri stars, and the P Cygni type of line profile for many lines. The presence of an abnormally strong neutral lithium absorption line at 6708 Å was first seen in T Tauri itself by Roscoe Sanford in 1947 (274) but he made no comment on the significance of this finding. The German astronomer Kurt Hunger (b.1921) visited Lick for a year from mid-1956 and worked with Herbig on T Tauri stars. It was at this time that the strong lithium in T Tauri and RY Tauri was confirmed (275). Walter Bonsack (b.1932), working with Greenstein at Caltech, then observed four T Tauri stars; all these and a T-Tauri-like star SU Aur had a very strong line of lithium (276). This result was extended to a further seven T Tauri stars by Bonsack in 1961 (277). He concluded that all T Tauri stars, if they show absorption lines at all, have a lithium abundance enhanced about a hundred times above the solar lithium-to-hydrogen ratio, based on the classic paper of Greenstein and Richardson (278) for the lithium abundance of the solar photosphere. The T Tauri lithium abundances (relative to calcium) were, however, comparable to the higher terrestrial value. A possible explanation, proposed by Bonsack and Greenstein, was lithium production on the surfaces of T Tauri stars, as the result of high energy cosmic ray protons inducing spallation reactions after being channelled down onto the star by a magnetic field. Such a model is no longer considered to be tenable. However, there may be no need to invoke lithium production in or on T Tauri stars, if the lithium content in the interstellar medium is also as high as observed in T Tauri spectra. The lower abundance in most main sequence stars is then due to lithium destruction at temperatures above a few million degrees, as a result of convective mixing of the surface stellar material into a star's interior (279).

The unusual star FU Orionis, which brightened by about six magnitudes in 1936 (280) to reach a nearly stable light-level near tenth photographic magnitude, may also be related to a pre-T-Tauri evolutionary phase of a newly-formed star. Although only H and K lines are in emission in the visible spectrum of FU Orionis, the very strong lithium line is a feature in common with T Tauri stars (281). An abundance analysis by Herbig (281) found lithium in FU Orionis about eighty times enhanced above the solar abundance.

Finally, Herbig in 1962 commented that T Tauri stars observed at high dispersion frequently show emission lines with shortward displaced absorption components (263). This line profile is especially visible at Hα and the K line and is characteristic of mass outflow as seen in P Cygni and many early-type stars, such as Wolf–Rayet stars (see section 9.2). T Tauri and RY Tauri both show this phenomenon and an extreme example is the luminous T Tau star LkHα120. Since Beals' mass outflow model (3) is generally regarded as being the correct interpretation of all P Cygni profiles, it was at first considered anomalous that gravitationally contracting stars should be losing and not accreting mass; however, in reality there is no contradiction between a slow quasi-equilibrium contraction accompanied by a more rapid mass loss due to material being driven off the star in an extended outer envelope.

Carlos Varsavsky in Argentina first estimated a mass loss rate for T Tauri of about 10^{-7} solar masses per year (282). Figures of this order of magnitude were subsequently confirmed by Leonard Kuhi (b.1936) from observations, mainly on the Lick 3m telescope, of the P Cygni profiles of six T Tauri stars (283). These rates imply a substantial fraction of the mass leaves a star during its contracting phase. As Kuhi pointed out, 'the effects of such large mass loss on the individual stars must be rather drastic' (283). However, material lost from T Tauri stars is expected to comprise only a small fraction of the overall interstellar medium.

9.11 The barium stars

Lines of ionised barium, especially the resonance line at 4554 Å, have been recorded by many observers in stellar spectra. For example, W.W. Morgan plotted the Rowland intensity of the 4554 Å line as a function of spectral type for dwarfs and giants from spectral type A0 to K5 (284), and this showed the much greater strength of this feature in the giants for types later than F8. Paul Merrill, in studying the spectra of red stars in 1926, had recorded that 'the ionised line λ4554.04 is weak in class M, strong in class N, and very strong in class S, being in fact one of the distinctive features of class S' (285). The presence of abnormally strong barium lines in several A stars, subsequently recognised as being metallic-line stars, was noted by Cora Burwell at Mt Wilson in 1938 (286). These different observations thus indicated both a considerable luminosity dependence for ionised barium lines, and also some marked abundance abnormalities in certain special stars.

A third class of star with abnormal strength of the 4554 Å line was announced by Bidelman and Keenan in 1951 (287). They described them as 'a group of G and K-type stars which show very strong lines of BaII but which, however, do not appear to be supergiant stars'. They named this new group the 'BaII stars' and showed they were G or K red giants which showed, in addition to much strengthened ionised barium lines, enhanced features of CH, CN, C_2 and ionised strontium. The brightest member of five stars listed was ζ Capricorni. This star was observed the following year by Roy Garstang (b. 1925) who found that in addition to barium and strontium, seven other elements heavier than iron were all present with unusual strength (288). Bidelman and Keenan concluded that the barium stars 'cannot be fitted into the spectral-class–luminosity-class array defined by normal stars'; that they appear to be related to the carbon stars and possibly also to the S stars, and that the strength of the 4554 Å line in S stars and BaII stars is not purely due to unusual conditions of ionisation or excitation.

In attempting to classify the barium stars, their relationship to other red giants of abnormal composition was an important consideration. Thus Bidelman preferred to place the barium stars as an early-type extension of the carbon C stars because the bands of C_2, CH and CN are all stronger than in the spectra of normal stars (289), while Keenan emphasised the similarity between the barium and the CH stars. He wrote: 'There is clearly a similarity between the BaII and the CH stars; the spectroscopic difference between the two groups lies primarily in the relative enhancement of CH bands as against BaII lines... The difference is one of degree, but the degree is considerable' (290). However, as Keenan also noted, there are important differences as well; in particular 'the BaII stars have much lower radial velocities [than the CH stars] on the average'.

An extensive curve-of-growth analysis of one of the barium stars, HR 2392, was undertaken by E.M. and G.R.Burbidge in 1957 (291). Carbon was found to have a slightly greater abundance than in κ Geminorum, a normal G8 III star used as a standard for the analysis. The abundances of elements from sodium to germanium were normal. On the other hand,

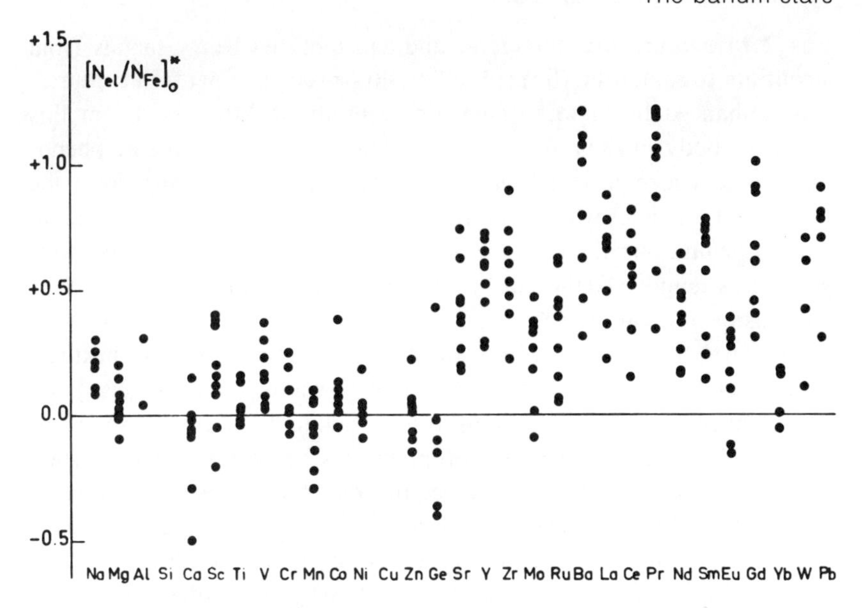

Fig. 9.15 Element abundances in barium stars, from a compilation by G. Cayrel de Strobel in 1976.

most of the elements heavier than strontium were enhanced by about a factor of ten; this included barium and the rare earths. The overabundant elements were nearly all those which can be synthesised deep inside red giant stars by the slow process of neutron captures, and the Burbidges were able to interpret the abundance anomalies in HR 2392 as the result of mixing of this material from the star's interior into the observable atmosphere. They suggested that these abundance anomalies may be the same as in the S stars, which differ only in temperature and pressure but not composition.

Brian Warner (b.1939) at the University of London surveyed the spectra of all known barium stars in 1964, by then twenty in number (if the high velocity star HD 26 is included) (292). He was able to devise a two-dimensional classification system based on medium dispersion slit spectra, using the normal temperature-dependent spectral type (in the range G5 to K5) and barium overabundance parameter on a scale from 1 to 5, Ba index 5 being the most extreme overabundance. Thus ζ Cap was classified K0-Ba3 on the Warner system.

Using high dispersion coudé spectra from the Radcliffe Observatory's 1.8m telescope in South Africa, Warner analysed the composition of nine of the barium stars (292). His results included HR 2392 and were in general agreement with the analysis of this star by the Burbidges: namely carbon was 2 to 5 times overabundant, and most of the heavy metals from strontium to samarium (formed by the slow process of neutron captures) were enhanced by factors of ten or generally slightly less. From this extensive body of data Warner concluded that the barium star phenomenon was a normal evolutionary stage of stars in the mass range from one to four solar masses which had undergone convective mixing of material from a helium-burning shell into the outer envelope. These results thus generally strengthened the conclusions reached by the Burbidges from their analysis of only one star.*

This was the position in 1965; the main development over the next few years was the discovery of many more barium stars, especially by MacConnell and his colleagues from Curtis Schmidt objective prism plates taken in Chile (293). When this program was begun thirty barium stars were known. Its completion increased this number to 150, while a further ninety stars showed marginal barium characteristics. These stars were found to have the same luminosity as normal G and K red giants. The statistics showed that about 1 per cent of all red giants in the same part of the Hertzsprung–Russell diagram are barium stars.

9.12 The discovery of CH stars When Philip C. Keenan and William W. Morgan at Yerkes Observatory proposed their new C classification for carbon stars in 1941, they noted a rare subgroup of five carbon stars 'characterised by unusually great CH intensity and almost complete absence of atomic lines in the blue region of the spectrum' (294). The main CH molecular feature they referred to was the Fraunhofer G-band covering much of the spectral region 4215–4325 Å in the blue, and observed in the majority of later-type stars, including the sun. The identification of CH as the principal origin of the G band was made by H.F.Newall, F.E. Baxandall and C.P. Butler in 1916 (295).

Keenan's classical paper in 1942 assigned the appellation 'CH star' to these five objects (296). He noted that in the blue part of the spectrum these stars 'could not readily be classified because their atomic lines are so unusually weak that only H and K can be recognised even on spectrograms of moderate scale'. Most of these stars had early R spectral types in the Harvard system. Keenan estimated equivalent spectral types from the visual region spectra as being late G to early K, corresponding to temperatures of 4600 to 3600 K. CH stars are therefore among the hottest of the carbon stars.

* Barium stars are now known to be binary stars. The peculiar abundances probably originated in the companions prior to mass exchange between the two stars.

Roscoe F.Sanford (1883–1958) at Mt Wilson added a further three CH stars to Keenan's list in 1944 (297). For three of these stars he was able to measure the strength of the D-line absorption due to interstellar clouds containing sodium atoms. This absorption gets stronger the more distant the star, and can thus give a rough estimate of distance and hence of luminosity. By this method Sanford found the CH stars to have the luminosities of typical giant stars. The evidence for their giant luminosities is discussed in a review by Keenan in 1958 (290).

The list of known CH stars was extended to eleven by W.P. Bidelman in 1956 (289). Of these at least ten are very high velocity objects, moving at 100 to 300 km/s relative to the sun, and hence belonging to the class of very old stars known as Population II. The discovery of a faint (eleventh magnitude) CH star in the Population II globular cluster ω Centauri by G.A. Harding (298) confirmed their membership of this population. Two of the stars in Bidelman's list, namely HD 26 (discovered by Keenan and Keller in 1951 (299)) and HD 201626 (discovered by Ruth Northcott in 1953 (300)) have become two of the best known members of the CH class, as a result of abundance analyses by G. Wallerstein and J.L. Greenstein in 1963 (301). They found most metals deficient by factors of respectively 5 and 30 relative to the standard star ε Virginis, but the carbon-to-iron ratio was 5 times that of ε Vir. Heavy metals such as barium and the rare earths were enhanced by about twenty times in both stars – the strong ionised barium lines in HD 26 had been remarked upon by Keenan and Keller (299), and are reminiscent of the barium stars.

Andrew McKellar at Victoria analysed the C^{12}/C^{13} isotope ratio in several carbon stars, including five CH stars in 1949 (302). This was one of the earlier reliable studies of stellar carbon isotope ratios and showed that four stars were rich in C^{13} (C^{12}/C^{13} from 3.0 to 6.3), whereas in one star this isotope was undetectable, and presumably therefore the ratio was near the normal solar value of about 80. No C^{13} was found in either HD 26 or HD 201626 by Wallerstein and Greenstein. On the other hand, another CH star, HD 209621, was found to be C^{13}-rich by Climenhaga (b. 1916) (303); the metal abundances in this star followed the same pattern as before with the iron-group elements deficient, the rare earths (relative to iron) enhanced (304).

9.13 Symbiotic stars

The recognition of symbiotic stars as a class of variable stars with peculiar spectra dates from 1933 when Merrill described the spectra of AX Persei, RW Hydrae and CI Cygni (305). He noted that these stars 'belong to that interesting group of objects whose spectra exhibit bright lines of ionized helium in combination with dark bands of titanium oxide, features which ordinarily occur only near opposite ends of the sequence of stellar temperatures'. In addition to the lines of ionised helium (principally that at 4686 Å), other emission lines found were from neutral helium, hydrogen, doubly ionised nitrogen, ionised iron and several forbidden nebular lines including the green 'nebulium' doublet. Merrill remarked on the close similarity of these three spectra with that of Z Andromedae, in which emission lines were first detected at Harvard by Mrs Fleming (306).

Z Andromedae had already been studied in detail by H. H. Plaskett (307) in 1928 and had been commented on by Miss Cannon in her 1916 article on spectra having bright lines (52). Here she assigned the star the spectral type Ocp, i.e. a peculiar member of the Wolf–Rayet class Oc. Plaskett's discussion was far more extensive (307). He analysed 16 spectra from the 72-inch telescope at Victoria and measured the wavelengths of 111 emission lines. All but twelve were identified. He found two components; stellar lines from elements such as hydrogen, neutral helium, ionised magnesium, ionised iron and nebular lines including those of ionised helium, doubly ionised nitrogen and forbidden lines of ionised oxygen typical of gaseous nebulae. A small radial velocity shift between these two systems was measured and found to be probably variable. Plaskett found no evidence of any absorption features in Z Andromedae. However, he deduced that a fairly cool stellar object was present in the system from the colour temperature he measured from the slope of the continuum. His value of 5200 K posed a fundamental problem, if the nebular emission lines were to be produced by the fluorescent mechanism of H. Zanstra (1894–1972), which was known to operate in typical gaseous nebulae (308). This theory required that the ultraviolet radiation from a hot star (temperature about 33 000 K) was the origin of the nebula's excitation and ionisation.

The presence of a cool stellar component associated with Z Andromedae was confirmed by Frank Hogg (1904–1951) when he observed titanium oxide absorption bands in the spectrum (309). Merrill also claimed to have detected titanium oxide absorption a decade earlier, from the illustrations to Plaskett's article (310). However the bands were apparently of variable intensity. The brightness of this star shows nova-like outbursts followed by a more gradual decline. One such outburst occurred in 1939 when Struve and Elvey reported the P Cygni-type of emission line profiles and the absence of any underlying M-type absorption bands (311).

After Pol Swings returned for his second visit to Yerkes Observatory from his native Belgium in 1939, a remarkable series of articles on the

Fig. 9.16 Pol Swings.

spectra of peculiar stars appeared in the *Astrophysical Journal* as a result of his collaboration with Otto Struve. Seven of these papers recorded changes in the emission-line spectra of Z Andromedae and related stars, including AX Persei, CI Cygni and other nova-like stars (312). In particular, changes in the emission lines of Z And after the major outburst of 1939 and a lesser eruption in 1941 were followed in considerable detail. Both late-type absorption features and the highest excitation emission lines disappeared after an outburst, and many lines had P Cygni profiles characteristic of mass loss in an expanding shell. One of the surprises of this work was the discovery of emission lines in AX Per and CI Cyg arising from very high states of ionisation, including forbidden lines due to Fe^{6+} and Ne^{4+} and possibly even Fe^{9+}, implying a very high temperature source of ionising radiation. A detailed summary of all spectroscopic observations of Z Andromedae and of AX Persei was compiled by Cecilia Payne-Gaposchkin in 1946, including the results from several decades of Harvard objective prism observations (313).

Shortly after the detailed series of results by Swings and Struve, Merrill again returned to a discussion of the spectra of the four principal members (Z And, AX Per, CI Cyg, RW Hya) of this class (310). He now devised a general classification of stars showing both emission lines and late-type absorption spectra. Five groupings were defined, but in only one of these (group III) were the emission lines from gas of high excitation observed. He used the term 'combination' spectrum for this type of object, namely the combination of cool absorption features such as titanium oxide bands, and emission lines of ionised helium, oxygen, neon, iron and several other highly ionised species. Twelve stars were assigned to this group. With one exception they were all long-period, irregular or nova-like variables.

Merrill had first used the term 'combination spectrum' in 1940 (314) and was careful to distinguish this from 'composite spectrum' which implied the superposition of 'two more or less normal spectra of different types'. Later on he was to use the term 'symbiotic star' for one displaying a combination spectrum, derived from the biological terminology for the cooperative cohabitation of dissimilar species (315). Since the time of Merrill's lecture on 'Symbiosis in Astronomy' at the eighth Liège astrophysics colloquium in 1957, the term 'symbiotic star' has been generally favoured over 'combination spectrum' star (316).

The suggestion that symbiotic stars (as they are now known) are binaries was first made by Frank Hogg when discussing the spectrum of Z And at a meeting of the American Astronomical Society in 1933. In his published summary he wrote: 'The writer suggests that the system consists of a normal, possibly somewhat variable M giant and a variable very high temperature dwarf, . . . which excites a nebular envelope . . . The nebular shell may arise by ejection of matter in the nova-like outbursts of the star' (317). Direct evidence to support this binary hypothesis was not immediately forthcoming, even though it was a neat solution to the principal anomaly of these stars, namely the symbiosis itself. The binary nature of Z And was, however, implicit in the discussion of Swings and Struve (see their first paper of 1941 (312)).

Several other observers supported the binary hypothesis of Z Andromedae and other symbiotic stars over the next few years. This was the view of Marie Bloch (1902–79) and Tcheng Mao-Lin at the Lyon Observatory when they summarised the results of the spectrophotometry of Z And in 1951 (318). They also redetermined the colour temperature of the cool component, finding it to be 5070K. Lawrence Aller in particular carried out a thorough study of three symbiotic variables, at the time he visited Victoria from Michigan in 1951 (319). For Z And and CI Cyg he interpreted the line strengths of emission features based on Zanstra's method for determining the temperatures of central stars in planetary nebulae. His result based on the intensity of the ionised helium 4686 Å line was 130 000 K for the 'hot source' of CI Cygni, 90 000 K for that of Z Andromedae. Lower values were obtained using the Balmer lines. J. Gauzit also interpreted his data for AX Persei using a model consisting of a cool M star with a very hot companion, and a large gaseous envelope surrounded both stars (320).

A binary model for symbiotic stars was also supported after Kraft's discovery of the binary nature of the recurrent nova T Coronae Borealis (321). This also has an underlying M-type absorption spectrum and appears to be closely related to the symbiotic objects. T CrB proved to be a double-lined spectroscopic binary with a hot subluminous B star as the companion. However, the radial velocity variations for the classical symbiotic stars were not at first considered to arise from orbital motion. Merrill found periodicities of a few hundred days in several of these stars (316) but favoured the view that these velocity variations were related to the stars' intrinsic variability rather than orbital motion. The situation in 1965 was summarised in a review by J. Sahade, where the binary hypothesis is clearly favoured but not considered to be definitely proven (322).

Periodic Doppler displacements due to orbital motion were found for several symbiotic stars by A.A. Boyarchuk at the Crimean Astrophysical Observatory in 1966 and 1967. First he analysed the lines of AG Pegasi and derived a binary model based on a hot Wolf–Rayet star (type WN6) with a cool M3 giant, both stars being surrounded by a dense envelope of electron temperature 17 000 K. The period was 800 days (323).

A similar model with a very hot star of temperature 10^5K for the secondary component could account for Z Andromedae (324). In each case, broad high excitation emission lines of ionised helium, nitrogen (N^{2+} and N^{3+}) etc. came from the envelope of the hot star. Metallic emission lines such as ionised iron lines came from that part of the M star facing the hot companion, and their velocity varied in antiphase to those from the early-type star's envelope. Other lines, such as the green 'nebulium' lines came from the nebula surrounding both stars, while the Balmer lines could originate from all three sources. Radial velocity curves for four symbiotic stars (AG Peg, BF Cyg, RW Hya, R Aqr) were presented by Boyarchuk in 1968 (325). These all gave additional direct support to the binary hypothesis.

Several compilations of symbiotic and related stars have been published. Bidelman's catalogue of emission line stars of types later than B (326) included 23 stars with combination spectra, while Mrs Payne-Gaposchkin's treatise, *Galactic Novae*, tabulated thirty-two 'stars with combination spectra and related binaries' (327). Neither list was especially homogeneous. Boyarchuk (325) has attempted to tighten the definition of a symbiotic star: spectroscopically late-type absorption features of titanium oxide, neutral and ionised calcium etc, must be seen, and also narrow emission lines of singly ionised helium and oxygen or higher ionised species with a width not exceeding 100 km/s. He then lists twenty-one certain symbiotic stars, and a further sixteen candidates. A very extensive compilation of all known early-type emission line stars by Lloyd Wackerling in 1970 included thirty classified as symbiotic stars with the spectral type notation of Z (328).

9.14 The spectra of supernovae

9.14.1 A note on the paucity of bright supernovae Supernovae are exploding stars which for a brief period acquire luminosities higher than those of any other known stellar objects. The light curves are characterised by a very rapid increase to around absolute visual magnitude -18 followed by a slow decline of perhaps 5 magnitudes or more over the next year. In the last thousand years in our Galaxy, there have been three well-recorded cases of supernova explosions. The last was seen in 1604 by Kepler. In each case these 'new stars' became the brightest objects in the sky. The next galactic supernova will certainly be an object of intense spectroscopic study. However, all supernovae observed spectroscopically so far have been in galaxies external to our own Milky Way system. Only one has been a naked-eye object at maximum. This was S Andromedae in the Andromeda spiral galaxy. It reached apparent magnitude $m_v = 5.4$ in August 1885. Stellar spectrum photography was then still in its infancy, and all observations were made with visual spectroscopes. Since 1885, only five other supernovae up to 1965 were brighter than tenth apparent photographic magnitude, and none has been bright enough to observe at high dispersion. Only one of these five was observed regularly over an extended period following outburst (see section 9.14.3). For these reasons our knowledge of supernova spectra contains many gaps. The following summarises some of the few landmarks in the literature.

A list of 111 supernovae recorded from 1885 to 1962 has been compiled by Fritz Zwicky (329). Forty-two supernovae which were examined spectrographically up to 1972 appear in a tabulation by J. B. Oke and L. Searle (330). A nearly exhaustive bibliography of the entire supernova literature to 1967 has been compiled by M. Karpowicz and K. Rudnicki (b.1926) (331).

9.14.2 S Andromedae and Z Centauri S Andromedae in the spiral galaxy M31 was the brightest supernova observed since the advent of spectroscopy. It erupted about 17 August 1885 and was first observed spectroscopically on the night of 1 September. Fifteen observers reported visual observations during that month, but no photography of the spectrum was undertaken. If S And had been only 10 light years more distant* then spectrograms would very likely have been secured and our knowledge of supernovae might have taken quite a different early course.

S Andromedae was observed intensively throughout September 1885 with a few observers continuing their observations through October to early November when the object was down to ninth magnitude. The earliest reports on 1 September by C.A. Young (332), H.C. Vogel (333) and R. Copeland (334) all indicated a continuous spectrum, although Copeland remarked: 'only on close examination could slight condensations indicative of bright lines be detected'.

By 3 September several observers reported having possibly seen lines. Thus Huggins wrote: 'There was an apparent condensation from about D to b [the letters referring only to positions in the spectrum], which might be due to bright lines in that part of the spectrum.... I was not able to be certain on this point' (335). The report of N. von Konkoly on 4 September was the first to be explicit on the lines: 'One has the definite impression that one is seeing bright fields on a dark background and indeed such a bright band is to be seen in the red, the yellow, the green and the blue. If this is the case, then these bright fields would correspond to the H-lines C and F as well as to D_3, and in fact at an enormously high pressure' (336). He suggested that the spectrum may be that of a carbon star (Vogel type IIIb) but Vogel himself was quick to discount such an idea (337).

Several other observers probably saw very broad lines over the next week or so. E.W. Maunder at Greenwich described the spectrum on 11 September, stating the brightest part to be at 5482 Å 'where at times the brightening appeared so definite as to suggest the presence of a bright line' (338). This statement is typical of several reports over the next few days. By mid-September eight observers were fairly certain that they had seen very diffuse lines while a further five may have seen them, though never with certainty. Only Young and von Gothard never reported lines, though the latter admitted that lines could easily have been missed. One of the last reports, by T.W. Backhouse on 5 November, was among the most positive for lines: 'There can be no doubt that the spectrum of the new star is highly interrupted... [and] contains more than one definite bright line' (339).

Cecilia Payne-Gaposchkin reviewed all the visual observations of S Andromedae in 1936 (340). The consensus of opinion is that the star had an orange-red colour, and that its spectrum had very broad diffuse emission

* The distance to the M31 galaxy is about 2 million light years.

bands which may not have been present until about 4 September of that year.

The first supernova spectrum ever to be photographed was that of Z Centauri which is still the second brightest extragalactic supernova ever recorded ($m_{pg}(\max) = 8.0$). This object was in the irregular galaxy (NGC 5253) and discovered by Mrs Fleming on a Harvard objective prism plate in July 1895, as reported by E.C. Pickering (341) who believed it to be of spectral type R. According to Zwicky, this result caused much confusion until it was eventually discredited in 1936 (342). It is similar to von Konkoly's classification of S And, and presumably arose from the impression of broad diffuse bands in the spectrum of a star of reddish colour. It was only the re-examination of the original spectrum by W. A. Johnson (343) (Fig. 9.17) and by Cecilia Payne-Gaposchkin (344) in 1936 that resolved the true nature of the star as one with broad diffuse emission lines. The only other observer of Z Cen was W.W. Campbell at Lick with a visual spectroscope. He reported 'some evidence of bright lines…but the light was too weak to enable me to decide' (345).

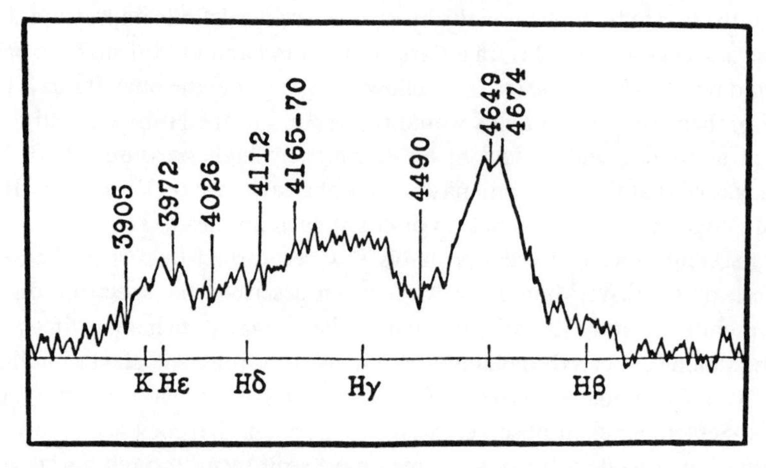

Fig. 9.17 Tracing of the Harvard spectrum of Z Centauri, the first supernova spectrum to be recorded.

9.14.3 Supernova spectroscopy to 1937 In the four decades 1896–1936 only three further supernovae were observed spectroscopically; in all cases the new observational data were sparse. Pease and Ritchey observed a faint supernova in a spiral galaxy, NGC 6946, in 1917 and obtained an objective prism plate: 'The continuous spectrum is strong and is crossed by what appears to be a series of bright bands' (346). In 1926 another faint supernova in a spiral system was observed both at Lick (347) by Shane and at Mt Wilson (348) by J.C. Duncan and S.B. Nicholson.

Four slitless spectrograms showed the same diffuse emission bands as before. Supernova 1936a, although it only reached photographic magnitude fourteen at maximum, was nevertheless the first to be observed with a slit spectrograph, by M.L. Humason (349) and W. Baade (1893–1960) (350) on the 100-inch Hooker telescope. The presence of a broad blue emission band at 4640 Å was confirmed by Baade as the dominant feature.

Fritz Zwicky (1898–1974) commented on the slow progress of supernova research in these years: 'This rather surprising lack of interest may perhaps be ascribed to a failure to realize the importance of supernovae, to occupation with other important investigations, to the uncertainty about the frequency of occurrence of these giant explosions and to the unavailability of suitable fast wide angle telescopes' (342). The absence of any known twentieth century supernovae brighter than magnitude eleven at maximum prior to 1937 must also have been a contributing factor.

Although only three supernovae were recorded spectroscopically in this period, Zwicky lists nineteen such outbursts in the same interval (342). The realisation that these stars occur in external galaxies (mainly spiral 'nebulae') necessitated assigning enormous luminosities once it was accepted that the spirals were outside our galactic system. Shapley estimated an absolute magnitude of -15 for S Andromedae in 1917 (351) if M31 were assumed to be an external star system, a figure which at that time seemed impossible, since it corresponded to about 10^8 times the solar luminosity. K. Lundmark (1889–1958) in Sweden, who with Hubble established the extragalactic nature of the spirals in 1925, was able to distinguish clearly between supernovae and novae (352). He used the terms 'upper-class' and 'lower-class' novae, the former being much rarer but some ten thousand times more luminous. Baade and Zwicky in 1934 introduced the term supernova (353) in place of 'upper-class nova'. The distinction was based on luminosity. It was not until 1938 that Baade pointed out: 'The totally different character of the spectra of common novae and supernovae puts it beyond doubt that we are dealing with two different classes of objects...' (354).

From 1933 to 1942 Zwicky at Caltech conducted a photographic supernova search using fast cameras, and as a result many new objects were found from about 1937, including two bright supernovae in 1937 (SN 1937c in galaxy IC 4182, and 1937d in NGC 1003). The first of these reached magnitude 8.2 and was the brightest supernova since Z Centauri in 1895. Both were observed spectroscopically by Rudolph Minkowski (1895–1976) who must be regarded as the greatest pioneer of supernova spectroscopy (355). His long series of spectra of these two objects, in the case of SN 1937c in IC 4182 35 spectra extending for 11 months from maximum, was the first systematic observing program to record in detail the changes in a supernova spectrum over such a long time interval. Minkowski divided the spectrum into two parts, since the behaviour above and below 5000 Å was quite different. Below 5000 Å in the blue several very broad intense emission bands showed relatively few spectral changes, but a curious progressive redwards shift of about 100 Å during the year of observations. The strongest blue band was at about 4650 Å. On the other hand in the yellow and red spectral region bands came and went with greater frequency, the spectrum was generally more complex and no progressive redshift was found. Minkowski concluded that all five earlier supernovae observed spectroscopically, with the possible exception of S Andromedae, showed closely similar spectroscopic behaviour.

The origin of the emission features completely defied identification and this has remained a considerable mystery ever since. However, Minkowski pointed out that the absence of hydrogen and helium lines was probably due to a very high degree of ionisation.

9.14.4 The classification of supernova spectra Primarily as a result of Zwicky's supernova search, the number of supernovae with spectroscopic observations increased considerably from 1937. At the beginning of that year there were only five supernovae studied spectroscopically; 5 years later there were at least seventeen. This sudden increase in data allowed Minkowski to place supernovae into two groups in 1941, which he introduced as follows:

> Spectroscopic observations indicate at least two types of supernovae. Nine objects... [these included SN1937c in IC4182] form an extremely homogeneous group provisionally called type I. The remaining five objects...are distinctly different; they are provisionally designated as type II. The individual differences in this group are large... The spectrograms of all objects of type I are closely comparable at corresponding times after maxima. Even at the earliest premaximum phase hitherto observed, the spectrum consists of very wide emission bands.... Spectra of type II have been observed from maximum until 115 days after. Up to about a week after maximum, the spectrum is continuous and extends far into the ultraviolet, indicating a very high color temperature. Faint emission is suspected near Hα. Thereafter, the continuous spectrum fades and becomes redder. Simultaneously, absorptions and broad emission bands are developed. The spectrum as a whole resembles that of a normal nova in the transition stage, although the hydrogen bands are relatively faint and forbidden lines are either

extremely faint or missing.... Supernovae of type II differ from those of type I in the presence of a continuous spectrum at maximum and in the subsequent transformation to an emission spectrum whose main constituents can be readily identified (356).

This passage thus clearly defined two quite distinct classes of supernova; although type II supernovae were not clearly identified until this time, their frequency of occurrence and luminosity at maximum are comparable to those of type I. Both the 1926 and 1936 supernovae were probably of this type, while Minkowski cited the 1940 supernova in NGC 4725 as a typical example. Although both the spectra and the light curves of type I form a very homogeneous group, the physical nature of these objects has proved much harder to decipher. On the other hand the most prominent lines in type II objects are due to the Balmer series and to doubly ionised nitrogen. Hubble described type II supernovae as 'gigantic ordinary novae' although any similarities are probably illusory (357). Although published spectroscopic data for fourteen type II supernovae from 1926 to 1972 are listed by Oke and Searle (330) none of these objects has an extensive series of observations comparable to Minkowski's type I supernova of 1937.

Zwicky reinitiated a collaborative supernova search from 1956 and many faint supernovae were discovered from that time, especially by M.L. Humason (358). As a result of spectroscopic data for newly discovered objects, Zwicky defined three further classes of supernova, III, IV and V (359) in addition to Minkowski's types I and II. However, the three new types were defined partly by reference to the light curve and mainly on the basis of only one definite member per type. Oke and Searle (330) could find no evidence to support more than the original two types of Minkowski, based primarily on the absence or presence of hydrogen lines. These authors also discussed spectroscopic observations of the many supernovae discovered since 1956. The type I supernova 1972e in NGC 5253 ($m_{pg} = 8.5$ at maximum) was the only bright object in their table with an extensive series of observations.

References

1. Cannon, A.J., *Harvard Ann.*, **28** (part II), 131 (1901).
2. Campbell, W.W., *Astron. and Astrophys.*, **13**, 448 (1894).
3. Beals, C.S., *Mon. Not. Roy. Astron. Soc.*, **90**, 202 (1929).
4. Wilson, O.C., *Astrophys. J.*, **95**, 402 (1942).
5. Halm, J., *Proc. Roy. Soc. Edinburgh*, **25**, 513 (1904).
6. Payne, C.H., *Harvard Bull.*, **855** (1928).

7. Beals, C.S., *Publ. Dominion Astrophys. Observ. Victoria*, **4** (no. 17), 269 (1930).
8. Edlén, B., *Observatory*, **55**, 115 (1932).
9. Payne, C.H., *Mon. Not. Roy. Astron. Soc.*, **92**, 368 (1932). See also *Zeitschrift für Astrophys.*, **7**, 1 (1933) and *Proc. Nat. Academy of Sci.*, **19**, 492 (1933).
10. Beals, C.S., *Observatory*, **56**, 196 (1933).
11. Perrine, C.D., *Mon. Not. Roy. Astron. Soc.*, **81**, 142 (1920).
12. Beals, C.S., *Publ. Dominion Astrophys. Observ. Victoria*, **6** (no. 9), 95 (1934).
13. Beals, C.S. and Plaskett, H.H., *Trans. I.A.U.*, **5**, 178 (1936).
14. Payne, C.H., *Trans. I.A.U.*, **5**, 178 (1936).
15. Beals, C.S., *Trans. I.A.U.*, **6**, 248 (1939).
16. Swings, P., *Astrophys. J.*, **95**, 112 (1942).
17. Hiltner, W.A. and Schild, R.E., *Astrophys. J.*, **143**, 770 (1966).
18. Huggins, W., *Mon. Not. Roy. Astron. Soc.*, **26**, 275 (1866).
19. Stone, E.J., *Mon. Not. Roy. Astron. Soc.*, **26**, 292 (1866).
20. Wolf, C. and Rayet, G., *Comptes Rendus de l'Académie des Sciences*, **62**, 1108 (1866).
21. Cornu, M.A., *Comptes Rendus de l'Académie des Sciences*, **83**, 1172 (1876).
22. Vogel, H.C., *Astron. Nachrichten*, **89**, 37 (1877).
23. Campbell, W.W., *Astron. and Astrophys.*, **11**, 799 (1892).
24. Huggins, W., *Astron. and Astrophys.*, **11**, 571 (1892).
25. Huggins, W. and Miller, W.A., *Proc. Roy. Soc.*, **15**, 146 (1867).
26. Sanford, R.F., *Publ. Astron. Soc. Pacific*, **59**, 87 and 344 (1947).
27. McLaughlin, D.B., *Publ. Astron. Soc. Pacific*, **58**, 159 (1946).
28. Lockyer, J.N., *Phil. Trans. Roy. Soc.*, **182A**, 424 and 439 (1891).
29. Adams, W.S. and Joy, A.H., *Publ. Astron. Soc. Pacific*, **45**, 249 and 301 (1933).
30. McLaughlin, D.B., *Publ. Astron. Observ. Uni. Michigan*, **9**, 13 (1949).
31. Stratton, F.J.M., *Publ. Solar Phys. Observ. Cambridge*, **4** (part 1) (1920).
32. Wright, W.H., *Publ. Lick Observ.*, **14** (part 2) (1925).
33. Wyse, A.B., *Publ. Lick Observ.*, **14** (part 3) (1940).
34. Baldwin, R.B., *Publ. Astron. Observ. Uni. Michigan*, **8**, 61 (1940).
35. Spencer Jones H., *Cape Observ. Ann.*, **10** (part 9) (1931).
36. Payne-Gaposchkin, C.H. and Menzel, D.H., *Harvard Observ. Circular*, **428** (1938).
37. Stratton, F.J.M., *Publ. Solar Phys. Observ. Cambridge*, **4**, 133 (1936).
38. McLaughlin, D.B., *Publ. Astron. Observ. Uni. Michigan*, **6**, 107 (1937).
39. Stratton, F.J.M. and Manning, W.H., *An Atlas of Nova Herculis 1934*, Publ. Solar Physics Observ., Cambridge, England (1939).
40. Harper, W.E., Pearce, J.A., Beals, C.S., Petrie, R.M. and McKellar, A., *Publ. Dominion Astrophys. Observ. Victoria*, **6**, 317 (1937).
41. Wyse, A.B., *Publ. Astron. Soc. Pacific*, **49**, 290 (1937).
42. McLaughlin, D.B., *Publ. Amer. Astron. Soc.*, **8**, 252 (1936).
43. Joy, A.H., Adams, W.S. and Dunham, T., *Publ. Astron. Soc. Pacific*, **48**, 328 (1936).
44. Sanford, R.F., *Astrophys. J.*, **102**, 357 (1945).
45. McLaughlin, D.B., *Publ. Amer. Astron. Soc.*, **10**, 310 (1943).

46. Weaver, H.F., *Astrophys. J.*, **99**, 280 (1944).
47. Larsson-Leander, G., *Stockholm Observ. Ann.*, **18** (no.3 and 4) (1954).
48. Meinel, A.B., *Astrophys. J.*, **137**, 834 (1963).
49. Saweljewa, M.W., *Astron. J.*, (USSR) **44**, 716 (1967).
50. Friedjung, M. and Smith, M.G., *Mon. Not. Roy. Astron. Soc.*, **132**, 239 (1966).
51. Doroschenko, W.T., *Astron. J.*, (USSR) **45**, 121 (1968).
52. Cannon, A.J., *Harvard Ann.*, **76** (no. 19), 3 (1916).
53. Pickering, E.C., *Harvard Ann.*, **27**, 1 (1890).
54. Lockyer, J.N., *Publication of the Solar Physics Committee*, 'Phenomena of New Stars', p. 5 (1914).
55. Newall, H.F., *Trans. I.A.U.*, **2**, 117 (1925).
56. Adams, W.S., *Trans. I.A.U.*, **1**, 95 (1922). Also ibid. *Astrophys. J.*, **57**, 65 (1923).
57. Russell, H.N., *Trans. I.A.U.*, **5**, 178 (1936).
58. Payne-Gaposchkin, C.H., *The Galactic Novae*, publ. North Holland Publishing Co., Amsterdam (1957).
59. McLaughlin, D.B., *Popular Astron.*, **46**, 373 (1938).
60. McLaughlin, D.B., *Publ. Astron. Observ. University of Michigan*, **8** (no. 12), 149 (1944).
61. Stratton, F.J.M., *Trans. I.A.U.*, **7**, 305 (1950).
62. Adams, W.S., *Trans. I.A.U.*, **1**, 100 (1922).
63. Secchi, A., *Comptes Rendus de l'Académie des Sciences*, **63**, 621 (1866).
64. Vogel, H.C., *Beobachtungen angestellt auf der Sternwarte des Kammerherrn von Bülow zu Bothkamp*, **2**, 29 (1873).
65. Secchi, A., *Sugli spettri prismatici de' Corpi Celesti*: Memoria del P. Angelo Secchi, Roma (1872).
66. Lockyer, J.N., *Proc. Roy. Soc.*, **57**, 173 (1895).
67. Campbell, W.W., *Astrophys. J.*, **2**, 177 (1895).
68. Fleming, W.P., *Harvard Ann.*, **56** (no. 6), 165 (1912).
69. Maury, A.C., *Harvard Ann.*, **28** (part I), 1 (1897). See p. 100 for γ Cas.
70. Scheiner, J.S., *Spectralanalyse der Gestirne*, published by W. Engelmann, Leipzig, p. 276 (1890). See also: Scheiner, J.S., *Astron. Nachrichten*, **122**, 320 (1889).
71. Merrill, P.W., *Lick Observ. Bull.*, **7** (no. 237), 162 (1913).
72. Merrill, P.W., Humason, M.L. and Burwell, C.G., *Astrophys. J.*, **61**, 389 (1925).
73. Merrill, P.W., Humason, M.L. and Burwell, C.G., *Astrophys. J.*, **76**, 156 (1932) (Paper II).
74. Merrill, P.W. and Burwell, C.G., *Astrophys. J.*, **78**, 87 (1933); ibid. **98**, 153 (1943); ibid. **110**, 387 (1949).
75. Merrill, P.W. and Burwell, C.G., *Astrophys. J.*, **112**, 72 (1950).
76. Henize, K.G., 'The Michigan-Mt. Wilson survey of the southern sky for Hα emission stars and nebulae', Ph.D. thesis, Uni. of Michigan (1955). See also Henize, K.G., *Astrophys. J. Suppl.*, **30**, 491 (1976).
77. Struve, O., *Astrophys. J.*, **73**, 94 (1931).
78. Morgan, W.W., *Astron. J.*, **51**, 21 (1944).
79. Slettebak, A., *Astrophys. J.*, **110**, 498 (1949).
80. Kourganoff, V., *Liège Astrophysics Symposium*, **20**, 117 (1958).
81. Curtiss, R.H., *Publ. Astron. Observ. Uni. Michigan*, **2**, 1 (1916).
82. Curtiss, R.H., *Publ. Astron. Observ. Uni. Michigan*, **2**, 36 (1916).

83. Curtiss, R.H., *Publ. Astron. Observ. Uni. Michigan*, **2**, 39 (1916).
84. Curtiss, R.H., *Publ. Astron. Observ. Uni. Michigan*, **4**, 163 (1932).
85. McLaughlin, D.B., *Publ. Astron. Observ. Uni. Michigan*, **4**, 175 (1932). See also: ibid. *Astrophys. J.*, **85**, 181 (1937).
86. Lockyer, W.J.S., *Mon. Not. Roy. Astron. Soc.*, **93**, 362 and 619 (1933).
87. Merrill, P.W., *Publ. Astron. Soc. Pacific*, **61**, 38 (1949).
88. Merrill, P.W. and Burwell, C.G., *Astrophys. J.*, **110**, 387 (1949).
89. Curtiss, R.H., *Publ. Astron. Observ. Uni. Michigan*, **3**, 16 (1923).
90. McLaughlin, D.B., *Astrophys. J.*, **88**, 622 (1938).
91. Struve, O. and Swings, P., *Astrophys. J.*, **97**, 426 (1943).
92. Merrill, P.W., *Astrophys. J.*, **115**, 145 (1952).
93. Morgan, W.W., *Publ. Yerkes Observ.*, **7** (part 3), 133 (1935).
94. Morgan, W.W., *Astrophys. J.*, **77**, 330 (1933).
95. Maury, A.C. and Pickering, E.C., *Harvard Ann.*, **28** (part I), 1 (1897). See p. 96, remark 69 on α^2 CVn.
96. Lockyer, J.N. and Baxandall, F.E., *Proc. Roy. Soc.*, **77A**, 550 (1906).
97. Baxandall, F.E., *Mon. Not. Roy. Astron. Soc.*, **74**, 250 (1914).
98. Baxandall, F.E., *Observatory*, **36**, 440 (1913) and ibid. *Mon. Not. Roy. Astron. Soc.*, **74**, 32 (1914).
99. Lunt, J., *Proc. Roy. Soc.*, **79A**, 118 (1907).
100. Morgan, W.W., *Astrophys. J.*, **75**, 46 (1932).
101. Ludendorff, F.W.H., *Astron. Nachrichten*, **173**, 1 (1906).
102. Belopolsky, A., *Astron. Nachrichten*, **196**, 1 (1913).
103. Guthnick, P. and Prager, R., *Veröffentlichungen der königlichen Sternwarte zu Berlin-Babelsberg*, **1**, 1 (1914).
104. Cannon, A.J., *Harvard Ann.*, **56**, 65 (1912); ibid. **56**, 115 (1912).
105. Cannon, A.J. and Pickering, E.C., *Harvard Ann.*, **91** to **99** (1918–24). The Henry Draper Catalogue.
106. Kiess, C.C., *Popular Astron.*, **25**, 656 (1917); ibid. *Publ. Astron. Observ. Uni. Michigan*, **3**, 106 (1919).
107. Shapley, H., *Harvard Bull.*, **798**, 2 (1924).
108. Payne, C.H., 'Stellar Atmospheres', *Harvard Observ. Monographs*, No. 1, p. 172 (1925).
109. Morgan, W.W., *Astrophys. J.*, **73**, 104 (1931).
110. Morgan, W.W., *Astrophys. J.*, **74**, 24 (1931).
111. Morgan, W.W., *Astrophys. J.*, **76**, 275 (1932).
112. Morgan, W.W., *Astrophys. J.*, **76**, 315 (1932); ibid. **77**, 77 (1933).
113. Bidelman, W.P., *Astrophys. J.*, **135**, 651 (1962).
114. Zeeman, P., *Zittingsverlagen der Akad. v. Wet. te Amsterdam*, **5**, 181 (1897). Accounts of this work in English are to be found in: *Phil. Mag.*, **43**, 226 (1897) and *Astrophys. J.*, **5**, 332 (1897).
115. Wright, W.H., *Lick Observ. Bull.*, **6**, 60 (1910).
116. Merrill, P.W., *Lick Observ. Bull.*, **7**, 162 (1913).
117. Minnaert, M.G.J., *Observatory*, **60**, 292 (1937).
118. Babcock, H.W., *Astrophys. J.*, **105**, 105 (1947).
119. Babcock, H.W., *Publ. Astron. Soc. Pacific*, **59**, 260 (1947). See also: ibid. *Astrophys. J.*, **114**, 1 (1951).
120. Babcock, H.W. and Burd, Sylvia, *Astrophys. J.*, **116**, 8 (1952).
121. Struve, O., *Trans. I.A.U.*, **7**, 287 (1950).
122. Gollnow, H., *Publ. Astron. Soc. Pacific*, **74**, 163 (1962).

123. Preston, G.W. and Pyper, D.M., *Astrophys. J.*, **142**, 983 (1965).
124. Babcock, H.W., *Astrophys. J. Suppl.*, **3**, 141 (1958).
125. Babcock, H.W., *Astrophys. J.*, **132**, 521 (1960).
126. Babcock, H.W., *Observatory*, **69**, 191 (1949).
127. Stibbs, D.W.N., *Mon. Not. Roy. Astron. Soc.*, **110**, 395 (1950).
128. Babcock, H.W., *Astrophys. J.*, **114**, 1 (1951).
129. Preston, G.W., *Publ. Astron. Soc. Pacific*, **83**, 571 (1971).
130. Deutsch, A.J., Ph.D. thesis, University of Chicago (1946), published in *Astrophys. J.*, **105**, 283 (1947).
131. Deutsch, A.J., *Trans. I.A.U.*, **8**, 801 (1952).
132. Struve, O. and Swings, P., *Astrophys. J.*, **98**, 361 (1943).
133. Burbidge, G.R. and Burbidge, E.M., *Astrophys. J. Suppl.*, **1**, 431 (1955).
134. Deutsch, A.J., *Publ. Astron. Soc. Pacific*, **68**, 92 (1956).
135. Pyper, D.M., *Astrophys. J. Suppl.*, **18**, 347 (1969).
136. Jaschek, M. and Jaschek, C., *Zeitschrift für Astrophys.*, **45**, 35 (1958).
137. Bertaud, C., *J. des Observateurs*, **42**, 45 (1959); ibid. **43**, 129 (1960).
138. Osawa, K., *Ann. Tokyo Astron. Observ.*, **9** (ser. 2), 123 (1965).
139. Bidelman, W.P., reported by A.E. Whitford in *Astron. J.*, **67**, 640 (1962). See p. 645.
140. Bidelman, W.P., *Publ. Astron. Soc. Pacific*, **72**, 471 (1960).
141. Aller, L.H. and Bidelman, W.P., *Astron. J.*, **67**, 571 (1962); ibid., *Astrophys. J.*, **139**, 171 (1964).
142. Garrison, R.F., *Astrophys. J.*, **147**, 1003 (1967).
143. Bernacca, P.L., *Contr. Osserv. Astrofis. Asiago,* no. 202 (1968).
144. Berger, J., *Contr. de l'Institut d'Astrophys., Paris*, sér. A, no. 217 (1956).
145. Stewart, J.C., *Astron. J.*, **61**, 13 (1956).
146. Hiltner, W.A., Garrison, R.F. and Schild, R.E., *Astrophys. J.*, **157**, 313 (1969).
147. MacConnell, D.J., Frye, R.L. and Bidelman, W.P., *Publ. Astron. Soc. Pacific*, **82**, 730 (1970).
148. Morgan, W.W., Keenan, P.C. and Kellman, E., *An Atlas of Stellar Spectra*, Univ. of Chicago (1943).
149. Slettebak, A., *Astrophys. J.*, **115**, 575 (1952).
150. Slettebak, A., *Astrophys. J.*,**119**, 146 (1954).
151. Parenago, P.P., *Soviet Astron.—A.J.*, **2**, 151 (1958).
152. Sargent, W.L.W., *Astrophys. J.*, **142**, 787 (1965) and ibid., *Magnetic and Related Stars*, ed. R.C. Cameron, p. 329 (1967).
153. Oke, J.B, *Astrophys. J.*, **150**, 513 (1967).
154. Burbidge, E.M. and Burbidge, G.R., *Astrophys. J.*, **124**, 116 (1956).
155. Slettebak, A., *Astrophys. J.*, **138**, 118 (1963).
156. Sargent, W.L.W., *Astrophys. J.*, **144**, 1128 (1966).
157. Baschek, B. and Searle, L., *Astrophys. J.*, **155**, 537 (1969).
158. Vogel, H.C. and Wilsing, J., *Publ. Potsdam Astrophys. Observ.*, **12** (no. 39), 1 (1899).
159. Titus, J. and Morgan, W.W., *Astrophys. J.*, **92**, 256 (1940).
160. Roman, N.G., Morgan, W.W. and Eggen, O.J., *Astrophys. J.*, **107**, 107 (1948).
161. Morgan, W.W. and Bidelman, W.P., *Astrophys. J.*, **104**, 245 (1946).

162. Weaver, H.F., *Publ. Astron. Soc. Pacific*, **58**, 246 (1946).
163. Greenstein, J.L., *Astrophys. J.*, **107**, 151 (1948).
164. Greenstein, J.L., *Astrophys. J.*, **109**, 121 (1949).
165. Miczaika, G.R., Franklin, F.A., Deutsch, A.J. and Greenstein, J.L., *Astrophys. J.*, **124**, 134 (1956).
166. Conti, P.S., *Astrophys. J.*, **156**, 661 (1969).
167. Böhm-Vitense, E., *Zeitschrift für Astrophys.*, **49**, 243 (1960).
168. Weaver, H.F., *Astron. J.*, **55**, 82 (1950).
169. Jaschek-Corvalan, M. and Jaschek, C.O.R., *Astron. J.*, **62**, 343 (1957).
170. Slettebak, A., *Astrophys. J.*, **121**, 653 (1955).
171. Abt, H.A., *Astrophys. J. Suppl.*, **6**, 37 (1961).
172. Abt, H.A., *Astrophys. J. Suppl.*, **11**, 429 (1965).
173. Abt, H.A. and Bidelman, W.P., *Astrophys. J.*, **158**, 1091 (1969).
174. Van't Veer-Menneret, C., *Ann. d'Astrophys.*, **26**, 289 (1963).
175. Conti, P.S., *Astrophys. J. Suppl.*, **11**, 47 (1965).
176. Conti, P.S., *Publ. Astron. Soc. Pacific*, **82**, 781 (1970).
177. Chaffee, F.H., *Astron. and Astrophys.*, **4**, 291 (1970).
178. Breger, M., *Astrophys. J. Suppl.*, **19**, 79 (1969).
179. Praderie, F., Thesis, Université de Paris (1967).
180. Michaud, G., *Astrophys. J.*, **160**, 641 (1970).
181. Bessel, F.W., *Astron. Nachrichten*, **22**, 145 (1844).
182. Clark, A.G., reported by G.P. Bond in *Astron. Nachrichten*, **57**, 131 (1862) and in *Mon. Not. Roy. Astron. Soc.*, **22**, 170 (1862).
183. Schaeberle, J.M., *Astron. Nachrichten*, **143**, 25 (1897).
184. Russell, H.N., *Publ. Amer. Astron. Soc.*, **3**, 22 (1918); see also ibid., *Popular Astron.*, **22**, 275 (1914).
185. Russell, H.N., *Astron J.*, **51**, 13 (1944).
186. Russell, H.N., 'Les novae et les naines blanches': III, Naines blanches. *Actualités scientifiques* No. 897, ed. A.J. Schaler, publ. by Hermann, Paris (1941).
187. Hertzsprung, E., *Astrophys. J.* **42**, 116 (1915).
188. Adams, W.S., *Publ. Astron. Soc. Pacific*, **26**, 198 (1914).
189. Adams, W.S., *Publ. Astron. Soc. Pacific*, **27**, 236 (1915).
190. Luyten, W.J., *Vistas in Astron.*, **2**, 1048 (1956).
191. van Maanen, A., *Publ. Astron. Soc. Pacific*, **29**, 258 (1917).
192. Adams, W.S. and Joy, A.H., *Publ. Astron. Soc. Pacific*, **38**, 121 (1926).
193. Lindblad, B., *Astrophys. J.*, **55**, 85 (1922).
194. Eddington, A.S., *Mon, Not. Roy. Astron. Soc.*, **84**, 308 (1924).
195. Fowler, R.H., *Mon. Not. Roy. Astron. Soc.*, **87**, 114 (1927).
196. Chandrasekhar, S., *Astrophys. J.*, **74**, 81 (1931); ibid., *Mon. Not. Roy. Astron. Soc.*, **91**, 456 (1931); ibid., **95**, 207 (1935).
197. Adams, W.S., *Proc. Nat. Acad. Sci.*, **11**, 382 (1925).
198. Moore, J.H., *Publ. Astron. Soc. Pacific*, **40**, 229 (1928).
199. Popper, D.M., *Astrophys. J.*, **120**, 316 (1954).
200. Greenstein, J.L. and Trimble, V.L., *Astrophys. J.*, **149**, 283 (1967).
201. Russell, H.N. and Atkinson, R. d'E., *Nature*, **127**, 660 (1931).
202. Joy, A.H., *Astrophys. J.*, **63**, 281 (1926).
203. Oosterhoff, P.Th., *Bull. Astron. Inst. Netherlands*, **6**, 39 (1930).
204. Öhman, Y., *Mon. Not. Roy. Astron. Soc.*, **92**, 71 (1931).
205. Kuiper, G.P., *Publ. Astron. Soc. Pacific*, **46**, 287 (1934).

206. Kuiper, G.P., *Publ. Astron. Soc. Pacific*, **47**, 279 (1935); ibid. *Astrophys. J.*, **87**, 592 (1938), *Astrophys. J.*, **89**, 548 (1939), *Astrophys. J.*, **91**, 269 (1940).
207. Kuiper, G.P., *Publ. Astron. Soc. Pacific*, **53**, 248 (1941).
208. Öhman, Y., *Stockholms Observ. Ann.*, **12**, 1 (1935).
209. Verweij, S., *Publ. Astron. Inst. of Amsterdam*, no. 5, p. 36 (1936).
210. Luyten, W.J., *Publ. Astron. Soc. Pacific*, **34**, 54 (1922).
211. Luyten, W.J., *Harvard Announcement Cards nos.*, **521**, **527**, **528** and **558** (1940–1).
212. Luyten, W.J., *Astrophys. J.*, **109**, 528 (1949).
213. Luyten, W.J., *American Scientist*, **48** (no. 1), 30 (1960).
214. Humason, M.L. and Zwicky, F., *Astrophys. J.*, **105**, 85 (1947).
215. Luyten, W.J., *Astron. J.*, **58**, 75 (1952).
216. Luyten, W.J., *White Dwarfs*, University of Minnesota, Minneapolis (1970).
217. Luyten, W.J., *Vistas in Astron.*, **2**, 1048 (1956).
218. Luyten, W.J., *Trans. I.A.U.*, **7**, 285 (1950). See p. 303.
219. Kuiper, G.P., *Astrophys. J.*, **89**, 548 (1939).
220. Luyten, W.J., *Astrophys. J.*, **116**, 283 (1952).
221. Morgan, W.W., quoted by Luyten, see reference (220).
222. Lynds, B.T., *Astrophys. J.*, **125**, 719 (1957).
223. Minkowski, R., reported by W.S. Adams, *Ann. Report of the Director of Mt. Wilson Observatory 1937–8*, p. 28 (1938).
224. Greenstein, J.L. and Matthews, Mildred S., *Astrophys. J.*, **126**, 14 (1957).
225. Eggen, O.J. and Greenstein, J.L., *Astrophys. J.*, **141**, 83 (1965); ibid., **142**, 925 (1965); ibid., **150**, 927 (1965).
226. Greenstein, J.L., *Handbuch der Physik*, **50**, 161 (1958).
227. Greenstein, J.L., *Stars and Stellar Systems*, vol. 6 (*Stellar Atmospheres*), edited J.L. Greenstein, chapter 19, p. 676 (1960).
228. Weidemann, V., *Zeitschrift für Astrophys.*, **57**, 87 (1963).
229. Weidemann, V., *Astrophys. J.*, **131**, 638 (1960).
230. Fleming, W.P., reported by A.J. Cannon, *Harvard Ann.*, **56**, 65 (1912), see p. 108.
231. Campbell, W.W., *Astrophys. J.*, **2**, 77 (1895).
232. Campbell, W.W., *Astrophys. J.*, **10**, 241 (1899).
233. Plaskett, J.S., *Publ. Dominion Astrophys. Observ. Victoria*, **4**, 111 (1928).
234. Wilson, R.E., *Lick Observ. Bull.*, **8**, 132 (1914).
235. Greenstein, J.L., *American Institute of Physics, Oral History Interview by P. Wright*, p. 17 (1974).
236. Greenstein, J.L., *Astrophys. J.*, **91**, 438 (1940).
237. Greenstein, J.L., *Astrophys. J.*, **97**, 252 (1943).
238. Greenstein, J.L. and Adams, W.S., *Ap.J.*, **106**, 339 (1947).
239. Merrill, P.W., *Publ. Astron. Soc. Pacific*, **56**, 42 (1944).
240. Greenstein, J.L. and Merrill, P.W., *Astrophys. J.*, **104**, 177 (1946).
241. Bidelman, W.P., *Astrophys. J.*, **109**, 544 (1949).
242. Bidelman, W.P., *Astrophys. J.*, **111**, 333 (1950).
243. Popper, D.M., *Publ. Astron. Soc. Pacific*, **54**, 160 (1942); ibid., **58**, 370 (1946); ibid., **59**, 320 (1947).

244. Bidelman, W.P., *Astrophys. J.*, **116**, 227 (1952).
245. Thackeray, A.D. and Wesselink, A.J., *Observatory*, **72**, 248 (1952).
246. Thackeray, A.D., *Mon. Not. Roy. Astron. Soc.*, **114**, 93 (1954).
247. Jaschek, M. and Jaschek, C., *Publ. Astron. Soc. Pacific*, **71**, 465 (1959).
248. Klemola, A.R., *Astrophys. J.*, **134**, 130 (1961).
249. Hill, P.W., *Mon. Not. Roy. Astron. Soc.*, **127**, 113 (1963); ibid., **129**, 137 (1965).
250. Cowley, A., Aller L.H. and Dunham, T., *Publ. Astron. Soc. Pacific*, **75**, 441 (1963).
251. Bidelman, W.P., *Astrophy. J.*, **117**, 25 (1952).
252. Ludendorff, F.W.H., *Astron. Nachrichten*, **173**, 1 (1906).
253. Berman, L., *Astrophys. J.*, **81**, 369 (1935).
254. Joy, A.H., *Astrophys. J.*, **102**, 168 (1945).
255. Adams, W.S. and Pease, F.G., *Publ. Astron. Soc. Pacific*, **27**, 132 (1915).
256. Slipher, V.M., *Bull. Lowell Observ.*, **3**, 63 and 66 (1918).
257. Cannon, A.J., *Harvard Circular*, **201** (1918); ibid., *Astron. Nachrichten*, **207**, 215 (1918).
258. Sanford, R.F., *Publ. Astron. Soc. Pacific*, **32**, 59 (1920).
259. Merrill, P.W., *Publ. Astron. Soc. Pacific*, **53**, 342 (1941).
260. Hubble, E.H., *Astrophys. J.*, **56**, 181 (1922).
261. Joy, A.H., *Publ. Astron. Soc. Pacific*, **44**, 385 (1932).
262. Joy, A.H., *Astrophys. J.*, **110**, 424 (1949).
263. Herbig, G.H., *Advances in Astron. and Astrophys.*, **1**, 47 (1962).
264. Herbig, G.H., *Quarterly J. Roy. Astron. Soc.*, **15**, 526 (1974).
265. Herbig, G.H., *J. Roy. Astron. Soc. Canada*, **46**, 222 (1952).
266. Herbig, G.H., *Astrophys. J.*, **125**, 612 (1957).
267. Ambartsumian, V., *Stellar Evolution and Astrophysics*, Academy of Sci., Armenian S.S.R. (1947).
268. Herbig, G.H., Dissertation, Uni. of California (1948).
269. Greenstein, J.L., *Astrophys. J.*, **107**, 375 (1948).
270. Struve, O. and Rudkjøbing, M., *Astrophys. J.*, **109**, 92 (1949).
271. Walker, M.F., *Astophys. J. Suppl.*, **2**, 365 (1956).
272. Henyey, L.G., le Levier, R. and Levee, R.D., *Publ. Astron. Soc. Pacific*, **67**, 154 (1955).
273. Herbig, G.H., *Mém. Soc. Roy. Liège*, [4] **18**, 251 (1958).
274. Sanford, R.F., *Publ. Astron. Soc. Pacific*, **59**, 134 (1947).
275. Hunger, K., reported by C.D. Shane, *Astron. J.*, **62**, 294 (1957).
276. Bonsack, W.K. and Greenstein, J.L., *Astrophys. J.*, **131**, 83 (1960).
277. Bonsack, W.K., *Astrophys. J.*, **133**, 340 (1961).
278. Greenstein, J.L. and Richardson, R.S., *Astrophys. J.*, **113**, 536 (1951).
279. Herbig, G.H., *Astrophys. J.*, **141**, 588 (1965).
280. Wachmann, A.A., *Beobacht. Zirkular der Astron. Nachrichten*, **21**, 60 (1939).
281. Herbig, G.H., *Vistas in Astron.*, **8**, 109 (1966).
282. Varsavsky, C., *Symposium on Stellar Evolution*, La Plata, p. 33 (1960)
283. Kuhi, L.V., *Astrophys. J.*, **140**, 1409 (1964).
284. Morgan, W.W., *Astrophys. J.*, **77**, 291 (1933).
285. Merrill, P.W., *Astrophys. J.*, **63**, 13 (1926).
286. Burwell, C.G., *Astrophys. J.*, **88**, 278 (1938).
287. Bidelman, W.P. and Keenan, P.C., *Astrophys. J.*, **114**, 473 (1951).

288. Garstang, R.H., *Publ. Astron. Soc. Pacific*, **64**, 227 (1952).
289. Bidelman, W.P., *Vistas in Astron.*, **2**, 1428 (1956).
290. Keenan, P.C., *Handbuch der Physik*, **50**, 93 (1958).
291. Burbidge, E.M. and Burbidge, G.R., *Astrophys. J.*, **126**, 357 (1957).
292. Warner, B., *Mon. Not. Roy. Astron. Soc.*, **129**, 263 (1965); see also ibid., *I.A.U. Symposium*, **26**, 300 (1966).
293. MacConnell, D.J., Frye, R.L. and Upgren, A.R., *Astron. J.*, **77**, 384 (1972).
294. Keenan, P.C. and Morgan, W.W., *Astrophys. J.*, **94**, 501 (1941).
295. Newall, H.F., Baxandall, F.E. and Butler, C.P., *Mon. Not. Roy. Astron. Soc.*, **76**, 640 (1916).
296. Keenan, P.C., *Astrophys. J.*, **96**, 101 (1942).
297. Sanford, R.F., *Astrophys. J.*, **99**, 145 (1944).
298. Harding, G.A., *Observatory*, **82**, 205 (1962).
299. Keenan, P.C. and Keller, G., *Astrophys. J.*, **113**, 700 (1951).
300. Northcott, R.J., *J. Roy. Astron. Soc. Canada*, **47**, 65 (1953).
301. Wallerstein, G. and Greenstein, J.L., *Astrophys. J.*, **139**, 1163 (1964).
302. McKeller, A., *Publ. Dominion Astrophys. Observ. Victoria*, **7**, 395 (1949).
303. Climenhaga, J.L., *Publ. Dominion Astrophys. Observ. Victoria*, **11**, 307 (1960).
304. Wallerstein, G., *Astrophys. J.*, **158**, 607 (1969).
305. Merrill, P.W., *Astrophys. J.*, **77**, 44 (1933).
306. Fleming, W.P., reported by E.C. Pickering, *Harvard Circ.*, **168**, 1 (1911).
307. Plaskett, H.H., *Publ. Dominion Astrophys. Observ. Victoria*, **4**, 119 (1928).
308. Zanstra, H., *Astrophys. J.*, **65**, 50 (1927).
309. Hogg, F.S., *Publ. Astron. Soc. Pacific*, **44**, 328 (1932).
310. Merrill, P.W., *Astrophys. J.*, **99**, 15 (1943).
311. Struve, O. and Elvey, C.T., *Publ. Astron. Soc. Pacific*, **51**, 297 (1939).
312. Swings, P. and Struve, O., *Astrophys. J.*, **91**, 546 (1940); **93**, 356 and **94**, 291 (1941); **95**, 152 and **96**, 254 (1942); **97**, 194 and **98**, 91 (1943).
313. Payne, C.H., *Astrophys. J.*, **104**, 362 (1946).
314. Merrill, P.W., *Spectra of long-period variable stars*, Uni. of Chicago Press, p. 105 (1940).
315. Merrill, P.W., *Astrophys. J.*, **111**, 484 (1950).
316. Merrill, P.W., *Mém. Soc. Roy. Sci. Liège*, (4) **20**, 436 (1958).
317. Hogg, F.S., *Publ. Amer. Astron.* Soc., **8**, 14 (1934).
318. Bloch, M. and Mao-Lin, T., *Ann d' Astrophys.*, **14**, 266 (1951).
319. Aller, L.H., *Publ. Dominion Astrophys. Observ. Victoria*, **9**, 321 (1954).
320. Gauzit, J., *Ann. d'Astrophys.*, **18**, 354 (1955).
321. Kraft, R.P., *Astrophys. J.*, **127**, 625 (1958).
322. Sahade, J., *Kleine Veröffentlichungen der Remeis Sternwarte, Bamberg*, **4** (no. 40), 140 (Third Bamberg Coll. on Variable Stars) (1965).
323. Boyarchuk, A.A., *Astron. Zhurnal*, **43**, 976 (1966) and **44**, 12 (1967). In translation in *Soviet Astron.*–A.J., **10**, 783 (1967) and **11**, 8 (1967).
324. Boyarchuk, A.A., *Astron. Zhurnal*, **44**, 1016 (1967). In translation in *Soviet Astron.–A.J.*, **11**, 818 (1968).
325. Boyarchuk, A.A., *I.A.U. Coll.* 'Non-Periodic Phenomena in Variable Stars', ed. L. Detre, p. 395 (1969).

326. Bidelman, W.P., *Astrophys. J. Suppl.*, **1**, 175 (1954).
327. Payne-Gaposchkin, C.H., see ref. (58), p. 217 (1957).
328. Wackerling, L., *Mem. Roy. Astron. Soc.*, **73**, 153 (1970).
329. Zwicky, F., *Handbuch der Physik*, **51**, 766, publ. by Springer-Verlag (1958). Continued by F. Zwicky in *Stars and Stellar Systems*, **8** (Stellar Structure), chap. 7, p. 367, publ. by University of Chicago Press (1965).
330. Oke, J.B. and Searle, L., *Ann. Rev. Astron. and Astrophys.*, **12**, 315 (1974).
331. Karpowicz, M, and Rudnicki, K., 'Preliminary Catalogue of Supernovae', *Publ. Astron. Observ. Warsaw University*, **15** (1968).
332. Young, C.A., *Sidereal Messenger*, **4**, 282 (1885).
333. Vogel, H.C., *Astron. Nachrichten*, **112**, 283 (1885).
334. Copeland, R., *Mon. Not. Roy. Astron. Soc.*, **47**, 49 (1886).
335. Huggins, W., *Nature*, **32**, 465 (1885). See also: ibid., *Observ.*, **8**, 333 (1885).
336. von Konkoly, N., *Astron. Nachrichten*, **112**, 286 (1885).
337. Vogel, H.C., *Astron. Nachrichten*, **112**, 387 (1885).
338. Maunder, E.W., *Mon. Not. Roy. Astron. Soc.*, **46**, 19 (1886).
339. Backhouse, T.W., *Mon. Not. Roy. Astron. Soc.*, **48**, 108 (1887).
340. Payne-Gaposchkin, C.H., *Astrophys. J.*, **83**, 245 (1936).
341. Pickering, E.C., *Harvard Circ.*, **4** (1895). See also: ibid. *Astrophys. J.*, **3**, 162 (1896).
342. Zwicky, F., see ref. (329), first part (1958).
343. Johnson, W.A., *Harvard Bull.*, **902**, 11 (1936).
344. Payne-Gaposchkin, C.H., *Astrophys. J.*, **83**, 173 (1936).
345. Campbell, W.W., *Astrophys. J.*, **5**, 233 (1897).
346. Ritchey, G.W., *Publ. Astron. Soc. Pacific*, **29**, 210 (1917).
347. Shane, C.D., *Publ. Astron. Soc. Pacific*, **38**, 182 (1926).
348. Duncan, J. and Nicholson, S., reported by M.L. Humason in ref. (349), (1936).
349. Humason, M.L., *Publ. Astron. Soc. Pacific*, **48**, 110 (1936).
350. Baade, W., *Publ. Astron. Soc. Pacific*, **48**, 226 (1936).
351. Shapley, H., *Publ. Astron. Soc. Pacific*, **29**, 216 (1917).
352. Lundmark, K., *Mon. Not. Roy. Astron. Soc.*, **85**, 865 (1925). See p. 887.
353. Baade, W. and Zwicky, F., *Publ. Nat. Acad. Sci.*, **20**, 254 (1934).
354. Baade, W., *Astrophys. J.*, **88**, 285 (1938).
355. Minkowski, R., *Astrophys. J.*, **89**, 156 (1939).
356. Minkowski, R., *Publ. Astron. Soc. Pacific*, **53**, 224 (1941).
357. Hubble, E., *Publ. Astron. Soc. Pacific*, **53**, 141 (1941).
358. Humason, M.L., *Publ. Astron. Soc. Pacific*, **72**, 208 (1960) and **73**, 175 (1961).
359. Zwicky, F., see ref. (329), second part (1965).
360. Warner, B., *Mon. Not. Roy. Astron. Soc.*, **137**, 119 (1967).

10 Quantitative analysis of stellar spectra

10.1 Introduction The successful analysis of the absorption lines in stellar spectra to obtain the abundances of the different elements became a feasibility for stars of a range of spectral types from the early 1940s. In this chapter the development of the subject from about 1940 to 1965 is reviewed. In addition, the developments in stellar atmosphere theory, in instrumental techniques and in laboratory data which made abundance analyses possible are discussed here, from the position reached at the end of Chapter 7.

Several important preliminary problems first had to be settled before substantial progress was possible. Those relating to instrumental techniques and laboratory data are deferred until section 10.8, and we treat here the problems that arose in the theory which was necessary to interpret stellar spectra. Firstly, the whole issue of stellar temperature measurements from the continuous spectra, or spectral flux gradients, was in a mess in the 1920s. This problem was gradually sorted out during the 1930s decade once it was realised that stars do not really radiate like black bodies. This is the subject of section 10.2.

Secondly, for the analysis both of the flux gradients and of spectral lines, some sort of model was necessary to describe the structure of stellar atmospheres. The development of model atmospheres from McCrea's first work in 1931 is therefore vital to the overall discussion. In particular, Wildt's discovery in 1939 of the H^- ion as the main opacity source in cooler stars was an important advance that at once cleared up a major obstacle in the interpretation of the spectrum of the sun and other cooler stars.

The basic principles for calculating the structure of a stellar atmosphere had already been laid by Schwarzschild, Milne and Eddington. In 1906 Schwarzschild had written his famous paper demonstrating radiative energy transport, rather than convection, to be the mechanism of energy transfer outwards through the solar atmosphere. He thus developed the theory of radiative equilibrium (1).

Schwarzschild in 1906 had also used the concept of hydrostatic equilibrium, in which the weight of the material above any layer must be

Fig. 10.1 Karl Schwarzschild.

exactly balanced by the pressure in that layer. Eddington in his book *The Internal Constitution of the Stars* developed this concept further and gave a fundamental equation governing atmospheric structure in hydrostatic equilibrium (2). He also gave here his approximate solution to the grey atmosphere problem, that is, to deducing the temperature structure of an atmosphere in which the optical opacity does not depend on wavelength. Although a hypothetical situation, this simplifying assumption at least allowed a unique solution to be found, in which the fourth power of the temperature varies linearly with the optical depth.

Then Milne in 1922 introduced the term 'local thermodynamic equilibrium' (LTE) (3), which can be considered as a clever astrophysical trick, in that the Saha and Boltzmann equations for ionisation and excitation are assumed to be valid even though the intensity of the radiation inside the solar atmosphere must depart from the isotropic Planck radiation of a black body. The validity of this assumption has been hotly debated ever since, but at least it made the investigation of solar and stellar atmospheric models possible in the days prior to computers, which would certainly not have been the case otherwise. Even with computers, the assumption of LTE was adopted in nearly every stellar abundance analysis prior to 1970.

Since theoretical model atmospheres could be used to predict stellar fluxes, the flux gradients (i.e. slope of the flux with wavelength) or colour temperatures determined by the observers could be directly compared with theory. Colour temperatures are those derived from the flux gradients and were previously obtained simply from comparison with the Planck curve pertaining to a black-body radiator. The colour temperatures of the models however differed markedly from their effective temperatures, which can be regarded as a more fundamental but observationally much harder to determine parameter. Effective temperature is related to the total amount of flux radiated by a star at all wavelengths from unit area of its surface. This theoretical result meant that black-body colour temperatures could not simply be used to interpret stellar spectra. Either the flux gradients had to be interpreted using a model atmosphere, or the temperatures had to be obtained by other means, such as from the spectral lines themselves.

10.2 Stellar colour temperatures from 1925

In the first decades of the present century the temperatures of stars had been deduced most notably by Wilsing and Scheiner at Potsdam (4) from visual observations, and by Rosenberg at Tübingen photographically (5). These temperatures were all colour temperatures obtained by fitting a Planck black-body energy distribution to the observed gradient. Although the Harvard sequence of spectral types was clearly shown to be a temperature sequence, serious disagreements between the Potsdam and Tübingen results, especially for the hotter stars, seemed to cast doubt on the validity of the Planck formula for stellar radiation. This was the conclusion of Alfred Brill when he again reviewed stellar temperature determinations in 1932 (6): 'The disagreement between the original temperature scales of Wilsing and of Rosenberg can be explained by the fact that the Planck law is no longer obeyed over the relatively wide spectral region of the visual and the photographically sensitive radiations'.

This concise statement had by 1931 both observational and theoretical support respectively from the work of C.S.Yü at Lick and of W.H. McCrea in Edinburgh (see section 10.3). Yü's photographic spectrophotometry in 1926 (7) had drawn attention to the large Balmer discontinuity in the spectra of some hotter stars which represented a substantial deviation from the Planck law for black bodies (see Chapter 8). Like Rosenberg, he also derived colour temperatures in the blue photographic region which were much too high for the atmospheres of early type stars (e.g. for Vega he obtained $T_c = 16\,000$ K).

Another extensive program of photographic temperature determinations was undertaken from 1923 by Professor Ralph Sampson

Fig. 10.2 Hans Kienle.

(1866–1939), the Astronomer Royal for Scotland. The observations were objective prism spectra of the brightest stars taken with a 6-inch Cooke astrograph, using mainly Ilford Panchromatic plates which would record the spectra as far as the Hα line in the red for the cooler stars (8). The spectra were traced on a novel recording microphotometer developed at the Royal Observatory.

Two papers measured the continuum gradients in the spectra of some eighty stars, relative to the gradient of Polaris (9). The gradients were then converted to what Sampson termed 'effective temperatures' (in practice they are colour temperatures) by adopting Capella as a black-body standard source at 5500K.* Sampson's results suffered from the same problems as those of earlier investigators, such as Rosenberg. By assuming a black-body spectrum, the earlier type stars were too hot (e.g. Vega 13 000 K). In one respect his method differed from earlier studies; Sampson recognised that departures from the black-body law existed, and hence

* The relative gradients were defined by $\Delta\phi = d\ln\,(F_\lambda/F_\lambda\,(\text{standard}))/d(\frac{1}{\lambda})$ where F_λ is the flux in a given star and F_λ (standard) that in a standard star such as Polaris. The 'colour temperature' T_c is the temperature of a black body with a gradient equal to the mean gradient of a star observed in a given spectral region. It is given by $\phi = \frac{c_2}{T_c}/(1 - e^{-c_2/\lambda T_c})$ where $c_2 = \frac{hc}{k} =$ 14 320 μ. Sampson called these 'effective temperatures', a term now reserved for the temperature of a black body with the same total emitted surface flux integrated over all wavelengths as that of the star.

used a stellar model of a black-body radiator surrounded by overlying layers that modify the spectrum. He believed solar limb darkening was a manifestation of this effect and corrected his stellar gradients on this basis.

The situation at the time of Sampson's work concerning stellar temperature measurements from the continuous spectrum was nicely summed up by William Greaves at Greenwich.

> The position in 1925 was that the whole subject was dominated by the belief that the radiation emitted by a star was distributed in wavelength roughly according to Planck's black-body formula. This belief is now known to be erroneous, and it is somewhat astonishing that it should have been held as widely as it was, since an observational fact was available which is conclusively fatal to it. The 'Balmer Discontinuity' has now become an astrophysical commonplace.... (10).

A new era in spectrophotometry began in the 1930s. The relative gradients were measured photographically and then converted to absolute gradients by the careful calibration of the observations by means of a standard lamp. This work laid the basis for the accurate determination of effective temperatures of stellar atmospheres which were obtained later, once reliable models could be constructed. At the time the work was done however, colour temperatures were still in common use, and there was still the lingering belief that these represented the 'true temperature' of the layers where the line spectrum is formed. The first absolute measurements came from William M.H. Greaves (1897–1955) and his colleagues at Greenwich, followed by Robley C. Williams (b.1908) at Ann Arbor, by Hans Kienle (1895–1975) and colleagues at Göttingen and by Daniel Barbier and Daniel Chalonge on the Jungfraujoch.

The Greenwich measurements were commenced in 1926 by Greaves with Charles Davidson and E. G. Martin. They measured gradients relative to the mean of nine A0 stars as standards, which were then calibrated using an acetylene burner (11). Later a tungsten filament lamp was employed (12) which was in turn calibrated at the National Physical Laboratory so as to be operated at a colour temperature of 2360K. The mean gradient for A0 stars reported in 1934 was 0.99μ corresponding to a colour temperature of 18 000 K, which was much hotter than the 10 000 K found by Fowler and Milne (13) and by Miss Payne (14) from Saha's ionisation theory. A system of 25 standard B and A stars was established on the Greenwich system, and the gradients in a further 38 stars of types B to G were measured.

The program at Göttingen under Kienle involved 50 Å/mm objective

Fig. 10.3 Kienle's objective prism camera at Göttingen, used for obtaining the continuous energy distributions of stars.

prism spectra for thirty-nine stars of types B0 to M with data covering the region from 3700 to 6850 Å (15). The results were presented as monochromatic magnitudes at about thirty intermediate wavelengths. In this respect they differed from the Greenwich presentation where only blue and red data points were used (this was one of many criticisms that Kienle had for the Greenwich data (16)). As at Greenwich these magnitudes were expressed relative to the mean energy distribution of A0 stars. In a further contribution by Kienle, Wempe and Beileke (17) the absolute spectrum of an A0 star was carefully calibrated using a tungsten filament lamp as an artificial star, which in turn was calibrated against a black body in the laboratory. The result was higher gradients for A0 stars ($\phi = 1.07\mu$), corresponding to a lower temperature than obtained by Greaves, but still too hot to agree with ionisation temperatures. What is more, the results were detailed enough to show a gradient decrease in going from red to blue, implying that the colour temperature of an A0 star changes from about 12 200K at 5550 Å, to about 23 500K at 4350 Å. This was therefore further direct evidence of departures from a black-body spectrum, in this instance in the shape of the Paschen continuum longwards of the Balmer jump.

The Michigan program was undertaken by Robley C. Williams at Ann Arbor from 1936. The aim of this limited but very carefully executed program was solely to establish the gradient for A0 stars, taking Vega as the primary standard and six other stars of the same spectral type as secondary standards (18). The Michigan $37\frac{1}{2}$-inch reflector was equipped with a tungsten filament standard lamp fitted to the top end of the tube to provide a calibration. The lamp revolved around the secondary mirror so as to illuminate uniformly an annulus of the primary. The lamp in turn was calibrated with a thermopile in the GEC laboratory at Nela Park, Ohio. The prismatic spectra had quite high dispersions in the blue (26.7 Å/mm at 4000 Å), falling to 133 Å/mm at 6000 Å. These were traced on a Moll microphotometer, and after calibrating the photographic response of each plate, and correcting the stellar magnitudes for extinction, the results were expressed as logarithmic intensities at eleven wavelengths in the stellar spectrum from 4040 to 6370 Å.

In Williams' first paper, a gradient for α Lyrae of $\phi_0 = 1.16\mu$ at 5000 Å was determined, corresponding to a colour temperature of 14 300K (18). This was increased to 1.19μ from a study of seven standard stars including α Lyrae (19). A third paper concluded that the differences between the Greenwich, Göttingen and Ann Arbor results must be due to problems of lamp calibrations, and did not lie in the astronomical observations themselves (20). To test this idea Williams took his lamp to Europe in 1939 and had it calibrated in four laboratories there, including the National Physical Laboratory at Teddington and the University Observatory in Göttingen. The results were mainly in good agreement, and did not support the suggestion of major discrepancies in stellar temperatures arising from this cause (21).

The pioneering Balmer jump and ultraviolet spectrophotometry of Barbier and Chalonge has already been discussed (see Chapter 8 and reference (22)). These were at that time the only observations to measure flux gradients on both sides of the Balmer jump. Barbier and Chalonge used a hydrogen discharge lamp at the telescope to calibrate the stellar spectra. The hydrogen lamp was in turn compared with a tungsten filament lamp, which they twice took to Göttingen (in 1936 and 1938) to calibrate against Kienle's black bodies (23). The most striking colour temperature differences were obtained from the Jungfraujoch observations between the blue and ultraviolet regions. For A0 stars in the blue region (3700 to 4500 Å) they found $\phi_{\text{blue}} = 1.00$ giving a colour temperature of 16 500K; while in the ultraviolet (3150 to 3700 Å) the gradient was $\phi_{\text{UV}} = 1.39$, resulting in 10 500K. This was certainly the most decisive observational evidence of non-Planckian continua in stellar spectra. What is more, these data were now being corroborated with theoretical results from the new technique of 'model stellar atmospheres'.

10.3 Early model stellar atmospheres

The theoretical interpretation of the observed energy distribution in the solar spectrum was a problem considered by Lindblad and by Milne during the early 1920s (3,24). Lindblad used the temperature distribution in the solar atmosphere derived by assuming a grey absorption coefficient, that is, one not depending on wavelength, and the results of this model gave moderately good agreement with the observations (25). The best agreement with limb-darkening observations could, however, be achieved by systematically lowering the continuous absorption in the spectral region from 4265 to 5955 Å. This gave a considerable blue peak to the theoretical energy distribution. Because the effect of absorption lines in reducing the flux was known to be substantial, the results were considered to agree quite satisfactorily with the observations.

The problem was to find a suitable source for this nearly constant opacity. The hydrogen atom alone was evidently not this source. This was shown dramatically by William H. McCrea (b. 1904) in Edinburgh, who calculated the first non-grey stellar models in 1931 (26). He was also the first to use the term 'model stellar atmosphere'. McCrea considered the simple case of a pure hydrogen atmosphere in which the continuous absorption arose solely from the grey contribution of free electrons and from the opacity of neutral hydrogen atoms excited to any level n. Here he used the formula of Kramers (27) who predicted an opacity κ_n varying as λ^3/n^5 for photon wavelengths λ less than the appropriate ionisation limit for each level, and of course falling abruptly to zero for longer wavelengths. For a series of models of various effective temperatures he computed the emergent flux as a function of wavelength (Fig. 10.4). The ionisation edges

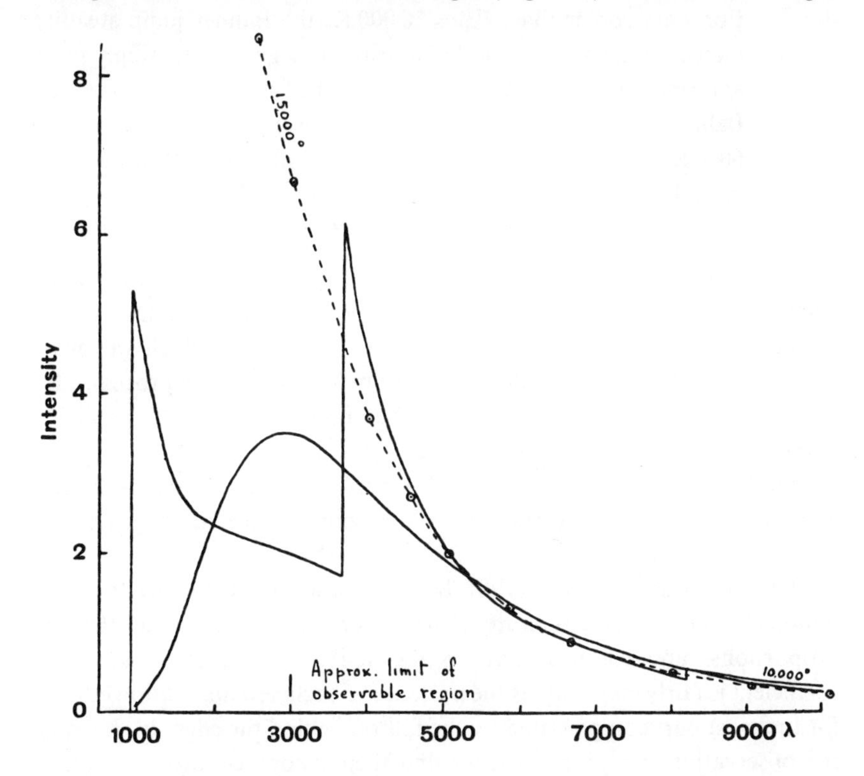

Fig. 10.4 The first model atmosphere fluxes calculated by W.H. McCrea in 1931; the results for a pure hydrogen atmosphere at 10 000 K are compared with black-body curves at 10 000 K and 15 000 K.

of hydrogen atoms in the second level gave the familiar Balmer jump. This is simply interpreted as being due to the sudden increase in the absorption at the wavelength 3650 Å. For wavelengths less than this, the opacity of the atmosphere is such that we only see to a shallower and relatively cooler depth where the radiated flux is less.

McCrea's conclusions were as follows:

1) For A and B stars the ultraviolet spectra were very different from Planck curves, but on the redward side of the Balmer discontinuity a Planck curve could be found that matched the results reasonably well.
2) For a B star with $T_{\mathrm{eff}} = 15\,000$ K, the colour temperature that matched the observations was about 18 000 K.

3) For an A star with $T_{eff} = 10\,000$ K, the colour temperature of 15 000 K matched the observations for $\lambda > 5000$ Å, but a higher colour temperature was needed between 4000 and 5000 Å.
4) For stars cooler than $T_{eff} = 10\,000$ K, the Balmer jump steadily increased in size and the theoretical fluxes gave increasingly poor agreement with the observations. For the sun, where the observed Balmer jump is almost non-existent, the Planck curve at about 6000 K from a grey atmosphere gave a far better (though still imperfect) fit to C.G. Abbot's observations (28) than to McCrea's fluxes for which atomic hydrogen was the source of opacity.

The failure of atomic hydrogen to model the observations for cooler stars led Biermann, Unsöld and Pannekoek to consider models with a substantial contribution of metals to the opacity, in which (as with hydrogen) photons are absorbed by the atom in an ionisation process. The composition of the solar atmosphere thus became a crucial ingredient. Although Russell's analysis of 1929 (29) had indicated the great preponderance of hydrogen, the result (based on one saturated hydrogen line) was sufficiently uncertain to allow some choice in the assumed composition in these models.

Thus Ludwig Biermann (1907–86) at Göttingen, calculated the continuous opacity from a mixture of hydrogen and metals in the Russell proportions, and claimed that 'in the visible region the absorption coefficient is fairly constant, as the presence of each new absorption edge is for the most part compensated by the fall-off behind the edge. In this way the observations mentioned earlier find their theoretical atomic explanation' (30).

Albrecht Unsöld in Kiel, Germany, first repeated McCrea's calculations for pure hydrogen atmospheres (31).* Like Biermann he considered the effect of metals on the opacity. In 1934 Unsöld rejected Russell's observational result that hydrogen atoms are about a thousandfold more abundant than all metals together.

> But the sureness of this [Russell's] procedure would be very low. For the most important elements the transition probabilities of the visible lines were only very roughly estimated and the excitation [population] ratios were not able to be specified exactly, and hence errors of the order of a factor 10 in the end result would seem quite possible (32).

* Unsöld's colour temperatures were all much higher than McCrea's and did not vary monotonically with effective temperature.

Unsöld instead adopted a model whose composition by mass was only about one third hydrogen and two-thirds metals (especially iron, silicon, magnesium) or by number hydrogen-to-metals only about 14 to 1, in an attempt to reproduce the required opacity. His results for the colour temperatures calculated from these models' fluxes now seemed to agree better with the effective temperatures, obtained by integrating the fluxes over all wavelengths.

It was these calculations and results that Unsöld included in the first edition of his influential treatise *Physik der Sternatmosphären* in 1938 (33). This work and its revised edition of 1955 dominated the theory of stellar atmospheres for the next several decades. Unsöld summed up this section of the earlier edition of his book thus:

> The frequency dependence of the opacity calculated for stellar material gives a qualitatively correct representation of continuous stellar spectra and their departures from the Planck radiation law. From a quantitative viewpoint, the predictions of the theory are fairly insecure because of the known unreliability of the calculation of metallic absorption coefficients and their frequency dependence. On the observational side a final determination of the absolute colour temperature scale of the stars is still lacking (33).

Antonie Pannekoek's analysis was similar to those of Biermann and Unsöld. He used a hydrogen-to-metals ratio of 1000 : 1 following Russell. He smoothed out 'the quasi-infinite number' of metallic absorption edges into an absorption continuum which was added to the atomic hydrogen opacity and the grey component due to free electrons (34). For stars with T_{eff} over 10 000 K the hydrogen opacity dominated and the colour temperature was 18 000 to 20 000 K, which seemed to agree with the high values measured at Greenwich by Greaves and his colleagues. Near 7000K the two temperatures were about equal and going cooler still, the colour temperatures were substantially less than the effective ones, a result similar to Unsöld's. In a later review of the stellar temperature scale, Pannekoek made it clear that he considered an effective temperature of about 10 500K was established for class A0. He cited evidence obtained from the analysis of the eclipsing binary β Aurigae by Russell, Dugan and Stewart which led to a direct determination of the effective temperature of this A0 star (35). He pointed out that the colour temperature from his models could be nicely explained using this effective temperature for A0 stars.

We can summarise these early studies of non-grey model atmospheres with the following thoughts. For early-type stars there was some success in explaining the anomalously high colour temperatures. For solar-type stars a fairly constant absorption coefficient was sought. In spite of the general belief that the metallic elements must somehow be responsible, the agreement of the fluxes with observations was only superficial. In fact, as Strömgren has noted: '... the observations agreed much better with a constant κ_ν' (36).

10.4 Rupert Wildt and the negative hydrogen ion

It was Rupert Wildt (1905–76) at Princeton who laid the confused results for the cool models with metallic absorption to rest. This German astronomer had migrated to the United States in 1935; he spent the years 1936–41 at Princeton and it was here in 1939 that he solved the missing opacity problem in the sun and cooler stars. His solution was that the rather unstable negative hydrogen ions, H^-, are the main source of 'fogginess' in the solar atmosphere. This brilliant result had an immediate effect, and almost overnight the whole course of solar atmospheric theory was changed (37).

In fact, knowledge of the existence and stability of H^- had been around for a decade prior to Wildt's discovery, but it seems that this knowledge was confined to physical rather than astrophysical circles. Hans Bethe (b. 1906) in Munich discussed the electron affinity of hydrogen in 1929 and showed that hydrogen with two electrons was stable with a positive heat of formation and ionisation energy of $I = 0.74$ electron volts (38). The Norwegian physicist Egil A. Hylleraas (1898–1965) independently came to the same conclusion in 1930, and deduced a value $I = 0.70$eV (39).

In June 1938 Wildt presented a paper at a symposium at Yerkes Observatory on 'Electron affinity in astrophysics' (40). In this work he applied Saha's equation to find the concentration of H^- ions in the upper solar atmosphere. The electrons came primarily from ionised metals, and these occasionally associated with the plentiful neutral hydrogen atoms to form the H^- ion. The calculations revealed 'the surprising fact that, in the higher levels [of the sun's atmosphere], the negative hydrogen ions are more abundant than the neutral hydrogen atoms excited to the ground state of the Balmer series.' At this time it was not certain whether the H^- ion had stable excited states, but the absence of H^- absorption lines in the solar spectrum seemed to indicate to Wildt that only the ground state was stable, and this subsequently proved to be correct. Several physicists at this time were attempting to calculate the photon absorption coefficient of the H^- ion by the photoionisation process. Clearly any photon of energy exceeding the ionisation potential might be absorbed; that is with wavelength less than 17 600 Å in the infrared (corresponding to $I = 0.70$ eV). At the time

Fig. 10.5 Rupert Wildt.

these calculations were only partially successful, but in a footnote to Wildt's published conference paper, a clear hint of the direction of his thoughts is given. Referring to the calculations of C.K. Jen (b. 1906) in 1933 (41) he wrote:

> Utilizing Jen's data, it has been shown that the addition of the H^- absorption to the metallic absorption produces for the solar atmosphere that desired independence of the absorption coefficient from wavelength which failed to emerge from Unsöld's and Pannekoek's theory.... This would mean the removal of a serious discrepancy between theory and observation and, in fact, the identification, by their continuous absorption spectrum, of the negative hydrogen ions so abundant in the solar atmosphere.

The important conclusion of this footnote was expanded into Wildt's classic paper of June 1939 in the *Astrophysical Journal* (37). Wildt stressed the fairly flat distribution of the H^- absorption coefficient, in so far as this was then known (from the calculations of Massey and Bates (42)) and as the observations of the solar spectrum required. The absorption by H^- was due both to bound–free photoionisation and to the free–free component (the latter arising from close encounters between electrons and hydrogen atoms). In addition, the small Balmer jump observed in solar-type stars could be explained from the dominance of the H^- opacity at 3650 Å over the contribution from neutral atomic hydrogen. The huge Balmer discontinuity shown in McCrea's pure hydrogen solar model thus disappeared, and with it also vanished one of Unsöld's main grounds for asserting the very low solar hydrogen-to-metals ratio (50:1 or less), based in part on the small observed Balmer jump.

The first reliable absorption coefficients for negative hydrogen ions came from the calculation of H^- photon absorption cross-sections by S. Chandrasekhar (b. 1910) at Yerkes Observatory, based on quantum mechanics. These results appeared in a series of papers from 1944 to 1946 (43) and showed for the bound–free absorptions a bell-shaped curve as a function of wavelength with a very broad maximum peaking at about 8500 Å, but still some 10 000 Å broad at half the maximum value. The free–free contribution rose steadily into the infrared. When both H^- contributions were added, an S-shaped function resulted, with a peak in the opacity at 8500 Å followed by a broad minimum at 1.6 μm (Fig. 10.6).

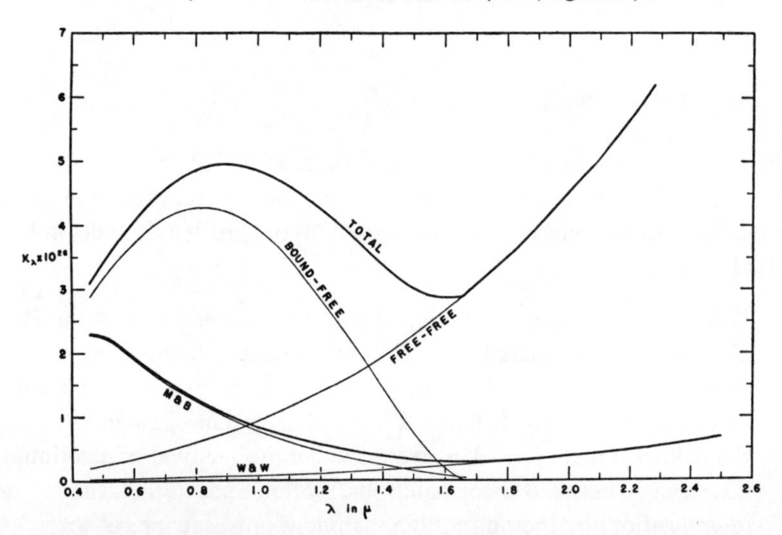

Fig. 10.6 The continuous absorption coefficient of H^- ions calculated by Chandrasekhar and Breen in 1946, and showing the total absorption as the sum of bound–free and free–free components.

In 1923 Milne, and also at nearly the same time Ragner Lundblad in Sweden, had shown how to deduce empirically the shape of the continuous opacity function for the solar atmosphere, by analysing the intensity distribution in the sun's spectrum (3, 44). This was applied by G.F.W.

Mulders in Utrecht as early as 1935, with solar data extending as far as 2.3 μm in the infrared. Mulders' opacity curve for the sun (Fig. 10.7) showed with great clarity 'a maximum near 9000 Å and again a minimum between 17 000 and 20 000 Å' which we can now ascribe to the long sought-for Wildt–Chandrasekhar H^- opacity source (45). A more detailed analysis by Daniel Chalonge and Vladimir Kourganoff (b. 1912) in Paris in 1946 (46) was able to confirm these results of Mulders (48) (see also the results of R. Peyturaux (b. 1925) (47) with the analysis of new data from a PbS cell for the solar near infrared region) and to give support to the calculations of Chandrasekhar for the H^- ion.

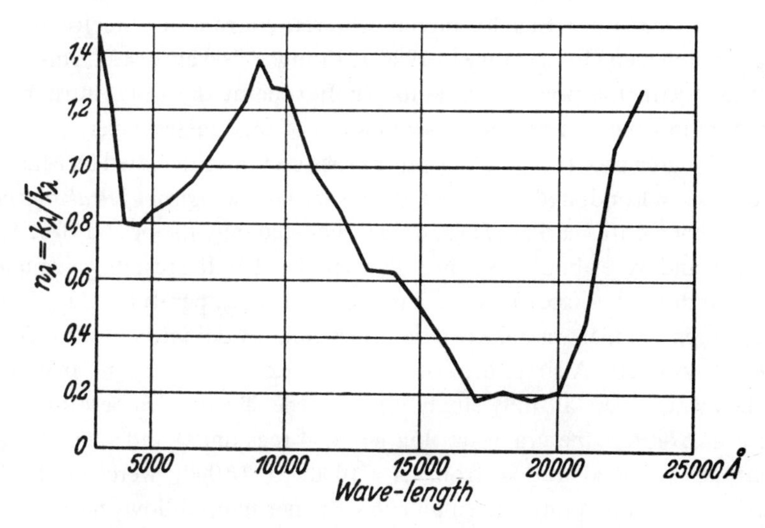

Fig. 10.7 The continuous absorption coefficient in the solar atmosphere deduced by G.F.W. Mulders in 1935.

Unsöld more than anyone had tackled the problems of the continuous solar opacity in the 1930s. When the second edition of *Physik der Sternatmosphären* appeared in 1955, he had this to say on H^-:

> As already indicated, the work of Wildt signified a complete revolution, not only for our ideas about the mechanism of how the solar spectrum originates, but also for those on the gas pressure in the sun's atmosphere. Relative to the earlier models, this had to be increased by about a factor of 30 as a result of the high hydrogen-to-metals ratio, for which B. Strömgren provisionally arrived at a value of 8000 to one (48).

10.5 Early model atmospheres in the 1940s after Wildt's discovery The result by Bengt Strömgren (1908–87) that Unsöld referred to (see section 10.4) was the construction of the first model atmospheres to incorporate the new H^- opacity. Strömgren's main goal was to construct solar models for different overall hydrogen-to-metals ratios, in order to study the strengths of Fraunhofer lines in the solar spectrum. The paper is important for being the first to use the new opacity; it is also a delight to read for the clarity of its exposition. Several of the results, after considerable subsequent refinement, have since become standard tools in spectral analysis. Strömgren's solar models were unfortunately not published in an astronomical journal, but in a memorial volume to his father, Elis Strömgren, who retired from the directorship of the Copenhagen Observatory in 1940 at the age of seventy (49). This fact, together with the political situation in Europe at the time, must have greatly hindered widespread dissemination of this classic paper.

Strömgren used the H^- absorption cross-sections calculated by Massey and Bates in London (42) which had appeared in the *Astrophysical Journal* only a few months earlier. These values included only the bound–free H^- opacity and were about a factor of two smaller than the results given later by Chandrasekhar (see (50) for comparison). The temperature structure of Strömgren's model was based on the approximate Eddington solution for a grey atmosphere. A simplification was that electrons came only from the metals which were all fully singly ionised. The effective temperature was fixed at 5740K,* the gravity at $\log g = 4.44$ (cgs units) and several high hydrogen-to-metal ratios from $A = 1000$ to $16\,000$, were used. As Strömgren pointed out, the small solar Balmer jump follows naturally if one adopts both the high hydrogen abundance and the H^- ion as the main continuous opacity source, which will dominate over the neutral hydrogen opacity even on the shorter wavelength side of the jump.

After solving for the H^- concentration using Saha's equation, Strömgren derived tables for the mean continuous opacity and gas pressure as functions of electron pressure and temperature. These were then used to find the continuous opacity and gas pressure as a function of depth in the sun, by solving the equation of hydrostatic equilibrium. Having thus calculated the model using the theory already derived by Schwarzschild, Milne, Eddington and others, Strömgren applied the Voigt profiles tabulated by Hjerting (52) to calculate line profiles for the solar spectrum.

An essential feature of his method was the use of an approximate formula that could take into account any arbitrary depth variation of the

* This is the value obtained by Milne (3) and very close to that by Minnaert (5748 K) (51) from Abbot's measures of solar radiation.

ratio η of line-to-continuous absorption coefficient within the atmosphere. The principles of this technique had been worked out by Strömgren in an earlier paper (53) and were applicable provided large changes in $\eta(\tau)$ with optical depth did not occur. It should be noted that Strömgren's method was just one of a variety of approximate methods developed for finding the fractional depression R of the continuous flux when line opacity is present in a stellar atmosphere. The simplest methods of all were based on the Schuster–Schwarzschild model, in which the continuous spectrum originates from the underlying photosphere, the lines in the separate reversing layer; or on the Milne–Eddington model, where the ratio η is considered to be depth-independent. These widely used models represent two extremes, with the real case somewhere in-between. Strömgren's method was a technique for tackling this in-between case when the function $\eta(\tau)$ was explicitly known. The weighting function technique of Unsöld in 1932 will not be described here (54); it too was able to calculate the depression R from $\eta(\tau)$ and a model. Both Strömgren's and Unsöld's methods were widely used in the 1940s and 1950s in the time interval between the introduction of model atmospheres and the introduction of electronic computers.

Some of Strömgren's qualitative results for solar lines are of interest. Because the continuous opacity is dominated by a species derived from hydrogen, and because the strength of weak neutral lines is proportional to η, the ratio of line-to-continuous opacity, Strömgren showed that the equivalent width of a weak line of a neutral metal is proportional to the ratio of the abundance of the metal to hydrogen. For a strong line of an ionised metal however, such as the K line of calcium, the strength was determined by the abundance ratio of calcium to all the metals. In this case the hydrogen abundance has no first order effect on the equivalent width, because of the combined effects of collisional damping and of the continuous opacity which just compensate each other.

Strömgren thus saw that for lines whose transition probabilities or oscillator strengths were known and also if the collisional damping constant could be estimated if the line is strong, then the theoretical equivalent width could be calculated. He did this for several lines and by changing the abundances of the element concerned, produced the first absolute solar curves of growth, from which the abundances (expressed as a ratio relative to hydrogen) could be obtained directly as a function of the equivalent width. He saw in addition, that the overall ratio of hydrogen-to-metals A could be determined by simultaneously considering both weak and strong lines of a given element. In effect he was choosing the A-value of that model which reproduced the observed collisional damping due to collisions with neutral hydrogen atoms.

Strömgren's results for three elements (sodium, potassium and calcium), from a total of only seven lines, gave abundances within a factor of about three of our currently accepted values. (Considering the gross simplifications, the agreement is fortuitous). His hydrogen-to-metals ratio, found from the method described for the K line and the weak 6573 Å neutral calcium line, was $A = 8000$.* This last value was somewhat higher than had been given by Russell in 1929 and hundreds of times greater than proposed by Unsöld in his 1934 solar model, and is also close to the modern value.

In a second paper in 1944, Strömgren and his colleagues in Copenhagen computed a series of model atmospheres for stars of spectral type A5 to G0 (55). This series or grid of models considered gravities from $\log g = 3.0$ to the main sequence value of 4.5 and hydrogen-to-metal ratios from 2500 to 16 000. The main difference between these models and the early solar models was the inclusion of the atomic neutral hydrogen continuous opacity as well as that of H^-. The H^- was still the principal opacity source in the F stars, but the free electrons could now come mainly from slightly ionised hydrogen rather than from the metals. For still hotter stars of A spectral type, Strömgren showed how the neutral hydrogen opacity became dominant, as the H^- ion sheds its extra electron at the higher temperatures.

These techniques were extended to the still hotter domain of B stars by Strömgren's student, Mogens Rudkjøbing (b. 1915) (56). Here the problem was specifically the modelling of the atmosphere of the B0 dwarf star τ Scorpii, which Unsöld had analysed using the so-called 'Grobanalyse' or coarse analysis technique (see section 10.9.1). The coarse method ignored the structure of the atmosphere and replaced quantities such as temperature, pressure and opacity by their mean values over all depths. Three surface gravities and two effective temperatures were used, to cover the likely parameters for τ Sco's atmosphere. Of course H^- was now no longer an opacity source; instead the photoionisation of atomic hydrogen and helium and scattering by free electrons were included. Rudkjøbing calculated the Balmer jumps, blue and ultraviolet spectral gradients, and ionised silicon (SiIV) line strengths from his models and compared these with the observations.

Lawrence Aller at Harvard in the 1940s also calculated models for early-type stars. His models had a grey temperature structure and were

* This value of A did not include the abundant but largely neutral elements carbon, nitrogen and oxygen. Inclusion of these elements would bring A down to almost 10^3, close to Russell's value.

calculated specifically for the analysis of the spectrum of the two A0 dwarfs, Sirius and γ Geminorum (78) (see section 10.9.3).

The main thrust of Strömgren's solar model in 1940 was to examine the behaviour of absorption lines in an atmosphere dominated by the H^- opacity. In 1945 Guido Münch (b. 1912) at Yerkes used the latest bound–free H^- opacity of Chandrasekhar to show that this opacity source gave reasonable agreement with solar observations, at least in the range from 4000 to 16 000 Å. Münch examined the energy distributions from the centre of the solar disk (the intensity spectrum), from the average of the whole disk (the flux spectrum) and the monochromatic solar limb darkening. All three were consistent with Chandrasekhar's H^- opacity being the principal contributor in the wavelength range mentioned (57). Münch also recalculated the structure of the solar atmosphere in 1947, using Chandrasekhar and Breen's H^- cross-sections with the free–free component included, and a temperature structure based on Chandrasekhar's exact solution of the grey atmosphere. The opacities were generally about 20 per cent above those of Massey and Bates used by Strömgren, while the pressures in the line-forming region were typically 20 per cent less (58).

10.6 Empirical and theoretical solar models and the line-blanketing problem

For the observational astronomer, the importance of model stellar atmospheres has been and is to provide a means of interpreting his observations and to determine temperatures, gravities and element abundances from them. For the theoretical astronomer, the observations of a close-up star such as the sun have provided an excellent test to keep the physics used in the models as close as possible to physical reality. We have therefore a system of interchange, whereby solar observations have indirectly, through the models, become the linch-pin for interpreting stellar spectra of widely differing types. In this section we outline some of the main developments in the late 1940s and 1950s in constructing solar models.

One of the first tasks was to demonstrate that the new H^- opacity of Wildt gave plausible results for the solar atmosphere. After Münch's work in 1945 (57) (see section 10.5), D. Chalonge and V. Kourganoff set out in 1946 to determine empirically the shape of the continuous opacity function, and confirm the H^- ion as a major source of opacity for the sun (46). Then came Daniel Barbier, who also confirmed H^- as the main opacity source (59) and then went on to deduce the solar temperature structure by analysing the extensive limb darkening observations of C.G. Abbot at the Smithsonian Astrophysical Observatory (60). He tabulated the first empirical $T(\tau)$ relationship for the sun.* Of the several empirical determinations of this function that followed in the next few years, especially notable was that of Erika Böhm-Vitense at Unsöld's institute in Kiel. She used the central intensities in strong spectral lines, where the opacity is high, in an attempt to map the temperatures at very shallow depths (61). Her rather cool boundary temperature for the sun was only 3800 K. She also obtained a new value for solar effective temperature by integrating the flux over all wavelengths, including contributions in the ultraviolet and infrared. The value was 5780 K, a fundamental datum frequently used in subsequent analyses.

On the theoretical side, the principal issue was to be able to predict the effect on the temperature structure ($T(\tau)$) of the non-greyness of the continuous absorption and of the thousands of line absorptions in the solar spectrum. If only the continuous opacity was considered, then the problem was relatively simplified. Guido Münch tackled this problem in 1947 (62). He used a method proposed by Chandrasekhar (63) in which a non-grey solution was approximated by a grey $T(\tau)$ provided the optical depth τ was defined in terms of an absorption coefficient flux-weighted over all wavelengths. Münch's result was a model with the surface layers about 55 K cooler than in the grey model proper.

However, it became clear that the correct way of tackling the non-grey problem was to find the temperature structure that gave a constant nett outwards radiative energy flux at all depths. If energy is transported by radiation, then this radiative equilibrium condition allowed $T(\tau)$ to be computed by a numerical iteration, using mathematical methods first proposed by Strömgren (64) and by Unsöld (65).

The first analysis to obtain a theoretical non-grey solar model by the iterative method to achieve flux constancy was by Raymond Michard at the Institut d'Astrophysique in Paris (66). He corrected an initial grey $T(\tau)$ for the sun, and found a flux constant model which was about 180 K cooler at the surface than the initial grey temperature structure.

The question of how to tackle the effect of the line opacities on the temperature structure is far more intractable, in part because of the problem of how to treat statistically the vast number of lines in the solar spectrum, of different strengths, and different ionisation and excitation levels and distributed non-uniformly throughout the spectrum. The effect of lines on the temperature structure was named by Milne in 1928 as

* T is the temperature and τ is the optical depth. τ is zero at the surface while $\tau = 1$ is a relatively deep layer. A photon emitted at depth τ and travelling along the outward normal has a probability of $e^{-\tau}$ of leaving the photosphere.

'line blanketing' (67). Milne showed how a relatively simple theory predicted a decrease in the limb darkening as a result of the line blanketing, due to the lines partially blocking the outward flux and causing a steepening of the temperature gradient near the surface. Early observations of the fraction of flux (η) removed by the lines ranged from about 10 to 16 per cent (averaged over all wavelengths) and these figures were, as both Milne and Lindblad showed in the early 1920s (3,25), in the right range to account for the slightly reduced limb darkening as compared with the limb darkening coefficient $u = 0.6$ expected from a line-free grey atmosphere.* However, as Lindblad pointed out, the variation of the continuous opacity with wavelength also influences the limb darkening, so the problem of disentangling these two competing influences on the observed value of u was not possible so long as the nature of the solar continuous opacity remained unknown.

Chandrasekhar considered line blanketing further in 1935, during his time at Trinity College, Cambridge. He showed that, in the case of an initially grey temperature structure, the effect of lines formed by a purely absorptive process was not only to steepen the temperature gradient near the surface: in addition the surface temperature was substantially cooler, and the temperatures deeper in the sun were all hotter, provided the line opacity persisted to these greater depths (68).† This of course was just the result later found from the empirical models, such as those of de Jager in 1952 (69) and of Erika Böhm-Vitense in 1954 (61) with surface temperatures of 4300 K and 3800 K respectively.

One requirement for the theoreticians was to have reliable values of η in each wavelength region of the solar spectrum. G.F.W. Mulders in 1935 measured the best solar values available, from the Revised Rowland Table of solar lines by C.E. St John and his colleagues (70). The mean value (over all wavelengths) was $\eta = 0.083$, rather less than accepted previously. His plotted results clearly showed how the line absorption occurs almost exclusively in the blue and ultraviolet regions of the solar spectrum (45) (Fig. 10.8). R. Michard showed later, using the Utrecht solar atlas, that Mulders' ultraviolet blanketing was in fact underestimated; he found a mean solar blanketing coefficient of $\eta = 0.124$ (71).

The question now is how can this blanketing data be used to predict the solar temperature structure. Münch's investigation of 1946 went further than Chandrasekhar's, because the strong blanketing in the ultraviolet of

* u is defined by the directional dependence of the surface intensity $I_0(\theta) = I_0(0)(1 - u + u\cos\theta)$.

† Chandrasekhar used the so-called Milne–Eddington model in which the ratio of line to continuous opacity is taken to be depth-independent.

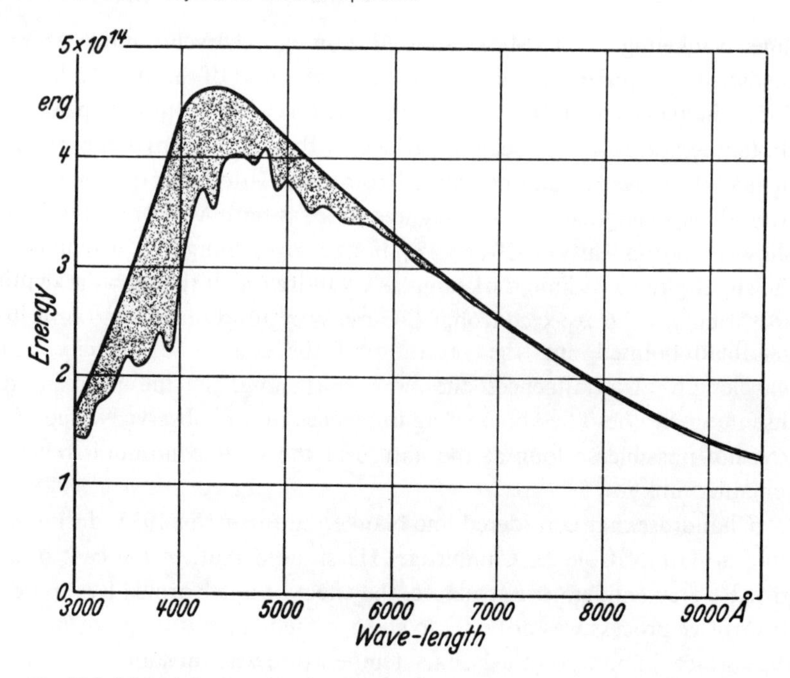

Fig. 10.8 Mulders' plot showing solar line blanketing, 1935. The shaded area represents the flux removed by the effect of the many absorption lines in the solar spectrum.

Mulders was used (72). This gave a solar surface temperature of about 4600 K, which seemed to agree rather well with K.O. Wright's excitation temperatures, which should also reflect the superficial line-forming layers (73) (see Chapter 7). The full treatment however requires not only the wavelength dependence of the blanketing, but also its depth dependence to be taken into account, and a flux-constant model is then achieved by iteration. If one could observe the flux spectrum below the surface the lines would generally weaken which can be predicted using Saha's equation.* Dietrich Labs (b. 1921) (74) then a student of Unsöld's in Kiel, and Jean-Claude Pecker (b. 1923) (75) in Paris tackled this difficult problem almost simultaneously. Both of these investigations involved a statistical treatment of putting the many thousands of real lines into groups represented by a few stronger fictitious lines distributed in different wavelength intervals. As Labs has pointed out, this involves solving a complex set of integral equations for radiative equilibrium, and in the days of mechanical

* The line weakening is partly due to increased ionisation at the higher temperatures, as given by the Saha equation. In addition, in the deeper layers the intensity more nearly approaches the line-free isotropic radiation field of a black body.

calculators operated manually by turning a handle, this was no mean feat! Labs' result was a marked cooling of the outermost solar layers only, to about 4300 K, but almost no effect on $T(\tau)$ for depths deeper than τ of about 0.01. He also remarked 'the precision of our improved surface temperature is not all that high.' Pecker also found this lower surface temperature (4200 K in his case), but his model in addition showed a rise in temperature near optical depth of 0.15 caused by the backwarming effect of the line blanketing, as predicted by Chandrasekhar. For greater depths the temperatures were practically the same as in the model without line blanketing.

The fullest theoretical treatment of solar line blanketing during this early period came from Karl-Heinz Böhm (b. 1923), another in the remarkable series of students that came from Unsöld's school from the early 1950s onwards (76). He iterated a line-blanketed solar model to flux constancy and claimed good agreement with the empirical model determined by his wife, E. Böhm-Vitense (61), between optical depths 0.05 and 1.5. He stressed the unreliability of his method at shallower depths.

10.7 Successive refinements to stellar model atmospheres 1940–65

For the analysis of stellar spectra, a variety of approaches of different complexity were used. The simplest viewpoint was to consider a star as having a sharply bounded photosphere radiating a continuous Planck spectrum of characteristic temperature T_{eff}. Surrounding this was a reversing layer in which the lines were formed, and which could be approximated by single mean parameters for temperature and electron pressure. This was the so-called Schuster–Schwarzschild model, which formed the basis for the method of 'coarse analysis' used by Unsöld for his pioneering work on the B0 star τ Sco (77).

The next possibility was to take approximate account of the atmospheric structure by using the grey $T(\bar{\tau})$ solution, which, for example, was the method of Lawrence Aller in 1942 for his analysis of the A0 dwarfs γ Geminorum and Sirius (78). Here the optical depth $\bar{\tau}$ was calculated from the mean of the non-grey absorption coefficient, the usual method being to take the weighted mean given by S. Rosseland (79).*

Since such models are not in general in radiative equilibrium, the next refinement was to iterate the $T(\bar{\tau})$ relationship by one of the standard numerical techniques for achieving flux constancy. This is particularly

* The Rosseland mean opacity is a harmonic mean weighted by the monochromatic derivatives ($\partial B_\lambda/\partial T$) prevailing at the depth where the opacities are evaluated.

important for earlier type stars in which the very non-grey neutral hydrogen opacity is dominant. Such flux constant models were first computed for early B stars by Anne B. Underhill (b. 1920) in Copenhagen (though further iteration beyond the first was deemed unnecessary) and by J.-C. Pecker in Paris in 1950 (80, 81) and by many other astronomers since.

Another refinement is some of the early models was the inclusion of convective transport of energy in a stellar atmosphere. Although Schwarzschild had shown that energy flowed outwards by radiation in the outer layers of the sun (1), he had also considered the stability of the radiative equilibrium. When the temperature gradient became too steep then convection would occur and at least some of the energy flux would be transported by this means. At the same time, the temperature gradient was reduced in the convective zone, thus altering the structure of the atmosphere and in principle this could affect the observed spectrum. Schwarzschild was unable to say, however, whether or not convection might play a rôle in the deeper layers. That the convective transport of energy does occur in these deeper layers of the sun's atmosphere was first shown theoretically by Unsöld in 1930: 'From numerical calculations it results that in the solar atmosphere, which should contain a considerable percentage of hydrogen, there is a zone which is roughly bounded by optical depths $\tau = 2$ and 30, which must contain convection currents. These are the cause of the solar granulation. . . .' (82). The convection zone in the sun in fact begins at optical depth about unity and is associated with the region in which hydrogen is partially ionised.

The problems of flux constant models that include line blanketing are much more difficult, if all the lines are to be accounted for, even though a statistical approximation is generally used. But there is one easy way out applicable to stars whose temperature structure is similar to the solar one. And that is to take the empirical solar temperature structure as determined by limb darkening and scale it according to the ratio of effective temperatures of the star to be modelled to that of the sun. Such an approach gives a stellar $T^* (\tau)$ relationship equal to $(T_{\mathrm{eff}}{}^*/T_{\mathrm{eff}}{}^{\odot})\,T^{\odot}(\tau)$ where $T^{\odot}(\tau)$ is derived from solar limb-darkening observations. Such a method was used by Roger Cayrel (b. 1925) and Jun Jugaku in 1963 (83) for the calculation of a grid of dwarf star models with effective temperatures from 3600 to 7200 K and abundances of metals from solar to one hundredth the solar content. Clearly this is an elegant solution to the stellar blanketing problem as the effect of the lines on the structure is automatically taken care of, and no iteration to flux constancy is required. However, the weakness of the method is evident, if stars with much more or much less than solar blanketing are considered. In the case of the Cayrel-Jugaku grid, the solar abundance stars at 3600 K were predicted to have their weak iron lines about a thousand times stronger than the same lines in the hottest metal-poor stars modelled.

For early-types stars scaled solar models are of course inapplicable, but here the line blanketing is due primarily to the strong Balmer lines, and also, for stars hotter than about A0, to the Lyman lines and some strong ultraviolet lines of ionised carbon, nitrogen and silicon. Since the principal lines giving rise to blanketing are few in number, an approximate treatment of their influence on restructuring the temperature distribution of a model can be undertaken. This was done by Eugene H. Avrett (b. 1933) and Stephen E. Strom (b. 1942) (84) at the Harvard-Smithsonian Observatory. Their models cover the range of effective temperatures from 10 000 to 20 000 K and gravities of $\log g = 3.0$ and 4.0 (cgs units). The results presented show the effect of various physical assumptions on the observed fluxes and on the Balmer jump, and for this reason are particularly instructive. Avrett and Strom considered in turn, models computed with a grey $T(\tau)$ structure, secondly those with a $T(\tau)$ corresponding to radiative equilibrium in the absence of line blanketing, thirdly they added the blanketing of the overlapping wings of higher members of the Balmer and Lyman lines of hydrogen, and finally they considered the additional line blanketing due to the cores of various strong ultraviolet lines of ionised carbon, nitrogen and silicon and of the first five Lyman lines. Since these line cores are all deep, the models simply removed all the flux in appropriate wavelength bands. The hydrogen wing opacities generally raised the level of the Balmer continuum ($\lambda < 3650$ Å) and reduced the size of the Balmer jump. The further addition of the strong line cores had a quite marked effect by raising the Balmer continuum, but especially at the shorter wavelengths for the hotter stars. The Balmer jump however remained unaltered.

Soon afterwards a comprehensive set of Balmer-line-blanketed models for stars of spectral type B8 to F2 was computed by Dmitri Mihalas (b. 1939) (85). To treat the line-blanketing problem, the first eighteen Balmer lines were considered, but lines below the Balmer jump were not included (the flux in the far Balmer continuum is in any case not large). From a grid of twenty models including dwarfs and evolved stars, fluxes were computed, from which such observable parameters as colours on the UBV system of photoelectric photometry and the Balmer jumps were found. These results led to a new calibration of the effective temperature scale for A stars based on the theoretical colours, provided a correction was made to allow for the effect of line blocking by the metallic lines. For the fundamental standard star Vega (A0V) Mihalas determined a value

$T_{eff} = 9790$ K from the colour indices, while the flux distribution and Balmer jump gave 9600 K. These models were still only an approximation since only Balmer line blanketing was treated; the observations they were compared with still required perfecting, especially the absolute flux calibration of Vega as it stood at that time. However, further developments can be regarded as fine tuning, and with the Mihalas calibration the difficulties in interpreting the flux gradients of A stars, that were so prominent in the 1920s, can be regarded as overcome.

10.8 The analysis of stellar spectra: four basic prerequisites The quantitative analysis of stellar spectra had its first tentative beginnings in the 1920s in the work of Cecilia Payne with her 1925 thesis and monograph *Stellar Atmospheres* (14), followed by that of Russell and Adams who analysed seven stars in 1928 (86) (see Chapter 7). After Minnaert and his Utrecht colleagues had developed the idea of the curve of growth for the solar spectrum, and after the application of this technique to stars by Pannekoek (87), by Struve and Elvey (88), by Louis Berman (89) and by Leo Goldberg (90), the stage was now set for the first detailed analysis of stellar spectra to obtain quantitative compositions. The essential requirements which opened the way forward were firstly, accurate stellar temperatures; secondly, reliable stellar line strength measurements; thirdly, accurate line identifications; and fourthly, sufficiently precise oscillator strengths for all the spectral lines to be analysed. Substantial progress was made in the first three of these prerequisites in the 1930s; the oscillator strength problem took longer to solve (lack of good data is still in the 1980s a major handicap), but significant strides forward were made, especially after the Second World War.

10.8.1 The effect of adopted temperature on derived abundances The first essential requirement was accurate stellar temperatures. The temperatures could come from the line spectra themselves as ionisation or excitation temperatures, or from the continuous spectra as colour temperatures. As we have seen, McCrea's model atmospheres were the first step in reconciling the colour temperatures with the ionisation temperatures (26) and successive improvements in stellar models rendered increasingly accurate temperatures. The importance of temperature in interpreting line strengths is illustrated by the example of a medium strength (equivalent width say a few hundred mÅ) neutral iron line in a solar-type star absorbing from a low-lying energy level. Doubling the iron abundance might typically increase this line's equivalent width by only 25 per cent, which on noisy photographic spectrograms may be less than the precision of measurement. On the other hand, lowering the temperature by only $1\frac{1}{2}$ per cent or about 260 K would produce the identical line strength increase. Measuring stellar temperatures to this accuracy is not easy, though it can be achieved with care from the flux gradients. For this reason alone it can be seen that the accuracies of stellar abundances are unlikely to be better than a factor of two, a figure also given by Unsöld for his analysis of τ Sco (albeit of B spectral type) and by many others since. Certainly many stellar temperatures are now known to a precision of twice this, or around 100 K, but then the random errors in the line strengths also have to be contended with.

10.8.2 Equivalent widths of lines by microdensitometry Reliable line strengths were the second requirement, and the development of recording microdensitometers, in some cases that could directly produce intensity records from the density, or blackening, in the non-linear photographic emulsion, was the important instrumental development. The Moll microphotometer (as it was then called) at Utrecht enabled Minnaert to produce the famous Utrecht solar atlas (see section 7.21). Another example of an instrument of considerable sophistication, that would simultaneously monitor a wedge calibration spectrum and use this to produce a direct intensity plot of a stellar spectrum on a second plate, was built at the University of Michigan Observatory in the late 1930s (91) (Fig. 10.9). With this instrument Albert Hiltner (b. 1914) and Robley Williams in 1946 produced their *Photometric Atlas of Stellar Spectra*, the first such atlas which presented tracings of the spectra of eight bright stars* covering the range of spectral types from B8 to M2 and luminosity classes from Ia to V (92). The atlas was produced from McDonald coudé prism plates, and covered the entire visible range from 3920 to 6720 Å. It provided a valuable reference for studying line profiles, line strengths and for distinguishing blended lines in stars of different types (Fig. 10.10). In the words of Chalonge: 'This atlas will become the indispensable tool of all those who deal with stellar spectroscopy' (93), while for Minnaert it was 'a continuous joy to "read" these records and to recognize many features, well known from verbal descriptions but now, for the first time, seen in a graphical representation' (94).

The need for high resolution spectrophotometric atlases for bright representative stars of different spectral types was recognised by Roger

* The stars were Rigel (B8Ia), Vega (A0V), Sirius (A1V), Deneb (A2Ia), α Per (F5Ib), Procyon (F5IV), Arcturus (K2p) and Betelgeuse (M2Ib).

Fig. 10.9 The Hiltner and Williams direct intensity microdensitometer at Michigan, 1940.

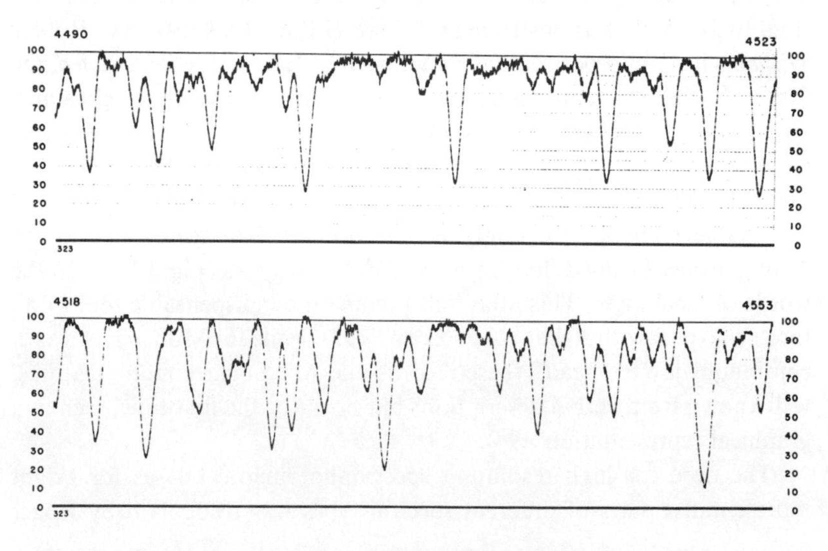

Fig. 10.10 A sample page from the Hiltner–Williams 'Photometric atlas of stellar spectra' showing part of the spectrum of α Persei.

Griffin at the University of Cambridge. In the early 1960s he obtained an extensive set of spectrograms of the metal-poor red giant Arcturus (spectral type K2p) on the Mt Wilson coudé spectrograph. The 114-inch camera was used to achieve dispersions of about 1.0 or 1.5 Å/mm in respectively the third order blue region or the second order red. Twenty-seven plates covering the spectral range from 3600 to 8825 Å were used; the calibrated microdensitometer tracing displayed the spectrum of Arcturus graphically with a resolution of about 28 mÅ in the blue or 40 mÅ in the red (95). This was the first star for which a spectral atlas has been produced comparable in quality to the Utrecht solar atlas of Minnaert and his colleagues. It is the standard reference for K-star spectrophotometry.

10.8.3 Line identification in standard stars The third requirement for any extensive analysis was a means of accurate line identifications in standard stars. Obviously it was essential to know the element, ionisation stage and transition that gives rise to any line being measured. Moreover, these data were necessary for detecting blends of two or more lines, which on even the best stellar spectra could appear as single. For this reason laboratory data alone were not enough, and needed to be supplemented with stellar identifications for stars of different spectral type. The prime source for cooler stars has always been Rowland's tables, which were revised in 1928 by St John and his colleagues (70) and listed 21 835 lines together with their wavelengths and Rowland intensities (including those in the infrared supplement)* of which fifty-seven per cent were 'referred to their sources'. With the publication of the Utrecht solar atlas in 1940, the opportunity arose for a second revision incorporating equivalent width measurements instead of the old Rowland intensities. Minnaert described in 1963 how the second revision came into being:

> These graphs contain an almost inexhaustible wealth of information. In order to put this material to use, the first step was obviously the derivation of a catalogue of Fraunhofer lines, with strengths expressed in equivalent widths.... Houtgast and other Utrecht astronomers had started the systematic measurements of equivalent widths already during the war. We had the good fortune, that Mrs Moore and Dr. Babcock were just then preparing a new edition of Rowland's table and that our intensities could be incorporated in this great enterprise. How many letters were exchanged in the course of this collaboration! It took more time

* The infrared region was revised in 1947 by H.D. Babcock and C.E. Moore (96), and the ultraviolet region was revised in 1948 (97).

> than we expected. At each consecutive meeting of the IAU we had the pleasure to meet and to discuss the progress of the work, but also we felt bitterly the regret that it was not yet finished and the anxiety that we would perhaps never reach the goal. Thanks to Mrs Moore, who carried the heaviest burden, the goal is practically attained. The ledgers are ready and the proofsheets are being corrected, what we have felt ever and ever again in the course of these years, that is the deepest admiration for Henry Rowland, who accomplished a similar enterprise 70 years ago, with so much less technical means, and whose work is still now a marvel of perfection (98).

The result, two years later, was the *Second revision of Rowland's preliminary table of solar spectrum wavelengths* by Charlotte Moore, Marcel Minnaert and J. Houtgast (99) with some 24 000 solar lines from 2935 Å to 8770 Å, 73 per cent of them with identifications – the cool-star spectroscopist's bible. If we detect some impatience in Minnaert's words of 1963, he didn't mention then that the Utrecht equivalent widths were published in 1960 'in advance of the complete work, because it is felt that such data are urgently needed in view of abundance determinations and comparisons with stellar spectra' (100). The ultraviolet part of this catalogue was based on new plates taken and reduced in Dublin at the Dunsink Observatory by H.A. Brück and his colleagues, to overcome the poor ultraviolet resolution of the original Utrecht atlas.

For line identifications in the spectra of all celestial sources other than solar-type stars, the standard reference work is Charlotte Moore's *Multiplet Table of Astrophysical Interest* (101) in its two editions of 1933 and, after revision, of 1945. Pol Swings (102) described the origin and contents of this well-known work:

> Since 1933 one book could always be found on the desk or within reach in the office of any solar or stellar spectroscopist; the book was always worn out and patched up. It was the first edition of Miss Moore's multiplet table, probably the most helpful table for astrophysicists in recent years. The original table had not really been planned for publication. An extension of Russell's famous "List of Ultimate and Penultimate Lines of Astrophysical Interest",* it had been prepared by Miss Moore in 1931 for her own use while working on the identification of atomic lines in the solar and sunspot spectra. Her manuscript was published in 1933 at the request of numerous colleagues; but the book was soon out of

* See (103).

> print, and, as a result of many laboratory investigations, it has become out of date for certain elements. When the need for a revision and extension became apparent, a program was planned at Princeton for the preparation of a revised edition. This program has required years of labor and has now been brought to a successful conclusion. The revised edition will save thousands of hours of labor to many astrophysicists, who will owe a great debt of gratitude to the author.
>
> The work contains two parts: a *Table of Multiplets*, and a *Finding List* of all the lines in the multiplet table. Each part is again subdivided into two sections – one on permitted lines (more than 23 000 lines), the other on forbidden lines (more than 2550 lines) (102).

However line identifications and wavelength measurements in the spectra of bright standard stars have also played an important role. This type of work was initiated by Vogel and his colleagues at Potsdam in the 1890s, with the introduction of quantitative measurements from stellar spectrograms. The extensive publication by Julius Scheiner (104) catalogued the lines in 47 stars (including the sun) covering a wide range in spectral types, and wherever possible he gave elements identifications and comparisons of the stellar and solar wavelengths. In the 1930s and 1940s this type of cataloguing work was carried on by numerous workers; Paul Merrill has published a useful list of over one hundred references of this type up to 1954 (105). Here only a few of the most notable stellar line catalogues are mentioned.

The compilation by Theodore Dunham, Jr. (1897–1984) of 1440 lines in the F5 supergiant α Persei deserves special mention, both because of its relatively early date and also because of the high dispersion and quality of the spectrograms. Dunham presented the α Persei catalogue for his doctoral thesis at Princeton under Russell in 1927 and it was published in 1929 (106) soon after the first revision of Rowland's tables by St John and colleagues. The observational material was obtained on the coudé prism spectrograph of the 100-inch telescope at Mt Wilson, with a dispersion of 2.9 Å/mm at Hγ. The catalogue listed wavelengths, intensities, identifications, excitation potentials and multiplet designations. It can only be regarded as a model of how to present an exhaustive description of a high dispersion stellar spectrum.

For B stars, the list of 296 lines in the B0 dwarf τ Scorpii by Struve and Dunham in 1933 is worth recalling, because this star was later to play such an important role in spectral analysis (107). For A stars the description of the spectra of thirteen 'type stars' by W.W. Morgan in 1935 became the standard work (108). Several hundred lines and identifications were listed for each star. The only publication of greater completeness that predates Morgan's work on A stars is the study of γ Geminorum (also in Morgan's list) by Sebastian Albrecht (1876–1960) at the Dudley Observatory, Albany N.Y. (109). For F stars Pannekoek, Greenstein and K.O. Wright all published line lists in the 1940s; but the work of J.W. Swensson at the McDonald Observatory on Procyon (F5IV) was especially notable since it included as many as 3600 lines (110). For solar type stars the Rowland table and its revisions served all practical purposes. However, Pannekoek measured the blue spectra of seven stars of types F3 to G6 from spectra obtained on a visit to the Dominion Astrophysical Observatory in Canada in 1929 (111), but the results were not published until 20 years later.

For stars cooler than the sun, Charlotte Moore at Princeton catalogued and analysed lines in the solar sunspot spectrum in 1932 (112). She determined that the spots are typically about 1000 K cooler than the normal solar photosphere and hence spectroscopically resemble a K-type star (a fact noted by Secchi as early as 1869 (113)). She catalogued 6312 sunspot lines from observations with the Mt Wilson 150-foot solar tower telescope for this work. Also at Princeton, Sidney Guy Hacker (b. 1908) continued in Dunham's footsteps with a catalogue of 3883 lines in the K2 giant Arcturus, once again based on Mt Wilson coudé spectra. About 3000 of these lines were identified as coming from 28 different atomic elements and 5 diatomic molecules (114). (For a discussion of the catalogue see (115).) Finally, for the M stars, we come to the work of Dorothy N. Davis, for her dedication surely the most remarkable of all those who have worked in this field. Her two huge compendia of stellar lines were for Antares (M1Ib) in 1939 (116) and for β Pegasi (M2II-III) in 1947 (117). For the latter star 10 000 lines were assigned wavelengths, intensities and, where possible, identifications. The lines came from 45 atomic elements and at least 15 diatomic molecules. This still represents the lengthiest study of any astronomical spectrum other than that of the sun.

10.8.4 The need for absolute oscillator strengths The fourth essential ingredient for the analysis of stellar spectra was the need for the oscillator strengths (or *f*-values) for the lines of different elements observed in stellar spectra. The need for both high quality oscillator strengths and for a large quantity of lines has always been acute. In 1934 only a handful of lines had known oscillator strengths as can be seen from the meagre table and references given by Mitchell and Zemansky in their textbook of that year (118).

Some of the early measurements of relative intensities of emission lines in arcs by R. Frerichs in Germany (119) and by G.R. Harrison in the U.S. (120) suffered from self-absorption and they were not on an absolute scale. Emission-line spectrography of a copper alloy arc was used much later by C.W. Allen and A.S. Asaad in London (121). At Mt Wilson, Robert and Arthur King began an extensive program in 1935 of absorption spectroscopy using a white light source shining through an electric furnace containing low density metallic vapour (122). Once again these results were for relative oscillator strengths and could therefore be used for determining stellar excitation temperatures (from the Boltzmann law) but not abundances. Absolute calibration was in principle possible, provided the metallic vapour pressures at the furnace temperatures were known, which was a major difficulty, or by use of the *f*-sum rule of W. Kuhn (123).*

Another technique, known as the 'hook method' used the method of anomalous dispersion in gases, and involved measuring the refractive index of a gas at wavelengths in the neighbourhood of an absorption line. This method had been used by D. Roschdestwensky in St Petersburg for the D lines in sodium vapour as early as 1910 (124).

Apart from laboratory measurements of oscillator strengths, a theoretical evaluation of absolute oscillator strengths is possible if wave functions can be calculated. In London D.R. Bates (b.1916) and Agnete Damgaard simplified these calculations by using hydrogenic wave functions, and showed that reliable results were possible at least for the transitions between the higher levels of the lighter atoms (125). The atoms with more complex electronic structure, such as those of the iron group which have many lines in solar-type stars, could not be treated by this method.

In 1946, the International Astronomical Union established a subcommittee of Commission 14[†] for 'Intensity Tables' of spectral lines under William Meggers (1888–1966) at the US National Bureau of Standards. Its objectives were to encourage and coordinate quantitative work on line intensity problems, as well as to collect and compile data in a form useful to astrophysicists. The motivation was stated to be the 'extremely meagre and inhomogeneous' data on *f*-values (i.e. oscillator strengths) available at that time (126). The successive reports of this subcommittee from 1950 onwards give some indication of the enormous accumulation of new oscillator strength determinations since then, though rarely at a rate and of the

* The *f*-sum rule states that the sum of the oscillator strengths for all lines from a given configuration equals the number of optically active electrons.

† This was the IAU Commission for Standard Wavelengths. It later became the Commission for Fundamental Spectroscopic Data.

quality to satisfy those undertaking the quantitative analysis of astronomical spectra.

One of the new methods able to give absolute *f*-values was the atomic beam method introduced in Göttingen by Hans Kopfermann and Günter Wessel (127). Absorption line spectroscopy was undertaken by passing light through the beam, in which the density of absorbing atoms was calculated by condensing the beam onto a plate which was then weighed. A similar atomic beam apparatus was also established at Caltech a few years later (128). The instrument is not simple to operate and large amounts of data did not ensue.

To measure the oscillator strengths of ions or of highly excited atoms, high temperature laboratory sources were necessary. One such source was the whirling-water stabilised arc of W. Lochte-Holtgreven and H. Maecker at Kiel (129) in which temperatures from 10 000 to 50 000K were possible. Another high temperature source is the shock tube, developed most notably at the University of Michigan from 1952 by Otto Laporte and his colleagues (130). Temperatures of over 10 000K were also attainable.

In spite of these various new methods for determining oscillator strengths, the still meagre amount of data even on relative oscillator strengths was emphasised by Robert King in 1954 (131). He tabulated data on relative line strengths for determinations involving ten or more lines from levels of different excitation. Only nineteen elements and fewer than 1900 lines were included. Only one element, titanium, was listed with relative intensity data for an ionised state.

A critical evaluation of all available oscillator strength data was undertaken by Goldberg, Edith Müller (b. 1918) and Aller at Michigan in 1960, at the time of their detailed reanalysis of the composition of the sun (132). Forty-two elements were considered and the quality of the best *f*-values estimated in each case. Only eleven elements and about 200 lines of those they measured had satisfactory absolute *f*-values experimentally determined at this time; a further seventy-one lines for seven elements had theoretical oscillator strength determinations of a quality they judged to be satisfactory.

A massive program to rectify the inadequate quantity of relative intensity measurements for spectral lines was commenced by W.F.G. Meggers at the National Bureau of Standards in the 1930s. The program ran for about 25 years; seventy elements and 39 000 spectral lines were measured by photographic emission-line spectroscopy from an arc struck between copper electrodes containing in addition exactly one part in a thousand of the desired element (133). Charles Corliss (b. 1919) and W.R.

Bozman undertook the calibration of these intensities for 25 000 of the lines to produce absolute oscillator strengths; this involved solution of the Saha equation in the arc and comparison with absolute *f*-values in the literature to convert from a relative to an absolute scale (134). These became by far the most extensive data set on *f*-values in the literature. However the accuracy of the absolute values was not high; a standard deviation for any given line of between 70 to 100 per cent was claimed, a level of precision which many astronomers have found inadequate for absolute abundance work on stars.

For solar-type stars, iron is easily the best represented element in terms of line numbers. Yet only about fifty blue neutral iron lines with *f*-values (out of 200 or 300 readily measurable) were tabulated by Corliss and Bozman. A new list of over 2000 lines of neutral iron was produced by Corliss and Brian Warner (135). This was from arc measurements of Corliss in 1964 though the data were combined with other measurements in the literature.

The determinations of the lifetimes of the excited states of atoms and ions can also lead to values for the oscillator strengths; of various techniques for doing this, that of beam foil spectroscopy developed by Stanley Bashkin (b. 1923) in the early 1960s has resulted in rapid progress. Together with A.B. Meinel (b. 1922) at the University of Arizona he passed positive ions accelerated to between 0.5 and 2 MeV through a thin foil. The passage through the foil gives rise to excitation or further ionisation, and the subsequent decay by photon emission is observed optically downstream (136). The similarity to the highly excited spectrum of a nova was noted.

Beam foil spectroscopy has given much new data on absolute *f*-values for lines in ionised spectra. By 1969 nearly half of the experimental references in the National Bureau of Standards, (NBS) *Bibliography on Atomic Transition Probabilities* (137) for the spectra of ions are beam-foil measurements. The results have shown serious systematic errors in the much used data of Corliss and Bozman. An example of such errors is afforded by the almost simultaneous discovery of several experimental groups in 1969 that the absolute scale of neutral iron oscillator strengths was about ten times smaller than the previously accepted values. This was the conclusion of a group in Kiel using a stabilised arc burning in argon (138), of the Harvard group from shock-tube measurements (139) and of the Caltech group from beam-foil spectroscopy (140). The result necessitated a corresponding ten-fold increases in the accepted solar iron abundance, a fact that illustrates well how much the astronomer is dependent on laboratory physicists for fundamental spectroscopic data.

10.9 Four pioneers in stellar abundance analysis: Unsöld, Greenstein, Aller, Wright

10.9.1 Unsöld and τ Sco: the method of 'Grobanalyse' Without question, the abundance analysis of the B0 dwarf star τ Scorpii by Albrecht Unsöld in 1941 represents a major step forward in the analysis of starlight (77). It was not the first attempt to determine stellar composition, by any means, (Payne, Russell and Adams took this honour), but it was by far the most detailed, was based (mainly) on high dispersion spectra, and above all the input physics (especially curve of growth theory) was applied with an assuredness that not even Russell in 1928 could hope to match. It is interesting to know why Unsöld chose an early B-type star for his first analysis of a stellar spectrum, and not a solar-type object in spite of the wealth of information already available on the solar spectrum and the physics of the solar atmosphere. He wrote: 'Previous investigations especially by O. Struve, show that early type stars promise considerable advantages with respect to their quantitative analysability. It is just the presumably abundant light elements – barely observable in the sun – which are represented here in several ionisation stages through numerous lines.' He continues by explaining how it is easier to solve the Saha equation in these circumstances to get the total abundance of an element, and notes that the results are less sensitive to the adopted temperature than for solar-type atmospheres. τ Scorpii is a bright southern ($m_v = 2.8$, declination $-28°$) sharp-lined B dwarf and was observed by Unsöld and Struve on the newly commissioned 82-inch reflector at McDonald Observatory during a 6 month visit that Unsöld spent at Yerkes and McDonald Observatories in 1939. Both coudé (dispersion 2.64 Å/mm at Hγ) and cassegrain (54 Å/mm at Hγ) spectra were obtained, the latter to study the effect of dispersion on the measured equivalent widths* and to extend the spectra to 3324 Å in the ultraviolet. The coudé exposure times were long, all over 5 hours. The spectrograms were traced on the Yerkes microdensitometer and Unsöld described the extremely laborious point-by-point hand rectification of the line profiles to convert from density to intensity, followed by the measurement of equivalent widths using a simple area integrating tool, the planimeter. About 200 lines were measured in this way, most of them from several spectra at both dispersions.

In the second paper on τ Sco Unsöld proceeded with his analysis. This was based on the method of 'Grobanalyse' or coarse analysis, based on assigning mean values of temperature and pressure to the line-forming

* The low dispersion line strengths were systematically 20 to 25 per cent larger.

Fig. 10.11 Albrecht Unsöld.

reversing layer of the Schuster–Schwarzschild model. The thickness of this layer was taken to be H, and the results expressed as 'the number of atoms above the photosphere' or log NH for each element. Theoretical curves of growth were computed for lines formed by pure absorption using an approximate formula first proposed by Minnaert which relates the flux depression below the continuum (R) to the abundance of atoms in the appropriate energy level for a given transition.* The Minnaert formula had the advantage of giving finite residual fluxes ($1\text{-}R_c$) in the centres of strong lines which, in the case of τ Sco, Unsöld estimated to be only about 45 per cent below the continuum. By using an approximation to the Voigt profile for the absorption cross-sections, the theoretical equivalent widths could thus be found by integrating the flux depression R over each line. The procedure was similar to that used by Menzel in 1936 (141). Depth-dependence of the line opacities in the reversing layer was therefore not taken into account. The cross-section for each line is proportional to the oscillator strength, and provided this fundamental parameter is known for each observed transition, then the theoretical curve gives directly a determination of log ($N_{r,s}H$) from the observed equivalent width expressed as log (W/λ). Here $N_{r,s}H$ is a column density of atoms in excitation level s

* The Minnaert formula is $\frac{1}{R} = \frac{1}{R_c} + \frac{1}{\tau_\lambda}$ where the optical depth in the line is $\tau_\lambda = \alpha_\lambda N_{r,s}H$. R_c is the maximum flux depression (centre of strong lines) and α_λ is the absorption cross-section per absorbing atom.

and ionisation state r. Unsöld used exclusively theoretically estimated oscillator strengths based on the formulae of Russell, Goldberg and others for the relative strengths of a line within a multiplet, and he used the so-called hydrogen-like transitions for a calibration between multiplets. For strong-line transitions he simply put $f \approx 1$.

Provided the temperature and electron pressure were known, then the Boltzmann and Saha equations together could be solved to find the total abundance of any element, log *NH*. For the electron pressure Unsöld used the Inglis–Teller formula and the strength of the Stark-broadened Balmer lines. The Saha equation was now applied to the ionisation equilibrium of several ions observed in two ionisation states to give the ionisation temperature of $T_{ion} = 28\,150 \pm 750$ K. This is the fundamental temperature determination used for τ Sco to obtain NH from $N_{r,s}H$ for all elements. Unsöld thus avoided interpreting the flux gradient or colour temperature of τ Sco (which is just as well since the star is considerably reddened by interstellar dust).

The final results were given for nine light elements including hydrogen (hydrogen, helium, carbon, nitrogen, oxygen, neon, magnesium, aluminium, silicon). These showed the proportions by number of hydrogen: helium: other elements of 85:15:0.24. In conclusion Unsöld notes the good agreement for τ Sco with Russell's (29) solar composition and in particular 'the enormous abundance of hydrogen' was now confirmed. Since Strömgren had arrived at the same conclusion for the sun by taking the H^- opacity into account (49) the study of τ Sco provided indirect support for the H^- opacity in cooler stars, as Unsöld explicitly noted.

This was the first coarse abundance analysis of any star using the curve of growth technique with the exception of Louis Berman's work on R CrB (89). It is perhaps surprising that Unsöld's reputation for stellar spectral analysis rests on the results for only one star (other than the sun). The explanation lies in his enormous contributions to the theoretical foundations of spectral analysis and the methods to be used, as expounded in the two editions of *Physik der Sternatmosphären* and in numerous papers in the *Zeitschrift für Astrophysik*. After the war a remarkable sequence of students came to Kiel to work under him. They all benefited from his incredibly broad and deep insight into the physics of stellar atmospheres. Many of these students applied Unsöld's techniques to the analysis of further stars – Gerhard Traving (b.1920) for 10 Lac (142), Kurt Hunger for Vega (143), Bodo Baschek for the subdwarf HD 140283 (144) are examples. The Kiel school has remained in the 'top league' for this type of work, though as we shall see, not without fierce competition from the West Coast Americans, from Paris and from Greenwich.

10.9.2 Greenstein and the differential analysis of F stars I next consider early contributions of Jesse Greenstein to the abundance analysis of stars, during the time he spent at Yerkes Observatory in the 1940s under Struve. In a series of papers from 1942 to 1949 he developed the important technique of differential abundance analysis. His first paper on abundances was for the peculiar hydrogen-deficient spectroscopic binary **υ** Sagittarii (145). Greenstein showed that helium was about one hundred times as abundant as hydrogen in this star (compared to the normal helium-to-hydrogen ratio of 0.1) (see section 9.9). However, the results were not analysed differentially, they were based on medium dispersion spectra only, and the abundances for iron were left in the form of the number of neutral atoms or ions above the photosphere.

From the point of view of analysis technique, the methods he used for the bright southern F0 supergiant Canopus are more significant (146). That he observed this star at all from McDonald is in itself remarkable, since it is always less than 7 degrees above the southern horizon.* The dispersion of his spectra was high and varied between 25 and 4 Å/mm. He measured the equivalent widths of about 300 lines in the blue region which were used in the analysis. The crux of the differential method that Greenstein now introduced was based on the realisation that the *relative* abundances of elements in Canopus to the sun could be determined by comparing the strengths of the same lines in the two stars. The advantage was that line oscillator strengths were not required, and these were previously one of the sources of uncertainty in abundance work. In effect, using the sun as a standard star implies that the solar atmosphere becomes the furnace from which the oscillator strengths are determined; however, the conversion of solar line strengths into *f*-values is not explicitly carried out, but only into a parameter X_0, the optical depth in the line centre of the reversing layer on a Schuster–Schwarzschild model. Greenstein obtained the X_0 values for solar lines from the Menzel–Baker–Goldberg solar curve of growth (147) based on C.W. Allen's Mt Stromlo equivalent widths (148).

These X_0 values for solar lines are proportional to the generally unknown oscillator strengths. After adjusting them for the known temperature difference of the sun and Canopus, Greenstein showed how to construct a curve of growth for lines in the spectrum of Canopus (Fig. 10.12).

* Curiously, Unsöld (τ Sco), Greenstein (υ Sgr, α Car) and Aller (Sirius) all included southern stars in their first abundance analyses. At this time there was no coudé spectrograph south of the equator, and the spectral observations were all poached by the observers at McDonald and Mt Wilson!

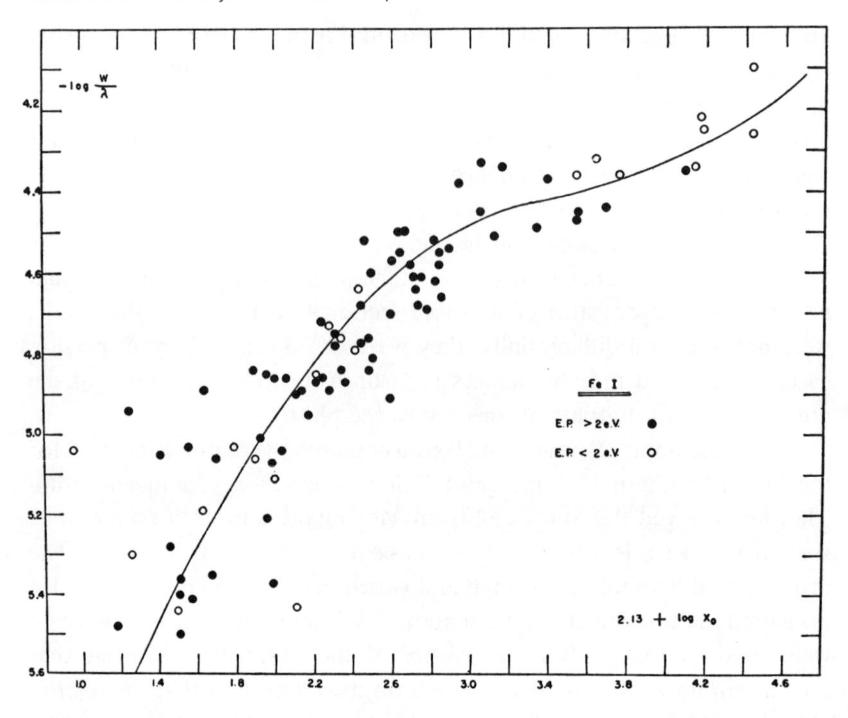

Fig. 10.12 Greenstein's differential curve of growth for iron lines in Canopus, 1942.

This is the principle of the differential curve of growth analysis. The measured horizontal shift gives the relative amount of a given absorber in the Canopus reversing layer to that in the sun. Greenstein did not, in this paper, develop the differential method to the extent of deriving element abundances. Instead he assumed the same overall composition for Canopus and the sun and simply derived the degree of ionisation for each element in the Canopus atmosphere. He thus did not push the potential of his method to its ultimate conclusion. As an afterthought, at the end of this paper, he reverted to using the unreliable theoretical oscillator strengths to obtain element abundances which he expressed relative to carbon for five other elements, including hydrogen (from Balmer line wings). In some ways, this reversion to an absolute analysis in the last paragraph spoilt the beauty of the paper as a whole. If he had followed Unsöld* for τ Sco, Greenstein would have corrected for ionisation and deduced overall (i.e. over all ionisation states) element abundances relative to the sun; if he had followed Strömgren in 1940, he would have deduced abundances relative to hydrogen via the hydrogen-dependence of the continuous opacity rather

* Unsöld's paper on τ Sco was not yet published but evidently Greenstein was aware of the results as these are referred to.

than using the Balmer lines. Strömgren's important paper on the sun was only referenced in the last paragraph, so it can be assumed the bulk of the Canopus analysis was completed before Greenstein had access to these works. In the meantime K.O. Wright at Victoria independently developed the method of differential analysis and published element abundances relative to the sun for three F stars (see section 10.9.4.).

In 1948 Greenstein published a differential curve-of-growth analysis of four F stars (ρ Pup, θ UMa, Procyon, α Per) and of the metallic-line star τ Uma (149). Here the method of differential coarse analysis is developed to its logical conclusion with the sun as a standard. The H and H^- continuous opacities are explicitly included, and the Saha equation is used with measured values of electron pressure and ionisation temperature,* to derive the overall element abundances from the abundances of individual ions or neutral atoms. The results for elements i were expressed as log $(Z_i/Z_i(\odot))$ where Z_i is the mass of element i per gram of stellar material. Furthermore, the pros and cons of the theoretical curves of growth based on the Schuster–Schwarzschild and the Milne–Eddington models were discussed; neither fitted the empirical solar curve of growth particularly well. However, the theoretical Milne–Eddington (ME) curves of Pannekoek and van Albada (150) agreed well with that calculated by Strömgren using a variable η with depth (49) (see section 10.5), and were used for Greenstein's work.

It is interesting to note that Miss Helen Steel at Harvard also produced a differential curve of growth for α Persei in 1945 from which she obtained an excitation temperature of 4400 K from neutral iron lines (151). However, her analysis from this point on made use of oscillator strengths, so from the point of view of technique, it broke no new ground. However, her analysis of α Persei preceded Greenstein's (149, 152) by over 2 years. Greenstein obtained a more realistic excitation temperature of 5100 K. Surprisingly, he does not reference the Harvard work nor comment on the cooler temperature found by Miss Steel.

The final results of his differential work Greenstein gave as follows:

> ....We see that the hydrogen abundance in the F stars is compatible with that in the sun within a factor of 2.† We have already shown that, on the whole, the metals show about the same relative abundances as in the sun...For the first time we may with some confidence say that the spectrum of a star could be predicted in

* See section 9.7.1 for further comments on τ UMa and the solution for electron pressure and ionisation temperature.

† This was determined only approximately from Balmer wing profiles.

> detail from that of the sun. Since we have a satisfactory source of opacity, the color temperature, the Balmer discontinuity, and the absolute intensities of the lines of hydrogen and the metals are all predictable and, with the exception of the first two, have now been proved consistent with observation (149).

A subsequent paper added further elements to the abundance results of these stars, but the method was the same (153).

10.9.3 Aller's abundance analyses Lawrence Aller's first curve of growth analysis was undertaken at Harvard under the tutelage of Menzel and Goldberg, who had already made substantial contributions to curve of growth theory and practice (see Chapter 7). His first results were for the A0 dwarfs, Sirius and γ Geminorum in 1942 (78). The observational material came from Mt Wilson, Lick and McDonald Observatories. This analysis is of course very early on, coming less than a year after Greenstein's study of Canopus. Unsöld's work on τ Sco is not referenced, so presumably was still unavailable due to the war in Europe.

Two features of this work on A dwarfs are of especial interest; firstly, Aller followed Greenstein and plotted differential curves of growth using the sun as a standard. The Menzel–Baker–Goldberg solar curve of growth from Allen's equivalent widths, as in Greenstein's case, was used. A differential curve for a star of spectral type considerably different from the solar type is only plotted with difficulty, as the stronger lines easily measurable in the sun are exceedingly weak in A0 stars, while the weaker solar lines are absent. Lines as weak as log $W/\lambda = -6.0$ were measured in Sirius. Aller derived excitation temperatures of about 6000 K for both stars from the differential curves of growth, and he noted that these values followed the trend established for the sun and Canopus of being considerably cooler than the effective temperature. The analysis goes on to find the numbers of various neutral atoms or ions above the A stars' photospheres relative to the sun, but the overall element abundances are not derived.* In this respect he thus went as far as Greenstein for Canopus.

The other aspect of this work which is innovative is the use of model atmospheres with a grey $T(\tau)$ structure. The models are not used to compute theoretical metal line strengths, so these are not 'model atmosphere' or 'fine' analyses in the currently accepted jargon. However, the wavelength dependence of the thickness of the Schuster–Schwarzschild reversing layer due to the non-greyness of the continuous absorption, is taken into account, following the proposal Unsöld had earlier outlined

* Aller gave a rough calculation to find $(H/Fe) = 3000$ by weight in Sirius.

(154), and the mean electron pressure is evaluated to compare with the results from the Stark-broadened Balmer lines.

The second abundance paper by Aller (now at Indiana University) is an important analysis of the O9.5 dwarf 10 Lacertae (155). This is a fully fledged coarse absolute analysis closely following Unsöld's work for τ Scorpii, including the use of the same oscillator strengths prescribed by Unsöld. Aller's introductory remarks justifying this analysis seem almost apologetic:

> Unsöld's study of τ Scorpii provides us with useful data..., but it seemed worthwhile to apply similar methods to at least one other representative main-sequence star. Accordingly, for comparison with the planetary nuclei, I have chosen 10 Lacertae, an O-type, main-sequence object, which falls close to the temperature and level of ionisation of representative planetary nuclei.

His motivation was thus to compare the compositions of stars with planetary nebulae. The study of these latter objects had been Aller's principal interest at Harvard.

The results for 10 Lac were presented for eight elements, for which the relative numbers in the reversing layer showed remarkable similarity with τ Sco; in fact Aller considered the agreement 'probably fortuitous in view of the uncertainties'. The hydrogen-to-helium ratio in 10 Lac was nearly 12:1, in this case somewhat greater than Unsöld gave for τ Sco (5.6:1).

Aller's third paper in his stellar abundance analyses is the most significant from point of view of method. This is the first 'fine analysis', meaning that stratification of the stellar atmosphere as given by a model, is used to interpret the line strengths. The term 'curve of growth analysis' has frequently been used in the literature as the opposite of a model atmosphere analysis. This was explicitly the meaning of this term as used by Aller when he gave a review 'On the Curve of Growth Method' in 1964 (156). This terminology is unfortunate and the phrases 'coarse analysis' or 'Grobanalyse' should be reserved for the case where atmospheric structure is replaced by mean values of temperature, electron pressure etc. As Strömgren showed empirically in 1940, for example, the concept and use of the curve of growth is equally valid in the case of a stratified atmospheric model.

The star Aller analysed in the first stellar 'fine analysis' in 1948 was γ Pegasi, a B2.5 IV object that had already received considerable attention, in particular from Gerhard Miczaika at the Heidelberg Landessternwarte and Anne Underhill at Yerkes. Miczaika's analysis of γ Peg was from medium dispersion (54 Å/mm at Hγ) cassegrain spectra on the 72cm Waltz telescope at Heidelberg (157). In view of the systematic effects of dispersion on measured equivalent width first found by Unsöld (77), the use of lower dispersions than 10Å/mm is now regarded as inadequate for abundance studies.* Miczaika's analysis, however, followed Unsöld's 'Grobanalyse' method, and his results for five light elements in their proportion to hydrogen agreed (within a factor of three) with the composition of τ Sco and 10 Lac. Anne Underhill catalogued 877 blue lines in γ Pegasi (159) from high dispersion McDonald coudé plates. She then computed model atmospheres for this and other B stars for the purpose of calculating theoretical Balmer Hγ line profiles to compare with the observations (160). These models had the grey $T(\tau)$ structure which Chandrasekhar had shown to be applicable to the non-grey case, provided a flux-weighted mean opacity was used to determine optical depth (63). The equivalent widths of Hγ and in addition of lines of neutral helium (4471 Å) and of ionised silicon (SiII and SiIII) were computed from the models using Strömgren's method (53). She showed how the gravity of a star could be determined from the Hγ wing profile or the silicon line strengths. γ Peg had a value 'intermediate between the two values chosen' for the gravities of the models ($\log g = 4.09$ and 2.0).

Now we return to Aller's study of γ Pegasi which followed shortly afterwards (161). He began with a coarse analysis: 'For a preliminary reconnaissance of the problem, the method employed by Unsöld in his discussion of τ Scorpii is probably as good as any that has been devised'. He thus derived 'numbers of atoms above the photosphere' on the Schuster–Schwarzschild model for ten light elements, as well as the electron pressure, and adopted an effective temperature of 20 000K before starting the model analysis. The model atmosphere was calculated closely following Miss Underhill's method, but the theoretical strengths of several lines of ionised carbon, nitrogen and oxygen were now included, as well as the wing profile of the Balmer Hδ line. This profile gave a gravity, while the line-by-line comparison of the calculated and observed strengths of carbon, nitrogen and oxygen showed by how much the abundances put into the theoretical calculation had to be adjusted to fit the observations. The final results gave the oxygen-to-hydrogen ratio as $\sim 10^{-4}$, with carbon-to-hydrogen and nitrogen-to-hydrogen slightly smaller. 'The high ratio of hydrogen to the elements of the oxygen group indicated by the Unsöld procedure is substantiated, although the relative proportions of carbon, nitrogen, and oxygen are not the same'. Aller ascribed to the depth-dependence of the excitation the relatively small differences between coarse and fine methods. The table summarises his results for γ Peg:

* This was K.O. Wright's conclusion in 1964 for example (158).

	Coarse (Unsöld procedure)	*Fine (Strömgren–Underhill procedure)*
log (C/H)	−4.47	−4.84
log (N/H)	−4.78	−4.69
log (O/H)	−4.18	−3.94

The method of fine analysis as introduced by Strömgren for the sun and by Aller and Underhill for stars was in principle a major advance in technique. In practice it did not render the coarse methods obsolete overnight; far from it, coarse analysis continued as the principal method throughout the 1950s and as an important technique in the 1960s. Aller himself was one of the main protagonists of coarse analysis. In 1966 he wrote: '. . . . one may legitimately ask if the whole curve-of-growth [i.e. coarse] procedure is obsolete. The answer depends on the type of star, the kind of data available and the computing facilities at hand. Certainly, an elaborate procedure is justified only if good data are available at the outset' (156). And later in the same article: 'The curve of growth has served as a useful tool although sometimes results obtained by it have been given greater credence than they deserved. Used wisely, it can continue as a powerful means of reconnaissance of stellar atmospheres'.

10.9.4 K.O. Wright and the analysis of four solar-type stars This discussion of abundance analyses in the 1940s would not be complete without mention of K.O. Wright's analysis of four F and G stars including the sun. (The stars were the F-type supergiants γ Cygni and α Persei and the F dwarf, Procyon.) These analyses were initially undertaken around 1939 and 1940 (see (162) for preliminary results) and submitted as a PhD thesis to the University of Michigan in 1940. The work was therefore either about simultaneous with, or even slightly preceding, Unsöld's τ Sco analysis, which was submitted to the *Zeitschrift für Astrophysik* in June 1941. However Wright did not publish his detailed results until 1947 for the reasons he himself explained:

> The results were revised and prepared for publication in December 1942 but printing was deferred until the end of the war. Recent advances in the methods of studying curve-of-growth phenomena have required a further complete revision of the analysis. [The discussion of the observations and reductions has] been changed only when additional information required such a revision but the remainder of the work was re-written in 1946 and 1947 (163).

The observational material was very extensive, being based on 83 high dispersion cassegrain spectrograms on the Victoria 72-inch telescope,

Fig. 10.13 Kenneth Wright.

including plates of the blue sky and moon, so that the sun could be analysed in the average light from the whole disc (the flux spectrum), as is necessarily the case for all stars.

The 1942 studies consisted of an absolute coarse analysis of the sun and the three stars based on the Schuster–Schwarzschild model and Goldberg's theoretical oscillator strengths. Abundances of neutral atoms and ions were given as column densities in the reversing layer, and excitation temperatures were measured from the lines of ionised scandium, neutral iron, and neutral and ionised titanium. On the other hand when the data were reworked in 1946–7 the analysis was treated differentially for the three stars relative to the sun. Unfortunately the opportunity of a true differential analysis, in which the same line in two stars is measured with identical instruments and reduction procedures, was not seized at this point, as Wright preferred to use either Allen's solar disc-centre equivalent widths (148) or his own measured from the Utrecht solar atlas. One new aspect of these analyses was the observation of different shapes for the curves of growth from neutral and ionised species. This was in part due to higher turbulent velocities from the ions, by about 50 per cent. These three differential analyses were completed and published in June 1947, some months before Greenstein had sent his manuscript on F stars to the *Astrophysical Journal* (149). Thus although Greenstein is often cited as the founder of differential analysis,* Wright derived the first differential

* Greenstein plotted the first differential curve of growth in 1942 (see section 10.9.2), but did not at that time derive differential element abundances.

element abundances. In practice, the honours should be shared.

Wright's 1946 results were presented for twenty-one elements in the form of log N, where N is the number of atoms plus ions of a given element above unit area at continuous optical depth 0.3. The hydrogen abundance was not specified, so the results are not so easily compared with Greenstein's, even though two stars were in common. Wright's results can be put in the form of $\log (N_{el}/N_{Fe})_* - \log (N_{el}/N_{Fe})_{\odot}$, that is the relative element-to-iron ratio in the star to the sun. When this is done then the well-determined relative element abundances of Wright are mainly within 0.3 in the logarithm (a factor of 2) of Greenstein's. Since this is comparable to the uncertainties, this rather weak test would indicate agreement. For elements represented by only a few lines (e.g. barium, zirconium, lanthanum) Greenstein's $(N_{el}/N_{Fe})_*$ ratios were, for Procyon, three to five times smaller than Wright's.

10.9.5 Concluding remarks on abundance analyses in the 1940s In the 1940s the techniques of stellar abundance analysis were first put into practice. Most analyses were coarse, either absolute or differential. The temperatures came mostly from the lines themselves as ionisation (especially for the hotter stars) or excitation temperatures (especially cooler stars). Such temperatures had the advantage of pertaining to the line-forming region, but were generally not so precise as would be obtained from the photoelectrically measured stellar colours that became available in the 1950s. The interpretation of the colours was greatly facilitated by new atmospheric models, both for computing theoretical flux gradients and for deriving the temperature in the shallower line-forming regions from that in the deeper continuum-forming layers which determined the colours.

At the Zürich symposium on *The Abundances of the Chemical Elements in the Universe* in 1948, Struve summed up the meeting with this comment:

> Perhaps the most striking result of the discussions is the remarkable degree of uniformity that has been observed in the most widely different astronomical sources. The sun, the main-sequence stars of type F and the He stars like Tau Scorpii and even the O-type star 10 Lacertae have all approximately the same composition. The observers tell us that the precision with which the abundances are now determined corresponds to a factor of perhaps a little better than 2 or 3. Even more surprising is the fact that Strömgren's result for interstellar matter and Menzel's for the planetary nebulae also indicate a composition that is essentially the same as that of normal stars... The first conclusion of this symposium is the establishment

> of a list of what we might call the normal abundances of the elements in the universe. These abundances seem to have the character of a universal law of nature (164).

Compare this with the statement of Sir William Huggins when he later described his earliest spectroscopic observations of 1863 (see Chapter 4):

> One important object of this original spectroscopic investigation of the light of the stars and other celestial bodies, namely to discover whether the same chemical elements as those of our earth are present throughout the universe, was most satisfactorily settled in the affirmative; a common chemistry, it was shown, exists throughout the universe (165).

So what, we might ask, had been achieved in nearly a century of hard effort since Huggins' time? Simply the difference in the trust that could be placed on a qualitative statement founded on the visual description of stellar spectra, and a quantitative, though still somewhat imprecise, statement based on photographic measurements and physical theory. Perhaps this was the main achievement of the first century of stellar spectroscopy from the time of Huggins – not so much a change in our overall view of the chemistry of the universe, but more a change in the conviction with which that view could now be asserted.

10.10 Abundance analyses from 1950

10.10.1 Overview 1950–65: who analysed which stars when, where and how? 'Perhaps the most exciting problem in stellar spectroscopy today' wrote Otto Struve in 1959,

> is the determination of abundances of chemical elements in the atmospheres of stars. We now realize that these abundances are not the same in all stars, and many of the most striking differences have been attributed to evolutionary processes. But I believe that we are beginning to wonder whether chemical abundances are *exactly* the same in *any two stars* (166).

Had Struve forgotten his words of a decade earlier? (see section 10.9.5). I think not; the 1940s was a period that saw rather approximate coarse analyses of normal abundance stars. The 1950s and beyond was, firstly, the period of refining these techniques using models, and secondly, it was the age of the analysis of peculiar stars. With more reliable abundances, with the considerable increase in the number of stars analysed, and the inclusion of stars of non-solar composition in observing programs, the conclusions of Struve in 1959 reflected the majority viewpoint of that time. Nor was Struve revoking his remarks of 1948; the general similarity in composition of the vast majority of stars that had been analysed (say by 1965) was still their most striking feature.

The chemically peculiar stars on the other hand are rare, and moreover, their compositions in many cases (the Ap stars excepted) can be regarded as a perturbation of the normal or solar composition. Indeed, one can take a view (albeit controversial) that emphasises the uniformity of stellar compositions, such as Unsöld's in 1969, though many astronomers would argue that here the case for uniformity is overstated. 'Even the oldest stars of the disk population have, in the main, a metal abundance which cannot be distinguished from that of the sun or the younger, spiral arm population – that is, all abundances in the disk and spiral-arm populations are the same, to within an error of, at most, 50 per cent' (167). Certainly Unsöld acknowledged the reality of low metal abundances in the very old halo population of stars and marked abundance peculiarities in certain 'anomalous' stars, but both these classes of objects are, he argued, so rare that they should not divert us from a theory that embraces the most striking feature for the remaining 95 per cent of all stars – their chemical uniformity. In spite of Unsöld's view, we now return to the spirit of Struve's words in 1959 and discuss some of the excitements and abnormalities he was referring to.

In 1953 the Centre National de la Recherche Scientifique organised a colloquium in Paris on stellar classification. At this meeting Cornelis de Jager (b. 1921) from Utrecht summarised the stellar abundance work undertaken up to that time:

> At present 36 more or less extensive lists of abundances in stellar atmospheres have been published, comprising one O-type star, 23 B-type stars, 3 of type A and 9 of type F... A few stars have been discussed by more than one author, for example γ Pegasi has been analysed five times, α Persei three times and γ Cygni twice. Of these stars, five B stars have been analysed using model atmospheres and the remainder by means of Unsöld's method, or the curve of growth (168).

de Jager cited nine astronomers who had contributed to this field for stars of Population I. A major theme in the 1950s was whether the fine analyses using models gave more reliable results. In 1953 de Jager's comment based mainly on Aller's fine analyses, was '... although we have a preference for stratified models, at present we cannot yet really say that the abundances obtained from stratified models necessarily have a greater value than those from homogeneous models' (168).

A later review on the subject of the determination of cosmic abundances was published by Minnaert in 1957 (169). Most of the important references to abundance analyses in the period 1944 to 1957 are cited here. By this time (i.e. 1957) model atmospheres had been applied to the analysis of τ Sco and 10 Lac by G. Traving at Kiel (170, 142), to the B1 supergiant ζ Persei by Roger Cayrel at the Institut d'Astrophysique in Paris (171), by Aller and his colleagues at Michigan for τ Sco (172), as well as in Aller's classic 1949 study of γ Pegasi already cited.

Aller at this time (1956–9) was involved in an extensive program on the analysis of B star spectra, in collaboration with J. Jugaku and G. Elste at Michigan. Four stars, including two supergiants ε CMa and 55 Cyg* and two dwarfs, τ Sco and γ Peg were included in this program. A feature of the eight papers Aller published on B star atmospheres is the increasing use of model atmospheres; thus the work on ε CMa (174) was a coarse analysis based on both Schuster–Schwarzschild and Milne–Eddington models. The work on γ Pegasi was a very detailed model atmosphere reanalysis using the latest models by A. Underhill (175). Aller's analysis of τ Sco with Jugaku and Elste (172) was the third analysis and second model atmosphere analysis published for this star. G. Traving was thus able to compare the results of Unsöld, of himself and of the Michigan team (142). He concluded that the small differences in element abundances came primarily from systematic differences in the equivalent width measurements rather than analysis technique. In his monograph in 1961 *The Abundance of the Elements* (176), Aller was able to summarise nearly two decades of work at Kiel, Michigan and in Paris on the composition of B stars. The uniformity of composition from star to star and with the interstellar medium was the most striking feature. These were all young objects of Population I yet also resembled the sun in their composition, indicating little change in the element abundances in the Galaxy in the last 4000 million years.

An extensive grid of fifty non-grey model atmospheres had been computed in 1957 by C. de Jager at Utrecht (Holland) and L. Neven (Royal Observatory, Uccle, Belgium), and was frequently referred to by those analysing spectra (177). The de Jager–Neven atmospheres covered the range from 4000 to 25 000 K in surface temperature, from 10 to 10^5 cm/s^2 in surface gravity and 'normal' abundances were adopted. The detailed

* 55 Cyg was earlier analysed by H.H. Voigt (b. 1921) (173) using the coarse analysis method developed by Unsöld for τ Sco.

results discussed the continuous fluxes with gradients and Balmer jumps calculated, as well as the theoretical profiles of Balmer and ionised helium lines and the calculated equivalent widths of seventy-eight lines of helium, carbon, nitrogen, oxygen and heavier elements – all important parameters vital to the observational spectroscopist (Fig. 10.14).

Unsöld's view at this time of fine analysis using model atmospheres is

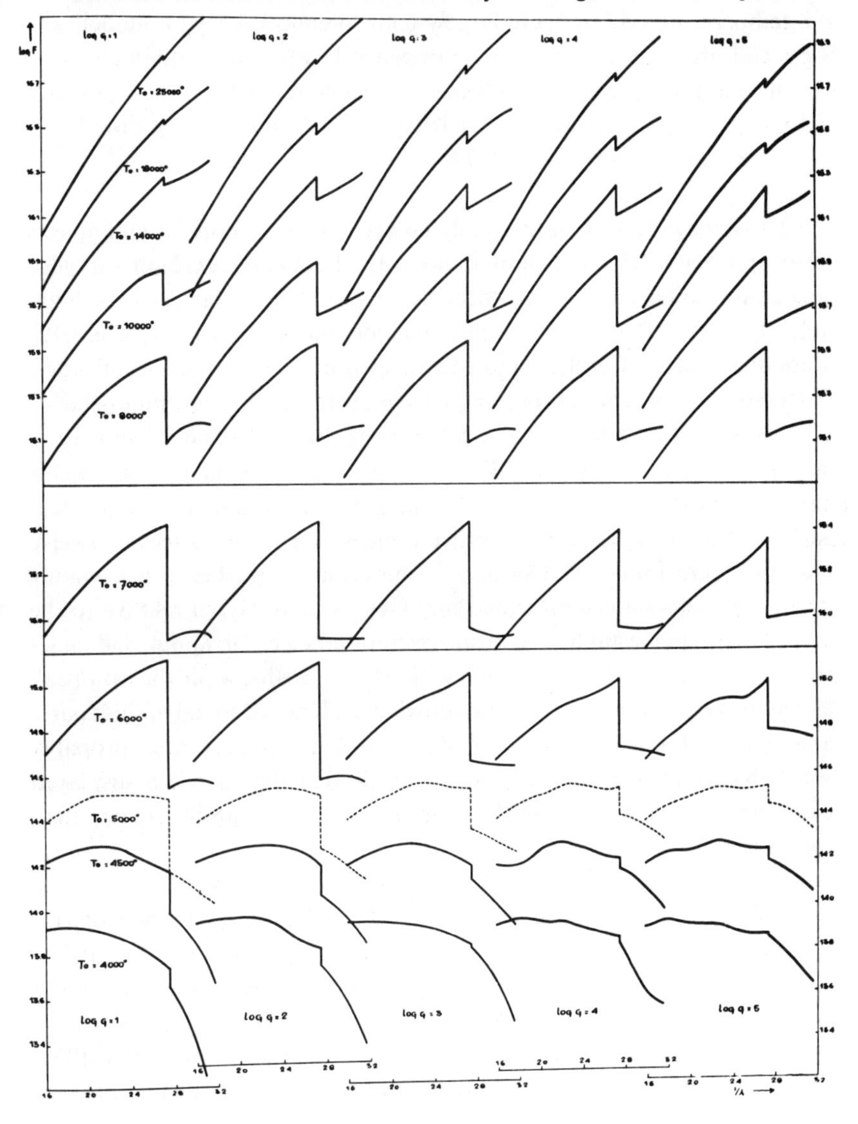

Fig. 10.14 Continuous spectral fluxes computed from the de Jager and Neven model stellar atmospheres, 1957.

interesting, since all the analyses from the Kiel group from 1941 to 1955 were coarse. He wrote in 1958:

> It is evident that a grey model offers very little advantage compared with a calculation using constant average values, where moreover the errors introduced by neglecting the variations both of temperature *and* pressure largely tend to cancel. Using a model atmosphere at all makes sense in general only if one computes at once a non-grey model with every possible refinement. That, by the way, is the reason why we [at Kiel] have hesitated a long time in using model atmospheres (178).

The stars with the first fine analyses were all young Population I objects with approximately solar abundances. In the 1950s attention was also turned to the peculiar stars, especially the peculiar A stars and the high-velocity stars of Population II, which includes the subdwarfs. Of several Ap stars analysed by Geoffrey and Margaret Burbidge (then at Cambridge, England), the differential coarse analysis of the bright europium star α^2 CVn was the first and the most extensive (179). The standard star was γ Geminorum with a normal A0IV spectrum. The McDonald coudé spectra covered nine different phases of the magnetic variation and the equivalent widths of as many as 1200 lines were measured. Abundances of twelve elements were found relative to γ Geminorum. A further eleven mainly heavier elements not represented in γ Gem were analysed relative to the sun. This paper established mean overabundances of about 830 times (relative to the sun) for lanthanum and the rare earths, while the iron-peak elements were about normal. The actual abundances derived in this coarse analysis need not be quoted in detail and are in any case probably somewhat imprecise; the qualitative result was that α^2 CVn displayed overabundances of selected elements orders of magnitude greater than found in normal stars.

The other major interest in abundance work in the 1950s was the analysis of the high-velocity stars of Population II. Interest in these objects had been aroused by Nancy Roman's 1950 classification of F, G and K stars into strong and weak-lined groups, with the high-velocity stars being confined to the latter group (180) (see section 8.8.4). Indeed more attention was directed to this type of star by abundance analysts in the period 1950-65 than to any other individual class. They are discussed in detail in section 10.10.2.

In March 1959 de Jager sent out a questionnaire to a number of observatories; he wished to ascertain what types of stars, of which luminosity class and belonging to which population, were being investigated at high dispersion ($\leqslant 20$ Å/mm) using what theoretical approach.

> The purpose of sending this questionary [sic] was to obtain insight into the detailed investigations going on into high dispersion stellar spectra, to learn at what observatories investigations are underway that eventually could lead to obtaining detailed model atmospheres, and to know at least roughly, in what way one tries to reach this latter goal (181).

There were twelve replies, covering stellar and solar work, though evidently the results discussed were investigations of all kinds, not simply abundance analyses. The respondents* mentioning stellar investigations were E.R. Mustel at the Crimean Astrophysical Observatory (studies of Be stars), Unsöld at Kiel (fine abundance analyses), Aller at Michigan (coarse and fine analyses of O and B stars) and also at Mt Wilson (abundances in subdwarfs and high-velocity stars), J.-C. Pecker in Paris (metallic-line and helium-rich stars), Mrs M. Hack at the Milano-Merate Observatory (curve of growth analyses of B and F stars), Miss A. Underhill at Victoria (atmospheric structure of Wolf–Rayet and O stars), K.O. Wright also at Victoria (analyses of ζ Aurigae-type stars† and K stars) and from de Jager himself for the Uccle and Utrecht observatories (use of model atmospheres for investigating B-star spectra). The response was certainly not complete (no reply from Yerkes-McDonald for example) but nevertheless it gives some insight into the wide range of high dispersion spectral studies in progress at that time.

The standard compilation of abundance analyses carried out up to the end of 1965 is the list given by R. Cayrel and G. Cayrel de Strobel (182). The Cayrels give forty references to determinations of the iron-to-hydrogen ratio in 154 stars. A total of 204 determinations are quoted with some stars being analysed several times by different workers. The Cayrel list is deliberately selective and is not intended as a comprehensive guide to all stellar abundance determinations to that date. Thus only two references are given to work prior to 1957 in the main table of their paper, and no references are given to any stars earlier in type than F0, except for a few very metal-poor subdwarfs whose anomalous A spectral types reflect their low metallicity rather than high temperature. Thus none of the early classic papers by Unsöld and Aller on B stars for example is listed (they are referred to in the text) nor is any of the work on Ap and Am stars by the

* The respondents for each institution, de Jager notes, are not necessarily the investigators.

† These are later-type supergiants with a hot dwarf companion.

Burbidges, van't Veer-Menneret and Conti. However, Minnaert's 1957 review fills many of these gaps (169). Surprisingly, the first ever stellar abundance analysis based on the curve-of-growth method from measurements of equivalent widths, namely Louis Berman's 1935 analysis of R Coronae Borealis, is not referred to in either of these reviews (89). The early curve of growth analyses of very late-type stars undertaken in Japan by Yoshio Fujita and subsequently by Yasumasa Yamashita (183, 184) are also unreferenced by Minnaert and the Cayrels. However, by selecting only the most reliable abundance analyses of later-type (i.e. F, G and K) stars, the Cayrel compilation has enhanced its usefulness for statistical purposes and for example, for the calibration of data from photoelectric photometry.

One of the most thorough abundance analyses of any star appearing in the Cayrel catalogue is the study of the G8 red giant, ε Virginis, by the Cayrels themselves (185). As many as 1400 lines were measured from nine Mt Wilson high dispersion coudé spectrograms. The abundances of 28 different elements were measured and for nearly every element the results showed a value close to that of the sun. Thus ε Virginis was shown to be a young star with an almost identical composition to the sun. It has since become a useful standard for differential analyses of other red giant stars.

10.10.2 Abundance analyses of stars of the halo population In 1915 Walter Adams called attention to three faint high proper motion A stars with peculiar spectra (186).* They had 'exceptionally narrow and well defined' hydrogen lines and the 4481 Å line of ionised magnesium, normally prominent in stars of this spectral type, was weak or absent. These stars were later classified as early F in the Henry Draper Catalogue, but Annie Cannon made only a passing reference to the weak K line in one of them (HD 140283) (187). Adams and Joy again made a brief reference to stars of this rare type in 1922 (188) and in 1935 the list was extended to six stars described as 'intermediate white dwarfs', since their trigonometric parallaxes indicated low luminosities that appeared to place them fainter than the normal main sequence stars in the (Hertzsprung–) Russell diagram, yet brighter than the white dwarfs such as Sirius B.

While at Yerkes, Gerard Kuiper found many more of this type of star, which he termed 'subdwarfs'. He showed these were all, or nearly all, high-velocity stars, that they comprised only a few per cent of dwarfs of middle spectral types and that they apparently were subluminous by an average of 1.9 magnitudes (189). The interest in these stars was greatly increased after

* The stars were HD 19445, HD 84937, HD 140283 since recognised as subdwarfs of the halo population.

W. Baade's identification of two populations of stars in the Andromeda galaxy in 1944 (190), and his conclusion was that Kuiper's subdwarfs belong to the Population II of our own Galaxy, along with for example the globular clusters.

More detailed spectroscopic studies of high-velocity stars began about 1950. Thus at Princeton, Martin (b. 1912) and Barbara Schwarzschild compared the spectra of three high-velocity F stars with those of lower space motion (191). They found the G band of CH stronger in the high-velocity group, but the neutral iron lines weaker, indicating that the ratio iron-to-hydrogen may be about half the value in the high-velocity stars. About the same time Nancy Roman at Yerkes devised her weak and strong-line classification criteria for F and G stars and showed that the high-velocity stars occur only in the weak-line group (180). Most of these high-velocity stars (velocity in excess of 70 km/s relative to the sun) were not subdwarfs or halo stars, which represent only the most extreme members of the weak-line group.

In Europe, the study of the spectra of high-velocity stars was practically simultaneous with the work cited of the Schwarzschilds and Roman in the United States. Thus Wilhelmina Iwanowska and her colleagues found a general line weakening of metal lines from a study of 22 later-type high-velocity stars (192). On the other hand G.R. Miczaika in Heidelberg was unable to establish any definite metal-line peculiarities for high-velocity red giants but he was able to confirm the presence of weaker CN bands and stronger CH (193).

As a result of the interest in subdwarfs at this time, Joseph Chamberlain (b. 1928) and L.H. Aller analysed the spectra of two of Adams' original stars, HD 19445 and HD 140283, in 1951 (194). This was the first curve of growth analysis for stars of the galactic halo population. Chamberlain and Aller found effective temperatures for both stars of 6300 K from the hydrogen line profiles, showing that they must be much cooler than normal A stars, and of the same temperature as dwarfs of type F. Their most interesting conclusion concerned the abundances of calcium and iron, the former based on the 4227 Å line of neutral calcium, the latter on a differential curve of growth for neutral and ionised iron lines using the sun as a standard. For HD 140283 calcium was nearly a factor of one hundred deficient, iron nearly a factor of ten; for HD 19445 the deficiencies were somewhat less but still substantial. Thus the subdwarfs were not early A stars with weak hydrogen lines, they were shown to be late F stars with weak metal lines, or hydrogen-rich stars (relative to their metal content)!

Shortly after the Chamberlain–Aller analysis of these subdwarfs two further papers came out which served to broaden our perspective on high-velocity stars. Philip Keenan and Geoffrey Keller (b. 1918) at Perkins Observatory considered the spectral classification of 108 suspected high-velocity stars (195), taken mainly from Miczaika's catalogue of stars of large space motion (196) (see section 8.8.4). They were able to demonstrate CN weakening for high-velocity G6 to K4 giants, while for earlier spectral types (F5 to G5, dwarfs and giants) the atomic lines were generally weaker and the G band of CH was stronger. Several of Nancy Roman's earlier conclusions (180) were confirmed. A second paper by Nancy Roman was explicitly directed to a discussion of seventeen of the high-velocity subdwarfs (197). She showed they were middle to late F main sequence stars but with abnormally blue colours. UBV photometry was discussed here for the subdwarfs for the first time. She also showed that their galactic orbits were generally highly eccentric; they thus belong to the halo population along with RR Lyrae stars and the globular clusters. The 1954 paper on subdwarfs was part of a much larger compilation of about 600 high-velocity stars that Nancy Roman published in 1955 (198). One of the objects in this high-velocity catalogue, HD 161817, was analysed by coarse differential methods by the Burbidges, who were now working at Caltech in Greenstein's group. The result was only moderate metal deficiency of two to three times (199).

The analysis of UBV photoelectric photometry (200,201), the use of narrow-band photoelectric scans of the continuous spectra (202) and the computation of model atmospheres (203) and interior models (204) for metal-poor subdwarf atmospheres all led to more precise analyses of the physical parameters and composition of subdwarfs from about 1954 onwards. In 1959 Bodo Baschek at Kiel re-analysed the spectrum of HD 140283, first with a coarse analysis, then using a model atmosphere (144). The mean deficiency of the heavier elements relative to the sun was a factor of 200. An extensive analysis of HD 140283, HD 19445 and another similar subdwarf, HD 219617, was completed by Aller and Greenstein the following year (205). These were differential analyses relative to the sun and the result for HD 140283 showed a hundred-fold deficiency of the metals. Aller later ascribed the small discrepancy between his work and Baschek's to differences in the measured equivalent widths, in the method of analysis and in the adopted effective temperature for this star (206). Baschek had based his temperature for HD 140283 (5940 K) on broad-band photometry, while Aller and Greenstein used a narrow-band scan with a correction for the effects of the absorption lines undertaken by W.G. Melbourne at Caltech (202). Melbourne's scan analysis resulted in a temperature of 5450 K, which now placed HD 140283 in the domain of the G stars, in fact cooler than the sun! Using these lower temperatures, the subdwarfs showed only slight or no subluminosity when compared with normal main sequence stars.

Other abundance analyses of subdwarfs in the 1960s confirmed this picture of them being very metal-poor high-velocity main sequence stars. Thus Arnold Heiser at Yerkes obtained element deficiencies of 200 times or more for HD 25329 (207), though a more thorough re-analysis of this star by Bernard Pagel (b. 1930) and A.L.T. Powell at the Royal Greenwich Observatory (208) lowered this figure to only 20 times deficient. This illustrates that large systematic errors were probably a feature of some of the earlier abundance work. Baschek carried out a differential model atmosphere analysis of another high-velocity dwarf (star 10367 in R.E. Wilson's General Catalogue of radial velocities) and found the heavy elements depleted by a factor of 40 relative to the sun (209). The star has an apparent visual magnitude of 11.2 and required exposure times of over 5 hours on the 100-inch Mt Wilson telescope! The spectra of two southern stars with mild Population II characteristics were analysed by I.J. Danziger working in Greenstein's group (210). One of these stars, γ Pavonis, showed a moderate deficiency of the iron group, but an over-deficiency of heavier elements.

The composition of halo population red giants was also a topic of active interest. As early as 1922, Lindblad had commented on the peculiar CN in three red giants in the globular cluster M13 (211). D.M. Popper found that CN-band weakness was a common feature of several red giants in the globular clusters M3 and M13 (212).

After Greenstein moved to the California Institute of Technology in 1948 he established a major program there for stellar abundance analysis. A survey of some of the early work then in progress was given by Greenstein in 1954 (213). A few years later, in 1957, he published a study of eight high-velocity red giants with Keenan (214). These stars had on average about two-thirds of the metal contents of low-velocity red giants, but ten times less of the molecule CN. These results were generally similar to those in another study by the Schwarzschilds with Searle and Meltzer at Princeton of six high-velocity red giants (215). A more detailed analysis of three halo red giants was commenced in 1954 by H.L. Helfer (b. 1929), G. Wallerstein and Greenstein at Caltech (216). Two of these stars were in globular clusters (M13 and M92), while a third was a high-velocity field star (HDE 232078) (Fig. 10.15). The analyses were coarse and differential relative to the sun. The metal deficiencies were from twenty to one hundred times in these stars, similar to the corresponding values for the subdwarfs. Three further halo red giants, all in the field, were analysed by George Wallerstein (b. 1930) with Greenstein and their colleagues as part of a continuation of

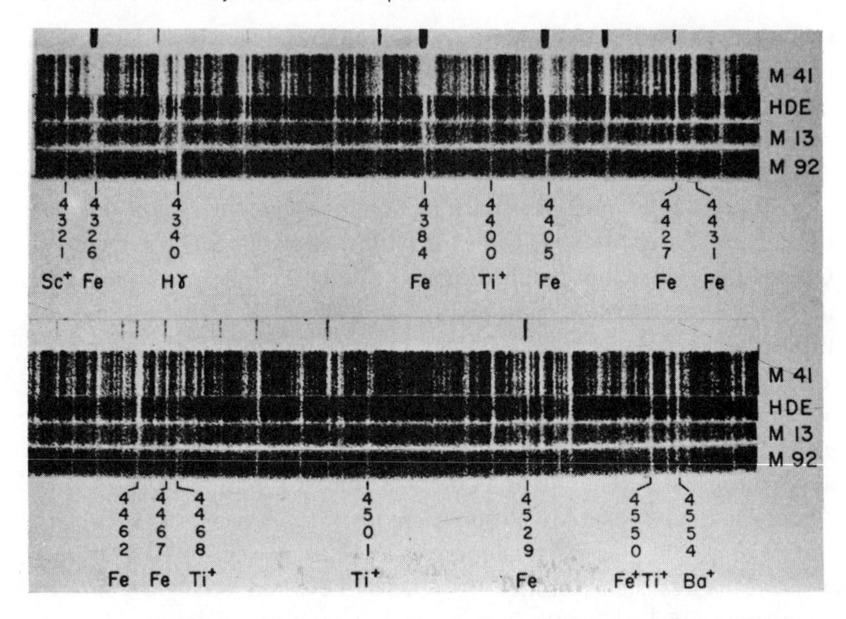

Fig. 10.15 Spectra of four stars, including three of the halo population. These were the first giant stars in the galactic halo to have their spectra analysed by curve of growth methods, by Helfer, Wallerstein and Greenstein in 1954.

this program (217) (Fig. 10.16). All three showed extreme metal deficiencies of 500 times or more. One star HD 122563, was especially interesting. The elements heavier than zinc were some fifty times more deficient than the iron group of elements, which were in turn 800 times deficient relative to the composition of the sun. This famous star was reanalysed by Pagel in 1965, also differentially relative to the sun (218). This reanalysis confirmed the large deficiency of iron and other metals by at least a factor of 200; however the relative distribution of heavy elements was more nearly solar. Of the

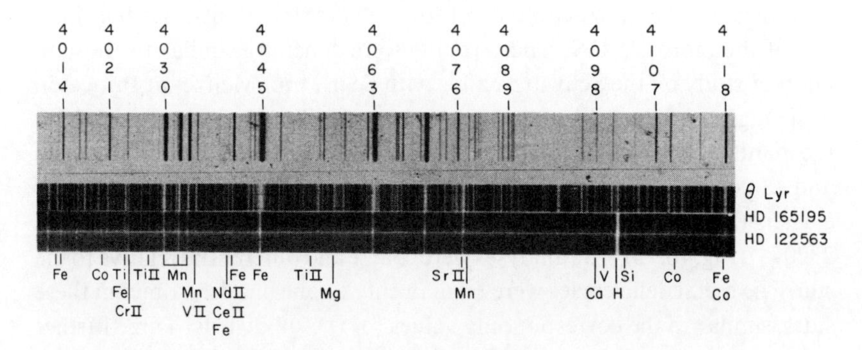

Fig. 10.16 Spectra of two extremely weak-lined halo giants are compared with θ Lyrae, a normal K0II star (from Wallerstein *et al.* ref. (217)).

elements heavier than iron, Pagel found only barium and cerium overdeficient, by a factor of about ten.

The intense interest in the composition of stars such as HD 122563 and the halo subdwarfs arose partly because of the highly influential treatise in 1957 of the Burbidges, W.A. Fowler and F. Hoyle in which stars are proposed as the sites for heavy element nucleosynthesis (B^2FH) (219). Many of the elements heavier than iron are believed to form by successive neutron captures onto nuclei of Fe^{56} inside red giant stars. According to a model of galactic evolution put forward by O.J. Eggen, D. Lynden-Bell (b. 1935) and A.R. Sandage in 1962 (220) the most metal-poor stars are also the oldest. Halo stars such as HD 122563 should form early on in the life of the Galaxy. This model of galactic evolution, together with the mechanisms for heavy-elements synthesis expounded in B^2FH, predicts overdeficiencies relative to iron for slow neutron-capture process elements, such as barium, in very old halo stars. Just such an age effect appeared to have been found in the case of HD 122563.

10.10.3 G dwarfs analysed by Wallerstein and others A program to analyse the spectra of a large sample of G dwarf stars was undertaken by George Wallerstein and his colleagues between about 1958 and 1961 working at first with Greenstein at Caltech and then at Berkeley. Since this was the first large-scale abundance survey to treat many stars with the same analysis techniques, it deserves a special mention. The spectra of G dwarfs are especially suited to abundance work, since they contain many sharp lines of a range of different elements and the composition of the stars is believed to reflect that of the interstellar gas clouds from which they were born. What is more, all these abundances were determined by the coarse differential method developed by Greenstein; the sun is thus an ideal standard. Nearly all the spectra were at 15 Å/mm and taken with the Mt Wilson 100-inch coudé spectrograph.

Much of the motivation for this work was directed towards a study of whether there is a general age-abundance relationship for stars in the Galaxy, as the theory of stellar nuclear synthesis with gradual heavy element enrichment of the interstellar gas would predict. An investigation of the relative distribution of the heavy elements, especially in the metal-poor stars, was a secondary aim of the program.

One of the early conclusions of this program came from a study of three of the solar-type dwarfs in the young galactic cluster, the Hyades (221). These were found to be possibly 20 per cent richer in heavy elements than the sun (a result of quite marginal significance in view of the random errors of such analyses, which frequently amount to a factor of 2). The age of the

Hyades stars was taken to be one billion (1×10^9) years, about five times less than the sun's age. Therefore, at most, only very slight enrichment of the interstellar gas from which stars form could have occurred in about the last 5 billion years.

Wallerstein's G dwarf program was eventually extended to a total of 31 disc population stars, selected for their wide range of kinematic and photometric properties (222). The low or zero rate of enrichment of the interstellar gas was confirmed; in fact some of the older, slightly evolved, stars were even marginally metal-rich. One firm conclusion was that the higher velocity disc stars (which on the basis of their kinematics may be slightly older than the sun, but not as old as the halo subdwarfs), for example 85 Pegasi (223), were nearly always metal-poor relative to the sun. In the case of 85 Peg, most metals were deficient by a factor of about three. Such deficiencies were considerably less marked than had been found for the halo subdwarfs, indicating that the rate of interstellar heavy element enrichment was only large very early in our Galaxy's evolution, when the halo stars were forming from clouds before or during the rapid collapse to the galactic disc.

Wallerstein also discussed the relative abundances of eleven heavy elements including iron, the best observed. In general, there was a remarkable uniformity in the element-to-element ratios from star to star, even in stars deficient in metals relative to hydrogen. Exceptions appeared to be manganese which in 85 Peg and other metal-poor stars was found to be overdeficient (222, 223); in addition seven stars showed an excess of the elements magnesium, silicon, calcium and titanium relative to iron. The nuclei of these elements contain an integral number of α-particles. In 1957 the B^2FH paper had proposed synthesis of such elements by α-particle additions to Ne^{20}, so it seemed plausible to seek an explanation in terms of this α-process.

An important result of having a large number of G dwarfs with known metal abundances, as well as several subdwarfs analysed by others, was the confirmation of a relationship between ultraviolet excess and abundance. The ultraviolet excess arises from the overall weakening of the many absorption lines which occur mainly in the ultraviolet part of the spectrum and can be conveniently derived from broad-band UBV photometry. Wallerstein's results established the first empirical relationship between these two parameters for G dwarfs (224, 222).

One of Wallerstein's conclusions, that there has been only very slight enrichment of the interstellar gas in the last several billion years, received independent support in 1963 from the work of John Hazlehurst (b. 1934) (225) and Bernard Pagel (226). Both these authors carried out separate abundance analyses of δ Eridani, a K0 subgiant star believed to be one of the oldest objects in the solar neighbourhood, from its position in the Hertzsprung–Russell diagram. Both analyses came to similar conclusions, that δ Eri has a solar composition, or is at most only very slightly deficient in metals. Hazlehurst estimated an age for δ Eri of 13 billion years, which in hindsight may be about two times too old. But even if this is so, the general implications of this work would not be negated.

10.10.4 Lithium in the sun and other dwarf stars In 1860 Kirchhoff reported the probable absence from the solar spectrum of the red line of lithium, which was familiar in the laboratory spectra of flames: 'From my observations, according to which no dark line in the solar spectrum answers to the red line of lithium, it would follow with probability that in the atmosphere of the sun lithium is either absent, or is present in comparatively small quantity' (227).

The lithium line of 6708 Å was first observed in the solar sunspot spectrum in 1907 by Hale and Adams (228) who, however, incorrectly identified it as calcium. This error was corrected in 1916 by Arthur King, also at Mt Wilson (229), who in addition showed from laboratory studies that the line has multiple structure with two or three close components.

When Andrew McKellar found a strong line of lithium in the cool carbon star WZ Cas in 1940, this was the only astronomical body other than the sun known to contain this element (see section 8.9.2) (230). By this time Harold Babcock and Charlotte Moore had found a very weak lithium absorption line in the general solar disc spectrum (231) and this was included in their tabulation of the red and infrared solar lines (96).

Greenstein and Robert Richardson (b. 1902) were the first to measure the equivalent width of this weak line of the solar disc (232). Their result was only 3.6 mÅ, from which a lithium abundance one hundred times less than that on earth was deduced. The abundance ratio lithium-to-hydrogen was of the order 10^{-11}. Such a small abundance was however already anticipated, as Russell had remarked in 1929 on the 'extreme rarity of Li' when he analysed the lines of the sunspot spectrum as part of his extensive treatise on the composition of the sun. Since the solar lithium content was anomalously low, Greenstein and Richardson proposed that the slow depletion of lithium in the solar atmosphere would occur by mixing of the surface layers with deeper hotter material in the sun, for lithium is destroyed by protons above about 3 million degrees.

In the 1940s and 1950s unusually strong lithium was found in a variety of stellar spectra. After McKellar's discovery of lithium in WZ Cas, Sanford found another strong-lithium carbon star, WX Cyg (233). Later

Keenan and Teske (b. 1930) showed lithium to be consistently enhanced in S stars (234) and Hunger found a very strong lithium line in two T Tauri stars (235). On the other hand, no lithium could be found in the interstellar medium by Lyman Spitzer (236).

All these results made it clear that the lithium content of stars shows large variations from object to object, with high values only in special types of star and an anomalously low value in the sun when compared with the terrestrial abundance. Walter Bonsack at Caltech explored the lithium line in 46 normal G and K stars, both dwarfs and giants (237). This was the first quantitative work on lithium abundances for any stars other than the sun. The lithium line was generally found to be a very weak absorption feature in most of Bonsack's stars, often on the limit of being measurable on his Mt Wilson coudé spectra. The stars analysed showed a total range of over a thousand in their lithium-to-vanadium ratios; there was a clear correlation of this ratio with spectral type but not with luminosity. Thus the later K stars were substantially poorer in lithium than the sun (however only one dwarf other than the sun had a measurable lithium line). Bonsack interpreted his results as being possibly due to deeper convective mixing in the cooler stars. This would bring lithium into progressively hotter interiors for the later spectral type stars, resulting in faster destruction of lithium atoms by protons.

Bonsack's work stimulated a lot of interest in the lithium abundances of later-type stars in the 1960s, especially after Bonsack and Greenstein measured lithium abundances in five T Tauri stars a hundred times above the solar value (238). George Herbig at Lick took up the challenge of explaining the wide variations in stellar lithium content from star to star. By 1963 he had observed 50 dwarfs from G0 to G8, including six in the Hyades cluster (age 500 million years) and one in the Pleiades (age 60 million years) (239). These young stars had high lithium abundances, around thirty times that of the sun. He proposed that the observations support a slow destruction of stellar lithium by convective mixing on a long timescale in main sequence stars, even though the theory of stellar structure was unable to predict convective zones penetrating deep enough into the interior.

Herbig's ideas were developed in much greater detail in an extensive article in 1965 treating one hundred nearby dwarf stars with spectral types from F5 to G8 (240). His hypothesis was that the lithium content of main sequence stars decays exponentially with time, the decay constant being sharply dependent on the stellar mass and hence spectral type. For the sun its value was estimated to be 1.32 billion years, but for cooler stars deeper convection gave smaller values (i.e. faster depletion of lithium). The initial lithium content when a star arrives on the main sequence would be high, possibly comparable to that in T Tauri stars.

Some of the best evidence for this hypothesis came from a study of the lithium abundances in Hyades dwarfs, which show a steady decline in abundance for spectral types later than G1, with a hundred times the solar lithium content, to less than the sun's abundance for the cooler stars (241). In addition, Herbig cited lithium abundance correlations with kinematics and H and K emission line strengths, both of which are believed to depend on stellar age.

Some contradictory evidence to Herbig's hypothesis has nevertheless been found, both by Herbig and by others. Peter Conti (242), Michael Feast (243) and George Wallerstein (244) all found old evolved subgiants* with a high lithium content. Herbig found a strong lithium line in χ Her, a star with rather a high space motion, and also an anomalous discrepancy in the lithium abundances of two otherwise identical G stars in the visual binary system ξ UMa (240). Moreover, he found a deficit of lithium-poor G dwarfs in his sample which would imply that his star selection procedure had inexplicably chosen relatively few stars older than 2 or 3 billion years in age. In spite of these problems, Herbig maintained that his data gave general support to his hypothesis that cool dwarf lithium abundances depend on a star's age.

10.10.5 HD 33579 and the first spectral analysis of an extra galactic star The two Magellanic Clouds, in the southern hemisphere skies, are the two nearest systems of stars to our own Galaxy, the Milky Way. Both are small irregular galaxies which are satellites to the Galaxy. The distances are both about 170 thousand light years which makes the most luminous stars in the Magellanic Clouds just bright enough for high dispersion spectroscopy with large telescopes.

Systematic spectrographic and photometric studies of the supergiant stars in the Clouds were undertaken by M.W. Feast, A.D. Thackeray and A.J. Wesselink from the mid-1950s on the Radcliffe Observatory reflector. A preliminary paper showed that these stars were generally of early spectral type with extremely sharp lines, especially of ionised iron (245). HD 33579 in the Large Magellanic Cloud was visually the brightest of these stars ($V = 9.11$) and a spectral type of A2 was assigned. The HD Catalogue had given a type of B to this star without decimal subdivision, with the comment that the spectrum is 'almost continuous'; presumably the narrow lines were nearly invisible on the low resolution objective prism plates.

* Conti's subgiant was δ Eridani—see section 10.10.3.

Further studies of the Cloud supergiants were carried out by Feast and his colleagues. Thus in 1958, a total of 130 stars was investigated and a Hertzsprung–Russell diagram drawn for several with known spectral types and photometric indices (246). These stars were between ninth and about twelfth apparent visual magnitude and predominantly of spectral type A0 or earlier. The spectral classifications were from medium dispersion plates from the 2-prism spectrograph on the Radcliffe telescope (49 or 86 Å/mm at Hγ).

Then in 1960 the Radcliffe astronomers published an important catalogue of 155 of the brightest stars in the Magellanic Clouds (247). MK spectral classifications were assigned and the luminosity class notation Ia-O adopted for stars which showed evidence of being brighter than standard stars of class Ia in the Galaxy.

In the case of HD 33579 its spectrum did not match that of any MK standard; the ionised metallic lines indicated type A2, the hydrogen lines A3. The final classification given was A3: Ia-O (e), the colon indicating some uncertainty in the decimal subdivision, the O the high luminosity of a super-supergiant or 'hypergiant', the (e) the presence of emission at Hα (248) but not at other wavelengths. Although HD 33579 is the brightest star in the Clouds at visual or blue wavelengths, the brightest bolometrically (over all wavelengths) are O-type supergiants with P Cygni line profiles indicating mass loss.

Prior to 1965 all the stars with abundance analyses were local objects in the immediate solar neighbourhood, the great majority being bright stars within one thousand light years, which corresponds to about one per cent of the overall diameter of our Galaxy. The question of the chemical uniformity not just of the solar neighbourhood, but of the galaxies within the universe at large, is clearly a fundamental one which raises further issues, such as whether the rates of element production by nuclear synthesis reactions in stars have been the same in different places. The abundance analyses of Magellanic Cloud stars are therefore of considerable importance.

The first high dispersion observations of HD 33579 in the Large Magellanic Cloud were secured by Antoni Przybylski using the coudé spectrograph on the Mt Stromlo 74-inch telescope. His 5-hour exposure on Christmas Eve, 1963, was at 10.2 Å/mm. He made a preliminary coarse analysis of this spectrum and compared the element abundances with those of the galactic supergiant α Cygni (249). In 1965 a higher quality spectrum exposed over three nights was obtained and the coarse analysis repeated (250). The result confirmed the preliminary finding that the metal content was about two-thirds of that of α Cygni, a normal star of approximately solar composition. Since the difference between this and a solar composition is probably less than the uncertainties of the analysis (especially so in view of the rather uncertain temperature and high luminosity of HD 33579, and because of the possibility that its lines may be partially filled by emission) Przybylski concluded that the heavy element content relative to hydrogen of the Large Magellanic Cloud was comparable to that of the Galaxy. Even though the Large Magellanic Cloud is apparently a less evolved galaxy than our own (it still contains many young blue supergiant stars and clouds of interstellar gas), and is of low mass compared to the Milky Way Galaxy, heavy element synthesis has proceeded to a comparable point in both systems.

References

1. Schwarzschild, K., *Göttinger Nachrichten, Mathematisch–Physikalische Klasse*, p. 41 (1906).
2. Eddington, A.S., *Internal Constitution of the Stars*, Cambridge (1926).
3. Milne, E.A., *Phil. Trans. Roy. Soc.*, **223A**, 201 (1922).
4. Wilsing, J. and Scheiner, J., *Publ. Astrophys. Observ. Potsdam*, **19**, 1 (1909), and also: Wilsing, J., Scheiner, J. and Münch, W., *Publ. Astrophys. Observ. Potsdam*, **24**, 1 (1919).
5. Rosenberg, H., *Abhandlungen der Kaiserliche Leopold-Carol Akademie der Naturforscher, Nova acta*, **101** (Nr. 2) (1914).
6. Brill, A., *Handbuch der Astrophysik*, **V** (Part I), 128 (J. Springer-Verlag, Berlin) (1932). See p. 158.
7. Yü, C.-S., *Lick Observ. Bull.*, **12**, 104 (1926).
8. Sampson, R.A., *Mon. Not. Roy. Astron. Soc.*, **83**, 174 (1923).
9. Sampson, R.A., *Mon. Not. Roy. Astron. Soc.*, **85**, 212 (1925) and ibid. **90**, 636 (1930).
10. Greaves, W.M.H., *Vistas in Astron.*, **2**, 1309 (1956).
11. Greaves, W.M.H., Davidson, C.R. and Martin, E.G., *Observations of Colour Temperatures of Stars*, His Majesty's Stationery Office (1932).
12. Greaves, W.M.H., Davidson, C.R. and Martin, E.G., *Mon. Not. Roy. Astron. Soc.*, **94**, 488 (1934).
13. Fowler, R.H. and Milne, E.A., *Mon. Not. Roy. Astron. Soc.*, **83**, 403 (1923).
14. Payne, C.H., *Harvard Coll. Observ. Monographs*, no. 1: Stellar atmospheres (1925).
15. Kienle, H., Strassl, H. and Wempe, J., *Zeitschrift für Astrophys.*, **16**, 201 (1938).
16. Kienle, H., *Vierteljahrschrift der Astron. Gesellschaft*, **67**, 397 (1932).
17. Kienle, H., Wempe, J. and Beileke, F., *Zeitschrift für Astrophys.*, **20**, 91 (1940).
18. Williams, R.C., *Publ. Observ. Uni. of Michigan*, **7**, 93 (1939).
19. Williams, R.C., *Publ. Observ. Uni. of Michigan*, **7**, 147 (1939).
20. Williams, R.C., *Publ. Observ. Uni. of Michigan*, **7**, 159 (1939).

21. Williams, R.C., *Publ. Observ. Uni. of Michigan*, **8**, 37 (1940).
22. Barbier, D. and Chalonge, D., *Ann. d'Astrophys.*, **4**, 30 (1941).
23. Barbier, D., Chalonge, D., Kienle, H. and Wempe, J., *Zeitschrift für Astrophys.*, **12**, 178 (1936) and Kienle, H., Chalonge, D. and Barbier, D., *Ann. d'Astrophys.*, **1**, 396 (1938).
24. Lindblad, B., *Uppsala Universitets Åarsskrift*, (1920).
25. Lindblad, B., *Nova Acta Regiae Societatis Sci. Upsaliensis*, (ser. IV) **6** (no. 1), 1 (1923).
26. McCrea, W.H., *Mon. Not. Roy. Astron. Soc.*, **91**, 836 (1931).
27. Kramers, H.A., *Phil. Mag.*, **46**, 836 (1923).
28. Abbot, C.G., *Astrophys. J.*, **34**, 197 (1911); and Abbot, C.G., Fowle, F.E. and Aldrich, L.B., *Smithsonian Misc. Coll.*, **74** (no. 7) (1923).
29. Russell, H.N., *Astrophys. J.*, **70**, 11 (1929).
30. Biermann, L., *Veröffentlichungen der Universitäts-Sternwarte Göttingen*, (Nr. 34) **3**, 45 (1933).
31. Unsöld, A., *Zeitschrift für Astrophys.*, **8**, 32 (1934).
32. Unsöld, A., *Zeitschrift für Astrophys.*, **8**, 225 (1934).
33. Unsöld. A., *Physik der Sternatmosphären*, Springer-Verlag, Berlin (1938).
34. Pannekoek, A., *Mon. Not. Roy. Astron. Soc.*, **95**, 529 (1935).
35. Pannekoek, A., *Astrophys. J.*, **84**, 481 (1936).
36. Strömgren, B., *Astrophysics*, edited by J.A. Hynek, (publ. McGraw-Hill) p. 192 (1951).
37. Wildt, R.W., *Astrophys. J.*, **90**, 611 (1939).
38. Bethe, H., *Zeitschrift für Physik*, **57**, 815 (1929).
39. Hylleraas, E.A., *Zeitschrift für Physik*, **60**, 624 (1930).
40. Wildt, R.W., *Astrophys. J.*, **89**, 295 (1939).
41. Jen, C.K., *Phys. Rev.*, (2) **43**, 540 (1933).
42. Massey, H.S.W. and Bates, E.R., *Astrophys. J.*, **91**, 202 (1940).
43. Chandrasekhar, S., *Astrophys. J.*, **100**, 176 (1944); ibid. **102**, 233 and **102**, 395 (1945); Chandrasekhar, S. and Breen, F.H., *Astrophys. J.*, **104**, 430 (1946).
44. Lundblad, R., *Astrophys. J.*, **58**, 113 (1923).
45. Mulders, G.F.W., *Zeitschrift für Astrophys.*, **11**, 132 (1935).
46. Chalonge, D. and Kourganoff, V., *Ann. d'Astrophys.*, **9**, 69 (1946); Chalonge, D., *Physica*, **12**, 721 (1946).
47. Peyturaux, R., *Ann. d'Astrophys.*, **15**, 302 (1952).
48. Unsöld, A., *Physik der Sternatmosphären* (Springer-Verlag, Berlin) 2nd edition, p. 165 (1955).
49. Strömgren, B., *Festschrift für Elis Strömgren*, edited by K. Lundmark, publ. by Einar Munksgaard, Copenhagen, p. 218 (1940).
50. Chandrasekhar, S., *Astrophys. J.*, **102**, 395 (1945).
51. Minnaert, M.G.J., *Bull. Astron. Inst. Netherlands*, **2**, 75 (1924).
52. Hjerting, F., *Astrophys. J.*, **88**, 508 (1938).
53. Strömgren, B., *Astrophys. J.*, **86**, 1 (1937).
54. Unsöld, A., *Zeitschrift für Astrophys.*, **4**, 339 (1932).
55. Strömgren, B., *Det Kgl. Danske Videnskabernes Selskab, Mat.–Fys. Meddelelser*, **21** (no. 3), 1 (1944) = Publikat. og mindre Meddelelser fra Kϕbenhavns Observ., no. **138** (1944).
56. Rudkjϕbing, M., *Publikat. og mindre Meddelelser fra Kϕbenhavns*

Observatorium, no. **145** (1947).
57. Münch, G., *Astrophys. J.*, **102**, 385 (1945).
58. Münch, G., *Astrophys. J.*, **106**, 217 (1947).
59. Barbier, D., *Ann. d'Astrophys.*, **9**, 173 (1946).
60. Abbot, C.G., *Smithsonian Ann.*, **3** (1913); ibid. **4** (1922).
61. Böhm-Vitense, E., *Zeitschrift für Astrophys.*, **34**, 209 (1954).
62. Münch, G., *Astrophys. J.*, **107**, 265 (1948).
63. Chandrasekhar, S., *Astrophys. J.*, **101**, 328 (1945).
64. Strömgren, B., *Astron. J.*, **53**, 107 (1948).
65. Unsöld, A., *Zeitschrift für Astrophys.*, **24**, 363 (1948).
66. Michard, R., *Ann. d'Astrophys.*, **12**, 291 (1949).
67. Milne, E.A., *Observ.*, **51**, 88 (1928).
68. Chandrasekhar, S., *Mon. Not. Roy. Astron. Soc.*, **96**, 21 (1936).
69. de Jager, C., *Rech. Astron. de l'Observ. d'Utrecht*, **13**, (part 1) (1952).
70. St John, C.E., Moore, C.E., Ware, L.M., Adams, E.F. and Babcock, H.D., *Carnegie Inst. Washington Publ.*, no. **396** (1928).
71. Michard, R., *Bull. Astron. Inst. Netherlands*, **11**, 227 (1950).
72. Münch, G., *Astrophys. J.*, **104**, 87 (1946).
73. Wright, K.O., *Astrophys. J.*, **99**, 249 (1944).
74. Labs, D., *Zeitschrift für Astrophys.*, **29**, 199 (1951).
75. Pecker, J.-C., *Ann. d'Astrophys.*, **14**, 152 (1951).
76. Böhm, K.H., *Zeitschrift für Astrophys.*, **34**, 182 (1954).
77. Unsöld, A., *Zeitschrift für Astrophys.*, **21**, 1 (1941); ibid. **21**, 22 (1941); ibid. **21**, 229 (1942); ibid. **23**, 75 (1942).
78. Aller, L.H., *Astrophys. J.*, **96**, 321 (1942).
79. Rosseland, S., *Mon. Not. Roy. Astron. Soc.*, **84**, 525 (1924).
80. Underhill, A.B., *Publik. og mindre Meddelelser fra Kϕbenhavns Observatorium*, no. **151** (1950) = Det Kgl. Danske Videnskabernes Selskab, Mat.-Fys. Meddelelser, **25** (no. 13), 1 (1950). See also Underhill, A.B., *Astrophys. J.*, **111**, 203 (1950).
81. Pecker, J.-C., *Ann. d'Astrophys.*, **13**, 433 (1950).
82. Unsöld, A., *Zeitschrift für Astrophys.*, **1**, 138 (1930).
83. Cayrel, R. and Jugaku, J., *Ann. d'Astrophys.*, **26**, 495 (1963).
84. Avrett, E.H. and Strom, S.E., *Ann. d'Astrophys.*, **27**, 781 (1964).
85. Mihalas, D., *Astrophys. J. Suppl.*, **13**, 1 (1965).
86. Russell, H.N. and Adams, W.S., *Astrophys. J.*, **68**, 9 (1928).
87. Pannekoek, A., *Proc. Acad. of Sci. Amsterdam*, **34**, 755 (1931).
88. Struve, O. and Elvey, C.T., *Astrophys. J.*, **79**, 263 (1934).
89. Berman, L., *Astrophys. J.*, **81**, 369 (1935).
90. Goldberg, L., *Astrophys. J.*, **89**, 623 (1939).
91. Williams, R.C. and Hiltner, W.A., *Publ. Observ. Uni. Michigan*, **8**, 45 (1940).
92. Hiltner, W.A. and Williams, R.C., *Photometric Atlas of Stellar Spectra*, Uni. of Michigan (1946).
93. Chalonge, D., *Ann d'Astrophys.*, **9**, 100 (1946).
94. Minnaert, M., *Astrophys. J.*, **104**, 331 (1946).
95. Griffin, R.F., *A Photometric Atlas of the Spectrum of Arcturus λ3600 to 8825 Å*, Cambridge Philosophical Society (1968).
96. Babcock, H.D. and Moore, C.E. The solar spectrum λ6600 to 13 495

Å, *Carnegie Inst. Washington*, publ. no. **579** (1947).

97. Babcock, H.D., Moore, C.E. and Coffeen, M.F., *Astrophys. J.*, **107**, 287 (1948).
98. Minnaert, M., Fourty [sic] years of solar spectroscopy, in *The Solar Spectrum*, p. 3, Utrecht Observatory Symposium, ed. C. de Jager. D. Reidel (1965).
99. Moore, C.E., Minnaert, M.G.J. and Houtgast, J., Second revision of Rowland's preliminary table of solar spectrum wavelengths. *National Bureau of Standards Monograph*, no. **61** (1965).
100. Minnaert, M., Preliminary photometric catalogue of Fraunhofer lines 3164 to 8770 Å. *Rech. Astron. de l'Observ. d'Utrecht*, **15**, 1 (1960).
101. Moore, C.E., *Multiplet Table of Astrophysical Interest*, Princeton Uni. Observ. and Mt Wilson Observ., Carnegie Inst, of Washington (1933); ibid., Revised multiplet table of astrophysical interest, *Contrib. from Princeton Uni. Observ.*, no. **20** (1945).
102. Swings, P., *Astrophys. J.*, **102**, 511 (1945).
103. Russell, H.N., *Astrophys. J.*, **61**, 233 (1925).
104. Scheiner, J., *Publ. Astrophys. Observ. zu Potsdam*, **7** (part 2) (no. 26), 167 (1895).
105. Merrill, P.W., Lines of the chemical elements in astronomical spectra, *Carnegie Inst. Washington*, publ. no. **610** (1958).
106. Dunham, T., *Contrib. Princeton Uni. Observ.*, no. **9** (1929).
107. Struve, O. and Dunham, T., *Astrophys. J.*, **77**, 321 (1933).
108. Morgan, W.W., *Publ. Yerkes Observ.*, **7** (part 3) (1935).
109. Albrecht, S., *Astrophys. J.*, **72**, 65 (1930).
110. Swensson, J.W., *Astrophys. J.*, **103**, 207 (1946).
111. Pannekoek, A., *Publ. Dominion Astrophys. Observ.*, **8** (no. 5), 141 (1949).
112. Moore, C.E., *Astrophys. J.*, **75**, 222 (1932); ibid., **75**, 298 (1932).
113. Secchi, A., *Comptes Rendus de l'Académie des Sciences*, **68**, 959 (1869).
114. Hacker, S.G., *Contrib. Princeton Uni. Observ.*, no. **16** (1935).
115. Hacker, S.G., *Astrophys. J.*, **83**, 140 (1936).
116. Davis, D.N., *Astrophys. J.*, **89**, 41 (1939). See also: ibid., **87**, 335 (1938).
117. Davis, D.N., *Astrophys. J.*, **106**, 28 (1947).
118. Mitchell, A.C.G. and Zemansky, M.W., *Resonance Radiation and Excited Atoms*, Cambridge Univ. Press, p. 145 (1934); reprinted (1961).
119. Frerichs, R., *Ann. der Physik*, **81**, 807 (1926).
120. Harrison, G.R., *J. Optical Soc. America*, **17**, 389 (1928).
121. Allen, C.W. and Asaad, A.S., *Mon. Not. Roy. Astron. Soc.*, **117**, 36 (1957); ibid., **117**, 622 (1957).
122. King, R.B. and King, A.S., *Astrophys. J.*, **82**, 377 (1935).
123. Kuhn, W., *Zeitschrift für Physik*, **33**, 408 (1925).
124. Roschdestwensky, D., *Zeitschrift für Wissenschaftliche Photographie*, **9**, 37 (1910); also in: ibid., *Ann. der Physik*, **39**, 307 (1912).
125. Bates, D.R. and Damgaard, A., *Phil. Trans. Roy. Soc.*, **242A**, 101 (1949).
126. Meggers, W.F.G., *Trans. I.A.U.*, **7**, 152 (1950).
127. Wessel, G., *Zeitschrift für Physik*, **126**, 440 (1949); Kopfermann, H.

and Wessel, G., *Zeitschrift für Physik*, **130**, 100 (1951).
128. Bell, G.D., Davis, M.H., King R.B. and Routley, P.M., *Astrophys. J.*, **127**, 775 (1958).
129. Lochte-Holgreven, W., *Observ.* **72**, 145 (1952); ibid., *Reports Progress in Physics*, **21**, 312 (1958); Maecker, H., *Zeitschrift für Physik*, **129**, 108 (1951).
130. Hollyer, R.N., Hunting, A.C., Laporte, O., Schwarcz, E.H. and Turner, E.B., *Phys. Rev.*, **87**, 911 (1952). See also Turner, E.B., *Proc. Nat. Sci. Foundation Conference on stellar atmospheres*, Indiana Univ., ed. M. Wrubel, p. 52 (1954).
131. King, R.B., *Proc. National Sci. Foundation conference on stellar atmospheres*, Indiana Univ., ed. M. Wrubel, p. 41 (1954).
132. Goldberg, L., Müller, E.A. and Aller, L.H., *Astrophys. J., Suppl.* **5**, 1 (1960). See also: Aller, L.H., *The Abundance of the Elements*, Interscience Publishers Inc., New York, p. 118 (1961).
133. Meggers, W.F., Corliss, C.H. and Scribner, R.F., *National Bureau of Standards Monograph*, no. **32** (1961).
134. Corliss, C.H. and Bozman, W.R., *National Bureau of Standards Monograph*, no. **53** (1962).
135. Corliss, C.H. and Warner, B., *Astrophys. J. Suppl.*, **8**, 395 (1964). See also: ibid., *J. Res. National Bureau Standards* **70A**, 325 (1966) for extension of results into the ultraviolet.
136. Bashkin, S. and Meinel, A.B., *Astrophys. J.*, **139**, 413 (1964).
137. Wiese, W.L., Smith, M.W. and Miles, B.M., Nat. Stand. Ref. Data System – *Nat. Bur. Standards*, **22**; Atomic transition probabilities (vol. II) (1969).
138. Garz, T. and Kock, M., *Astron. and Astrophys.* **2**, 274 (1969).
139. Huber, M. and Tobey, F.L., *Astrophys. J.*, **152**, 609 (1968); Grasdalen, G.L., Huber, M. and Parkinson, W.A., *Astrophys. J.*, **156**, 1153 (1969).
140. Whaling, W., King, R.B. and Martinez-Garcia, M., *Astrophys. J.*, **156**, 389 (1969).
141. Menzel, D.H., *Astrophys. J.*, **84**, 462 (1936).
142. Traving, G., *Zeitschrift für Astrophys.*, **44**, 142 (1957).
143. Hunger, K., *Zeitschrift für Astrophys.*, **36**, 42 (1955).
144. Baschek, B., *Zeitschrift für Astrophys.*, **48**, 95 (1959).
145. Greenstein, J.L., *Astrophys. J.*, **91**, 438 (1940).
146. Greenstein, J.L., *Astrophys. J.*, **95**, 161 (1942).
147. Menzel, D.H., Baker, J.G. and Goldberg, L., *Astrophys. J.*, **87**, 81 (1938).
148. Allen, C.W., *Mem. Commonwealth Solar Observatory*, **1**, (no. 5) (1934).
149. Greenstein, J.L., *Astrophys. J.*, **107**, 151 (1948).
150. Pannekoek, A. and van Albada, G., *Publ. Astron. Inst. University of Amsterdam*, no. **6** (part 2) (1946).
151. Steel, H.R., *Astrophys. J.*, **102**, 43 (1945).
152. Greenstein, J.L. and Hiltner, W.A., *Astrophys. J.*, **109**, 265 (1949).
153. Greenstein, J.L., *Astrophys. J.*, **109**, 121 (1949).
154. Unsöld, A., *Zeitschrift für Astrophys.*, **21**, 229 (1942).
155. Aller, L.H., *Astrophys. J.*, **104**, 347 (1946).
156. Aller, L.H., *I.A.U. Symposium*, **26**, 99 (1966).
157. Miczaika, G.R., *Zeitschrift für Naturforschung*, **3a**, 241 (1948).

158. Wright, K.O., *I.A.U. Symposium*, **26**, 15 (1966).
159. Underhill, A.B., *Astrophys. J.*, **107**, 337 (1947).
160. Underhill, A.B., *Astrophys. J.*, **107**, 349 (1947).
161. Aller, L.H., *Astrophys. J.*, **109**, 244 (1949).
162. Wright, K.O., *Publ. American Astron. Soc.*, **9**, 276 (1939); ibid. **10**, 34 (1940).
163. Wright, K.O., *Publ. Dominion Astrophys. Observ., Victoria*, **8** (no. 1), 1 (1947).
164. Struve, O., *Trans. I.A.U*, **7**, 487 (1950).
165. Huggins, W., Publications of Sir William Huggins' Observatory vol .II (as footnote to article in *Phil. Trans. Roy. Soc.*, **154**, 413 (1864)), publ. W. Wesley and Son (1909).
166. Struve, O., *Sky and Telescope*, **19**, 7 (1959).
167. Unsöld, A., *Science*, **163**, 1015 (1969).
168. de Jager, C., 'Principes fondamentaux de classification stellaire', *Colloque international du C.N.R.S.*, no. **55**, 141 edited by E. Schatzman, Paris (1955) = Overdruk Sterrewacht Utrecht, no. **17** (1955).
169. Minnaert, M., *Mon. Not. Roy. Astron. Soc.*, **117**, 315 (1957).
170. Traving, G., *Zeitschrift für Astrophys.*, **36**, 1 (1955).
171. Cayrel, R., Thèse de doctorat d'état, Paris (1957), published in : *Ann. d'Astrophys. Suppl.*, no. **6** (1958).
172. Aller, L.H., Elste, G. and Jugaku, J., *Astrophys. J. Suppl.*, **3**, 1 (1957).
173. Voigt, H.H., *Zeitschrift für Astrophys.*, **31**, 48 (1952).
174. Aller, L.H., *Astrophys. J.*, **123**, 117 (1956).
175. Aller, L.H. and Jugaku, J., *Astrophys. J.*, **127**, 125 (1958); ibid; *Astrophys. J. Suppl.*, **4**, 109 (1959).
176. Aller, L.H., *The Abundance of the Elements*, Interscience Publishers Inc., New York, p. 113 (1961).
177. de Jager, C. and Neven, L., *Rech. Astron. de l'Observ. d'Utrecht*, **13** (no. 4) (1957).
178. Unsöld, A., *Mon. Not. Roy. Astron. Soc.*, **118**, 3 (1958).
179. Burbidge, G.R. and Burbidge, E.M., *Astrophys. J. Suppl.*, **1**, 431 (1955).
180. Roman, N.G., *Astrophys. J.*, **112**, 554 (1950).
181. de Jager, C., *Comm. de l'Observ. Royal de Belgique*, No. **157**, I.A.U. Colloquium: 'The empirical determination of stellar photospheric structure', ed. C. de Jager and L. Neven (1959).
182. Cayrel, R. and Cayrel de Strobel, G., *Ann. Review Astron. and Astrophys.*, **4**, 1 (1966).
183. Fujita, Y., *J. Japanese Phys. Soc.*, **2**, 204 (1947); ibid., *Publ. Astron. Soc. Japan*, **4**, 81 (1952).
184. Yamashita, Y., *Publ. Astron. Soc. Japan*, **8**, 142 (1956).
185. Cayrel, G. and Cayrel, R., *Astrophys. J.*, **137**, 431 (1963).
186. Adams, W.S., *Astrophys. J.*, **42**, 172 (1915).
187. Cannon, A.J. and Pickering, E.C., *Harvard Ann.*, **96**, 228 (1921).
188. Adams, W.S. and Joy, A.H., *Astrophys J.*, **56**, 262 (1922).
189. Kuiper, G.P., *Astron. J.*, **53**, 194 (1948).
190. Baade, W., *Astrophys. J.*, **100**, 137 (1944).
191. Schwarzschild, M. and Schwarzschild, B., *Astrophys. J.*, **112**, 248 (1950).
192. Iwanowska, W., Frackowiak, M. and Kažymierczak, M., *Bull. Astron.*

Observ. Torun, no. **9**, 25 (1950).
193. Miczaika, G.R., *Zeitschrift für Astrophys.*, **27**, 1 (1950).
194. Chamberlain, J. and Aller, L.H., *Astrophys. J.*, **114**, 52 (1951).
195. Keenan, P.C. and Keller, G., *Astrophys. J.*, **117**, 241 (1953).
196. Miczaika, G.R., *Astron. Nachrichten*, **270**, 249 (1940).
197. Roman, N.G., *Astron. J.*, **59**, 307 (1954).
198. Roman, N.G. *Astrophys. J. Suppl.*, **2**, 195 (1955).
199. Burbidge, E.M. and Burbidge, G.R., *Astrophys. J.*, **124**, 116 (1956).
200. Schwarzschild, M., Searle, L. and Howard, R., *Astrophys. J.*, **122**, 353 (1955).
201. Wildey, R.L., Burbidge, E.M., Sandage, A.R. and Burbidge, G.R., *Astrophys. J.*, **135**, 94 (1962).
202. Melbourne, W.G., *Astrophys. J.*, **132**, 101 (1960).
203. Swihart, T.L. and Fischel, D., *Astrophys. J. Suppl.*, **5**, 291 (1960) and Swihart, T.L., *Astrophys. J.*, **132**, 915 (1960).
204. Reiz, A., *Astrophys. J.*, **120**, 342 (1954).
205. Aller, L.H. and Greenstein, J.L., *Astrophys. J. Suppl.*, **5**, 139 (1960); ibid., *Astrophys. J.*, **132**, 520 (1960).
206. Aller, L.H., see ref. (176), p. 208 (1961).
207. Heiser, A.M., *Astrophys. J.*, **132**, 506 (1960).
208. Pagel, B.E.J. and Powell, A.L.T., *Roy. Observ. Bull.*, no. **124** (1966).
209. Baschek, B., *Zeitschrift für Astrophys.*, **61**, 27 (1965).
210. Danziger, I.J., *Astrophys. J.*, **143**, 527 (1966).
211. Lindblad, B., *Astrophys. J.*, **55**, 85 (1922) (see p. 116).
212. Popper, D.M., *Astrophys. J.*, **105**, 204 (1947).
213. Greenstein, J.L., *Proc. N.S.F. Conference on Stellar Atmospheres*, Indiana Univ., ed. M. Wrubel, p. 38 (1954).
214. Greenstein, J.L. and Keenan, P.C., *Astrophys. J.*, **127**, 172 (1957).
215. Schwarzschild, M., Schwarzschild, B., Searle, L. and Meltzer, A., *Astrophys. J.*, **125**, 123 (1956).
216. Helfer, H.L., Wallerstein, G. and Greenstein, J.L., *Astrophys. J.*, **129**, 700 (1954).
217. Wallerstein, G., Greenstein, J.L., Parker, R., Helfer, H.L. and Aller, L.H., *Astrophys. J.*, **137**, 280 (1963).
218. Pagel, B.E.J., *Roy. Observ. Bull.*, no. **104** (1965); see also ibid., *J. Quantitative Spectroscopy and Radiative Transfer*, **3**, 139 (1963).
219. Burbidge, E.M., Burbidge, G.R., Fowler, W.A. and Hoyle, F., *Rev. Modern Physics*, **29**, 547 (1957).
220. Eggen, O.J., Lynden-Bell, D. and Sandage, A.R., *Astrophys. J.*, **136**, 748 (1962).
221. Helfer, H.L., Wallerstein, G. and Greenstein, J.L., *Astrophys. J.*, **132**, 553 (1960); Wallerstein, G. and Helfer, H.L., *Astrophys. J.*, **129**, 347 (1959).
222. Wallerstein, G., *Astrophys. J. Suppl.*, **6**, 407 (1961).
223. Wallerstein, G. and Helfer, H.L., *Astrophys. J.*, **129**, 720 (1959).
224. Wallerstein, G. and Carlson, M., *Astrophys. J.*, **132**, 276 (1960).
225. Hazlehurst, J., *Observ.*, **83**, 128 (1963).
226. Pagel, B.E.J., *Observ.*, **83**, 133 (1963).
227. Kirchhoff, G., *Phil. Mag.*, **19** (ser. 4), 195 (1860).
228. Hale, G.E. and Adams, W.S., *Astrophys. J.*, **25**, 75 (1907).

229. King. A.S., *Astrophys. J.*, **44**, 169 (1916).
230. McKellar, A., *Publ. Astron. Soc. Pacific*, **52**, 407 (1940); ibid., *Observ.*, **64**, 4 (1941).
231. Babcock, H.D. and Moore, C.E., reported by A. McKellar, *Observ.*, **64**, 4 (1951).
232. Greenstein, J.L. and Richardson, R.S., *Astrophys. J.*, **113**, 536 (1951).
233. Sanford, R.F., *Astrophys. J.*, **99**, 145 (1944); ibid., **111**, 31 (1950).
234. Keenan, P.C. and Teske, R.G., *Astrophys. J.*, **124**, 499 (1956); Teske, R.G., *Publ. Astron. Soc. Pacific*, **68**, 520 (1956).
235. Hunger, K., *Astron. J.*, **62**, 294 (1957).
236. Spitzer, L., *Astrophys. J.*, **109**, 548 (1949).
237. Bonsack, W., *Astrophys. J.*, **130**, 843 (1959).
238. Bonsack, W. and Greenstein, J.L., *Astrophys. J.*, **131**, 83 (1960).
239. Herbig, G.H., *Astron. J.*, **68**, 280 (1963).
240. Herbig, G.H., *Astrophys. J.*, **141**, 588 (1965).
241. Wallerstein, G., Herbig, G.H. and Conti, P.S., *Astrophys. J.*, **141**, 610 (1965).
242. Conti, P.S., *Observ.*, **84**, 122 (1964).
243. Feast, M.W., *Mon. Not. Roy. Astron. Soc.*, **134**, 321 (1966).
244. Wallerstein, G., *Astrophys. J.*, **145**, 759 (1966).
245. Feast, M.W., Thackeray, A.D. and Wesselink, A.J., *Observ.*, **75**, 216 (1955).
246. Feast, M.W., Thackeray, A.D. and Wesselink, A.J., *Observ.*, **78**, 156 (1958).
247. Feast, M.W., Thackeray, A.D. and Wesselink, A.J., *Mon. Not. Roy. Astron. Soc.*, **121**, 337 (1960).
248. Henize, K.G., *Astrophys. J. Suppl.*, **2**, 315 (1956).
249. Przybylski, A., *Nature*, **205**, 163 (1965).
250. Przybylski, A., *Mon. Not. Roy. Astron. Soc.*, **139**, 313 (1968).

11 Some miscellaneous topics in stellar spectroscopy: individual stars of note, stellar chromospheres and interstellar lines

11.1 Introduction A miscellany of topics is covered in this final chapter. The first section discusses the spectra of a few individual peculiar stars of unusual interest, which appear not to be members of the broader groups of peculiar stars discussed under the headings of Chapter 9.* A large number of stellar spectral idiosyncrasies could have been included here, but I have restricted the discussion to only four objects. Several outstanding candidates were omitted, including β Lyrae (probably the star with the greatest number of spectroscopic references in the entire literature), since it has been referred to, albeit briefly, in earlier chapters; and also the spectroscopic binary and supergiant ζ Aurigae, with its atmospheric eclipses.

Secondly, two major topics of importance are treated which fell outside the headings in earlier chapters. The Wilson–Bappu effect encompasses the calcium emission lines from the chromospheres, or hot turbulent outer layers, of the cooler stars. The analysis of these lines has become a major method of determining stellar luminosities and distances.

Finally, the topic of interstellar absorption lines in stellar spectra is reviewed. This is an appropriate topic for a book on stellar spectroscopy, since it is the analysis of the starlight which reveals the presence of interstellar clouds along the line of sight to the star. Moreover, this subject brings us to the advent of far ultraviolet spectroscopy from above the atmosphere; the dawn of this new era in the mid-1960s makes an

* To be more precise, the stars discussed appeared for many years to have spectra both unique and unclassifiable. Neither of these adjectives may be appropriate today for any of the stars included in section 11.2.

appropriate place to conclude this review of 150 years of stellar spectroscopy since the time of Fraunhofer.

11.2. Some individual stars of note

11.2.1 The spectrum of P Cygni The star P Cygni is an old and very slow nova whose unusual (though not unique) spectral characteristics have earned it a special place in the history of stellar spectroscopy. A Dutch chartmaker and geographer, Willem Janszoon Blaeu (1571–1638) observed it as a third magnitude star in August 1600, still in the pre-telescopic era, presumably some months after its nova outburst. It stayed at third magnitude for 6 years before fading slowly. A second outburst occurred about 1655 once again followed by a slow fading some years later. From 1665 it was in a transitory stage with frequent brightness fluctuations around the naked-eye limit of about magnitude six. It has been near fifth magnitude with only small brightness variations since 1715. (See (1) for full details of the light variations.)

The first reference to the spectrum of P Cygni came from E.W. Maunder at Greenwich. He used a single-prism spectroscope in October 1888 and found a single bright line at 4858 Å, which we can presume to be Hβ (2). The star was recorded on two objective prism plates at Harvard soon afterwards, as it appears in the Draper Memorial Catalogue of 1890 (3) where Mrs Fleming gave it spectral type Q (which then signified of miscellaneous or unclassifiable type) with the brief additional remark: 'spectrum nearly continuous, traversed by bright hydrogen lines'.

The key characteristic that distinguishes the P Cygni spectrum are the double lines throughout the spectrum, a somewhat broad emission line bordered by a usually narrower absorption feature immediately on its violet edge (Fig. 9.4). This property was discovered by James Keeler using a visual spectroscope on the giant Lick refractor in 1889, when he noted the 'appearance of the bright lines in the green, which are described as having dark borders on their more refrangible edges.... As only one observation of P Cygni was made with suitable apparatus, the appearance may have been merely a mistaken impression' (4). It was clearly an observation requiring the utmost skill. But early spectrographic observations at Harvard by Miss Maury (5) and at Potsdam by Vogel and Wilsing (6) soon confirmed Keeler's discovery. The German observers concluded from plates taken in October 1895: 'The spectrum [is] very similar to that of β Lyrae, since bright and dark lines appear in pairs next to each other.' They evidently believed the star to be double and gave two separate spectral types for the absorption and emission spectra. Both they and Miss Maury gave wavelengths of absorption and emission features, mainly of hydrogen and neutral helium. Belopolsky found the dark lines to be displaced towards the violet by about 1.5 Å (but for $H\beta$ and $H\gamma$ lines, by about twice this amount) whereas the bright lines were at their normal wavelengths (7), and this was later to become an important clue for the interpretation of the spectrum. However, Frost rejected a velocity interpretation of the absorption lines' blue displacement, because it appeared to vary from line to line:

> ... the velocity derived for $H\beta$ would be greater by nearly 50 km [per second] than that for $H\delta$. This amount is obviously much in excess of even the large accidental errors unavoidable in the measurement of such broad lines; and at once negatives the possibility that these are Doppler effects (8).

Paul Merrill also observed P Cygni between 1907 and 1913; he was the first observer to suggest that small temporal variations in line positions might be occurring (9). For many years the question of spectral changes in P Cygni was controversial. At Harvard, Boris P. Gerasimovič (1889–1937) even claimed the absorption features to have once been on the red sides of the emission for some Balmer lines, from a re-examination of old plates taken in 1887 and 1888 (10). C.T. Elvey for one remained sceptical (11). Meanwhile high resolution coudé spectra sometimes showed doubled hydrogen absorption lines, as noted at Mt Wilson by Walter Adams and Theodore Dunham (12) and then by O.C. Wilson (13). This last paper confirmed the suspicion of spectral changes but no periodicity could be assigned.

The breakthrough in understanding P Cygni came in 1929 when C.S. Beals published his well-known paper 'on the nature of Wolf–Rayet emission' (see section 9.2) (14). He wrote:

> Both P Cygni and η Carinae have been numbered among the novae... This similarity with novae, considered in connection with the absorption on the violet edges of emission lines and the variation in width of P Cygni lines with wave-length, suggests that the peculiarities in the spectra of these stars is [sic] due to the ejection of gaseous material in a manner similar to that suggested for Wolf–Rayet stars.

The absorption lines were produced by the light coming directly from the stellar photosphere being absorbed or scattered in that part of the shell on the line of sight to the observer. Beals soon realised that the complex absorption line structures gave information on stratification in the shell (15). This theme was taken up by Struve, who studied Yerkes spectra of P Cygni from the giant refractor in a detailed analysis (16). As well as discussing the spectral effects due to line formation in a geometrically diluted radiation field, he discovered an 'obvious correlation between velocity [of the shell's expansion] and ionization potential'. Lines of lesser ionisation potential (e.g. the Balmer lines) had higher Doppler velocities away from the star (up to nearly 200 km/s) and originated in the outer layers; on the other hand silicon (SiIV) and nitrogen (NIII) absorptions were found expanding at only about 50 km/s in layers closer to the star where the ionising ultraviolet light was more intense. Thus Struve concluded the shell was not only expanding but accelerating outwards as it did so. As predicted on Beals' theory, the emission lines on the other hand, although they also arise from the gaseous shell, have all about the same mean Doppler velocity which is close to that of the underlying star itself. Because the shell is transparent, all parts expanding in all directions are seen, giving rise to Doppler-broadened but undisplaced emission lines.

Many observers have since confirmed the validity of this model. One of the most detailed studies came from Beals himself in 1950 (17). He too showed how the expansion velocity depended on the ionisation potential of the element. Because the emission tended to be stronger for the lower potential species, these must originate from the larger more extensive outer layers (indeed, for the doubly ionised nitrogen lines, the emission is not observed, only a violet-displaced absorption) indicating that the sense of the velocity change must be an outward acceleration. Beals also studied the spectra of sixty-nine other stars* that showed the P Cygni characteristic line profiles (17). These were sometimes only seen at Hα, and often showed variations on the basic pattern of emission plus violet displaced absorption. These different profiles were classified into eight types, four of them unique to stars of the P Cygni type, and interpreted on the basis of various models for the expanding shell.

Finally, we bring the story of P Cygni up to the year 1968 with M. de Groot's extensive analysis from thirty-six high resolution spectra (19). This work took into account the splitting of the lower ionisation absorption lines, in particular the Balmer lines, into as many as three components: 'The triple structure of the absorption lines can be explained by supposing that absorption takes place in three different shells', whose respective radii thicknesses and velocities were deduced. What is more, the radial velocity

* The existence of other stars with spectra of the P Cygni type had been recognised for some time. Annie Cannon listed eleven in 1916, though all were considerably fainter than P Cygni itself (18).

of the outermost shell was found to pulsate with a 114-day period, between peak values of about – 180 and – 240 km/s, thus accounting for some of the mysterious spectral variations reported by earlier observers.

11.2.2 η Carinae The spectrum of the star η Carinae is one of the more unusual emission-line spectra of any celestial object. The star itself has had a remarkable history of apparent magnitude changes. Prior to 1837 it was between second and fourth magnitudes, but brightened in that year to zeroth magnitude, making it one of the brightest stars in the sky. From about 1859 to 1870, η Carinae faded slowly to magnitude eight. In the next century from 1870 there have been several small amplitude fluctuations of less than a magnitude, and since about 1940 there has also been a steady rise in brightness to the present level near magnitude six. η Carinae is surrounded by a small elongated outer envelope of nebulosity, no more than 10 arc seconds in extent. The much larger Carina gaseous nebula also surrounds the star.*

The first account of the visual spectrum of the stellar object came from A. le Sueur in 1870. His observations on the Great Melbourne telescope showed that many bright emission lines were present (21) (see section 4.15 for a fuller discussion). Le Sueur identified the Balmer lines Hα and Hβ but was unsuccessful in assigning correct identifications to any other features. For example, the displacement of a bright yellow line from the normal wavelength of the sodium D lines was suspected, but this observation had to await another quarter century before the observed D_3 line could be identified with neutral helium (see section 4.11).

Photographic spectra of η Carinae were recorded by the Harvard observers in Peru with the 13-inch Boyden telescope from 1892 to 1899 and described by Annie Cannon in 1901 (22) (Fig. 11.1). Both dark and bright lines were in evidence: 'Taking only the dark lines into consideration, the spectrum would be classed as F5G. Considering the bright lines, there is a marked resemblance to the spectrum of Nova Aurigae... On some plates, the dark lines appear inconspicuous, and nearly the whole spectrum is crossed by bright lines. On the other plates, dark lines are distinctly seen on the edges of shorter wave-length of the bright bands' (22). The interesting points are the changes in the spectrum and the report of the P Cygni-type of line profiles. The fact that changes had occurred from an essentially absorption-line spectrum in 1892–3, to a predominantly emission-line spectrum by 1895, was emphasised is a discussion of the Harvard observations by F.E. Baxandall in 1919 (23). Later on B.J. Bok also

* See E. Gaviola's discussion of the surrounding nebulosity (20).

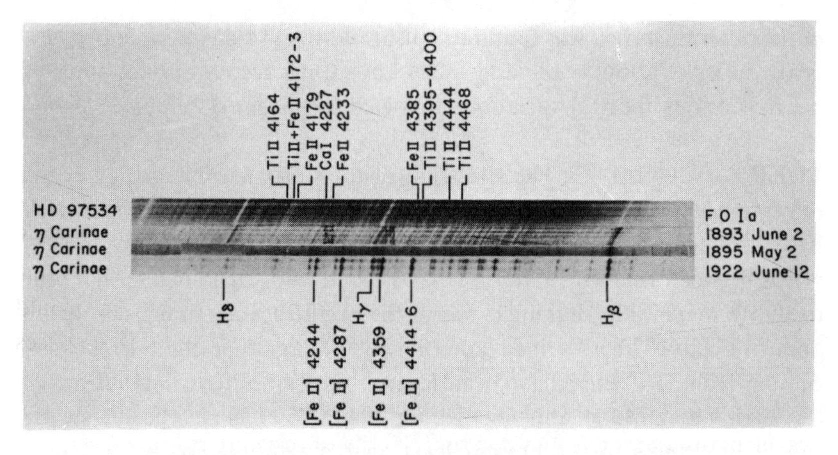

Fig. 11.1 Early Harvard spectra of η Carinae.

discussed this early Harvard material; he measured as many as 70 absorption lines in an 1893 spectrum, but in 1895 these were generally much weaker and many strong emission lines were present (24).

The η Carinae spectra observed in this early period have been rediscussed extensively by Nolan Walborn (b. 1944) and Martha Liller (b. 1931) (25). They point out that the Harvard spectra began shortly after a one magnitude brightness increase from 1887 to 1889 followed by a slow decline in the 1890s. The spectral development resembled that of a nova during this decline. On the other hand, the emission-line spectrum seen by le Sueur in 1870 was markedly different from the first of the Harvard observations.

Two spectra of η Carinae were obtained at the Royal Observatory at the Cape in 1899 and described by Sir David Gill in 1901 (26). One of these was an objective prism spectrum; the second was the first recorded slit spectrogram and exposed over four consecutive nights. Gill reported 'a very remarkable bright-line spectrum' but there is no mention of dark (absorption) lines. Since the slit spectrograph was equipped with an iron spark for a wavelength comparison, accurate wavelengths could be measured for the first time. The wavelengths and intensities of thirty-two blue lines on the slit spectrogram were tabulated; only three Balmer lines were identified.

Some progress in the problem of line identifications was made by Joseph Moore and Roscoe Sanford in 1913. They used two spectra at 36 Å/mm dispersion (at Hγ) obtained in Santiago as part of the D.O. Mills radial velocity expedition to Chile from Lick Observatory (27). They found many bright lines but none of the dark ones that had been reported from Harvard. Many of the emission lines were identified with the ionised lines of iron, titanium and chromium, but a considerable number still remained unidentifiable. Similar conclusions were reached by Joseph Lunt at the Cape in 1919 (28). He obtained two spectra, exposing each for 9 hours over 3 nights; he confirmed the presence of iron and chromium lines in addition to those of hydrogen; he was less certain of the presence of titanium. Many lines still defied identification and he concluded 'that we have to deal with a spectrum containing numerous lines, some of them very strong, with which we are unacquainted terrestrially'. Lunt was also the first since Annie Cannon to report the possible presence of absorption lines, 'but they could not be identified with known lines, and they may be merely interspaces between the bright lines' (28).

The next major advance came in 1928 when Paul Merrill, using the wavelengths and intensities of Moore and Sanford and of Lunt, identified many of the lines previously of unknown origin, with forbidden transitions between metastable states of singly ionised iron (i.e. the [FeII] spectrum) (29). Merrill noted at the close of his paper that 'the present investigation was suggested by Bowen's success in identifying several of the strong nebular* lines with transitions from metastable states in oxygen and nitrogen atoms' (29).

An extension of the line identifications into the red part of the spectrum was undertaken by H. Spencer Jones (30). He measured the spectrum on a red objective prism plate that had been obtained in 1919 at the Union Observatory in Johannesburg. It was the first spectrogram of η Carinae to show the very great strength of Hα, and also the first to record a weak line at 5875 Å, and Spencer Jones tentatively assigned this to the D_3 line of helium. However evidently D_3 in 1919 was much fainter than in 1870, when le Sueur was able to see it in a visual spectroscope.

Miss D. Hoffleit (b. 1907) made a study of 40 objective prism plates taken at Arequipa between 1892 and 1930 and was able to identify several dozen new lines in the spectrum (31). She also discussed the evidence for changes in the line intensities, and confirmed that these had taken place in the period 1892–1902, but not thereafter.

From 1950 several very thorough new investigations of the spectrum of η Carinae were undertaken. For example, E. Gaviola at the Cordoba Observatory in Argentina obtained 70 slit and 33 slitless grating spectra on the 1.5 m reflector from 1944 to 1951; measurements were made on 21 of these (32). A total of 903 emission lines and 76 absorption lines from 3029 to 6966 Å were measured. The emission lines were often asymmetric and multiple showing two, three or more components. He found the Hα line to be brighter than all other lines together. His spectra showed only slight

* i.e. lines in planetary and diffuse nebulae; see section 8.3.

evidence for spectral changes with the exceptions of the neutral helium lines which were definitely variable. All but forty-one of the lines in Gaviola's list were identified; apart from the Balmer and neutral helium lines, the vast majority were ascribed to singly ionised metals such as iron, chromium and titanium.

Another extensive η Carinae program was commenced by A.D. Thackeray at Radcliffe Observatory, Pretoria, at about the same time (33). In this first paper Thackeray also gave an extensive line list, and for the first time the η Car spectrum was catalogued into the infrared to 8870 Å. This was in spite of his use of a prism spectrograph on the 74-inch telescope; the dispersion varied from 10 Å/mm in the ultraviolet to 400 Å/mm in the infrared. Nevertheless, some strong infrared lines were found, six of them being unidentified. A few doubly ionised elements were identified from lines in the spectrum.

As with Gaviola's study, multiple structure was present in many emission lines and the strongest ones also showed P Cygni profiles with violet absorption components shifted by -450 km/s, and thus supporting Annie Cannon's description of the spectrum half a century earlier. Similarities with the spectrum of RR Telescopii, a slow nova, were noted. This was the first of several papers Thackeray produced on the η Carinae spectrum. After the completion of the coudé spectrograph at Radcliffe Observatory in 1959 (the first in the southern hemisphere), Thackeray was able to obtain much higher quality infrared spectra of η Carinae at 31.2 Å/mm on Kodak IN plates (34). Many new infrared lines were catalogued; some were identified with neutral nitrogen but several of the strongest still defied identification. The coudé spectrograph was also used to explore the ultraviolet region in 1967 (35) at which time the lines of several elements were found to have changed strength since the earlier observations.

When the Mt Stromlo coudé spectrograph was completed on the 74-inch telescope in 1961 by Dunham, observations of η Carinae were the very first to be obtained. The dispersion of 10.2 Å/mm was higher than on the Radcliffe coudé. Aller and Dunham published a new line list for the blue spectral region based on this material and noted further small changes in the spectrum, including the weakening since 1953 of redward displaced emission components of some of the strong iron lines (36). Mt Stromlo coudé observations were continued by A.W. Rodgers (b. 1932) and Leonard Searle during 1964 and 1965 (37). These observers also obtained photoelectric scans of the continuous radiation to about 1 μm with a scanning spectrophotometer. This period marks the transition from the cataloguing and identification of spectral lines, which had taken nearly a

Fig. 11.2 Roscoe Sanford.

century of effort, to one of analysis and modelling of the source. Rodgers and Searle emphasised the probable non-stellar origin of the radiation and the inhomogeneities in the electron temperature and density of the emitting material. The greater part of the volume could be described by an electron temperature of 20 000 K and number density of three million electrons per cubic centimetre, from an analysis of forbidden lines of nitrogen and sulphur. They also discussed the possibility that the continuum may arise from synchrotron radiation due to high energy electrons interacting with a magnetic field.

Our understanding of the underlying object of η Carinae advanced substantially in the late 1960s. G. Neugebauer (b. 1932) and J.A. Westphal (b. 1930) at Caltech showed the continuum rises steeply into the infrared. Their data extended to a wavelength of 19.5 μm, where η Car is the brightest object in the whole sky (38). At the same time, B.E.J. Pagel at the Royal Greenwich Observatory found evidence for a very large reddening of the radiation, from an analysis of the observed emission line intensities of forbidden iron lines (39). The evidence pointed to a model in which the large excess of infrared radiation came from a thick dust shell completely surrounding a central hot star. The thermal radiation from an outer dust shell at about 250 K would account for the large infrared excess at 19.6 μm while scattering of the visible radiation by the dust would also produce the circumstellar reddening. The dust model was first proposed by Bernard Pagel (40) and since then by many others (see for example (41)). It is possible that the variations of the apparent visual magnitude since the early nineteenth century have not been accompanied by changes in the total luminosity at all wavelengths, but only by changes in the amount of dust absorption together with thermal reradiation in the infrared part of the spectrum.

Table 11.1 *Published line lists for η Carinae*

Author(s)	*Reference*	*Wavelength region (Å)*	*Number of lines*	*Number identified*
Gill, D	(26)	4060–5020	39	3
Moore, J.H. and Sanford, R.F.	(27)	4100–5170	76	
Lunt, J.	(28)	4060–5020	103	30
Merrill, P.W.	(29)	4060–5020	94	75
Spencer Jones, H.	(30)	3930–6570	66	59
Hoffleit, D.	(31)	3900–5600	44	44
Gaviola, E.	(32)	3029–6966	903	862
Thackeray, A.D.	(33)	3677–8863	436	322
Thackeray, A.D.	(34)	6666–9122	143	88
Thackeray, A.D	(35)	3076–4244	325	284
Thackeray, A.D.	(35)	4863–6882	157	137
Aller, L.H. and Dunham, T.	(36)	3455–5019	432	396

11.2.3 He^3 in 3 CenA 3 CenA is a mid-type B star which was studied by Bidelman in 1960 from plates taken at the Chile Station of the Lick Observatory (see section 6.11). He recognised that the spectrum 'contains a considerable number of lines that are not generally seen in stellar spectra' (42). Lines of ionised phosphorus were the most unusual, and although not unique to 3 CenA, they were far stronger than in any other star then known.* The companion (3 CenB) is a normal B8 dwarf.

Even more unusual is the discovery of a very high abundance of the light helium isotope He^3 in the atmosphere of this star. The solar system ratio He^3/He^4 is of the order of 10^{-4}. In B stars this ratio can be measured spectroscopically from the fairly large redwards isotope shift in several of the neutral helium lines, amounting to 0.50 Å for the 6678 Å line from pure He^3. In 1961 Sargent and Jugaku found shifts corresponding to nearly pure He^3 in ten helium lines from coudé spectra in both the blue and red spectral regions (43). From wavelength measurements they estimated He^3/(He^3 + He^4) to be 0.84. A variety of nuclear reactions were proposed that might have been able to account for the excess of He^3.

An abundance analysis of 3 CenA was carried out at this time by

* Bidelman found strong ionised phosphorus lines in κ Cancri (B8p) shortly afterwards.

Jugaku, Sargent and Greenstein (44). This showed that phosphorus was by no means the only enhanced element; krypton was 1300 times overabundant, and gallium possibly 8000 times. Some elements on the other hand were deficient relative to hydrogen, including helium.

3 CenA is the best known and first discovered of the P-Ga subclass of Bp stars. Several other stars of this class have since been found and some (e.g. ι Ori B) are also He^3 stars. As with other Bp and Ap stars, it is probable that the element diffusion hypothesis can best explain the observed peculiarities. Eight definite cases of He^3 stars were known in 1979 (45) and they appeared to be intermediate in temperature between the early B-type helium-strong stars and the later B helium-weak stars.

11.2.4 Przybylski's star, HD 101065 The unusual spectrum of HD 101065 was discovered by Antoni Przybylski (1913–85) at Mt Stromlo Observatory in 1960 as part of a program to observe the spectra of high proper motion stars in the southern hemisphere (46). Although the star is given a B5 spectral type in the Henry Draper Catalogue, Przybylski described it as 'a G0 star with high metal content'. After a preliminary analysis, he concluded 'that the abundance of metals will prove to be many times higher than that of normal main sequence stars.... many lines should be attributed to rare earths'. This initial statement was based on an inspection of the extremely rich absorption line spectrum rather than a quantitative analysis. Later Przybylski showed that the metal enhancement was highly selective, and some elements of high solar abundance (such as iron) were markedly deficient in this peculiar star, making it, overall, metal-poor.

By 1963 it was clear that HD 101065 was extremely bizarre in its composition. Although the star is only slightly hotter than the sun (from its colours), Przybylski found that 'lines of holmium are stronger than lines of any other element.... This behaviour of the holmium lines is surprising since holmium is one of the elements that has not yet been observed with certainty in stellar spectra and could not be seen at all in the spectrum of the sun' (47). There were over 100 identifiable ionised holmium lines in a blue coudé spectrogram, but none of iron; six of the fifteen strongest blue lines were identified as holmium features (48).

An abundance analysis by the coarse curve of growth method in 1966 confirmed a high rare earth element content (49). Most rare earths were enhanced by a factor of about a thousand. On the other hand, calcium was markedly deficient. Iron lines still could not be identified, but the red lithium line was probably present. A coudé spectrum obtained by Warner on the Radcliffe telescope confirmed that lithium is over two orders of magnitude overabundant (50).

Przybylski has commented that 'while spectral peculiarities are superimposed on a more or less normal spectrum in other abnormal stars, there are scarcely any normal features in the spectrum of this star' (49). HD 101065 has a unique spectrum; that it may be a cool analogue of the Ap class of star has been suggested by several authors. An abundance analysis by G. Wegner and A. Petford confirmed very large rare earth overabundances and the presence of iron was reported for the first time (51). Abundances for forty-nine elements were given, which represents one of the largest number of elements detected in the spectrum of any star other than the sun.

In his most recent discussions on HD 101065, Przybylski has clearly favoured a diffusion process to account for the strange composition of his star; he does not believe that the abnormalities can be directly ascribed to nuclear reactions (52).

11.3 Emission lines at H and K and the Wilson–Bappu effect

Solar-type stars have many thousands of photospheric absorption lines in their visible spectra, of which the H and K lines of ionised calcium in the far violet are the strongest. Emission lines were not known to occur in any stars of spectral type G and K (other than the sun) until about 1900, when G. Eberhard (1867–1940) at Potsdam discovered weak emission cores in the centres of the strong H and K absorption lines of Arcturus, on heavily exposed plates (at 16 Å/mm dispersion in the violet). The discovery was reported by Eberhard and Schwarzschild in 1913 (53) by which time two further K stars, Aldeberan and σ Geminorum, were also found to show these so-called 'H and K reversals'. The reversal in σ Geminorum was especially strong (Fig. 11.4). Extremely weak reversals were also known in the solar spectrum itself, since their discovery by Henri Deslandres in Paris in 1892 (54) who had assigned the labels H_1 and K_1 for the main solar absorption lines; H_2, K_2 for the emission reversals and H_3, K_3 for a very weak absorption feature in the emission, giving each emission line the appearance of a double peak.

The emission (H_2, K_2) now found at Potsdam in three K stars was much stronger than in the solar spectrum. On the other hand, other stars of the same spectral type appeared to have no reversals.

In 1897 Deslandres moved to the Meudon Observatory (at that time an institution independent of the Paris Observatory) and it was here in the

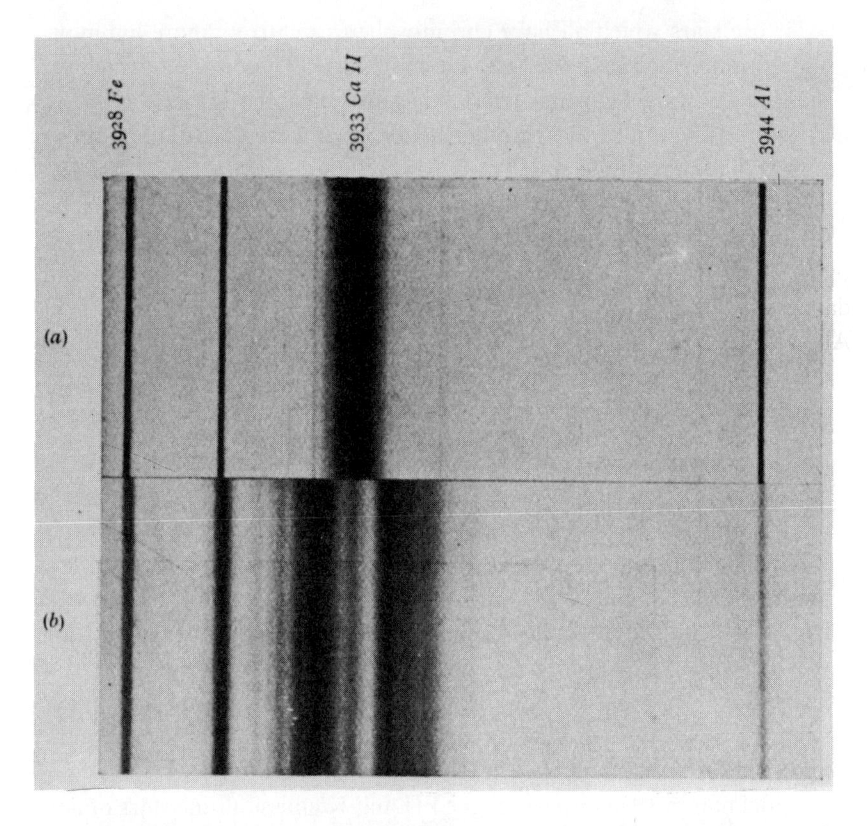

Fig. 11.3 K_2 and K_3 lines of ionised calcium in the sun and Arcturus photographed by H.A. Deslandres.

early 1920s that he and V. Burson made further discoveries of H and K emission-line stars. The observing program began in 1920 and the first results a year later reported seven K stars and one G star with emission, including the original three stars of the Potsdam observers (55). Soon afterwards ε Geminorum (type G) and α Orionis (type M) were added to the list, and in both these stars the H_3 and K_3 absorption components were 'wide and readily visible' (56).

The main results of the Meudon program were reported in 1922 for a total of twelve late-type stars showing H_2 and K_2 emission, including three with H_3 and K_3 absorption also visible (β Dra was the third such star) (57). The emission had always been recognised as arising from the chromosphere (the hot tenuous outer layers) rather than the more quiescent lower-lying photosphere, because the emission reversals of the sun were stronger in sunspots and prominences, and these are associated with chromospheric activity. Deslandres and Burson make here an interesting comment:

> Those stars which all have chromospheres relatively more luminous and important than the Sun, are all giants... The stars... which have the widest chromospheric lines also have the greatest luminosity; and if on the other hand one notices that the Sun, of absolute magnitude equal to 5.2, is certainly a dwarf star, it seems as if the intensity of the chromospheric lines is closely linked to the luminosity of the star (57).

The authors go on to speculate that a new method of measuring spectroscopic parallaxes for late-type stars based on the H and K emission reversals should be possible and might even represent an improvement of the recently established spectroscopic parallax technique of Adams and Kohlschütter at Mt Wilson.

This conclusion was not based on a clear distinction between emission line width and intensity, as the quotation shows. However, the prophecy of a new technique for spectroscopic parallaxes was to prove quite correct. Meanwhile, the number of stars known to show the H and K reversals was greatly increased, mainly from discoveries at Mt Wilson by Joy and R.E. Wilson (58) who catalogued 445 G, K and M stars with calcium emission, of which ninety-five (generally supergiants) also showed the K_3 central absorption. They noted that the use of ultraviolet-transmitting optics, including a grating to replace prisms, had contributed to the much increased discovery rate.

Bidelman in 1954 also catalogued H and K emission-line stars of later spectral type, as part of a very extensive catalogue and bibliography of all known emission-line stars showing either bright hydrogen or bright calcium lines (59). Most of the 426 H and K emission-line stars catalogued also appeared in the Joy and Wilson tabulation (58), but they were listed under four categories, those of spectroscopic binaries, giants (including subgiants and supergiants), dwarfs and Cepheids.

A systematic study of H and K emission lines was commenced at Mt Wilson by Olin C. Wilson (b. 1909) in 1938 using 10 Å/mm coudé spectra from the 100-inch telescope. The program was continued after the war and the first results in 1954 showed that the total width (measured in km/s) of the K_2 line correlated well with the Mt Wilson spectroscopic absolute magnitudes (60).

A full discussion of the data for 185 stars of type G,K and M was published in a classic paper by Olin Wilson, with the assistance of the Indian astronomer M.K. Vainu Bappu (1927–82) (61). These two astronomers worked together on the calcium emission reversals at the time Bappu was at Mt Wilson on a Carnegie Fellowship, from 1952 to 1954. In this paper the logarithm of the K_2-line emission width, after correction for instrumental broadening, was plotted against absolute visual magnitude on the Yerkes system. The points defined a straight line over a 15 magitude range of absolute magnitude. From this remarkable calibration they concluded that the absolute magnitude (and hence parallax or distance) of any late-type star could be fixed with a precision of half a magnitude from the width of K_2 emission.

The relationship between log $W(K_2)$ and M_v has become known as the Wilson–Bappu effect as a result of this well-known paper. In using this appellation, we should remember the groundwork done three decades earlier by Deslandres, who first suggested that such a correlation might exist. Wilson and Bappu came to the conclusion that turbulence in optically thin stellar chromospheres was the most likely source of the broadening of the ionised calcium emission lines, but no theoretical basis for their linear relationship was proposed. If turbulence is correct, Wilson pointed out that his results with Bappu imply a turbulent broadening varying as the one-sixth power of the star's luminosity (62).

Wilson presented an improved calibration of the linear coefficients for the Wilson–Bappu relationship in 1959 (63) using the known absolute magnitudes of the sun and Hyades giants. The probable error of a single measurement was shown to be 0.26 magnitudes, while the diagram also contained some intrinsic scatter estimated at 0.20 magnitudes.

This Wilson–Bappu relationship explicitly concerns the width of the calcium emission, not its intensity. In 1963 Wilson announced that the intensity of H and K emission in main sequence stars probably depends on their age (64). He reached this conclusion from studying spectra of stars in four open clusters (Hyades, Praesepe, Coma and Pleiades) as well as of the sun. The youngest stars, those in the Pleiades, had the strongest emission, the sun the weakest. He concluded that cool dwarf stars initially have strong emission reversals, but these slowly decline in intensity. Wilson considered this effect might be related to the average surface magnetic field strength, since for the sun, local regions of high chromospheric activity with strong reversals of H and K were also associated with a strong magnetic field. Further evidence for an age versus emission intensity relationship was presented by Wilson and Andrew Skumanich (b. 1929) from a sample of 114 main sequence field stars (65).

These ideas were incorporated by Wilson into a broader picture, when he presented evidence for a correlation between the calcium emission intensities and stellar rotation as well as the age (66). For this purpose he used the data on stellar rotational velocities and photoelectric photometry in the literature for a sample of 308 stars, mainly dwarfs of types F and G.

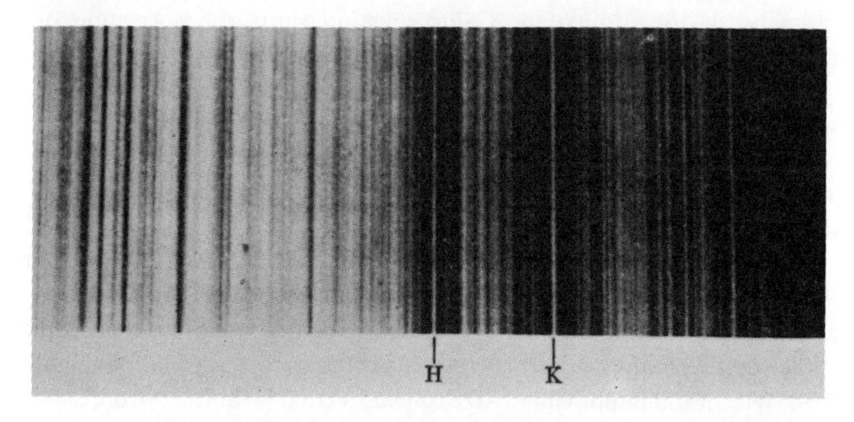

Fig. 11.4 H and K emission lines in σ Geminorum, photographed by Eberhard and Schwarzschild.

The faster rotators were shown to be clearly segregated from the slow ones in the Hertzsprung–Russell diagram in the sense that dwarfs cooler than colour index $b - y = 0.285$* were also slow rotators. These cooler stars also have hydrogen convection zones in their outer envelopes. The interaction between rotation and convection results in strong magnetic fields and also chromospheric activity, as observed in the calcium emission cores. This in turn results in magnetic braking of the rotation by the mechanism proposed by Schatzman (67) (see section 6.31). As the rotation is slowed, the chromospheric activity also decays, thus explaining the observed relationship of the calcium emission intensity with age. On the other hand, stars earlier in spectral type than F4 have no hydrogen convection zones and therefore no chromospheric activity even if they are young. They remain as fast rotators during their main sequence lifetimes if they happened to be born with a high rotational velocity.

Wilson's extensive observations of the H and K emission in late-type stars were made from Mt Wilson for mainly northern hemisphere stars. A program to redress the balance in the southern hemisphere was undertaken by Brian Warner when he observed the emission widths of the K lines of 200 bright southern G, K and M stars, using the coudé spectrograph at the Radcliffe Observatory, Pretoria (68). He showed that his absolute magnitudes using the Wilson–Bappu calibration of the emission widths were to a precision of 0.60 magnitudes from measurements of one plate.

* This index is on the Strömgren photometric system and corresponds to about spectral type F4.

11.4 Interstellar absorption lines In 1904 the Potsdam astronomer J. Hartmann announced the presence of a 'stationary CaII K line' in the spectrum of the spectroscopic binary, δ Orionis (69). He was referring to the fact that the sharp K line was observed to have a fixed velocity of 16 km/s whereas the much broader lines of hydrogen and helium displayed variable velocities of peak-to-peak amplitude about 200 km/s with a 5.7 day period. He concluded that the origin of the K line might not be stellar at all and that 'there is very probably a cloud of calcium vapour between us and the star, which causes the absorption of this ray'.

Stationary lines were soon found in other early-type spectroscopic binary stars. However, Hartmann's interpretation of the stationary line was not at first generally accepted. Vesto Slipher at the Lowell Observatory was one who did favour the interstellar origin (70). He predicted that interstellar sodium D lines should also be observable in early-type spectra, and in fact these were discovered a decade later by Miss Mary Heger at Lick from the spectra of the binary stars β Sco and δ Ori (71). On the other hand, several astronomers supported the concept of a circumstellar cloud of calcium vapour as the origin of the sharp lines at H and K. Oliver Lee devised such a model in 1913 for the binary α Camelopardalis (72), while the observations by Reynold K. Young at Victoria of the binary 12 Lacertae were especially influential in support of the circumstellar cloud model (73) on the basis of velocity variations he claimed to have found in the calcium lines.

The general acceptance of the interstellar hypothesis came when J.S. Plaskett was able to show that the radial velocity of the calcium lines from the spectra of about fifty O and B stars differed from and was usually less than the corresponding velocity of the star; systematic differences of up to 40 km/s sometimes arose (74). Such an effect would be impossible on the circumstellar hypothesis. Plaskett also showed the sharp calcium and sodium lines were always present in O-type stellar spectra, whether the star was known as a binary or not. He concluded that his data supported the presence of a widely distributed yet tenuous interstellar gas. Plaskett's evidence in favour of interstellar matter, was firmly supported by Eddington in 1926 in his influential Bakerian lecture to the Royal Society on 'Diffuse matter in interstellar space' (75). Eddington emphasised that the absorption occurs everywhere along the line of sight to a star, and that it might therefore be expected to be stronger for more distant stars. This hypothesis was tested by Struve at Yerkes. From eye-estimates of interstellar K-line intensities, firstly from 321 O to B3 stars (76), later from 1718 stars of these early types (77), he showed that the assumption that the line intensity depended only on distance was fully justified. He also studied the galactic distribution of the calcium intensities and demonstrated that stars with strong interstellar lines were confined to the plane of the Galaxy.

Struve next collaborated with his Soviet colleague, Boris Gerasimovič (then visiting Harvard College Observatory) to deduce the mean total density of interstellar gas (their result was of the order of 10^{-26} g/cm^3), and to demonstrate that the interstellar gas partakes in the differential rotation of the Galaxy in the same way as the stars (78). On average the absorbing gas has about half the distance and hence half the velocity of the star whose spectrum is observed. Both these results were confirmed from a thorough study by J.S. Plaskett and J.A. Pearce at Victoria, using interstellar calcium intensities for 314 stars (79).

Further data on the strengths of the interstellar calcium lines was obtained by E.G. Williams during his time at Mt Wilson Observatory in the early 1930s. For sixty-seven early-type stars he showed there was a statistical correlation between the K-line absorption intensity and the stars' colour excess or reddening (due to interstellar obscuring material, or dust) (80). This result was readily explained, since the more distant stars have both more gaseous calcium atoms and more dust particles along the line of sight. However, Williams concluded these phases were probably not coextensive in space. His results were the first to make use of microdensitometer tracings to determine K-line absorption intensities. He was able to apply this technique the following year to measure the K-line intensity of Nova Herculis and hence estimate a distance of 370 parsecs*, one of the few available distance methods for such stars (81) (Fig. 11.5).

The first analysis of the strength of interstellar lines using the new curve of growth technique of Minnaert was by Unsöld, Struve and Elvey (82). In fact, this was the first application of the curve of growth to any astronomical object other than the sun. The shape of the curve, and hence the ratio of the strengths of doublets (such as K to H), should depend on how the lines are broadened. Like Eddington (75, 83), the Yerkes astronomers believed that differential galactic rotation which sets up a linear velocity gradient in the absorbing gas with distance in the galactic plane, should be the principal cause of line broadening.

On the galactic rotation hypothesis, interstellar line strength should correlate with the mean radial velocity of absorbing material. Beals at Victoria looked for such a relationship (84) but found no clear-cut trend existed. What is more, he found several stars in which the interstellar K lines show a double or 'complex' structure. The evidence supported the idea of discrete interstellar clouds along the line of sight, with the random or

* About 1200 light years.

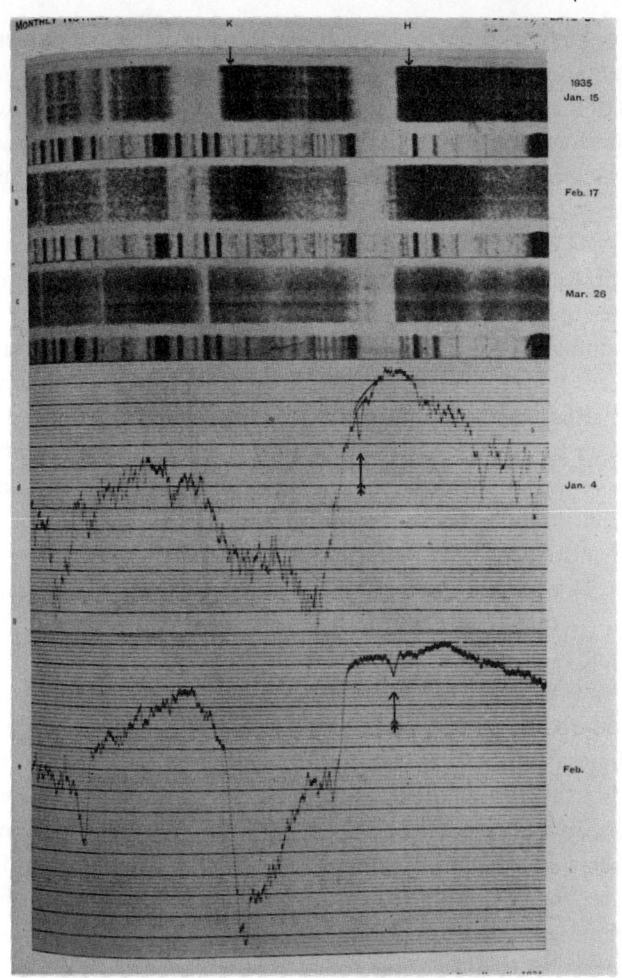

Fig. 11.5 The interstellar K-line in Nova Herculis, observed by E.G. Williams, 1935.

irregular motions of the clouds being the principal cause of line broadening by the Doppler effect. This view was strongly supported by a detailed study of the interstellar sodium D lines by Wilson and Merrill (85). They concluded:

> If it is agreed that the widening and displacement of the interstellar lines are manifestations of the Doppler effect, the only reasonable hypothesis . . . is that the interstellar gas exists in discrete masses, or clouds, and that these clouds, while partaking of the general galactic rotation, have additional relative motions.

For these results, Wilson and Merrill had continued the work of E.G.

Williams and measured interstellar line intensities from microdensitometer tracings, and this resulted in a Mt Wilson catalogue of the equivalent widths of interstellar calcium and sodium lines in several hundred stars by these authors with Roscoe Sanford and Miss Cora Burwell (86). Data on the radial velocity displacements and strengths of three further unidentified lines at 5780, 5797 and 6284 Å were also given. This was the first large body of interstellar line spectrophotometric data, described by Münch (87) as being of 'transcendental importance' since it replaced the earlier eye estimates of Struve and others.

Up until 1934 the H and K lines and the two D lines were the only interstellar lines studied; these four are easily the strongest found. In 1934 Merrill announced the discovery of four further interstellar lines in the yellow-red spectral region (88).* All these features were rather wide with diffuse edges, quite different in character from the extremely sharp interstellar lines studied hitherto. Another stationary line described as 'a vague feature near λ4427' was added by Merrill in 1936 to the interstellar list (89). This last feature was studied in more detail by Beals and Blanchet at Victoria (90) (Fig. 11.6). It is extremely broad, typically about 40 Å across and is observed as a shallow depression in the continuum, centred on about 4430 Å, for the more distant early-type stars. Its presence in forty-six stars was recorded at Victoria and its interstellar character established. Münch has described the problem of identifying the origin of this and the other diffuse interstellar bands as 'the most challenging problem of stellar spectroscopy' (87). Beals and Blanchet concluded with the comment: 'In the absence of a definite identification in terms of absorption by gaseous molecules, the possibility that λ4430 may be due to absorption by particles of solid material should not be entirely disregarded' (90). Fifty years later this suggestion is still a topic for active research.

The development of the coudé spectrograph on the 100-inch telescope (see (91) for discussion of instrumental developments and section 1.3.4) led to the discovery of several additional sharp but weak interstellar lines by Theodore Dunham and W.S. Adams, which were quickly identified. These included interstellar lines of ionised titanium (92), neutral potassium and calcium (93) and iron (94). Additional interstellar features at 3958, 4233 and 4300 Å were also reported by Dunham (93). All these features were identified as arising from interstellar molecules, the first two from CH^+ (95), the last from CH (96), following a suggestion of Swings and Rosenfeld (97). In addition, two weak lines near 3874 Å were identified as CN lines by McKellar (96) and by Adams (91). The increase in the known number of

* Three have wavelengths as given above; the fourth is at 6614 Å.

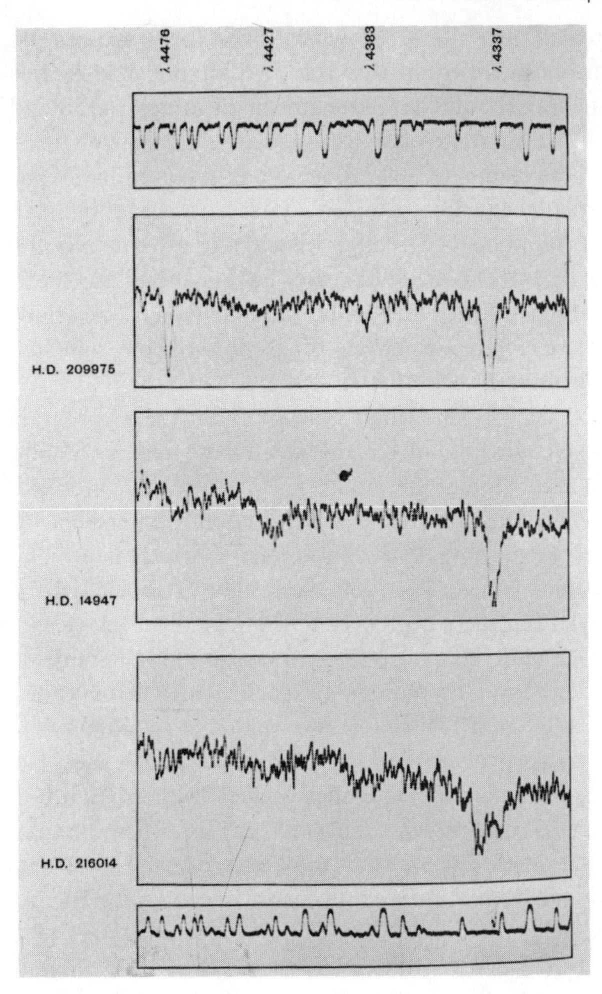

Fig. 11.6 The interstellar diffuse band at 4430 Å (Beals and Blanchet, 1938).

interstellar absorption features can be judged from a tabulation given by Greenstein in 1951; fifteen atomic and eleven molecular lines were all positively identified, while the origin of six diffuse bands remained a mystery except to say that they were definitely interstellar (98).

The Mt Wilson program to study the interstellar lines was continued on the 100-inch coudé spectrograph both during and after the war years by Walter Adams. He was able to obtain high dispersion spectra at 2.9 Å/mm of unrivalled quality, and it is the extensive results reported in two papers in 1943 and 1949 that now turned many of the preliminary findings of earlier workers into established and accepted facts (94, 99). The first of these papers treated fifty stars, of which over forty showed the complex interstellar lines with from two up to five components. Adams measured the interstellar CH and CH^+ strengths in his sample and found that these varied greatly from star to star in their relative intensities. His second paper extended these results to 300 OB stars and reported many more cases of complex interstellar lines. A few stars have weak interstellar K-line absorption components with high velocities relative to a local standard of rest, in one case (HD 169454) of nearly 100 km/s. This was the discovery by Adams of high-velocity interstellar gaseous clouds. Invariably the stars showing these clouds also have stronger low-velocity interstellar absorption components in their spectra.

The next decade and a half following the Adams program served to further consolidate but not greatly extend these findings. Guido Münch at Mt Wilson and Palomar Observatories was one who continued in Adams' footsteps; he studied the velocity distribution of interstellar clouds (100) and with H. Zirin (b. 1929) observed the interstellar lines in stars of high galactic latitude lying at least 230 parsecs above or below the plane of the Galaxy (101). For these objects complex interstellar lines were a common feature, often indicating high-velocity clouds along the line of sight.

After the Lick Observatory 120-inch (3 m) telescope became operational in 1959 that institution too was able to obtain the very high resolution spectra required to study faint interstellar lines. Herbig found several new molecular lines identified as due to either CH or CH^+ with this instrument (102). Instrumental developments elsewhere also gave new data on interstellar lines. The McMath solar telescope and spectrograph at Kitt Peak National Observatory could record spectra with 10 mÅ resolution when used with an image tube to observe bright stars. This was the technique used by W.C. Livingston (b. 1927) and C.R. Lynds (b. 1928), who found that the D_2 line of interstellar sodium in the spectrum of α Cygni could be resolved into five components (103) where previously only two components were known to exist. Another technique employed a Fabry–Pérot scanning interferometer, an instrument able to obtain very high resolution line profiles. L.M. Hobbs (b. 1937) used such a device on the new Lick telescope and again found the multiple components of the interstellar D_2 line of α Cygni; in addition he claimed to detect the hyperfine structure of two of the five main components, seen as a splitting into two components about 20 mÅ apart (104).

A new era in interstellar absorption line studies began in 1965. In this year Donald C. Morton (b. 1933) and Lyman Spitzer (b. 1914) at Princeton obtained the first results of their ultraviolet spectroscopy from Aerobee rockets (105). These were by no means the first rocket observations of stellar spectra at wavelengths below the transparency limit of the earth's atmosphere near 3000 Å. As early as 1955 discrete sources had been detected from an Aerobee rocket launched by the US Naval Research Laboratory in fairly broad photometric pass-bands (106). The Morton and Spitzer observations were the first, on the other hand, to have sufficient resolution to detect absorption lines in stellar spectra. The potential advantages of observing the interstellar lines in stellar spectra in the ultraviolet were earlier emphasised in a discussion by Spitzer and F.R. Zabriskie (b. 1933) (107).

Morton and Spitzer used two grating spectrographs that recorded the ultraviolet spectra at about 65 Å/mm dispersion and 1 Å resolution photographically on film, which was subsequently recovered by parachute. The first partially successful flight was in June 1965; the parachute failed and the camera and spectrograph in the payload were broken open on impact. Nevertheless some unfogged film was recovered with spectra of two B stars, δ and π Scorpii being successfully recorded (Fig. 11.7). Of the twenty-nine absorption lines identified, four were attributed to interstellar ionised carbon, neutral oxygen, ionised silicon and ionised aluminium in the wavelength region 1261 to 2000 Å (105, 108).

Further successful Aerobee flights followed; later in 1965 Morton recorded spectra of six O or early B stars in Orion (109). Four of the stars were found to have strong P Cygni line profiles with absorption components showing violet Doppler shifts of several thousand km/s. This indicated the process of circumstellar gas leaving these stars at high velocities. In addition, the very strong interstellar line of ground-state neutral hydrogen known as the Lyman-α line at 1216 Å was recorded spectrographically for the first time, in the stars δ and ζ Orionis. From the strength of this line Morton was able to estimate a column density of hydrogen atoms along the line of sight to these stars.

These observations of the line spectra of stars in the far ultraviolet region of the spectrum, unobservable on the surface of the earth, but seen now from the spectrographs carried aloft on high altitude rockets, brings this story of the first one and a half centuries of stellar spectroscopy from the time of Fraunhofer to a close. The rocket observations were able to record spectra for only a few minutes of their trajectory, and the signal-to-noise ratio in the photographs was not very high. Clearly the capability for very high resolution ultraviolet spectroscopy with good photometric precision that satellites afforded would revolutionise this branch of astronomy. Indeed, the third Orbiting Astronomical Observatory, or Copernicus satellite, achieved just this after its launch in August 1972. The interstellar absorption lines of a wide range of elements were shortly

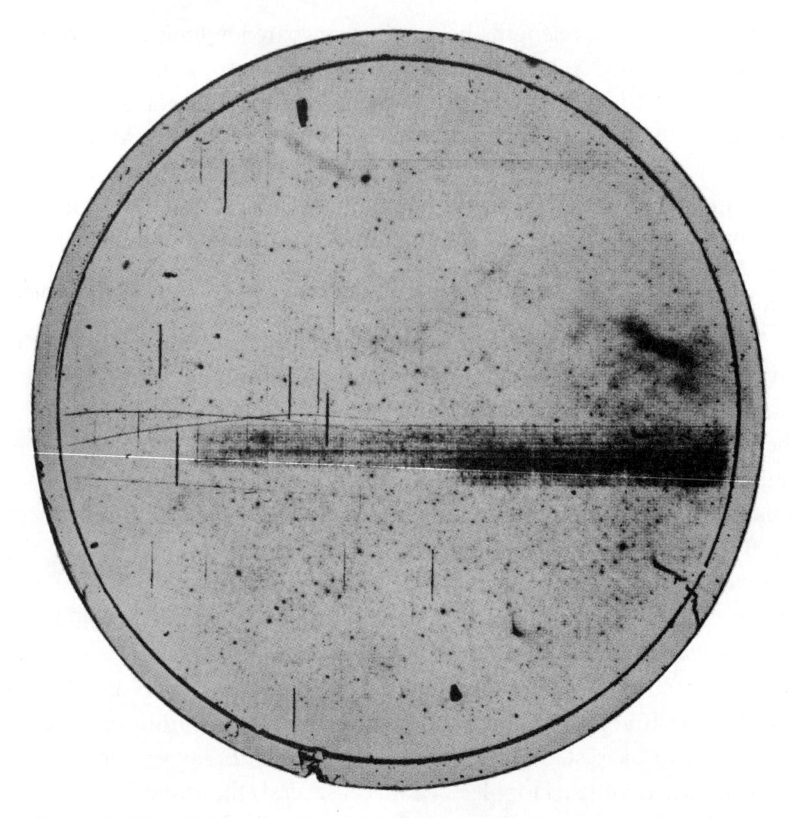

Fig. 11.7 Ultraviolet rocket spectra of δ Sco and π Sco by Morton and Spitzer in 1965. The π Sco spectrum extends over three-quarters of the diameter from 1260 to 2180 Å; that of δ Sco over the right-hand third of the diameter, from 1260 to 1720 Å. The dark vertical lines are zero-order (undispersed) spectra of stars in Libra.

afterwards recorded for the first time and the resolution was around 0.1 Å or less, considerably superior to that achieved by the rocket spectrographs (110). The discussion of the spectroscopic results from the era of satellite astronomy belongs, however, to the story of the second 150 years, which has only just now begun.

References

1. Müller, G. and Hartwig, E., *Geschichte und Literatur des Lichtwechsels der sicher veränderlichen Sterne (Leipzig)*, **2**, 444 (1920).
2. Maunder, E.W., *Mon. Not. Roy. Astron. Soc.*, **49**, 300 (1889).
3. Pickering, E.C., *Harvard Ann.*, **27**, 1 (1890).
4. Keeler, J.E., *Astron. and Astrophys.*, **12**, 361 (1893).

5. Maury, A.C. and Pickering, E.C., *Harvard Ann.*, **28** (Part I), 1 (1897), see p. 101.
6. Vogel, H.C. and Wilsing, J., *Publ. Astrophys. Observ. Potsdam*, **12**, 27 (1899).
7. Belopolsky, A.A., *Astrophys. J.*, **10**, 319 (1899).
8. Frost, E.B., *Astrophys. J.*, **35**, 286 (1912).
9. Merrill, P.W., *Lick Observ. Bull.*, **8**, 24 (1913).
10. Gerasimovič, B.P., *Harvard Bull.*, **852**, 16 (1927); ibid., **867**, 16 (1929).
11. Elvey, C.T., *Astrophys. J.*, **68**, 416 (1928).
12. Adams, W.S. and Dunham, T., reported by W.S. Adams in *Annual Report, Director, Mt Wilson Observ.*, p. 181 (1931).
13. Wilson, O.C., *Astrophys. J.*, **84**, 296 (1934).
14. Beals, C.S., *Mon. Not. Roy. Astron. Soc.*, **90**, 202 (1929).
15. Beals, C.S., *Observ.*, **57**, 319 (1934).
16. Struve, O., *Astrophys. J.*, **81**, 66 (1935).
17. Beals, C.S., *Publ. Dominion Astrophys. Observ. Victoria*, **9**, 1 (1950).
18. Cannon, A.J., *Harvard Ann.*, **76**, (No. 3), 19 (1916).
19. de Groot, M., *Bull. Astron. Inst. Netherlands*, **20**, 225 (1969).
20. Gaviola, E., *Astrophys. J.*, **111**, 408 (1950).
21. le Sueur, A., *Proc. Roy. Soc.*, **18**, 22 (1870); ibid., **18**, 245 (1870).
22. Cannon, A.J., *Harvard Ann.*, **28** (part II), 175 (1901).
23. Baxandall, F.E., *Mon. Not. Roy. Astron. Soc.*, **79**, 619 (1919).
24. Bok, B.J., *Popular Astron.*, **38**, 399 (1930).
25. Walborn, N.R. and Liller, M.H., *Astrophys. J.*, **211**, 181 (1977).
26. Gill, D., *Mon. Not. Roy. Astron. Soc.*, **61**, [66] (1901).
27. Moore, J.H. and Sanford, R.F., *Lick Observ. Bull.*, **8**, 55 (1913).
28. Lunt, J., *Mon. Not. Roy. Astron. Soc.*, **79**, 621 (1919).
29. Merrill, P.W., *Astrophys. J.*, **67**, 391 (1928).
30. Jones, H.S., *Mon. Not. Roy. Astron. Soc.*, **91**, 794 (1931).
31. Hoffleit, D., *Harvard Bull.*, **893**, 11 (1933).
32. Gaviola, E., *Astrophys. J.*, **118**, 234 (1953).
33. Thackeray, A.D., *Mon. Not. Roy. Astron. Soc.*, **113**, 211 (1953).
34. Thackeray, A.D., *Mon. Not. Roy. Astron. Soc.*, **124**, 251 (1962); see also: ibid., *Observ.*, **81**, 102 (1961).
35. Thackeray, A.D., *Mon. Not. Roy. Astron. Soc.*, **135**, 51 (1967).
36. Aller, L.H. and Dunham, T., *Astrophys. J.*, **146**, 126 (1966).
37. Rodgers, A.W. and Searle, L., *Mon. Not. Roy. Astron. Soc.*, **135**, 99 (1967).
38. Neugebauer, G. and Westphal, J.A., *Astrophys. J.*, **152**, L89 (1968); Westphal, J.A. and Neugebauer, G., *Astrophys. J.*, **156**, L45 (1969).
39. Pagel, B.E.J., *Nature*, **221**, 325 (1969); see also Ade, P. and Pagel, B.E.J., *Observ.*, **90**, 6 (1970).
40. Pagel, B.E.J., *Astrophys. Letters*, **4**, 221 (1969).
41. Davidson, K., *Mon. Not. Roy. Astron. Soc.*, **154**, 415 (1971).
42. Bidelman, W.P., *Publ. Astron. Soc. Pacific*, **72**, 24 (1960).
43. Sargent, W.L.W. and Jugaku, J., *Astrophys. J.*, **134**, 777 (1961).
44. Jugaku, J., Sargent, W.L.W. and Greenstein, J.L., *Astrophys. J.*, **134**, 783 (1961).
45. Hartoog, M.R. and Cowley, A.P., *Astrophys. J.*, **228**, 229 (1979).
46. Przybylski, A., *Nature*, **189**, 739 (1961).

47. Przybylski, A. and Kennedy, P.M., *Publ. Astron. Soc. Pacific*, **75**, 349 (1963).
48. Przybylski, A., *Acta. Astron.*, **13**, 217 (1963).
49. Przybylski, A., *Nature*, **210**, 20 (1966).
50. Warner, B., *Nature*, **211**, 55 (1966).
51. Wegner, G. and Petford, A., *Mon. Not. Roy. Astron. Soc.*, **168**, 557 (1974).
52. Przybylski, A., *I.A.U. Coll.*, **32**, 351 'Physics of Ap Stars' (1976).
53. Eberhard, G. and Schwarzschild, K., *Astrophys. J.*, **38**, 292 (1913).
54. Deslandres, H.A., *Comptes Rendus de l'Acad. des Sciences*, **115**, 22 (1892); ibid., **119**, 457 (1894).
55. Deslandres, H.A. and Burson, V., *Comptes Rendus de l'Acad. des Sciences*, **172**, 405 (1921).
56. Deslandres, H.A. and Burson, V., *Comptes Rendus de l'Acad. des Sciences*, **172**, 729 (1921).
57. Deslandres, H.A. and Burson, V., *Comptes Rendus de l'Acad des Sciences*, **175**, 121 (1922).
58. Joy, A.H. and Wilson, R.E., *Astrophys. J.*, **109**, 231 (1949).
59. Bidelman, W.P., *Astrophys. J. Suppl.*, **1**, 175 (1954).
60. Wilson, O.C., *Proc. Nat. Sci. Foundation Conference on Stellar Atmospheres*, Bloomington, Indiana Univ. p. 147 (1954).
61. Wilson, O.C. and Bappu, M.K.V., *Astrophys. J.*, **125**, 661 (1957).
62. Wilson, O.C., *Astrophys. J.*, **126**, 525 (1957).
63. Wilson, O.C., *Astrophys. J.*, **130**, 499 (1959).
64. Wilson, O.C., *Astrophys. J.*, **138**, 832 (1963).
65. Wilson, O.C. and Skumanich, A., *Astrophys. J.*, **140**, 1401 (1964).
66. Wilson, O.C., *Astrophys. J.*, **144**, 695 (1966).
67. Schatzman, E., *Ann. d'Astrophys.*, **25**, 18 (1962).
68. Warner, B., *Mon. Not. Roy. Astron. Soc.*, **144**, 333 (1969).
69. Hartmann, J., *Sitzungsberichte der Preussischen Akad. der Wissenschaften Berlin*, **14**, 527 (1904); ibid., *Astrophys. J.*, **19**, 268 (1904).
70. Slipher, V.M., *Lowell Observ. Bull.*, **2**, 1 (1909).
71. Heger, M.L., *Lick Observ. Bull.*, **10** (no. 326), 59 (1919).
72. Lee, O.J., *Astrophys. J.*, **37**, 1 (1913).
73. Young, R.K., *Publ. Dominion Astrophys. Observ. Victoria*, **1**, 219 (1921); ibid., *Observ.* **43**, 399 (1920).
74. Plaskett, J.S., *Mon. Not. Roy. Astron. Soc.*, **84**, 80 (1923).
75. Eddington, A.S., *Proc. Roy. Soc.*, **111(A)**, 424 (1926).
76. Struve, O., *Astrophys. J.*, **65**, 163 (1927).
77. Struve, O., *Astrophys. J.*, **67**, 353 (1928). See also: ibid., *Mon Not. Roy. Astron. Soc.*, **89**, 567 (1929).
78. Gerasimovič, B.P. and Struve, O., *Astrophys. J.*, **69**, 7 (1929).
79. Plaskett, J.S. and Pearce, J.A., *Publ. Dominion Astrophys. Observ. Victoria*, **5**, 167 (1933).
80. Williams, E.G., *Astrophys. J.*, **79**, 280 (1934).
81. Williams, E.G., *Mon. Not. Roy. Astron. Soc.*, **95**, 573 (1935).
82. Unsöld, A., Struve, O. and Elvey, C.T., *Zeitschrift für Astrophys.*, **1**, 314 (1930).
83. Eddington, A.S., *Mon. Not. Roy. Astron. Soc.*, **95**, 2 (1934).
84. Beals, C.S., *Mon. Not. Roy. Astron. Soc.*, **96**, 661 (1936).

85. Wilson, O.C. and Merrill, P.W., *Astrophys. J.*, **86**, 44 (1937).
86. Merrill, P.W., Sanford, R.F., Wilson, O.C. and Burwell, C.G., *Astrophys. J.*, **86**, 274 (1937).
87. Münch, G., *Stars and Stellar Systems*, **7**, 365: 'Nebulae and Interstellar Matter', edited by B. Middlehurst and L.H. Aller (1968).
88. Merrill, P.W., *Publ. Astron. Soc. Pacific*, **46**, 206 (1934).
89. Merrill, P.W., *Astrophys. J.*, **83**, 126 (1936).
90. Beals, C.S. and Blanchet, G.H., *Mon. Not. Roy. Astron. Soc.*, **98**, 398 (1938).
91. Adams, W.S., *Astrophys. J.*, **93**, 11 (1941).
92. Dunham, T. and Adams, W.S., *Publ. Amer. Astron. Soc.*, **9**, 5 (1937); Dunham, T., *Nature*, **139**, 246 (1937).
93. Dunham, T., *Publ. Astron. Soc. Pacific*, **49**, 26 (1937).
94. Adams, W.S., *Astrophys. J.*, **97**, 105 (1943).
95. Douglas, A.E. and Herzberg, G., *Astrophys. J.*, **94**, 381 (1941).
96. McKellar, A., *Publ. Astron. Soc. Pacific*, **52**, 187 (1940).
97. Swings, P. and Rosenfeld, L., *Astrophys. J.*, **86**, 483 (1937).
98. Greenstein, J.L., *Astrophysics – a Topical Symposium*, edited by J.A. Hynek, published by McGraw-Hill Book Co., chapter 3, p. 526 (1951).
99. Adams, W.S., *Astrophys. J.*, **109**, 354 (1949).
100. Münch, G., *Astrophys. J.*, **124**, 42 (1957).
101. Münch, G. and Zirin, H., *Astrophys. J.*, **133**, 11 (1961).
102. Herbig, G.H., *Astron. J.*, **65**, 491 (1960).
103. Livingston, W.C. and Lynds, C.R., *Astrophys. J.*, **140**, 818 (1964).
104. Hobbs, L.M., *Astrophys. J.*, **142**, 160 (1965).
105. Morton, D.C. and Spitzer, L., *Astrophys. J.*, **144**, 1 (1966).
106. Byram, E.T., Chubb, T.A., Friedman, H. and Kupperian, J.E., Jr., *Astron. J.*, **62**, 9 (1957).
107. Spitzer, L. and Zabriskie, F.R., *Publ. Astron. Soc. Pacific*, **71**, 412 (1959).
108. Stone, M.E. and Morton, D.C., *Astrophys. J.*, **149**, 29 (1967) presented further results for these stars.
109. Morton, D.C., *Astrophys. J.*, **147**, 1017 (1967).
110. Morton, D.C., Drake, J.F., Jenkins, E.B., Rogerson, J.B., Spitzer, L. and York, D.G., *Astrophys. J.*, **181**, L103 (1973).

Appendix I Table of solar lines designated by letters by Fraunhofer and others

Adapted from Appendix I in *Astronomical Physics* by F. J. M. Stratton, Methuen and Co., London (1925), p.183.

Letter	*Wavelength (Å)*	*Chemical origin*	*Authority for letter*[a]
x_4	8806.8	Mg	Ab.
x_3	8662.2	CaII	Ab.
x_2	8542.1	CaII	Ab.
x_1	8498.1	CaII	Ab.
Z	8227.	telluric H_2O	Ab.
A	7593.7	telluric O_2	Fr.
a	7160.	telluric H_2O	B. P.
B	6867.2	telluric O_2	Fr.
C	6562.8	Hα	Fr.
D_1	5895.9	NaI	Fr.
D_2	5890.0	NaI	Fr.
D_3[b]	5875.6	HeI	—
E	5269.6	FeI	Fr.
b_1	5183.6	MgI	B. P.
b_2	5172.7	MgI	B. P.
b_3	{5169.1 5168.9}	{FeII FeI}	B. P.
b_4	5167.3	MgI	B. P.
F	4861.3	Hβ	Fr.
d	4383.6	FeI	B. P.
G	4314.2	CH	Fr.
g	4226.7	CaI	B. P.
h	4101.7	Hδ	B. P.
i	4045.8	FeI	B. P.
H	3968.5	CaII	Fr.
K	3933.7	CaII	Ma.
L	3820.4	FeI	Be.
M	3727.6	FeI	Be.

Appendix I (*Contd.*)

Letter	*Wavelength (Å)*	*Chemical origin*	*Authority for letter*[a]
N	3581.2	FeI	Be.
O	{3440.6 / 3441.0}	{FeI / FeI}	Be.
P	3361.2	TiII	Be.
Q	3286.8	FeI	Es.
R	3179.3	CaII	Es.
r	3143.8	TiII	Ma.
S	3100.0	FeI	Ma.
s	3047.6	FeI	Co.
T	3020.7	FeI	Ma.
t	2994.4	FeI	Co.
U	2947.7	FeI	Co.

[a] Authorities:

Fr.	Fraunhofer, J., *Denkschrift. der Königl. Akad. der Wissenschaften zu München,* **5**, 193 (1817).
B. P.	Baden Powell, *British Association Report*, p. 1 (1839).
Be.	Becquerel, E., *Bibliothèque Universelle de Genève*, **40**, 341 (1842).
Es.	Esselbach, E., *Ann. der Physik*, **98**, 513 (1856).
Ma.	Mascart, E., *Ann. Scientifiques de l'Ecole Normale Polytechnique*, **1**, 219 (1864).
Ab.	Abney, W. deW., *Phil. Trans. Roy. Soc.*, **177**, 653 (1880).
Co.	Cornu, M.A. *Spectre Normal du Soleil* (1881).

[b] Not photospheric.

Appendix II Vogel's first spectral classification scheme of 1874

The following is a translation of Vogel's description of his first spectral classification scheme of 1874 which appeared in *Astronomische Nachrichten,* **84**, 113 (1874):

'I with to propose the following division, which should correspond to our present knowledge on the spectra of the fixed stars:

'Class I: Spectra in which the metallic lines are seen only very weakly or not at all, and the refrangible parts of the spectrum (blue, violet) especially stand out by their intensity.

'a) Spectra in which, apart from the very weak metallic lines, the hydrogen lines are visible and stand out through their width and intensity (most white stars such as Sirius and Vega belong here).

'b) Spectra in which the individual metallic lines are only quite weakly visible or are not to be seen at all, and the hydrogen lines are also missing (β, γ, δ, ε Orionis).

'c) Spectra in which the hydrogen lines are in emission, and apart from these lines, the D_3 line is also visible, being bright as well (so far only β Lyr and γ Cas known).

'Class II: Spectra in which the metallic lines stand out very clearly. The refrangible parts of the spectrum are dim in comparison with the previous class, and in the less refrangible parts weak bands sometimes appear.

'a) Spectra with very numerous metallic lines, which are easily recognisable, especially in the yellow and green, by their intensity. The hydrogen lines are mostly strong, but not so strikingly broadened as with class Ia. However, in some stars these are weak and in such cases weak bands are usually to be seen in the less refrangible parts as numerous closely packed lines (Capella, Arcturus, Aldeberan).

'b) Spectra in which, apart from dark lines and several weak bands, a number of bright lines appear (T Coronae), and also to be included in all probability are the stars observed by Wolf and Rayet in Cygnus as well as the variable star R Gem (although the faintness of the light of this last named means that absorption lines can only be suspected, it does indeed have dark bands in the red and yellow).

'Class III: Spectra in which apart from dark lines numerous dark bands appear in all parts of the spectrum, and the refrangible parts of the spectrum are conspicuously weak.

'a) Apart from dark lines, bands are to be recognised in the spectrum, of which the most conspicuous are dark towards the violet where they terminate sharply, whereas on the red side the bands are dim and appear washed out (α Her, α Ori, β Peg).

'b) Spectra in which dark very broad bands are to be found, and whose intensity profile is the other way round when compared to those of the previous subdivision, and in which therefore, the strongest bands are sharply terminated on the red side where they are darkest, whereas towards the violet they gradually fade away (so far only faint stars are known of this type: Schjellerup catalogue red stars no.78, 152, 273, etc.)'.

Appendix III Summary of replies to the 1910 questionnaire on spectral classification

(See section 5.10 for explanation of columns)

Name	1	2	3	4	5
1. W.S. Adams, Mt Wilson	Yes	i) Neglect of some metallic line criteria ii) More detail needed for A stars iii) Spectral type R should not be used.	None yet	Yes, but later	i) Enhanced lines ii) Colour indices
2. S. Albrecht, Cordoba	Yes	O, B stars need improved classification	None yet, wait	—	—
3. W.W. Campbell, Lick	Yes	Errors in Harvard spectral types	None yet, possibly never	Yes	Use of spectral regions from green to red.
4. Miss A.J. Cannon, Harvard	Yes	i) B,A etc. should become B0, A0 (Hertzsprung's suggestion) ii) Ma should become M0; Mb should become M5; Mc should possibly be given another letter as so different.	None at present	Yes, but care needed in eliminating atmospheric & instrument effects	i) Peculiar stars with strong Sr or Si lines ii) B stars with bright Balmer lines.

Appendix III (*Contd.*)

Name	1	2	3	4	5
5. A.L. Cortie, Stonyhurst Coll. Obs.	Yes, but retain Secchi's I, II, III, IV as well	Harvard system doesn't give names that readily describe spectral types, as for example Lockyer's system does.	Yes, in near future. Possibly some modification of Harvard system	Certainly not yet; too complicated.	None
6. H.D. Curtis, Lick	Yes	Further subdivisions in Draper class may later be needed.	Adopt Draper system immediately	No	—
7. R.H. Curtiss, U. of Michigan	Yes	i) Notation in Draper system arbitrary, order of letters is disturbed. ii) Peculiar and composite spectra not easily classified on Draper system.	Yes (which system not specified)	Yes	Variable radial velocity should be noted in classification.
8. H. Ludendorff and G. Eberhard, Potsdam	Yes	Order of letters in Draper scheme is not very nice, but now too late to alter.	Draper system should be recommended as temporary measure	Desirable, though difficult.	Visual end of spectra needs further study.
9. Mrs W.P. Fleming, Harvard	Yes	More detailed classification needed for classes M, N, R.	Not now, nor in near future.	Yes	Further subdivisions needed for classes M, N, and R.
10. E.B. Frost, Yerkes	Yes	i) Lack of logical sequence of letters in Draper system. ii) Low dispersion spectra fail to reveal details of slit spectra. iii) More detail needed for peculiar A stars. iv) However, no modifications recommended to Draper system as these would cause confusion.	Unwise now, or in near future.	Yes	(Reply to Q5 incorporated in that for Q2)

11.	M. Hamy, Paris	Yes	The Draper notation gives little feeling for the essential characteristics of each type.	Yes, now. Could be Draper system, though this is likely to be a temporary adoption.	No, too complicated.	More account should be taken of temperature as Lockyer tried to do.
12.	J. Hartmann, Göttingen	Yes, but also retain Secchi's system.	Draper system often too detailed; Secchi's coarser types sometimes better.	A combination of Secchi and Draper systems should be used.	—	i) Any system should be a pure classification; interpretation will come later. ii) Bright standard stars for detailed study should be named in each subdivision.
13.	E. Hertzsprung, Potsdam	—	i) Draper system not quite linear with colour index and should be modified so that it is exactly. ii) Change A to A0, etc. iii) Some decimal subdivisions on Draper system used far more often than others.	Yes, Draper system should be used now.	Yes	Absolute luminosity should be studied further and taken into account in classification.
14.	S.S. Hough, Cape Observ.	Yes	Poor nomenclature and arbitrary order of letters.	Adopt Draper system, but revise the notation.	—	i) No further criteria needed at present as too much detail undesirable. ii) Later on a system based on physical or chemical conditions might be justified.

Appendix III (*Contd.*)

Name	1	2	3	4	5
15. F. Küstner, Bonn	Yes	—	Adopt Draper system for the time being.	—	Red end of spectra should be investigated
16. H.C. Lord, Emerson McMillan Observatory.	Yes	More photographs of typical stellar spectra should be published.	Yes, now (presumably Draper classification)	No. Pictures of spectra are better than a line width symbol.	Useful to have set of pictures of each spectral type taken at different resolutions.
17. J. Lunt, Cape Observ.	—	i) Subdivisions of all types should be with letters a, b, c... (e.g. Ba to Be, instead of B0 to B9). ii) Only one typical star of each class should be named. iii) Further subdivision of class A is needed.	i) Adopt Draper system in near future after modifications & improvements. ii) Circulate photographs of typical stars of each type.	Yes	The fundamental criterion for classification should be a chemical one.
18. Miss A.C. Maury, Harvard	Yes. My own scheme differs mainly in nomenclature rather than substance.	Letters or numerals in order would be desirable. This is presumably an evolutionary order from types O to N.		Yes. My parallel series of a, b, c stars were unmistakable.	i) Peculiarities of A stars should be included. ii) Evolution important.
19. J.A. Parkhurst, Yerkes	Yes	Criteria for classifying F, G, K stars need to be more definitely stated.	—	—	Visual and UV parts of spectra should both be taken into account.

20.	E.C.Pickering, Harvard	Yes	i) Draper scheme needs extending for M stars. ii) Present notation needs improving for peculiar stars.	Not until laws of stellar evolution have been established.	Yes	Possibly spectral types could be objectively measured rather than estimated.
21.	J.S. Plaskett, Ottawa	Yes	Present Draper notation doesn't suggest anything in regard to character of spectra.	Yes, but not till Solar Union meeting in 1913.	Yes	Red end of spectra, including Hα, should be studied.
22.	H.N. Russell, Princeton	Yes	None, except some details of notation might be revised.	Yes, Draper system at once, provided there is general unanimity.	Yes	i) Study of yellow and green spectral regions. ii) More work with slit spectrographs on line widths. iii) Study of luminosity effects. iv) Spectral types based only on line spectrum, not colours etc.
23.	J. Scheiner, Potsdam	No	The Draper system is not simple, and the basic principles of the classification, line breadth and the appearance of certain elements, are not clearly distinguished. The second Vogel system is more elegant and simpler, but also would need substantial revision before being adopted.	No, not the Draper system. Nor is the time ripe for any system to be universally adopted.	—	Factors such as proper motion and parallax need further investigation.

Appendix III (*Contd.*)

Name	1	2	3	4	5
24. F. Schlesinger, Allegheny	Yes	i) Draper system only covers blue, UV region. ii) It is based on objective prism spectra, though slit spectra show more detail. iii) Neglects variations in continuous spectrum. iv) Neglects changes in wavelength of some lines with spectral type.	No system should be recommended either now or in near future.	Yes, using slit spectrograms, though this is of minor importance.	(See answer to Q.2.)
25. K. Schwarzschild, Potsdam.	Yes	Draper system has unfortunate choice of letters.	Not in near future	Hertzsprung has shown Maury's line width to be an important criterion.	Weak lines should be studied more at high dispersion.
26. W. Sidgreaves, Stonyhurst College	Yes	Names, such as Altairian, preferable to numbers and letters.	No	Yes	—
27. V.M. Slipher, Lowell	Yes	The letters should be in alphabetical order. Rename class O as A etc. so as to achieve this.	Yes, provided there is general agreement on a uniform system.	Yes	i) Spectrum from 4900 to 6900 Å should be studied further. ii) Continuous spectrum should be considered.
28. J. Wilsing, Potsdam	—	Many spectroscopists are already in favour of the Draper classification. Changing the letters is necessary, though this would be of no fundamental significance.	—	—	Characteristic type spectra should be acquired by each observatory.

Sources of illustrations

The sources of many of the illustrations are given in the Acknowledgements, or in the text. In other cases sources are given below. The notation (8:153) refers to reference (153) cited in Chapter 8.

Fig.	Source
1.1	(8:153)
1.2	(1:55)
1.3	(1:62)
1.5	(1:62)
2.1	(2:12)
2.2	M. Rohr, *Zeitschrift für Instrumentenkunde*, **46**, 273 (1926).
2.4	(2:16)
3.1	*Vistas in Astronomy*, **9**, 277, ed. A Beer, Publ. Pergamon Press, Oxford (1967).
3.2	*Smithsonian Annual Report* p. 294 (1904).
3.3	(3:17)
3.4	(3:20)
3.5	(3:22)
3.6	(3:24)
3.7	(3:25)
3.8	*Himmel und Erde*, **11**, 170.
3.9	(3:45)
3.10	(3:45)
4.1	(4:6)
4.2	(4:6)
4.3	*Astron. and Astrophys.*, **11**, 640 (1892).
4.4	J. Pohle, *P. Angelo Secchi*, Bachem Verlag, Cologne (1904).
4.5	As for 4.4.
4.6	A. Secchi *Sugli spettri prismatici delle stelle fisse*, Firenze (1867).
4.7	As for 4.6.
4.8	A. Secchi, *Atti dell 'Accademia Pontificia de' Nuovi Lincei* (1872).
4.9	As for 4.8
4.10	*Astrophys. J.*, **26**, 128 (1907).
4.13	(5:55)
4.14	(4:61)
4.15	*Astrophys. J.*, **27**, 1 (1908).
4.16	H.C. Vogel, *Beobachtungen der Sternwarte zu Bothkamp*, **I** (1872).
4.17	(4:80)
4.18	A.L. Cortie, *Astrophys. J.*, **53**, 233 (1921).
4.19	(4:91)
4.20	(4:90)
4.21	(8:153)
5.1	S.I. Bailey, *Astrophys. J.*, **50**, 236 (1919).
5.2	(5:56)
5.3	E.C. Pickering, Henry Draper Memorial, *Annual Report*, No. 2 (1888).
5.4	A.J. Cannon, *Astrophys. J.*, **34**, 314 (1911).
5.5	*Sky and Tel.*, **11**, 106 (1952).
5.6	(5:3)
5.7	(5:72)

5.8 (5:3)

6.1 O. Struve, *Sky and Tel.*, **22**, 132 (1961).
6.2 *Astrophys. J.*, **12**, 239 (1900).
6.3 (4:78)
6.4 As for 6.1.
6.5 H.D. Babcock, *Publ. Astron. Soc. Pacific*, **56**, 146 (1944).
6.6 *Ann. Solar Phys. Observ.*, Cambridge **1.**
6.8 (6:68)
6.9 (6:68)
6.10 (6.68)
6.11 (6:70)
6.12 G.H. Herbig, *Publ. Astron. Soc. Pacific*, **63**, 191 (1951).
6.13 *J. Brit. Astron. Assoc.*, **40**, 241 (1930).
6.14 (6:101)
6.15 *Astrophys. J.*, **83**, 1 (1936).
6.16 (8:153)
6.17 (8:153)
6.18 R.M. Petrie, *Publ. Astron. Soc. Pacific*, **63**, 218 (1951).
6.19 *Publ. Astron. Soc. Pacific*, **38**, 184 (1926).
6.20 *Roy. Astron. Soc. Occasional Notes*, **3**, 189.
6.21 R.M. Petrie, *Publ. Astron. Soc. Pacific*, **63**, 215 (1951).
6.23 *Proc. Roy. Soc.*, **99**, i (1921).

7.1 A.V. Nielsen, *Centaurus*, **9**, 219 (1963).
7.2 As for 7.1.
7.4 As for 7.1.
7.5 D.E. Osterbrock, *Sky and Tel.*, **51**, 91 (1976).
7.6 *Sterne und Weltraum*, **9**, 63 (1970).
7.7 'Professor Megnad Saha: His Life, Work and Philosophy'. ed. S.N. Sen. Calcutta (1954).
7.8 *J. Brit. Astron. Assoc.*, **45**, 175 (1935).
7.10 C.de Jager, *Quarterly J. Roy. Astron. Soc.*, **12**, 338 (1971).
7.11 (7:79)
7.12 (7:87)
7.14 (7:79)
7.15 (7:88)
7.16 (7:93)
7.17 S. Chandrasekhar, *Astrophys. J.*, **139**, 423 (1964).
7.18 (7:108)

8.1 *Monthly Not. Roy. Astron. Soc.*, **107**, 1 (1947).
8.2 D.E. Osterbrock, *Sky and Tel.*, **51**, 96 (1976).
8.3 (1:62)
8.4 (8:43)
8.5 J.H. Oort, *Quarterly J. Roy. Astron. Soc.*, **7**, 330 (1966).
8.6 *Mitt. Astron. Gesell.*, p. 43 (1963).
8.7 *Sterne und Weltraum*, **4**, 135 (1965).
8.8 (8:82) part 1.
8.9 (8:100)
8.11 (8:107)
8.12 D. Chalonge, *Ann. d'Astrophys.*, **19**, 263.
8.13 A. McKellar, *Publ. Astron. Soc. Pacific*, **70**, 129 (1958).
8.14 (8:16)
8.15 O. Struve, *Sky and Tel.*, **22**, 134 (1959).
8.16 (8:153)
8.17 *Popular Astron.*, **56**, 119 (1948).
8.18 D.A. MacRae, *Sky and Tel.*, **30**, 7 (1965).
8.19 As for 8.18
8.20 (8:185)
8.21 A.H. Joy, *Publ. Astron. Soc. Pacific*, **74**, 41 (1962).
8.22 (8:192)
8.23 (8:194)
8.24 (8:207)

9.1 (9:2)
9.3 (9:15)

9.4 R.H. Curtiss, *Publ. Astron. Observ. Univ. Michigan*, **3**, 22 (1923).
9.5 (9:74) (1949 paper).
9.6 (9:103)
9.7 (9:103)
9.8 (9:119) (1951 paper).
9.9 *Sky and Tel.*, **39**, 33 (1970).
9.10 (9:204)
9.11 J.L. Greenstein, *I.A.U. Symp.*, **10** and *Ann. d'Astrophys.*, Suppl. **8** (1959).
9.12 (9:236)
9.13 W.S. Adams, *Publ. Astron. Soc. Pacific*, **63**, 183 (1951).
9.14 (9:254)
9.15 G. Cayrel de Strobel, *I.A.U. Symp.* **72**, 38 (1976).
9.16 P.J. Ledoux, *Ciel et Terre*, **100**, 135 (1984).
9.17 (9:343)
10.1 *Sterne und Weltraum*, **5**, 189 (1966).
10.2 O. Heckmann, *Sterne und Weltraum*, **4**, 231 (1965).
10.3 (10:17)
10.4 (10:26)
10.5 *Sky and Tel.*, **51**, 156 (1976).
10.6 (10:43) (1946 paper).
10.7 (10:45)
10.8 (10:45)
10.9 (10:91)
10.10 (10:92)
10.11 O.C. Wilson, *Publ. Astron. Soc. Pacific*, **68**, 89 (1956).
10.12 (10:146)
10.14 (10:177)
10.15 (10:216)
10.16 (10:217)

11.1 (11:25)
11.2 *Publ. Astron. Soc. Pacific*, **63**, 183 (1951).
11.3 F.J.M. Stratton, *Astronomical Physics*, p. 132, Methuen and Co. (1925).
11.4 (11:53)
11.5 (11:81)
11.6 (11:90)
11.7 (11:105)

Index of names

Dates are given on pages with numbers in bold type.

Index of star names

References to non-stellar objects (eg. planets, star clusters, galaxies) can be found in the subject index.

Index of spectral lines

(a) Balmer and Lyman lines

(b) Fraunhofer lines

(c) Index of spectral lines by wavelength

Index of subjects